Proceedings of the International School on Applied Mathematics

SYMMETRIES AND NONLINEAR PHENOMENA

Other Publications in this Series

1. *Stochastic Processes Applied to Physics and Other Related Fields*, edited by B. Gomez, S. M. Moore, A. M. Rodriguez-Vargas, and A. Rueda (1983).

2. *Workshop on the Search for Gravitational Waves*, edited by E. Posada and G. Violini (1983).

3. *Non Conventional Energy Sources*, edited by G. Furlan, H. Rodriguez, and G. Violini (1984).

4. *Galaxies, Quasars and Cosmology*, edited by L. Z. Fang and R. Ruffini (1985).

5. *Reactor Physics for Developing Countries and Nuclear Spectroscopy Research*, edited by K. Lieb and G. Medrano (1986).

6. *Proceedings of the First Latin American School on Biophysics*, edited by Fayad. A. M. Rodriguez-Vargas, and G. Violini (1987).

7. *Methods and Applications of Nonlinear Dynamics*, edited by A. W. Saenz (1988).

8. *First International Course on Condensed Matter*, edited by D. Prosperi, S. Rosati, and G. Violini (1988).

Proceedings of the International School on Applied Mathematics

SYMMETRIES AND NONLINEAR PHENOMENA

CIF Series – Vol. 9

Paipa (Colombia),
Feb 22-26, 1988

Editors

D. LEVI
Dipartimento di Fisica
Universita'di Roma "La Sapienza"
P. zale A. Moro, 2
00185 Roma, ITALY

P. Winternitz
Centre de Recherches Mathématiques
Université de Montréal
CP 6128, Succ. "A"
Montreal, Quebec
H3C 3J7 CANADA

World Scientific
Singapore • New Jersey • London • Hong Kong

Published by

World Scientific Publishing Co. Pte. Ltd.
5 Toh Tuck Link, Singapore 596224
USA office: 27 Warren Street, Suite 401-402, Hackensack, NJ 07601
UK office: 57 Shelton Street, Covent Garden, London WC2H 9HE

British Library Cataloguing-in-Publication Data
A catalogue record for this book is available from the British Library.

SYMMETRIES AND NONLINEAR PHENOMENA
Proceedings of the International School on Applied Mathematics

ISBN-13 978-9971-5-0663-6
ISBN-10 9971-5-0663-7
ISBN-13 978-9971-5-0701-5 (pbk)
ISBN-10 9971-5-0701-3 (pbk)

PREFACE

It is a great pleasure for me to present this new volume of the CIF series, the ninth it it, which contains the proceedings of the Conference on Symmetries and Nonlinear Phenomena held last February in Paipa, a small town not too far from Bogota.

Before making some specific comments about the Conference and its place in the program of the Center, it is appropriate to notice how the series is now CIF series and no longer the ACIF one. This change reflects the formal foundation of the Centro Internacional de Fisica, which took place at the end of 1985 and whose effects on our publications have started to manifest, as is shown by the fact that this is the fourth title published this year.

The Conference was very successful. It covered a topic of great interest, covering many different aspects of the applications of group theory to the study of nonlinear phenomena. The reason why CIF included this Conference in its 1988 program is of a different nature. On one hand, the scientific arguments mentioned above (see also the Introduction) played an important role in this decision. However, one should not forget that one of the main aims of the Centro Internacional de Fisica is to promote scientific research in Latin America and, in particular, the Andean region. For this, the collaboration of scientists of the region both at regional level and with those of advanced countries is fundamental. The possibility of carrying on advanced research is in fact something which may contribute to reduce the risk of the brain drag. Conferences like this are powerful tools in order to reduce the scientific isolation of our region.

Even if the attendance to the Conference was probably lower than what could have been hoped for, nevertheless it started many scientific contacts whose results will only be seen in the future. As a matter of fact, the reduced attendance has also an obvious explanation, when one considers how the structure of CIF activities is changing, reducing the number of courses and increasing the weight of research. This may indeed provoke some initial difficulty among the usual users of CIF's activities, but we have no doubt that eventually the regional scientific community will adapt itself to this new trend.

By no means this change corresponds to denying the importance of what CIF (and earlier, ACIF) has accomplished in Latin America since the beginning of its activity (March 1982). The importance of this accomplishment can be seen clearly from some figures which can be quoted. Until last April, sixty-five events have been organized, not only in Colombia, but also in Argentiana, Brazil, Dominican Republic, Equador and Peru. This realization has involved the participation of 4216 scientists, of whom 548 were lecturers. Such an activity has certainly contributed to the improvement of the scientific atmosphere in this subcontinent and has opened interesting new perspectives to scientific research. However, the exposure of the regional scientific community to this interaction with international science, though important, cannot be considered a final result. Its meaning can only be found in the creation of actual possibilites of research which should privilege the horizontal cooperation of scientists of the region. This consideration, which has accompanied ACIF and CIF from the beginning of this activity is leading the Centro Internacional de Fisica to considerably increase its weight of research with respect to Courses. The first steps in this direction have been given with the realization of an annual research workshop on theoretical physics, which was held for the first time in 1985. This Conference corresponds to an analogous line, having been oriented toward the identification of possible joint research programs of the regional scientific community with colleagues from other regions. It can be observed that both the above-mentioned activities are of a theoretical nature. This does not correspond to the choice of CIF. On the contrary, our present programs give much weight to experimental research. This year we shall put in operation our first laboratory devoted to the biophysics of membrances. A number of other laboratories are in the program.

We hope that this activity will produce important fruits in the near future and look with similar interest to all the branches of the physical sciences in which we are working. Thus, let us conclude this introduction expressing the hope that the fruits of this Conference can be seen very soon.

Bogota, June 15, 1988 G. Violini

CONTENTS

Introduction.

The International School on Applied Mathematics: **Symmetries and Nonlinear Phenomena** held in Paipa, Colombia, February 22-26, 1988 brought together a group of experts on nonlinear equations and their applications in physics, and some local participants. The speakers presented reviews of topics to which they themselves had made important contributions and also results of new original research. The contents of the talks were written up and are contained in this volume. The individual articles manifest a considerable diversity in topics and applications, together with a definite unity in mathematical approach.

It is well known that many, if not all, of the fundamental equations of physics are nonlinear and that linearity is achieved as an approximation. One of the important developments in applied mathematics and theoretical physics over the recent years is that many nonlinear equations, and hence many nonlinear phenomena, can be treated as they are, without approximations, and solved by essentially linear techniques.

One of the standard techniques for solving linear partial differential equations is the Fourier transform. During the last 20 years, it was shown that a class of physically interesting nonlinear partial differential equations can be solved by a nonlinear extension of the Fourier technique, namely the inverse scattering transform. This reduces the solution of the Cauchy problem to a series of linear steps. This method, originally applied to the Korteweg-de Vries equation is now known to be applicable to a large class of nonlinear evolution equations in one space and one time variable, to quite a few equations in $2+1$ dimensions and also to some equations in higher dimensions. The corresponding nonlinear equations are called "completely integrable" and can be interpreted as the equations of motion for completely integrable infinite dimensional Hamiltonian systems. They have many interesting properties, like having Lax pairs, infinitely many conserved quantities and integrals of motion, soliton and multisoliton solutions, periodic and quasiperiodic solutions, recursion operators, Bäcklund transformations, etc.

Symmetry, group theory and differential geometry play an important role in the understanding of the structure of these equations, in generating the integrable equations and in finding their solutions. Many of the articles in this volume are devoted to the type of integrable system discussed above.

Most nonlinear equations, on the other hand, are not integrable and cannot be treated via the inverse scattering transform, nor its generalizations. They can of course be treated by numerical methods, which is the most common procedure. Interesting qualitative and quantitative features are often missed in this manner and it is of great value to be able to obtain at least particular exact analytic solutions of nonintegrable equations. Here again group theory comes to the rescue. Indeed, Lie group theory was originally created as a tool for solving ordinary and partial differential equations, be they linear, or nonlinear. New developments have also occurred in this area. Some of them have their origins in computer science. The advent of algebraic computing and the use of such computer languages as

REDUCE, MACSYMA, MAPLE, etc., have made it possible to write computer programs that construct the Lie algebra of the symmetry group of an equation, or at least greatly aid in its construction. Other important advances concern the theory of infinite dimensional Lie algebras, such as loop algebras, Kac-Moody and Virasoro algebras which frequently occur as Lie algebras of the symmetry groups of integrable equations in $2+1$ dimensions. Furthermore, practical and computerizable algorithms have been proposed for finding all subgroups of a given Lie group and for recognizing Lie algebras given by their structure constants.

The emphasis in this volume is precisely on these new developments in the theory of differential equations and nonlinear physical systems. The subjects treated can roughly be subdivided into the following mutually overlapping categories.

1. Completely integrable nonlinear dynamical systems, their Lax pairs, Hamiltonian formulations, recursion operators, etc. The articles by Fokas and Santini, by Magri, Morosi and Tondo and by Olver discuss the general schemes from which all properties of the integrable infinite-dimensional systems in $1+1$ or more dimensions can be obtained. Other articles are devoted to specific integrable systems, in particular integrable quantum spin chains (papers by Barouch, Fokas and Papageorgiou and by Fuchssteiner and Falck). The only known completely integrable nonlinear systems in n dimensions, where n is arbitrary, are generalizations of the sine-Gordon and wave equations. They are discussed in the article by Tenenblat.

2. Lie group theory and differential geometry as tools for solving nonlinear differential equations. Lectures that fall into this general category concern both integrable and nonintegrable equations. Kamran's contribution is an up to date review of the Cartan equivalence problem stressing its algorithmic features and its application to the classification of ordinary differential equations into equivalence classes. The general problem of variable separation in linear and nonlinear partial differential equations, and systems of equations, was reviewed by Miller, emphasizing its Lie algebraic aspects. A specific type of generalized symmetry, namely "reciprocal transformations", that go beyond Lie point symmetries, is discussed by Rogers. Saint-Aubin used the example of the CP^n sigma model to demonstrate the existence of symmetries that depend on the class of solutions that they are applied to. The occurrence of Kac-Moody-Virasoro algebras as Lie algebras of Lie groups of local point transformations leaving integrable nonlinear equations in $2+1$ dimensions invariant was reviewed by Winternitz.

3. Physical applications of nonlinear systems. Among these we mention the contribution of Holm who discussed, using numerous examples, mainly coming from fluid dynamics, the applicability and importance of various definitions of stability. Levi's contribution is motivated by the observed propagation of solitary waves through straits and in oceans. He presents a $2+1$ dimensional model, describing such waves, that in many interesting cases reduces to an integrable system. Wolf reviews the subject of nonlinear optics from the view point of Hamilton-Lie theory. The systematic use of group theory not only sim-

plifies calculations but provides a conceptual framework for the classification of optical aberrations.

4. Among the other topics covered in this volume are supersymmetry, as it occurs in nonrelativistic quantum mechanics: an original and important extension of the concept of symmetry group in quantum mechanics (see the first of Vinet's articles). Vinet also reviewed the group theoretical aspects of the Berry phase, playing an important role in quantum mechanics and quantum field theory. The article by Ortega and Caycedo is concerned with the existence and stability of positive periodic solutions of certain nonlinear eigenvalue problems. In his article, Zuluaga applies variational methods to determine conditions under which certain generalized Hammerstein integral equations have solutions.

The book is intended for graduate students in applied mathematics and theoretical physics, and also for experts in the field. The articles were written having clarity and pedagogical soundness in mind. Most of them contain extensive bibliographies which considerably increase their usefulness.

Minneapolis, September 8, 1988

D. Levi P. Winternitz

On the Time Evolution Equation of Spin Systems and Their Continuum Limit

E. Barouch, A.S. Fokas
V. Papageorgiou
Department of Mathematics and Computer Science
Clarkson University
Potsdam, New York 13676, U.S.A.

May 1988

Abstract

The recursion operator for the Landau-Lifshitz equation is presented and the relations between various discrete spin systems and their continuum limits are discussed.

1 Introduction

The anisotropic quantum spin Hamiltonian on a chain with nearest neighbor interaction (XYZ)is given by

$$H = -\frac{1}{2}\sum_{j=1}^{N}\left\{J_x S_j^1 S_{j+1}^1 + J_y S_j^2 S_{j+1}^2 + J_z S_j^3 S_{j+1}^3\right\} \tag{1}$$

where the operators S_j^ℓ $(\ell = 1,2,3)$ satisfy the Pauli commutation and anticommutation relations, i.e.

$$\left[S_n^j, S_m^k\right] = 2i\epsilon^{jk\ell} S_n^\ell \delta_{nm} \tag{2}$$

$$S_n^j S_n^k = \delta^{jk} + i\epsilon^{jk\ell} S_n^\ell \tag{3}$$

where $\epsilon^{jk\ell}$ is a totally antisymmetric tensor with $\epsilon^{1,2,3} = 1$. The equation of motion of the nth spin-vector is given by

$$\frac{d}{dt}\mathbf{S}_n = \mathbf{S}_n \wedge J_q\left(\mathbf{S}_{n+1} + \mathbf{S}_{n-1}\right) \tag{4}$$

where $\wedge$ is the standard vector product, $\mathbf{S} = (S_n^1, S_n^2, S_n^3)$, and J_q is diagonal matrix

$$J_q = diag(J_x, J_y, J_z). \tag{5}$$

Equation (4) admits a Lax pair. Consider the quantum operator L_n associated with the nth lattice site to be

$$L_n = \begin{bmatrix} \omega_4 + \omega_3 S_n^3 & \omega_1 S_n^1 - i\omega_2 S_n^2 \\ \omega_1 S_n^1 + i\omega_2 S_n^2 & \omega_4 - \omega_3 S_n^3 \end{bmatrix}, \tag{6}$$

with the Baxter parametrization

$$\omega_4 + \omega_3 = \rho sn(u + \eta, k) \tag{7.a}$$

$$\omega_4 - \omega_3 = \rho sn(u - \eta, k) \tag{7.b}$$

$$\omega_1 + \omega_2 = \rho sn(2\eta, k) \tag{7.c}$$

$$\omega_1 + \omega_2 = \rho k sn(2\eta, k) sn(u - \eta, k) sn(u + \eta, k) \tag{7.d}$$

satisfying the relation

$$J_x : J_y : J_z = \left[1 + ksn^2(2\eta)\right] : \left[1 - ksn^2(2\eta)\right] : \left[cn(2\eta)dn(2\eta)\right] \tag{8}$$

and $sn(u,k)$ is the standard Jacobi elliptic function.

Consider the system

$$U_{n+1} = L_n u_n \tag{9.a}$$

$$\frac{d}{dt}U_n = M_n u_n \tag{9.b}$$

and its compatibility condition

$$\frac{dL_n}{dt} = M_{n+1}L_n - L_n M_n. \tag{10}$$

The shift operator L_n is taken as the first member of the pair, and Sogo and Wadati [1] computed M_n by considering the evolution equation (4) in the explcit representation of dL_n/dt. They found the following structure for M_n:

$$M_n = M_n^4 I + \sum_{k=1}^{3} M_n^k \sigma^k \tag{11.a}$$

$$M_n^k = \sum_{j=1}^{3} F_j S_n^j S_{n-1}^j \tag{11.b}$$

$$M_n^k = G_k\left(S_n^k + S_{n-1}^k\right) + \sum_{i=1}^{3}\sum_{j=1}^{3} \epsilon^{ijk} H_k S_n^i S_{n-1}^j, \quad k = 1,2,3 \tag{11.c}$$

with the spectral functions F_j, G_k, H_k obtained explicitly.

2 The Landau-Lifshitz Equation

In a preprint from 1979, Sklyanin [2] took a heuristic continuum classical limit of (4) and obtained the Landau-Lifshitz (LL) equation

$$\mathbf{S}_t = \mathbf{S} \wedge \mathbf{S}_{xx} + \mathbf{S} \wedge J\mathbf{S}, \tag{12}$$

where J is the diagonal matrix

$$J = diag(J_1, J_2, J_3). \tag{13}$$

Let:

$$\mathbf{S}_n \to \mathbf{S}(x), \quad J_q \sim I + \frac{1}{2}\delta^2 J, \quad t \to \delta^{-2} t \tag{14}$$

$$\mathbf{S}_{n\pm1} \cong \mathbf{S}(x) \pm \delta \mathbf{S}_x(x) + \frac{1}{2}\delta^2 \mathbf{S}_{xx}. \tag{15}$$

Then as $\delta \to 0$, equation (4) yields equation (12). The Lax pair of the LL equation is given by Sklyanin [2] as

$$L = \sum_{j=1}^{3} S_j W_j \sigma_j, \tag{16}$$

$$V = \sum_{i,j,k=1}^{3} W_i \sigma_i S_j S_{k,x} \epsilon^{ijk} + 2\sum_{j=1}^{3} a_j S_j \sigma_j \tag{17}$$

with $a_1 = -W_2 W_3$ (and cyclic permutations), σ_j are the standard 2 x 2 Pauli spin matrices, and the elliptic functions W_j are given by

$$W_1^2 - W_3^2 = \frac{1}{4}(J_1 - J_3) \equiv \alpha \tag{18}$$

$$W_2^2 - W_3^2 = \frac{1}{4}(J_2 - J_3) \equiv \beta. \tag{19}$$

The LL Lax pair can be obtained as a limit of the quantum pair of Sogo and Wadati [1]. The importance of the Landau Lifshitz equation manifests itself in the reductions to other integrable systems. In particular Sklyanin considered the reduction $x \to R^{1/2}x$, $S_1 = -p/R$, $S_2 = \{1 - p^2/R^2\}^{1/2} sin\varepsilon$, $S_3 = \{1 - p^2/R^2\}^{1/2} cos\varepsilon$, $R(J_3 - J_2 \to \gamma$, $(J_3 - J_1)/R \to 1$, $(J_2 - J_1)/R \to 1$, $R \to \infty$. This reduction yields the sine-Gordon equation

$$\begin{aligned} p_t &= \varepsilon_{xx} - \tfrac{1}{2}\gamma sin 2\varepsilon \\ \varepsilon_t &= p. \end{aligned} \tag{20}$$

Furthermore, consider the reduction $t = R^{1/2}\tau$, $J_3 - J_1 = J_3 - J_2 = \gamma R$, $R \to \infty$ and define ψ as

$$\psi = R^{1/2} e^{i\gamma R \tau}(S_1 + S_2). \tag{21}$$

Substitution in the LL equation yields the nonlinear Schrödinger equation

$$i\psi_t = -\psi_{xx} - \frac{\gamma}{2}|\psi|^2\psi. \tag{22}$$

The central goal in "quantum inverse scattering" is the generation of a family of commuting Hamiltonians that provide a family of integrable evolution equations. A most natural method to generate the hierarchy of the LL evolution equation is the construction of a recursion operator [3]. The recursion operator for the LL equation, Φ_{LL}, is given by

$$\Phi_{LL} \doteqdot \Phi_{HM}^2 - \frac{1}{4}\pi\left((4A\mathbf{S}) \wedge (\mathbf{S} \wedge \cdot) - (D^{-1}\{\mathbf{S} \bullet (4A\mathbf{S}) \wedge (\mathbf{S} \wedge \cdot)\})\mathbf{S}_x - (D^{-1}\{\mathbf{S} \bullet (\mathbf{S} \wedge \cdot)_x\})(4A\mathbf{S} \wedge \mathbf{S})\right) \tag{23}$$

with

$$\Phi_{HM} \doteqdot -\frac{1}{2}\left[\mathbf{S} \wedge D - \{D^{-1}(\mathbf{S} \wedge \mathbf{S}_x \bullet \cdot)\}\mathbf{S}_x\right], \tag{24}$$

where

$$\pi\mathbf{a} \doteqdot -\mathbf{S} \wedge (\mathbf{S} \wedge \mathbf{a}) \text{ and } D = \frac{\partial}{\partial x}.$$

The existence of a recursion operator in bi-Hamiltonian form implies a hierarchy of integrable equations, as well as a hierarchy of Poisson structures. In particular the LL equation is a bi-Hamiltonian system with two compatible Hamiltonians. The hierarchy of evolution equations is obtained also in terms of the adjoint of Φ_{LL}, Ψ_{LL} which satisfies

$$\mathbf{S} \wedge (\Psi_{LL}\mathbf{p}) = \Phi_{LL}(\mathbf{S} \wedge \mathbf{p}) \equiv \Omega_{LL}\mathbf{p}, \tag{25}$$

for vectors $\mathbf{p}$ such that $\mathbf{p} \bullet \mathbf{S} = 0$, where Ω_{LL} is the second Hamiltonian operator. Explicitly we obtain:

$$\mathbf{S}_t = \mathbf{S} \wedge \Psi_{LL}^{n-1}(\alpha \mathbf{S} \wedge \mathbf{S}_x), \quad n = 1, 2, ..., \quad \alpha = const. \tag{26.a}$$

$$\mathbf{S}_t = \mathbf{S} \wedge \Psi_{LL}^{n-1}(\mathbf{c}), \quad n = 1, 2, 3, ... \tag{26.b}$$

and $\mathbf{c} = 0$ is a constant vector.

The original LL equation is obtained from n=2 of (26.b), while the next known equation of Date, Jimbo, Miwa and Kashiwara [4] is obtained from (26.a) with n=2. A similar study [5] was carried out for the XYh Hamiltonian given by

$$H = -\frac{1}{2}\sum_{j=1}^{N}\left\{J_x S_j^x S_{j+1}^x + J_y S_j^y S_{j+1}^y + h S_j^z\right\} \tag{27}$$

and both L_n and M_n were computed as well as the R matrix establishing its complete integrability.

The quantum equations of motion posses a heuristic classical continuum limit. However its complete integrability has not been established.

The time evolution of the quantum spin system is given by the Lax pair

$$\frac{d}{dt}\mathbf{S}_n = \mathbf{S}_n \wedge I_q\left(\mathbf{S}_{n+1} + \mathbf{S}_{n-1}\right) + hB\mathbf{S}_n \tag{28}$$

where the matrices I_q and B are given by

$$I_q = diag(J_x, J_y, 0) \tag{29.a}$$

$$B = \begin{bmatrix} 0 & 1 & 0 \\ -1 & 0 & 0 \\ 0 & 0 & 0 \end{bmatrix} \tag{29.b}$$

Equation (28) is completely integrable and its Lax pair has been derived as well as its hierarchy of mastersymmetries. Its continuum limit is given by

$$\frac{d}{dt}\begin{pmatrix} S_1 \\ S_2 \\ S_3 \end{pmatrix} = \begin{bmatrix} hS_2 - S_3(S_{2,xx} + j_2S_2) \\ -hS_1 + S_3(S_{1,xx} + j_1S_1) \\ S_1(S_{2,xx} + j_2S_2) - S_2(S_{1,xx} + j_1S_1) \end{bmatrix} \tag{30}$$

The R-matrix limit as well as the Lax pair limit has not been established, raising the suspicion that (30) is not completely integrable.

Acknowledgements

This work was partially supported by the Office of Naval Research under Grant Number N00014-76-C-0867, National Science Foundation under Grant Number DMS-8501325, and Air Force Office of Scientific Research under Grant Number 87-0310.

References

[1] Sogo, K.; Wadati, M.: Prog. Theor. Phys. **68** 85 (1982).

[2] Sklyanin, E.K.: On Complete Integrability of the Landau-Lifshitz Equation, Preprint LOMI, E-3-79, Leningrad, 1979.

[3] Barouch, E.; Fokas, A.S.; Papageorgiou, V.G.: The Bi-Hamiltonian Formulation of the Landau-Lifshitz Equation, INS #89, Clarkson University.

[4] Date, E.; Jimbo, M.; Kashiwara, M.; Miwa, T.: Landau-Lifshitz Equation: Solitons: Quasi Periodic Solutions and Infinitely Dimensional Lie Algebras, J. Phys. A. **16**, 221-236 (1983).

[5] Barouch, E.; Fuchssteiner, B.: Mastersymmetries and Similarity Equations of the XYh Model, Stud. Appl. Math. **73**, 221-237 (1985).

Conservation Laws for Integrable Systems

A.S. Fokas*
Department of Mathematics
Stanford University
Stanford, CA 94305

P.M. Santini
Universita Degli Studi - Roma
Istituto di Fisica "guglielmo Marconi"
Piazzale dell Scienze, 5
1-00185 Roma, ITALY

April 1988

Abstract

A unified approach to conserved quantities of integrable evolution equations in 1+1 and 2+1 is presented. The examples of Korteweg-deVries, nonlinear Schrödinger, Kadomtsev-Petviashvili, and Davey-Stewartson equations are used to illustrate the general theory.

*Permanent Address: Department of Mathematics and Computer Science, Clarkson University, Potsdam, New York 13676

1 Introduction

It is well known that equations integrable by the inverse scattering transofrm, both in 1+1 (one spatial and one temporal dimensions) and 2+1 (two spatial and one temporal dimensions) possess infinitely many conserved quantities.

For equations in 1+1 an effective approach to these conserved quantities is as follows: Associated with a given evolution equation there exists a recursion operator Φ which maps symmetries to symmetries. If the equation admits a Hamiltonian formulation, then the Hamiltonian operator Θ maps gradients of conserved quantities to symmetries. Thus, using Φ and Θ, (or alternatively the adjoint of Φ) a hierarchy of conserved gradients follows. Given a gradient function γ it is in principle possible to find its potential and thus obtain a hierarchy of conserved quantities. However, in practice, this procedure is rather cumbersome, and is thus desirable to seek a direct approach to finding conserved quantities.

There exist the following situation in the literature where an explicit construction of conserved quantities has been implemented:

(i) Let γ denote a conserved gradient for the Korteweg-deVries (KdV) or the Benjamin-Ono (BO) equation. Then $I = \int_R dx\gamma$ is the associated conserved quantity.

(ii) The mastersymmetry approach introduced in [1] yields an explicit construction of conserved quantities [2]-[6]. However, this approach has certain limitations discussed in [7].

It turns out that (i) and (ii) above are related. Indeed the origin of (i) is the existence of Galilean invariance which is, in a sense, a generator of the underlying mastersymmetry.

For equations in 2+1, the question of finding a recursion operator remained open until recently and thus the mastersymmetry formalism was the only feasible approach to obtaining conserved quantities.

We have recently derived the recursion operators associated with wide classes of integrable equations in 2+1 [8]-[10] and we have established the relation between recursion operators and mastersymmetries. Using this general theory we have been able to define explicitly the conserved quantities in terms of the recursion operator Φ, a starting conserved gradient function γ, and a suitable function T. In particular $I^{(n)} = \langle (\Phi^{\dagger})^n\gamma, T\rangle$, where $\dagger$ denotes the adjoint with respect to the bilinear form $\langle , \rangle$. For the KdV, $\langle a, b\rangle = \int_R dxab$, $T = 1$ and $I^{(n)} = \int_R dx\gamma^{(n)}$ are conserved quantities.

In this paper we first derive the equation for $I^{(n)}$ in general, and then illustrate its applicability for the KdV, nonlinear Schrödinger, Kadomtsev-Petviashvili and Davey-Stewartson equation.

2 General Notions and Results

Lemma 2.1

1. Let Φ be hereditary (Nijenhuis) [11] i.e.,

$$\Phi_d[\Phi v]w - \Phi\Phi_d[v]w \quad \text{is symmetric w.r.t. } v, w. \tag{2.1}$$

Then

$$[\Phi^n K_1, \Phi^m K_2]_L = \Phi^{n+m}[K_1, K_2]_L + \Phi^m \Sigma_{r=1}^n \Phi^{n-r} S_2 \Phi^{r-1} K_1 - \Phi^n \Sigma_{r=1}^m \Phi^{m-r} S_1 \Phi^{r-1} K_2, \tag{2.2}$$

where

$$S_i \doteq \Phi_d[K_i] + [\Phi, K_{i_d}], \tag{2.3}$$

[,] dentoes the usual commutator, d is an appropriate directional derivative, and the Lie bracket $[,]_L$ is defined by

$$[A, B]_L \doteq A_d[B] - B_d[A]. \tag{2.4}$$

Throughout this paper, m,n are nonnegative integers.

2. Let Θ be a Hamiltonian oeprator, i.e. Θ is skew symmetric with respect to an appropriate bilinear form $\langle,\rangle$ and Θ satisfies the Jacobi identity

$$\langle \Theta_d[\Theta A]B, C\ \rangle + \text{ cyclic permutation } = 0. \tag{2.5}$$

Then

$$[\Theta f, \Theta g]_L = \Theta grad\langle f, \Theta g\rangle + \Theta\left\{\left(f_d - f_d^\dagger\right)[\Theta g] - \left(g_d - g_d^\dagger\right)[\Theta f]\right\}, \tag{2.6}$$

where $\dagger$ denotes the adjoint of an operator with respect to $\langle,\rangle$ and the gradient of a functional I is defined by

$$I_d[v] = \langle gradI, v\rangle. \tag{2.7}$$

3. Let Φ be hereditary (Nijenhuis), Θ be a constant skew symmetric operator, and assume that

$$\Phi\Theta = \Theta\Phi^\dagger. \tag{2.8}$$

Then

$$(\Phi^m T)_d\Theta + \Theta(\Phi^m T)_d^\dagger = \Phi^m\left(T_d\Theta + \Theta T_d^\dagger\right) + \Sigma_{r=1}^m \Phi^{r-1}\Theta\left(\Phi^\dagger\right)^{m-r} S^\dagger, \tag{2.9}$$

where

$$S^\dagger = \Phi_d^\dagger[T] + \left[T_d^\dagger, \Phi^\dagger\right]. \tag{2.10}$$

Proof. The derivation of the above results is given in [10].

Theorem 2.1 Assume:

(i) Φ is hereditary (Nijenhuis).

(ii) $\Phi\Theta = \Theta\Phi^\dagger$, Θ constant skew symmetric operator.

(iii) γ is a gradient.

(iv) Φ is a strong symmetry for $\Theta\gamma$, i.e. $\Phi_d[\Theta\gamma] + [\Phi, (\Theta\gamma)_d] = 0$.

(v) $\Phi_d[T] + [\Phi, T_d] = \beta$, $T_d\Theta + \Theta T_d^\dagger = 0$.

(vi) $\Phi[\Theta\gamma, T]_L = \alpha\Theta\gamma$.

Then

$$\Phi^n(\Theta\gamma) = \frac{1}{\alpha + (n+1)\beta}\Theta \text{ grad } \langle\left(\Phi^\dagger\right)^{n+1}\gamma, T\rangle, \tag{2.11}$$

i.e.

$$I^{(n)} = \langle\left(\Phi^\dagger\right)^n\gamma, T\rangle \tag{2.12}$$

are conserved quantities for the hierarchy of equations generated by Φ.

Proof. Let $f = \gamma$, $\Theta g = \Phi^m T$ in (2.6) and use $\Theta_d = 0, \gamma_d = \gamma_d^\dagger$ to obtain

$$[\Theta\gamma, \Phi^m T]_L = \Theta \text{ grad}\langle\gamma, \Phi^m T\rangle - \left\{(\Phi^m T)_d\Theta + \Theta(\Phi^m T)_d^\dagger\right\}[\gamma]. \tag{2.13}$$

Equation (2.2) implies

$$[\Theta\gamma, \Phi^m T]_L = \Phi^m[\Theta\gamma, T]_L + \Phi^m\Sigma_{r=1}^n\Phi^{-r}S\Phi^{r-1}\Theta\gamma, \tag{2.14}$$

where we have used the fact that Φ is a strong symmetry for $\Theta\gamma$. Equation (2.9) implies

$$(\Phi^m T)_d\Theta + \Theta(\Phi^m T)_d^\dagger = \Sigma_{r=1}^m\Phi^{r-1}\Phi^{m-r}\Theta S^\dagger \tag{2.15}$$

since $T_d\Theta + \Theta T_d^\dagger = 0$, and $\Phi\Theta = \Theta\Phi^\dagger$. Substituting (2.14) and (2.15) in (2.13) and using $S = S^\dagger = \beta$, $\Phi[\Theta\gamma, T]_L = \alpha\Theta\gamma$ it follows that

$$\alpha\Phi^{m-1}\Theta\gamma \qquad = \Theta\text{grad } \langle\gamma, \Phi^m T\rangle - \beta m\Phi^{m-1}\Theta\gamma,$$

or

$$(\alpha + \beta m)\Phi^{m-1}(\Theta\gamma) = \Theta\text{grad}\langle\gamma, \Phi^m T\rangle = \Theta\text{grad } \langle(\Phi^\dagger)^m\gamma, T\rangle.$$

Remarks 2.1

(i) If Φ is factorizable in terms of compatible Hamiltonian operators then assumptions (i), (ii) are fulfilled.

(ii) The above theorem is useful for constructing conserved quantities of integrable equations provided that one can find γ and T satisfying assumptions (iii), (iv) and (v), (vi) respectively. In what follows we will present suitable such γ and T for several applications.

3 Applications in 1+1

3.1 The KdV Hierarchy

The Korteweg-deVries (KdV) equation

$$q_t = q_{xxx} + 6qq_x, \tag{3.1}$$

is associated with the hereditary operator

$$\Phi = D^2 + 4q + 2q_x D^{-1}, \quad D = \frac{\partial}{\partial x}, \quad D^{-1} f = \int_{-\infty}^{x} f(\xi) d\xi. \tag{3.2}$$

(i) It can be shown that Φ is hereditary [12].

(ii) Let $\Theta = D$, which is obviously constant and skew symmetric.

$$\Phi^\dagger = D^2 + 4q - 2D^{-1} q_x, \quad \text{thus} \quad D\Phi^\dagger = D^3 + 2(Dq + qD) = \Phi D.$$

(iii) Let $\gamma = q$, then since $\gamma' = 1, \gamma$ is a gradient.

(iv) Since Φ is invariant under x-translations, Φ is a strong symmetry for q_x.

(v) Let $T = 1$ and let prime denote the Frechét derivative, then $T' = 0$. Also

$$\Phi'[T] + [\Phi, T'] = 4, \quad \text{i.e.} \quad \beta = 4.$$

(vi) $[\Theta\gamma, T]_L = [q_x, 1]_L = D \bullet 1$, thus

$$\Phi[q_x, 1]_L = (D^2 + 4q + 2q_x D^{-1}) D \bullet 1 = 2q_x, \quad \text{i.e.} \quad \alpha = 2.$$

Thus theorem 2.1 implies:

Proposition 3.1 The functionals

$$I^{(n)} = \int_{-\infty}^{\infty} dx (\Phi^\dagger)^n q, \tag{3.3}$$

where Φ is defined in (3.2), are conserved quantities of the KdV hierarchy. For example $n = 0,\ \ 1$, imply

$$I^{(0)} = \int_{-\infty}^{\infty} dx q, \quad I^{(1)} = \int_{-\infty}^{\infty} dx (q_{xx} + 3q^2).$$

3.2 The NLS Hierarchy

The nonlinear Schödinger is a special reduction ($r = \pm q^*$) of

$$Q_t = \sigma Q_{xx} - 2qrQ, \quad Q \doteqdot \begin{pmatrix} 0 & q \\ r & 0 \end{pmatrix}, \quad \sigma \doteqdot \begin{pmatrix} 1 & 0 \\ 0 & -1 \end{pmatrix}. \tag{3.4}$$

(i) Rather than verifying that the associated recursion operator is hereditary we choose to derive this operator using the underline spectral problem. This derivation also implies that Φ is hereditary [13].

Consider the matrix eigenvalue problem (AKNS)

$$W_x = \lambda \sigma W + QW. \tag{3.5}$$

Equation (3.5) is compatible with

$$W_t = VW, \tag{3.6}$$

iff

$$Q_t = V_x - [Q, V] - \lambda[\sigma, V].$$

Considering the diagonal and off-diagonal parts of the above equations and noting that the product of off-diagonal matrices is diagonal and that $[\sigma, V] = [\sigma, V_0] = 2\sigma V_0$, we obtain

$$Q_t = V_{0_x} - [Q, V_D] - 2\lambda\sigma V_0 \tag{3.7}$$

$$0 = V_{D_x} - [Q, V_0], \tag{3.8}$$

where subscripts 0 and D denote off-diagonal and diagonal matrices respectively.

Let

$$V_0 = \Sigma_{j=0}^{n} \lambda^j V^j, \quad V^j \text{ off} - \text{diagonal}, \tag{3.9}$$

then (3.8) implies

$$V_D = D^{-1}\Sigma_{j=0}^{n} \lambda^j [Q, V^j]. \tag{3.10}$$

Hence, equation (3.7) yields

$$Q_t = \Sigma_{j=0}^{n} \lambda^j V_x^j - \Sigma_{j=0}^{n} \lambda^j \left[Q, D^{-1}[Q, V^j]\right] - 2\sigma \Sigma_{j=1}^{n+1} \lambda^j V^{j-1}.$$

Thus

$$V^n = 0, \quad V^{j-1} = \frac{1}{2}\sigma\left(V_x^j - \left[Q, D^{-1}[Q, V^j]\right]\right), \quad j = 1, \ldots, n-1, \quad Q_t = V_x^0 - \left[Q, D^{-1}[Q, V^j]\right]. \tag{3.11}$$

To solve the above equations, let

$$\Psi V_0 \doteq \sigma\left(V_{0_x} - \left[Q, D^{-1}[Q, V_0]\right]\right). \tag{3.12}$$

Thus

$$V^{j-1} = \frac{1}{2}\Psi V^j, \quad j = 1, \ldots, n-1, \quad V^n = 0, \quad Q_t = \sigma\Psi V^0. \tag{3.13}$$

Equation (3.8) with $V_0 = V^n = 0$ implies $V_{D_x}^n = 0$, and (3.11b) with $j = n$ implies

$$V^{n-1} = -\frac{1}{2}\sigma\,[Q, V_D^n]. \tag{3.14}$$

Equation (3.13a) implies $V^0 = \left(\frac{\Psi}{2}\right)^{n-1} V^{n-1} = -\frac{1}{2^n}\Psi^{n-1}\sigma\,[Q, V_D^n]$. Hence

$$Q_t = -\frac{1}{2^n}\sigma\Psi^n\sigma[Q, V_D], \quad V_{D_x} = 0. \tag{3.15}$$

Thus the eigenvalue problem (3.5) is associated with the equations

$$Q_t = \sigma\Psi^n\sigma[Q, V_D] = \Phi^n[Q, V_D], \tag{3.16}$$

where the action of Φ on off-diagonal matrices is given by

$$\Phi V_0 = D\sigma V_0 - \left[Q, D^{-1}[Q, \sigma V_0]\right], \tag{3.17}$$

and V_D is any diagonal matrix such that $V_{D_x} = 0$. Let $V_D = \sigma$, then $n = 0, 1, 2$ yield

$$\begin{aligned} Q_t &= \sigma Q \\ Q_t &= Q_x \\ Q_t &= \sigma Q_{xx} - 2q r Q. \end{aligned} \tag{3.18}$$

Remark 3.1 Since Q is a 2×2 off-diagonal matrix, symmetries and gradients of conserved quantities are 2×2 off-diagonal matrices. For example, it is easy to verify that σQ is a symmetry for Q_x:

$$[\sigma Q, Q_x]_L = (\sigma Q)'[Q_x] - (Q_x)'[\sigma Q] = \sigma Q_x - D(\sigma Q) = 0.$$

Let us verify that σQ_x is a gradient:

$$\langle (\sigma Q_x)'[V], U\rangle = \langle (\sigma D V) U\rangle\rangle = \int_{-\infty}^{+\infty} dx\ trace\ \sigma \begin{pmatrix} 0 & V_{1_x} \\ V_{2_x} & 0 \end{pmatrix} \begin{pmatrix} 0 & U_1 \\ U_2 & 0 \end{pmatrix} =$$

$$= \int_{-\infty}^{+\infty} dx (V_{1_x} U_2 - U_1 V_{2_x}) = \int_{-\infty}^{+\infty} dx (U_{1_x} V_2 - V_1 U_{2_x}) =$$

$$= \int_{-\infty}^{+\infty} dx\ trace \begin{pmatrix} 0 & U_{1_x} \\ -U_{2_x} & 0 \end{pmatrix} \begin{pmatrix} 0 & V_1 \\ V_2 & 0 \end{pmatrix} = \langle (\sigma D U) V\rangle.$$

In what follows we choose

$$\gamma = Q, \quad T = x\sigma Q.$$

Assumptions (i), (ii) of Theorem 2.1 are satisfied if Φ is given by (3.17) and $\Theta = \sigma$. Furthermore, Q is a gradient ($Q' = I$), and it is easily verified that Φ is a strong symmetry for σQ. $T' = x\sigma$, thus

$$T'\sigma + \sigma(T')^{\dagger} = x\sigma^2 - x\sigma^2 = 0.$$

$$\begin{aligned} \Phi'[T]V + [\Phi, T']V = &- \left[x\sigma Q, D^{-1}[Q, \sigma V]\right] - \left[Q, D^{-1}[x\sigma Q, \sigma V]\right] + \\ & D\sigma x \sigma V - \left[Q, D^{-1}[Q, xV]\right] - x\sigma\left\{D\sigma V - \left[Q, D^{-1}[Q, \sigma V]\right]\right\} \\ = &-x\left[\sigma Q, D^{-1}[Q, \sigma V]\right] + \left[Q, D^{-1}x[Q, V]\right] + V + xV_x \\ &- \left[Q, D^{-1}x[Q, V]\right] - xV_x + x\sigma\left[Q, D^{-1}[Q, \sigma V]\right] = V, \end{aligned}$$

i.e. $\beta = 1$. Also $[\sigma Q, x\sigma Q]_L = 0$, i.e. $\alpha = 0$. Thus:

Proposition 3.2 The functionals

$$I^{(n)} = \int_{-\infty}^{\infty} dx \quad \text{trace}\left((\Phi^\dagger)^n Q\right) x\sigma Q, \tag{3.19}$$

where Φ is given by (3.17) are conserved quantities for the hierarchy

$$Q_t = \Phi^m \sigma Q. \tag{3.20}$$

The NLS corresponds to $m = 2$ and $r = \pm q^*$.
$n = 0$ implies trace $Q\sigma Q = 0$.
$n = 1$ implies trace $\sigma Q_x \sigma Q = (qr)_x$.
Thus

$$I^{(1)} = \int_{-\infty}^{\infty} dx x (qr)_x = -\int_{-\infty}^{\infty} dx (qr).$$

4 Applications in 2+1

4.1 The KP Hierarchy

The Kadomtsev-Petviashvili (KP) equaiton

$$q_t = q_{xxx} + 6qq_x + 3\alpha^2 D^{-1} q_{yy}, \tag{4.1}$$

is associated with the following operator [8]

$$\Phi = D^2 + q^+ + Dq^+ D^{-1} + q^- D^{-1} q^- D^{-1}, \tag{4.2a}$$

where

$$q^\pm = q_1 \pm q_2 + \alpha(D_1 \mp D_2), \quad D_i \doteq \frac{\partial}{\partial y_i}, \quad q_i = q(x, y_i, t). \tag{4.2b}$$

It was shown in [9] that the above operator is hereditary. Also if $\Phi^\dagger$ denotes the adjoint of Φ with respect to the bilinear form

$$\langle f_{12}, g_{12} \rangle = \int_{R^3} dx dy_1 dy_2 f_{21} g_{12}, \quad f_{12} \doteq f(x, y_1, y_2), \tag{4.3}$$

then

$$\Phi D = D \Phi^\dagger.$$

The directional derivative of the operators $q^\pm$ is defined by

$$q_d^\pm[\sigma] f \doteq \int_R dy_3 \left(\sigma_{13} f_{32} \pm \sigma_{32} f_{13}\right). \tag{4.4}$$

Associated with the above operator there exist two starting symmetry operators

$$\hat{N} \doteq q^-, \quad \hat{M} \doteq Dq^+ + q^- D^{-1} q^-. \tag{4.5}$$

It was shown in [9] that:

1. Φ is a strong symmetry for $\hat{K} \bullet 1$, $\hat{K} = \hat{N}$ or $\hat{M}$.
2. $D^{-1}\hat{K} \bullet 1$ is an extended gradient function.

Let $T = \delta_{12} \doteqdot \delta(y_1 - y_2)$, where δ denotes the Dirac's distribution. Then $T_d = 0$. Also

$$\Phi_d[T]f + [\Phi, T_d]f = q^+[\delta]f + Dq^+[\delta]D^{-1}f + q^-[\delta]D^{-1}q^- D^{-1}f +$$

$$q^- D^{-1}q^-[\delta]D^{-1} = 4f,$$

since

$$q_d^+[\delta]f = \int_R dy_3\,(\delta_{13}f_{32} + \delta_{32}f_{13}) = 2f$$

$$q_d^-[\delta]f = \int_R dy_3\,(\delta_{13}f_{32} - \delta_{32}f_{13}) = 0.$$

Thus $\beta = 4$. Also,

$$[\hat{N}, \delta]_d = [q^-, \delta]_d = q^-[\delta] = 0$$

$$[\hat{M}, \delta]_d = [Dq^+ + q^- D^{-1}q^-, \delta]_d = Dq_d^+[\delta] + q_d^-[\delta]D^{-1}q^- + q^- D^{-1}q^-[\delta] = 2D$$

Hence

$$\Phi[\hat{M} \bullet 1, \delta]_d = \left(D^2 + q^+ + Dq^+D^{-1} + q^- D^{-1}q^- D^{-1}\right) 2D \bullet 1 = 2\hat{M} \bullet 1.$$

Thus $\alpha = 0$ if $\hat{K} = \hat{N}$ and $\alpha = 2$ if $\hat{K} = \hat{M}$.

Proposition 4.1 The functionals

$$I^{(n)} = \int_{R^2} dx dy_1 \gamma_{11}^n, \quad \gamma_{11}^n = (\gamma_{12}^n)_{2=1}, \gamma_{12}^n \doteqdot \left(\Phi^\dagger\right)^n D^{-1}\hat{K}, \tag{4.6}$$

where $\hat{K} = \hat{N}$ or $\hat{M}$, defined in (4.5) and Φ is given by (4.2), are conserved quantities of the KP hierarchies.

4.2 The DS Hierarhcies

The Davey-Stewartson (DS) equation

$$iq_t + \frac{1}{2}(q_{xx} + \alpha^2 q_{yy}) = q(\varphi - |q|^2)$$

$$\varphi_{xx} - \alpha^2\varphi_{yy} = 2|q|^2_{xx}, \tag{4.7}$$

is the reduction $r = q^*$ of the following equation

$$Q_t = \sigma\left(Q_{xx} + \alpha^2 Q_{yy}\right) + QA - AQ, \quad Q = \begin{pmatrix} 0 & q \\ r & 0 \end{pmatrix},$$
$$(D_x - JD_y)\, A = (D_x + JD_y)\, \sigma Q^2 \tag{4.7}$$

where $\sigma = diag(1, -1)$.

A. Derivation of the Associated Recursion Operator

We will show that the eigenvalue problem

$$W_x = JW_y + QW, \quad J = \alpha\sigma, \tag{4.8}$$

gives rise to the hereditary operator

$$\Phi = \sigma(P - Q^+ P^{-1} Q^+), \tag{4.9}$$

where

$$PF \doteqdot F_x - JF_{y_1} - F_{y_2}J, \quad Q^\pm F \doteqdot Q_1 F \pm FQ_2, \quad Q_i \doteqdot Q(x, y_i, t), \tag{4.10}$$

and Φ acts only on off diagonal matrices.

The above operator is derived in [9] : Associated with

$$W_x = \hat{Q}W, \quad \hat{Q} \doteqdot Q + JD_y,$$

we consider the flow

$$W_t = \hat{V}W.$$

The above equations are compatible iff the following operator equation is valid

$$\hat{Q}_t = \hat{V}_x - [\hat{Q}, \hat{V}]. \tag{4.11}$$

Let

$$(\hat{V}F)(x, y_1) \doteqdot \int_R dy_2 V(x, y_1, y_2) F(x, y_2). \tag{4.12}$$

Then

$$[\hat{Q}, \hat{V}]F = \int_R dy_2 \left(Q^- V + JV_{y_1} + V_{y_2} J\right) F_2, \quad \hat{V}_x F = \int_R dy_2 V_x F.$$

Also $\hat{Q}_t F = Q_t F = \int_R dy_2 \delta_{12} Q_{2_t}$, $\delta_{12} = \delta(y_1 - y_2)$. Hence, equation (4.11) implies the distribution equation

$$\delta_{12} Q_{2_t} = PV - Q^- V,$$

where P, Q^- are defined in (4.10). Letting $V = V_D + V_0$, where the subscripts denote diagonal and off-diagonal parts we obtain

$$\delta_{12}Q_{2_t} = PV_0 - Q^-V_D, \quad PV_D - Q^-V_0 = 0. \tag{4.13}$$

Let

$$V_0 = \Sigma_{j=0}^n \delta^j V^j, \quad V^j \ \text{off}-\text{diagonal}, \quad \delta^j = \left(\frac{\partial}{\partial y_1}\right)^j \delta. \tag{4.14}$$

Then (4.13b) implies

$$V_D = P^{-1}Q^-\Sigma_{j=0}^n \delta^j V^j.$$

Substituting for V_0, V_D in (4.13a) we obtain

$$\delta_{12}Q_{2_t} = \left(P - Q^-P^{-1}Q^-\right)\Sigma_{j=0}^n \delta^j V^j.$$

It can be shown that

$$\left[P - Q^-P^{-1}Q^-, \ \delta^j\right] V_0 = -2\alpha\delta^{j+1}\sigma V_0,$$

hence

$$\delta_{12}Q_{12_t} = \Sigma_{j=0}^n \delta^j \left(P - Q^-P^{-1}Q^-\right) V^j - 2\alpha\Sigma_{j=1}^{n+1}\delta^j\sigma V^{j-1}.$$

Thus

$$V^n = 0, \quad V^{j-1} = \frac{1}{2\alpha}\sigma\left(P - Q^-P^{-1}Q^-\right)V^j, \quad j = 1, ..., n-1, \quad \delta_{12}Q_{2_t} = \left(P - Q^-P^{-1}Q^-\right)V^0. \tag{4.15}$$

To solve the above equations, we define the operator Ψ acting on off-diagonal matrices by

$$\Psi \doteq \sigma\left(P - Q^-P^{-1}Q^-\right). \tag{4.16}$$

Thus

$$V^n = 0, \quad V^{j-1} = \frac{\Psi}{2\alpha}V^j, \quad j = 1, ..., n, \quad \delta_{12}Q_{2_t} = \sigma\Psi V^0.$$

If $V^n = 0$, (4.13b) implies $PV_D^n = 0$, then (4.15b) yields $V^{n-1} = -\frac{1}{2}\alpha\sigma Q^- V_D^n$. Thus

$$V^0 = \left(\frac{\Psi}{2\alpha}\right)^{n-1} V^{n-1} = -\frac{\Psi^{n-1}}{(2\alpha)^n}\sigma Q^- V_D^n.$$

Hence

$$\delta_{12}Q_{2_t} = -\frac{\sigma\Psi^n}{(2\alpha)^n}\sigma Q^- V_D = \frac{\sigma\Psi^n}{(2\alpha)^n}Q^+\sigma V_D = -\frac{1}{(2\alpha)^n}\Phi^n Q^- V_D, \tag{4.17}$$

where

$$\Phi\sigma = \sigma\Psi, \quad \text{i.e.} \quad \Phi = \sigma\left(P - Q^+P^{-1}Q^+\right), \tag{4.18}$$

and we have used $\sigma Q^- = -Q^+\sigma$.

B. Derivation of the DS Equation

Integrating equation (4.17) we obtain

$$Q_{1_t} = \int_R dy_2 \delta \sigma \Psi^n Q^+ \sigma V_D = -\int_R dy_2 \delta \Phi^n Q^- V_D, \tag{4.19}$$

where Ψ, Φ are defined in (4.16), (4.18), $Q^{\pm}$ are defined in (4.10) and V_D is any diagonal matrix such that $PV_D = 0$, P given by (4.10). The DS equation corresponds to $V_D = \sigma$ and $n = 2$. Letting$V_D = \sigma$ in (4.19) we find

$$Q_{1_t} = \int_R dy_2 \delta \sigma \Psi^n Q^+ I \tag{4.20}$$

(i) <u>$n = 0$</u>
$Q^+ I = Q_1 + Q_2$, thus $Q_t = \sigma Q$.
(ii) <u>$n = 1$</u>
We first note that

$$PF = F_x - JF_{y_1} - F_{y_2}J = \begin{cases} D - J(D_1 + D_2)F, & F \text{ diag} \\ D - J(D_1 - D_2)F, & F \text{ off}-\text{diag}. \end{cases}$$

$$\Psi Q^+ I = \Psi(Q_1 + Q_2) = \sigma\{P(Q_1 + Q_2) - Q^- P^{-1} Q^- (Q_1 + Q_2)\} = \sigma\{P(Q_1 + Q_2) - Q^- F\},$$

where

$$(D_x - J(D_1 + D_2))\, F = Q_1^2 - Q_2^2. \tag{4.21}$$

Thus

$$\sigma \Psi Q^+ I = (Q_1 + Q_2)_x - J(D_1 - D_2)(Q_1 + Q_2) - Q^- F,$$

or

$$\sigma \Psi Q^+ I = (Q_1 + Q_2)_x - J\left(Q_{1_{y_1}} - Q_{2_{y_2}}\right) - (Q_1 F - F Q_2). \tag{4.22}$$

Hence

$$\int_R dy_2 \delta \sigma \Psi Q^+ I = 2Q_{1_x} - Q_1 F_{11} + F_{11} Q_1. \tag{4.23}$$

However, equation (4.21) implies

$$\int_R dy_2 \delta\, (D_x - J(D_1 + D_2))\, F = 0,$$

or

$$(D_x - JD_1) \int_R dy_2 \delta F = 0,$$

or

$$(D_x - JD_1)\,F_{11} = 0, \tag{4.24}$$

where we have used $\delta(D_1+D_2) = D_1\delta$. Equation (4.23) with appropriate boundary conditions at infinity implies $F_{11} = 0$ and hence equaiton (4.23) yields

$$Q_t = Q_x.$$

(iii) $\underline{n=2}$

$$\Psi^2 Q^+ I = \Psi^2(Q_1+Q_2) = \sigma\left(P - Q^- P^{-1} Q^-\right)\Psi(Q_1+Q_2) = \sigma P\Psi(Q_1+Q_2) - \sigma Q^- G, \tag{4.25}$$

where

$$(D_x - J(D_1+D_2))G = Q^-\Psi(Q_1+Q_2) = Q^-\sigma\left(P - Q^- P^{-1} Q^-\right)(Q_1+Q_2) =$$

$$= -\sigma Q^+\left\{(Q_1+Q_2)_x - J\left(Q_{1_{y_1}} - Q_{2_{y_2}}\right)\right\} + \sigma Q^+ Q^- F.$$

Thus

$$(D_x - J(D_1+D_2))G = -\sigma Q^+\left\{(Q_1+Q_2)_x - J\left(Q_{1_{y_1}} - Q_{2_{y_2}}\right)\right\} + \sigma\left\{\left(Q_1^2 - Q_2^2\right)\right\}F.$$

Hence

$$(D_x - JD_1)G_{11} = -2\sigma\left(Q_1^2\right)_x. \tag{4.26}$$

Equation (4.25) yields

$$\Psi^2 Q^+ I = \sigma P\sigma\left(P - Q^- P^{-1} Q^-\right)(Q_1+Q_2) - \sigma Q^- G$$

$$= P^2(Q_1+Q_2) - PQ^- F - \sigma Q^- G$$

$$= P^2(Q_1+Q_2) - (Q_1F - FQ_2)_x + J(D_1 - D_2)(Q_1F - FQ_2) - \sigma(Q_1G - GQ_2)$$

$$= P^2(Q_1+Q_2) - (Q_1F - FQ_2)_x + J\left\{Q_1(D_1 - D_2)F - ((D_1 - D_2)F)Q_2\right\} + \left(Q_{1_{y_1}} - Q_{2_{y_2}}\right)F - \sigma(Q_1G - G$$

Thus

$$\int_R dy_2\,\delta\sigma\Psi^2 Q^+ I = 2\sigma\left(Q_{1_{xx}} + \alpha^2 Q_{1_{y_1y_1}}\right) - (Q_1G_{11} - G_{11}Q_1)$$

$$+\sigma J\int_R dy_2\left\{Q_1(D_1 - D_2)F - ((D_1 - D_2)F)Q_2\right\}, \tag{4.27}$$

where we have used $F_{11} = 0$. But $PF = Q_1^2 - Q_2^2$ implies

$$\delta(D_1 - D_2)PF = \delta(D_1 - D_2)(Q_1^2 - Q_2^2) = \delta P(D_1 - D_2)F,$$

thus

$$(D_x - JD_1)\int_{-\infty}^{\infty} dy_2\delta(D_1 - D_2)F = 2(Q_1^2)_{y_1}. \tag{4.28}$$

Let

$$-2A \doteq F_{11} - \alpha\int_R dy_2\delta(D_1 - D_2)G, \tag{4.29}$$

then equation (4.27) yields

$$Q_t = 2\sigma(Q_{xx} + \alpha^2 Q_{yy}) + 2QA - 2AQ,$$

while(4.26), (4.28) imply

$$-2\left(D_x - JD_{y_1}\right)A = -2\sigma(Q_1^2)_x - 2\alpha(Q_1^2)y_1 = -2\left(D_x + JD_{y_1}\right)Q_1^2.$$

Hence

$$Q_t = \sigma\left(Q_{xx} + \alpha^2 Q_{yy}\right) + QA - AQ$$

$$(D_x - JD_y)A = (D_x + JD_y)Q^2. \tag{4.30}$$

C. Conserved Quantities

Associated with the recursion operator Φ there exist two starting symmetry operators

$$\hat{N} = Q^-, \quad \hat{M} = Q^-\sigma = -\sigma Q^+. \tag{4.31}$$

It was shown in [10] that:

1. Φ is a strong symmetry for $\hat{K}\cdot I$, $\hat{K} = \hat{N}$ or $\hat{M}$.
2. σQ^-, Q^+ are extended gradient functions.

Let

$$T = \frac{x}{2}\sigma Q^+\delta I. \tag{4.32}$$

Then

$$T_d[f] = \frac{x}{2}\sigma Q_d^+[f]\delta I = \frac{x}{2}\sigma\int_R dy_3(f_{13}\delta_{32} + \delta_{13}f_{32}) = x\sigma f.$$

Thus

$$\sigma T_d + T_d^\dagger\sigma = 0.$$

Also it can be shown (see Appendix A of [10]) that

$$\Phi_d[T]f + [\Phi, T_d]f = f, \quad i.e.\ \beta = 1.$$

$$\Phi[\hat{K} \cdot I, T] = 0, \qquad i.e.\ \alpha = 0.$$

Hence Theorem 2.1 implies:

Proposition 4.2 The functions

$$I^{(n)} = \int_{R^2} dx dy_1 x \ \ \text{trace}\left(\Phi^n \hat{K} \cdot I\right)_{11} Q_1, \tag{4.33}$$

where Φ is defined by (4.9), $\hat{K} = Q^-$ or σQ^+, are conserved quantities of the DS hierarchy.

Acknowledgements

This work was partially supported by the Office of Naval Research under Grant Number N00014-76-C-0867, National Science Foundation under Grant Number DMS-8501325, and Air Force Office of Scientific Research under Grant Number 87-0310. A.S. Fokas is grateful to J. Keller and his group at the Department of Mathematics, for their hospitality during his sabbatical leave at Stanford University.

References

[1] A.S. Fokas and B. Fuchssteiner, Phys. Lett. A **86** 341 (1981).

[2] W. Oevel and B. Fuchssteiner, Phys. Lett. A **88**, 323 (1982).

[3] I.Y. Dorfman, Deformation of Hamiltonian structures and integrable systems, to appear.

[4] B. Fuchssteiner, Progr. Theoret. Phys. **70** 150 (1983).

[5] W. Oevel, A Geometrical approach to integrable systems admitting scaling symmetries, to appear.

[6] H.H. Chen, Y.C. Lee, and J.E. Lin, Phys. D **9** 439 (1983); Phys. Lett. A **91**, 381 (1982).

[7] A.S. Fokas, Stud. Appl. Math. **77** 253-299 (1987).

[8] A.S. Fokas and P.M. Santini, Stud. Appl. Math. **75** 179 (1986).

[9] P.M. Santini and A.S. Fokas, Commun. Math. Phys. **115** 375-419 (1988).

[10] A.S. Fokas and P.M. Santini, Recursion operators and bi-Hamiltonian structures in multidimensions II, INS 67, to appear in Commun. Math. Phys.

[11] B. Fuchssteiner, Nonlinear Anal. **3** 849 (1979).

[12] A.S. Fokas and B. Fuchssteiner, Lett. Nuovo Cimento **28** 299 (1980); B. Fuchssteiner and A.S. Fokas, Phys. D **4** 47 (1981).

[13] A.S. Fokas and R.L. Anderson, J. Math. Phys. **23** 1066 (1982).

Computer Algorithms for the Detection of completely integrable Quantum Spin Chains

Benno Fuchssteiner
Uwe Falck

Universität Paderborn
Germany

ABSTRACT

Algorithms to test and to detect complete integrability for quantum mechanical spin-1/2 chains are given and their implementations in MAPLE are described. Symmetry group generators and their corresponding recursion formulas (master symmetries) for several spin chains are given.

1. Introduction

Although completely integrable quantum spin chains originated in statistical mechanics their wide range of applications goes far beyond the boundaries of this special field. Nowadays they find applications in mathematical economy (e.g. interaction models) as well as in certain areas of pure and applied mathematics which are rather unrelated to statistical mechanics (e.g. algebraic geometry, knot-theory, partial differential equations, etc.).

Much of the work in this area centers around the deep contributions which lead to the solution of the Heisenberg anisotropic spin chain (or the so called XYZ-model) having the Hamiltonian

$$H_{XYZ}=\sum_{n=1}^{N} J_x\sigma^x{}_n\sigma^x{}_{n+1}+J_y\sigma^y{}_n\sigma^y{}_{n+1}+J_z\sigma^z{}_n\sigma^z{}_{n+1}. \tag{1.1}$$

Here, J_x , J_y and J_z are arbitrary coefficients and the $\sigma^r{}_n$, $n=1,\dots,N$; $r=x,y,z$ are a representation of the Pauli spin matrices fulfilling the following relations

$$\sigma^x{}_n\sigma^y{}_n=i\sigma^z{}_n \quad \textit{(and cyclic permutations of x,y,z)} \tag{1.2}$$

$$\sigma^r{}_n\sigma^r{}_n=1 \tag{1.3}$$

$$\sigma^r{}_n\sigma^{\bar{r}}{}_m=\sigma^{\bar{r}}{}_m\sigma^r{}_n \quad \textit{for } n,m=1,\dots,N \textit{ with } n\neq m. \tag{1.4}$$

Formally we assumed in (1.1) that

$$\sigma^r_{N+k}=\sigma^r_k \text{ for all } k \tag{1.5}$$

thus prescribing periodic boundary conditions.

Already the solution (i.e. the determination of the eigenvectors) of the completely isotropic case $J_x=J_y=J_z$ (the XXX-model), which was given in 1931 by Bethe [11] counts as one of the fundamental results in the theory of spin models. About 40 years later the work on the XYZ-model culminated in R.J. Baxters brilliant work which allowed him to solve the fully anisotropic case by linking it to the so called 8-vertex-model. Already the computation of the ground state energy [6] [7] was a considerable achievement, which then lateron was crowned [8],[9],[10] by the complete determination of the eigenvectors of the so called transfer matrix. In between, pioneering work was done by Kramers-Wannier [30] (introduction of the transfer matrix) Onsager [35] (solution of the two- dimensional Ising model) and by Lieb [32] (XXZ-model, 6-vertex model).

Since all this work is highly nontrivial and technically complicated a new level of understanding was reached when Takhtadzhan and Fadeev [47] introduced the QIST (quantum inverse scattering theory) and applied it successfully to the XYZ-model. This application was further extended by Sogo-Wadati [45] who also gave the correct Lax-pair for the anisotropic spin chain.

The wide applicability of results on spin-chains can be seen from the fact that even particular cases and their classical continuum limits are leading to highly interesting dynamical systems. For example, many completely integrable partial differential equations in 1+1 dimensions are reductions of the Landau-Lifshitz equation [31], which itself is the classical continuum limit of the XYZ-model. This applies among others to the nonlinear Schrödinger equation and the sine-Gordon equation [44] (in real life coordinates). Further interesting spin chains are those with external magnetic fields, for example the XYh-model

$$H_{XYh}=\sum_{n=1}^{N} J_x\sigma^x_n\sigma^x_{n+1}+J_y\sigma^y_n\sigma^y_{n+1}+h\sigma^z_n \tag{1.6}$$

which also is completely integrable [4],[12],[39] and which is equivalent to the two-dimensional Ising model [46],[3].

The last model belongs to an extended class of integrable models, namely those which can be diagonalized by the Jordan-Wigner transformation[34],[12]. But apart from these and the higher Hamiltonians of the XYZ-model not very many integrable spin chains are known.

Since all the remarkable results and striking applications of the interesting spin chains are mostly due to complete integrability, it certainly is of interest to ask whether other completely integrable spin chains can be found.

Certainly, there are completely integrable spin chains of higher order which are not yet known, but already the computations necessary to show the complete integrability of the XYZ-model are so involved that it seems completely hopeless to look for more complex examples of complete integrability in this area. In spite of the fundamental nature of complete integrability, up to now, there is no algorithm in order to determine whether or not a system has this property [25,p.212].

Such an algorithm for spin-1/2 chains and its computer implementation is the content of this paper. Our aim is a computeralgebra package doing the following:

Given a spin-1/2 Hamiltonian H then (if existent) the computer should give us:

a) The first nontrivial Hamiltonian commuting with the given H

and

b) A recursion formula for higher order Hamiltonians commuting with H.

In this aim we succeeded. So the Lie-Bäcklund symmetries for the dynamical system given by the spin Hamiltonian are computed by a package based on a formula manipulation system (in our case MAPLE [23]).

Surprisingly, the CPU-times for the determination of complete integrability for spin chains with nearest neighbor interaction, or nearest-nearest neighbor interaction, are rather moderate. This is unexpected because even the determination of the far less complex Lie-point symmetries for dynamical systems by computers are leading to rather impressive running times (see the pioneering work of F. Schwarz [40],[41],[42],[43]). The reason for this surprising behavior is that the algorithm proposed in this paper already uses very many features of the structural properties of complete integrability (especially the existence of mastersymmetries which corresponds, for finite dimensional systems, to the existence of angle-variables [21]). This is in contrast to the algorithm which usually is applied for the determination of Lie-point symmetries, and which works in full generality also for those systems which are not completely integrable.

The paper is organized in the following way: In section 2 a brief survey of the theoretical background leading to mastersymmetries is given on a general level. In section 3 the details for an approximate solution of the division problem in the Lie algebra of spin operators are described and in section 4 the algorithm for finding symmetries and recursion formulas is to be found. In 5 details on the implementation are reported and technical details together with typical running times are given. Section 6 contains a brief overview on results and "surprises". In the concluding section further problems and directions for further research are indicated.

2. Background

We consider dynamical systems of the form

$$A_t = K(A), \quad A \in M \tag{2.1}$$

where $K(A)$ is a C^∞-vector field on a manifold M. Generally, one is interested in one-parameter diffeomorphism groups on M leaving the flow (2.1) invariant. Alas, apart from trivial cases, an explicit determination of such a group almost never is possible. Therefore one considers infinitesimal descriptions of these groups. So one-parameter groups are described by their infinitesimal generators. A vector field $G(A)$ is an infinitesimal generator of a one-parameter symmetry group for (2.1) if (2.1) commutes with the flow

$$A_t = G(A), \tag{2.2}$$

which exactly happens when G and K commute with respect to the vector field Lie algebra. In case the manifold M is a vector space this vector field commutator is given by

$$[G,K] = [G,K](A) = \frac{\partial}{\partial \varepsilon}\Big|_{\varepsilon=0} \{K(A+\varepsilon G(A)) - G(A+\varepsilon K(A))\}. \tag{2.3}$$

Thus the infinitesimal aspect of determining the symmetry groups of (2.1) requires the determination of $K^@$,the commutant

$$K^@ = \{G \in \mathbf{L} \mid [K,G]=0\} \tag{2.4}$$

of K with respect to the Lie algebra $\mathbf{L}$. Of course K is a member of $K^@$. A hamiltonian system on a 2N-dimensional manifold [1] is completely integrable if $K^@$ (or a suitable subalgebra) is an N-dimensional abelian Lie algebra of hamiltonian vector fields. For infinite dimension the notion of complete integrability is not yet completely clear, usually one requires that $K^@$ is sufficiently large and abelian. Thus a test for complete integrability is reduced to the construction of sufficiently large abelian subalgebras of $K^@$.

In case of a quantum mechanical system the manifold under consideration is $M = L_{sa}(\mathbf{H},\mathbf{H})$ the set of all selfadjoint operators on a suitable Hilbert space $\mathbf{H}$, and given a fixed Hamiltonian $H \in L_{sa}(\mathbf{H},\mathbf{H})$, the dynamics on M is defined to be

$$A(t)_t = i[H, A(t)], \quad A \in L_{sa}(\mathbf{H},\mathbf{H}). \tag{2.5}$$

Here the bracket [,] is the usual commutator of operators in Hilbert space. Thus the vector field $K(A)$ is of the special form

$$K(A)=i[H\ ,\ A]. \tag{2.6}$$

A simple exercise shows that for all vector fields of the same type

$$G(A)=i[\hat{H}\ ,\ A] \tag{2.7}$$

we have

$$[G(A)\ ,\ K(A)]=0 \textit{ if and only if } [H\ ,\ \hat{H}]=0. \tag{2.8}$$

So the determination of these symmetry groups requires finding those $\hat{H}$ which commute with H . This amounts to the same as the description of the spectral properties of H.

Nevertheless, the determination of the one-parameter symmetry groups of (2.5) is not completely trivial although system (2.5) obviously is a linear flow. The difficulty stems from the fact that the manifold $M=L(\mathbf{H},\mathbf{H})$ is rather large in dimension. In order to overcome this difficulty we change the viewpoint. This change of viewpoint we demonstrate first for the most simple quantum mechanical systems, the **harmonic oscillator**. There the Hamiltonian is given by

$$H=\frac{1}{2}(p^2+q^2) \tag{2.9}$$

where p and q are selfadjoint with

$$[q\ ,\ p]=i. \tag{2.10}$$

The physical quantities are scaled such that $h=2\pi$. All irreducible representations of (2.9) are isomorphic, therefor, for the description of the dynamics of the system it is not at all necessary to fix the manifold $M=L(\mathbf{H},\mathbf{H})$ under consideration. So, in order to get rid of unnecessary quantities we now consider the dynamics, given by H, as a flow on the manifold given by all pairs of p,q such that (2.10) holds on a suitable Hilbert space. Then the dynamic

$$q_t=i[H\ ,\ q]=-ip \tag{2.11}$$

$$p_t=i[H\ ,\ p]=iq \tag{2.12}$$

can be written

$$\begin{bmatrix} q \\ p \end{bmatrix}_t = \begin{bmatrix} 0 & -1 \\ 1 & 0 \end{bmatrix} \begin{bmatrix} q \\ p \end{bmatrix} \tag{2.13}$$

in complete analogy to the classical harmonic oscillator. Only the p and q are not classical quantities any more but rather elements of an operator manifold. This change of manifold leads to a rather considerable change of the mathematical structure of the corresponding vector fields. Whereas for (2.5) all selfadjoint operators lead to vector fields, now for (2.12) vector fields have to be operator-valued functions in p and q, for example polynomials. For brevity we call from now on the pairs (p,q) **field variables.** It should be remarked that this change of viewpoint is rather dramatic because according to (2.5) the dynamic of every quantum dynamical system is linear whereas according to this new viewpoint the dynamic may be highly nonlinear. For example, if we take the Hamiltonian

$$H=\frac{1}{4}(p^4+q^4) \tag{2.14}$$

instead of (2.9) then the system given on the manifold of all possible (p,q) is

$$\begin{bmatrix} q \\ p \end{bmatrix}_t = \begin{bmatrix} 0 & -1 \\ 1 & 0 \end{bmatrix} \begin{bmatrix} q^3 \\ p^3 \end{bmatrix} \tag{2.15}$$

certainly a nonlinear flow now.

To sum up: By introduction of the field variables we have reduced the size of the manifold considerably thus taking into account the disadvantage that now the dynamics may be nonlinear. But there is an additional advantage coming from this viewpoint, namely that now we have a more precise idea what complete integrability may be for quantum systems. To be concrete, we call a quantum system **completely integrable** if it is completely integrable with respect to the last viewpoint. There complete integrability is rather precise since our new viewpoint makes out of the quantum mechanical system a classical flow (on a manifold of a rather considerable size).

In case of the spin chains the field variables are the σ^r_n, $r=x,y,z$, $n=1,\dots,N$. The manifold under consideration is the set of all tuples of the σ^r_n such that relations (1.2) - (1.4) are fulfilled. The vector fields will be the polynomials in these variables. We perform all computations in the Lie algebra given by commutators of operators instead in the Lie algebra of vector fields. This is justified because of (2.8).

Hence, for a given polynomial H in the field variables, we now look for a (suitably large) space of other polynomials in the field variables such that these constitute an abelian operator algebra commuting with H. Since we are interested in arbitrary large dimension the only way of describing these abelian commutants is by way of recursive definition of its generating elements.

There are very many standard methods for recursive generation of commutants, for example by inverse scattering theory, by application of hereditary operators or by the Hirota bilinear mechanism (see [2], [15], [17], [16], [26], [28], [27]). Unfortunately, it turns out that in these cases either these standard mechanisms do not work or they are not suitable for algorithmic approaches. Therefore we choose to perform the construction of the commutant by way of mastersymmetries, which is a simple and straightforward method having the additional advantage of being rather accessible for computer algebra methods. Mastersymmetries do exist for all known completely integrable systems. They were first discovered in case of the BO [14] (Benjamin-Ono-equation) and the KP [36] (Kadomtsev-Petviashvili eq.), they exist for the XYZ-model [20] and the XYh-model [5]. Systematic studies can be found in [18] and [38], [37].

Let **L** be a Lie algebra. A **mastersymmetry** M for $H \in \mathbf{L}$ is an element of **L** such that

$$[[M\ ,\ H]\ ,\ H]=0. \tag{2.16}$$

Thus

$$H_1=[M\ ,\ H] \tag{2.17}$$

obviously is a symmetry generator since $[H_1\ ,\ H]=0$. Because of the Jacobi identity

$$H_2=[M\ ,\ H_1] \tag{2.18}$$

also gives a symmetry. In case of an abelian $H^{@}$ one can continue this process of construction [20] to obtain a sequence

$$H_{n+1}=[M\ ,\ H_n],\quad n\in \mathbb{N} \tag{2.19}$$

of Lie algebra elements commuting with H. Thus commutation with M maps $H^{@}$ in $H^{@}$ and, under additional and reasonable assumptions, we can expect that $H^{@}$ is generated out of H by successive application of the commutator with M. So our only task is to find a Hilbert space operator M fulfilling (2.16).

Alas, for selfadjoint H this cannot be, apart for trivial M . This is easily seen:
By decomposition into selfadjoint and antiadjoint part and multiplication with a suitable complex number we may assume that M is antiadjoint and $H_1=[M\ ,\ H]$ is again selfadjoint. Hence there is a common spectral decomposition for H and H_1. Now take an eigenvector e (or a generalized eigenvector) to see that

$$<e,[H\ ,\ M]e>=<He,Me>-<e,MHe>=\lambda<e,Me>-\lambda<e,Me>=0. \tag{2.20}$$

Here λ is the real eigenvalue of H with respect to e . This shows that $[H\ ,\ M]=0$ and M itself must be an element of $H^{@}$. This argument also carries over to higher order [18] mastersymmetries by a well-known theorem of elementary algebra [29, p. 39].

In order to avoid this difficulty, mainly due to the fact that we tried to deal with honest operators, we have to make a slight detour.

In order to describe mastersymmetries in terms of commutators of operators we first extend the manifold under consideration in such a way that the Hamiltonian becomes a completely unbounded operator but nevertheless describes the same flow as before. Of course, then it must be possible to recover the original flow by reduction. This idea is carried out by assuming that the number of lattice points becomes infinite, i.e. the lattice is represented by $\mathbb{Z}$. And a Hamiltonian H (like the one given in (1.1)) now becomes formally

$$H=H_{XYZ}=\sum_{n\in\mathbb{Z}} J_x\sigma^x{}_n\sigma^x{}_{n+1}+J_y\sigma^y{}_n\sigma^y{}_{n+1}+J_z\sigma^z{}_n\sigma^z{}_{n+1}. \tag{2.21}$$

This does not change the flow under consideration because, by virtue of (1.2)-(1.4), we still have for the dynamics of the $\sigma^r{}_n$ the same equations as before

$$(\sigma^x{}_n)_t=J_z\sigma^y{}_n(\sigma^z{}_{n+1}+\sigma^z{}_{n-1})-J_y\sigma^z{}_n(\sigma^y{}_{n+1}-\sigma^y{}_{n-1}) \tag{2.22}$$

$$(\sigma^y{}_n)_t=J_x\sigma^z{}_n(\sigma^x{}_{n+1}+\sigma^x{}_{n-1})-J_z\sigma^x{}_n(\sigma^z{}_{n+1}-\sigma^z{}_{n-1})$$

$$(\sigma^y{}_n)_t=J_y\sigma^x{}_n(\sigma^y{}_{n+1}+\sigma^y{}_{n-1})-J_x\sigma^y{}_n(\sigma^x{}_{n+1}-\sigma^x{}_{n-1}) \tag{2.24}$$

from where the original flow can be completely recovered by introducing the periodic boundary conditions (1.5) .

So, the operator H now is not a Hilbert space operator anymore but rather an external derivative on the elementary polynomials in the field variables $\sigma^r{}_n$. Here, **elementary** means that only a dependence of finitely many $\sigma^r{}_n$ is given, H itself clearly is not elementary.

The Lie algebra where commutators are computed now shall be the set of all those derivations which still look like operators, i.e. the set of all finite sums of terms like

$$S(n)=\sum_{n\in\mathbb{Z}} \alpha(n)\sigma^{r_1}{}_n\sigma^{r_2}{}_{n+1}\sigma^{r_3}{}_{n+2}\cdots\sigma^{r_{k+1}}{}_{n+k}. \tag{2.25}$$

Here the $\alpha(n)$ are assumed to be polynomials in n. One should note that sums where higher order terms at the same lattice point, like $\sigma^r{}_n\sigma^{\tilde{r}}{}_n$ occur also can be rewritten in this form. This is a consequence of (1.2)-(1.3). By abuse of notation we still call this algebra the operator Lie algebra. Now, we try to find a mastersymmetry in this enlarged Lie algebra. The structure of this mastersymmetry then has to be such that it is not compatible with the reduction given by the periodic boundary conditions (1.5). This requirement gives a good deal of information about the structure of the mastersymmetry. To see this we introduce the notion of translation invariance. An element of the operator Lie algebra is said to be **translation invariant** if it is not changed by the substitution $n \to n+N$, N fixed but arbitrary. Hence $S(n)$ given in (2.25) is translation invariant if and only if the $\alpha(n)$ are independent of n. The importance of this notion is that by the replacement

$$\sum_{n\in\mathbb{Z}} \to \sum_{n=1}^{N} \tag{2.26}$$

these operators can be interpreted as honest Hilbert space operators if periodic boundary conditions are introduced. Furthermore this replacement is such that a homomorphism for commutators is given. Since after reduction the mastersymmetries cannot be represented as Hilbert space operators they have to be given by operators breaking the translation invariance. And we can assume that a mastersymmetry M has to be of the form

$$M(n)=\sum_{n\in\mathbb{Z}} n^k S_n+\sum_{n\in\mathbb{Z}} \alpha(n) Z_n \tag{2.27}$$

where $k\geq 1$ and $\alpha(n)$ is a polynomial of degree $\leq k-1$. Using the fact that

$$H_1=[M,H] \tag{2.28}$$

is a translation invariant symmetry we obtain from the substitution $n\to(n+N)$ that that

$$M(n+N)-M(n),\ N \textit{ fixed but arbitrary} \tag{2.29}$$

must be an element of $H^@$, hence a symmetry of the system. Using this argument successively, we find that

$$S=\sum_{n\in\mathbb{Z}} S_n \tag{2.30}$$

itself must be a symmetry of the system. Furthermore, under the assumption that all symmetries commute with translation invariance one obtains $k=1$. Thus the general form of a mastersymmetry will be

$$M=\sum_{n\in\mathbb{R}} nS_n+Z \tag{2.31}$$

where Z is a translation invariant operator and where

$$S=\sum_{n\in\mathbb{R}} S_n \tag{2.32}$$

is a translation invariant symmetry. For short we write (2.31) as

$$M=nS+Z. \tag{2.33}$$

3. Division in a Lie Algebra with Subgrading

Effective computation with polynomials usually needs arguments using the degree of given polynomials. Of similar importance in the algorithmic search of symmetries is the highest-order-term projection. In most situations where classical fields are considered one deals in some sense with a free algebra where polynomial degrees are available. The grading given by that usually is compatible with the Lie algebra structure since then the degrees are connected to eigenvalues of derivations. In these cases we speak of scalings, and projections onto parts with maximal scaling are, because of the Jacobi identity, Lie algebra homomorphisms. For spin chains the situations is not as clear since the underlying structure is not a free algebra anymore. This is due to the fact that representing the relations of Pauli matrices we have implemented a reduction on a free algebraic structure. Indeed, there is no way of finding a scaling symmetry in the whole algebra of possible spin chain hamiltonians. This fact is, among other reasons, responsible for the difficulties one encounters in treating critical exponents by means of similarity solutions (see [13] for the importance of similarity solutions).

Fortunately for finding computational strategies, by way of going down from higher order terms to lower order terms, full compatibility of "degrees" with the Lie algebra operations is not necessary. That means that we can work with a subgrading instead of a grading. Recall, that grading in an algebra **L** means that each algebra element A can be decomposed in a unique way

$$A=\sum_{k=0}^{n} A_k \tag{3.1}$$

such that the A_k are either zero or have homogeneous degree k. Furthermore, for elements A_k, B_j with degrees k and j, respectively, either the commutant $[A_k,B_j]$ vanishes or has degree $k+j$. Now, if $\phi(A)$ denotes the supremum of the degrees k occurring in this series and if P_k denotes the projection given by $A \rightarrow A_k$ then

i) $\phi:\mathbf{L}\rightarrow \boldsymbol{N}_0\cup\{-\infty\}$

and

ii) $P_{\{\cdot\}}$ is a measure on $\boldsymbol{N}_0$ with values in the linear projections on L.

Furthermore, the compatibility conditions of grading and algebraic structure imply for all $A,B\in \mathbf{L}$

$$\phi(0)=-\infty \tag{3.2}$$

$$\phi(A+B)\leq\max(\phi(A),\phi(B)) \tag{3.3}$$

$$\phi(A-P_{\phi(A)}(A))<\phi(A) \tag{3.4}$$

and

$$\phi([A,B])\leq\phi(A)+\phi(B). \tag{3.5}$$

And a **subgrading** is a pair ϕ, P such that i), ii) and (3.2) to (3.5) hold. In addition we require that for all $n\in \boldsymbol{N}_0$

$$\mathbf{L}_n=\{A\in \mathbf{L}|\phi(A)\leq n\} \tag{3.6}$$

are linear subspaces of $\mathbf{L}$ such that $\mathbf{L}_{n+1}\neq\mathbf{L}_n$. The highest-order-term projection is now defined by

$$ho(A)=P_{\phi(A)}(A). \tag{3.7}$$

An important step for an algorithm which determines symmetries is the solution of the following

DIVISION PROBLEM:
Given Lie algebra elements H and R and some $m\geq 0$ with $\phi(R)\leq m+\phi(H)$. Find all X such that

$$\phi(X)\leq m \tag{3.8}$$

and

$$[H,X]=R. \tag{3.9}$$

The finite dimensionality of the subspaces $\mathbf{L}_n$ ensures that there is an algorithm determining all possible solutions of this problem. One algorithm is given by taking X to be an arbitrary sum with respect to the elements of a basis $\mathbf{L}_m$ and equating the coefficients of $[H,X]$ (with respect to a basis in $\mathbf{L}_M$, $M=m+\phi(H)$) with those of R. Thus the variable coefficients in the representations of X are determined.

This procedure is the crucial part of the algorithm described in the next section. For efficiency, the problem is solved by successive approximation. The properties of the subgrading yield that up to rather high order interaction for spin chains this results in a program with reasonable running times.

Now, under mild additional assumptions, the determination of symmetries can be achieved.

Let some $H\in\mathbf{L}$ be given and assume that for some $H_1\in H^{@}$ the highest order term $ho(H_1)$ is given. Then a simple procedure can retrieve H_1 from its highest order term. First we put

$$R=-[H,ho(H_1)] \tag{3.10}$$

and

$$M=\phi(H_1)-1. \tag{3.11}$$

Obviously there is a solution

$$X=H_1-ho(H_1) \tag{3.12}$$

of the corresponding division problem because H_1 commutes with H.

Of course, this solution may not be unique since X is determined by (3.9) only modulo elements out of $H^{@}$. These additional elements will appear in our solution with free parameters and will yield additional symmetries. Similar arguments lead to the determination of mastersymmetries instead of symmetries.

4. The algorithms

First we have to fix our subgrading. This we do in the following way. Each operator T is decomposed according to different ranges in interaction,

$$T=T_0+T_1+T_2+T_3+T_4+...+T_n \tag{4.1}$$

where T_n consists only of operators with (n+1)-neighbor interaction. For example T_2 could be of the following form

$$T_2=\alpha\sigma^x{}_n\sigma^y{}_{n+1}\sigma^z{}_{n+2}+\beta\sigma^z{}_n\sigma^y{}_{n+2} \tag{4.2}$$

and so on. The quantities $P_{\{\cdot\}}$ and ϕ are then defined as

$$P_{\{n\}}(T)=T_n \tag{4.3}$$

$$\phi(T_n)=n. \tag{4.4}$$

The crucial part of our program package is a subroutine CS (commutator solution) which instead of the division problem solves the following

REDUCED DIVISION PROBLEM:
Given Lie algebra elements H and R and some $m\geq 0$ with $\phi(R)\leq m+\phi(H)$. Find all $X=ho(X)$ with $\phi(X)\leq m$ such that

$$ho[ho(H),X]=ho(R) \tag{4.5}$$

This routine is called upon by $CS(H,R,m)$ or $CS(H,R)$ since the parameter m by default is set equal to $\phi(R)-\phi(H)$. Details of this procedure are given later. The advantage of dealing with this reduced problem instead of the full problem is that the size of linear (and nonlinear equations later on) to be solved shrinks considerably. So the running time of the whole package is cut down to a reasonable performance.

Although this routine could be used to solve the division problem iteratively we use it directly in order to find the symmetries we are looking for.

PROCEDURE: SYM(H,S) :
Given a symmetry S such that there is a mastersymmetry of the form

$$M=nS+Z, \quad Z \textit{ translation invariant} \tag{2.33}$$

this procedure computes the symmetry

$$H_1=[M,H]=SYM(H,S). \tag{4.6}$$

Based on the observation that the highest order term of M can be assumed to be given by the highest order term of nS we know that the highest order term of H_1 is given by the highest order term of $[nS,H]$ and we can use the following algorithm of successive approximation:

Step 0:Put $H_1:=[nS,H]$, $m:=\phi(H_1)-1$

Step 1:Put $R:=[H_1,H]$. *If* $R=0$ *then RETURN* (H_1) *else GOTO Step 2.*

Step 2:Determine $X=CS(H,R,m)$ *If there is no solution then RETURN ("There is no symmetry of this form") else GOTO Step 3.*

Step 3:Put $H_1:=H_1+X$, $m:=m-1$ *and GOTO Step 1.*

Obviously, in *Step 0* the quantity H_1 is computed correctly in its highest order and each run computes H_1 correctly up to one order less. Hence the algorithm has to stop either after $\phi([nS,H])-1$ runs, giving the correct H_1, or it stops before by telling us that for the given S there is no, mastersymmetry of the form (2.33).

It looks as if the algorithm only works if a symmetry S is already known. But this is not so since we can always start with $S=H$. In order to determine the mastersymmetry itself we need the following procedure

PROCEDURE: GHO(H,R,E) :
The procedure GHO ('Given highest order') determines those X with

$$ho(X)=ho(E) \tag{4.7}$$

such that

$$[X,H]=R. \tag{4.8}$$

Step 0:Put $X:=E$, $m:=\phi(E)-1$

Step 1:If $[X,H]=R$ *then RETURN(X) else GOTO Step 2.*

Step 2:Put $X:=X+CS(H,[X,H]-R,m)$, $m:=m-1$ *If there is no solution then RETURN ("There is no solution") else GOTO Step 1.*

Of course, once we have implemented this algorithm there is no need for SYM anymore because

$$SYM(H,S)=GHO(H,0,[nS,H]). \tag{4.9}$$

Now, the determination of the first nontrivial mastersymmetry (if existent) is given by

$$MAS(H,H)=GHO(H,SYM(H,H),nH). \tag{4.10}$$

This means that first we use SYM to determine one nontrivial symmetry and then we determine via GHO the mastersymmetry which commutes the H we started with into this symmetry.

Of course, if this is successful then by further commutation with M we can compute as many higher order symmetries as we like, and these higher order symmetries then can be used to compute further mastersymmetries: Another, more tedious way, to compute symmetries would be successive use of SYM, i.e.

$$H_1=SYM(H,H) \tag{4.11}$$

$$H_2=SYM(H,H_1) \tag{4.12}$$

$$H_{n+1}=SYM(H,H_n). \tag{4.13}$$

It should be remarked that running through these routines we encountered some surprises which will be described in Section 6.

We end this section by briefly describing the crucial algorithm CS which gives the solution of the reduced division problem.

PROCEDURE: CS(H,R,M) (commutator solution):

Step 0:If $R=0$ then RETURN(0). If $\phi(R)<\phi(H)$ then RETURN ('Solution nonexistent')

Step 1:Find a suitable basis $B_1, \ldots ,B_N$ of homogeneous elements (i.e. $ho(B_i)=B_i$) with degree between $\phi(R)-\phi(H)$ and M.

Step 2:Look in a hashtable of possible commutators in order to sort out a suitable subset of those elements of $\{B_1, \ldots ,B_N\}$ which can contribute to the solution X. Call these elements $\{A_1, \ldots ,A_k\}$.

Step 3:Make the Ansatz

$$X=\sum \alpha_i A_i$$

and consider

$$ho[ho(H),X]=ho(R)$$

as a system of linear equations for the α_i *(and the so called free parameters).*

Step 4:Solve this system and put X equal to the sum which corresponds to this solution. If the solution of the system is not unique keep the additional parameters as free parameters for future use. If there is no solution, even if free parameters are sacrificed, then RETURN ("solution nonexistent").

It should be remarked that for CS it is essential that linear equations are solved in full generality (i.e. including the homogeneous solutions) and that the free parameters in these solutions are available for future use, namely when future applications of CS are necessary.

This treatment of free parameters also is essential when a systematic search for integrable spin chains shall be performed. In this case a Hamiltonian is taken where external free parameters appear as coefficiens in front of characteristic pieces of interaction terms. Then if CS runs into a deadlock the equations which are not solvable are considered as equations for the external free parameters which now appear in a quadratic way. And making these parameters available CS tries to find solutions for the Hamiltonians given by these new restrictions for the external free parameters.

5. Implementation

The program package is written in MAPLE [23], a formula manipulation system developed by the University of Waterloo. The choice for a formula manipulation system was mainly based on our desire for rapid prototyping and on the fact that for these systems very many sophisticated algorithms are available. The result of this choice is that the whole package is relativity short (about 400 lines of code). Although a more sophisticated use of the very many sophisticated tools available for these formula manipulation systems certainly should enable us to cut down the package to less than 150 lines of code (thus loosing a lot of transparency). In fact we only used very few of the tools of MAPLE in order to ensure portability to either other systems or standard languages like PASCAL or C.

A result of this programming policy is that the package still is relatively slow. Certainly by using more of those tools which are in the kernel of maple we could decrease the running time for complicated examples by a factor 10. Rewriting the whole package after the experience we have with this first version also should give a considerable increase in performance. In fact, the only part of the whole package which has been completely rewritten is the commutator program for spin operators and this gave a dramatic increase in performance for this part. First, the program was based on really commuting matrices like the Pauli-matrices (certainly the most stupid way to do it!), then we used hash tables for certain commutators and this gave an improvement by a factor around 50.

This experience led us to base very many parts of the program package on hash tables. A decision which is somewhat doubtful in its wisdom.

MAPLE was chosen over other systems for many reasons. Availability was the least important one of these reasons. In earlier work [22] we had experienced most favorable comparisons with respect to running time between MAPLE and other systems. Our general experience seems that systems based on C are indeed much faster than those based on LISP. In addition MAPLE appealed to us because of its transparent programming structure, and the fact, that the set of tools was relatively small and did not have this overwhelming abundance which can be found in systems like MACSYMA. The most important reason for giving MAPLE the preference over other systems was that we had the desire to run our programs on small sized computers efficiently.[1] Nowadays, we even have rather satisfactory results by running our programs on a Macintosh SE. Originally the program development was begun on a Cadmus 9520 with version 3.3 of MAPLE, lateron we switched to MAPLE 4.0 on a SUN 3/50 (under BSD UNIX 4.2). Really fastidious examples we usually run on a SUN 3/260.

The essential parts of the whole package are an efficient commutator program and the routine CS. Running a profile on how much time is consumed by different parts of the program one finds that for typical application of SYM 30% of the running time is consumed by the commutator program and 35% by the routine CS.

For CS the solver for systems of linear and quadratic equations is a crucial tool. After having available version 4.0 of MAPLE, we relied in this point completely on MAPLE, and as far as we can say, the solver works in an efficient and satisfactory way. This is in contrast to earlier versions. For the commutator program the difficulty had to be overcome how to deal with infinite sums of products of Pauli matrices. To do this we chose externally the following representation of for infinite sums:

Notation:	Real Meaning:
$\alpha * H(i,k)$	$\alpha \sum_{n\in\mathbb{Z}} \sigma^{(i)}{}_n \sigma^{(k)}{}_{n+1}$
$\alpha * H(s,0,j,r) + \beta * H(i,k)$	$\alpha \sum_{n\in\mathbb{Z}} \sigma^{(s)}{}_n \sigma^{(j)}{}_{n+2} \sigma^{(r)}{}_{n+3} + \beta \sum_{n\in\mathbb{Z}} \sigma^{(i)}{}_n \sigma^{(k)}{}_{n+1}$

1) At that time neither the AT-version of REDUCE did exist, nor the version for SUN-workstations was available.

Here the symbol H does indicate the infinite sum, its entries are the different spin matrices, and always the N-th entry denotes the spin matrix sitting on the lattice point $n+N$. The index 0 denotes the identity matrix, which together with the Pauli matrices forms a basis of 2x2-matrices. The indices 1,2,3 stand for x,y,z, respectively. One should observe that always, by rewriting, the sum can be written in such a way that the sub-indices start with n.

In this notation the Hamiltonians of the XYZ- and XYh-model, respectively have the following form

$$H_{XYZ}=J1*H(1,1)+J2*H(2,2)+J3*H(3,3) \tag{5.1}$$

$$H_{XYh}=J1*H(1,1)+J2*H(2,2)+h*H(3). \tag{5.2}$$

Internally, this notation is changed into hash tables and the commutators of basic elements are also stored as hash tables. Originally the notation was chosen in order to use the tools available in MAPLE for dealing with functions (procedures) and to work with remember tables. This idea has been abandoned in order to ensure portability to other systems (and to Pascal or C). In the future the whole package will be rewritten either in C or by using more sophisticated tools from MAPLE. Most probably both will be done in order to compare efficiency and performance.

The whole package is available as a test-version. (Warning: Comments and handbook are written in German).

At the end of this section let us say a few words about our general experience in working with algebraic formula manipulation systems.

The general advantages are obvious: Rapid prototyping and availability of sophisticated standard algorithms. Furthermore, quite often the syntax of a formula manipulation system is much closer then any other programming language to the day to day formulas a mathematician writes. These advantages enable even the unexperienced programmer to have good results after a little while.

Another point which is of utmost importance, at least for the applications we have in mind, is good performance with respect to running times and memory space economy. This point usually is underestimated and put aside by the hint that speed and memory space are increased so fast by technical progress, that this cannot be an issue anymore in the development of software. To our experience this is just not true since real fastidious applications of formula manipulation systems require so much in time and memory that they are in most cases still beyond reach. To save time and memory is not important for its own sake, but this is important for whether a problem is feasible or not.

A last point of great importance is that for a mathematician, or a scientist in general, it is absolutely necessary that at each stage he must be able to control what he is doing. Apart from scientific ethics this a question of practical value. This requirement implies the necessity that a user of a formula manipulation system must have, in principle, access to the source codes of all parts of the system he is using. Many developers claim that this is not really necessary since most of the libraries are known in source anyway, and that the remaining parts (kernel) have to be compared to a compiler where access is not really necessary. To our opinion this viewpoint is wrong although it has to be admitted that the reasons for not making all the sources completely available are understandable. Sometimes for having a good judgement about efficiency and possibilities of a computer algebra system a look on the sources even of the kernel is indispensable. (Apart from the fact that even the most careful developer cannot avoid to produce from time to time a bug in the kernel).

Last not least, but this is more a matter of convenience, a computer algebra system should not have too many tools and options because this will hurt its transparency as a programming language and its portability to small systems. Although most users are often impressed by the power and the copiousness of a system they are nevertheless confused and scared away by the same fact.

6. Results and Surprises

For convenience and brevity we use here the notation introduced in section 5.

a. The XYZ-model

Starting with the Hamiltonian of the XYZ-model

$$H0=\sum J_x\sigma^x{}_n\sigma^x{}_{n+1}+J_y\sigma^y{}_n\sigma^y{}_{n+1}+J_z\sigma^z{}_n\sigma^z{}_{n+1} \tag{6.1}$$

$$=J1*H(1,1)+J2*H(2,2)+J3*H(3,3)$$

the call $SYM(H0,H0)$ yields (after 9 sec systems time on a SUN 3/50) the known [33], [20] symmetry

$$H1:=4*I*J1*J3*H(1,2,3)-4*I*J1*J2*H(1,3,2)-4*I*J3*J2*H(2,1,3)+$$
$$4*I*J1*J2*H(2,3,1)+4*I*J3*J2*H(3,1,2)-4*I*J1*J3*H(3,2,1). \tag{6.2}$$

Further calls

$$H2:=SYM(H0,H1) \tag{6.3}$$

$$H(n+1)=SYM(H0,Hn) \tag{6.4}$$

yield higher order symmetries (these are to long to write down here). Time and memory necessary for these computations on a SUN 3/50 workstation are:

Symmetries for the XYZ-model on a SUN 3/50:

Call	words_used (total)	words_used (maximal)	time (sec)
SYM(H0,H0)	144000	106 000	9
SYM(H0,H1)	1021217	264 192	109
SYM(H0,H2)	2884004	309 248	341
SYM(H0,H3)	10953000	555 000	1279

The time is system time and one word is 32 bits. The corresponding system times on a Macintosh SE are roughly ten-times the times on a SUN 3/50. The necessary memory space on a MAC SE is roughly 50% of the space necessary on the SUN 3/50 and a SUN 3/260 needs only a quarter of the time cosumed by a SUN 3/50. It has to be admitted that computing, for example, $H4{:=}SYM(H0,H3)$ needs an awful lot of time and memory, but it has to be taken in account that $H4$ really is a monster representing interaction of up to 6 neighbors and being a sum of roughly 100 terms (each of them being an infinite sum). A good guess is that the explicit form of this hamiltonian will never appear explicitly in the literature.

In order to avoid such monsters we look for a recursion formula for the symmetries of the XYZ-model, i.e. for a mastersymmetry. The call MAS(H0,H1) gives in about 17 seconds the well known mastersymmetry [20].

$$M0{:=}n*(J1*H(1,1)+J2*H(2,2)+J3*H(3,3)). \qquad (6.5)$$

Now, the symmetries are easily accessible by commutation with $M0$. Since the symmetry SYM(H0,H1) was found by the program package under the assumption of the existence of a mastersymmetry of the form

$$M1{:=}nH1+Z \qquad (6.6)$$

we would like to find this mastersymmetry. But here we are in for a surprise $MAS(H0,H1)$ yields the result that there is no mastersymmetry of this kind. Indeed, $M0$ is the only mastersymmetry we can find for the XYZ-model. So, it seems, that mastersymmetries play for our algorithm the role of a nonexistent ghost who is nevertheless helpful if one believes in it. In order not to escape in metaphysical regions we give a short explication of this puzzling phenomenon.

Its occurrence is a consequence of the subtle differences between complete integrability on finite and infinite dimensional manifolds. For completely integrable systems on finite dimensional manifolds there is, at least locally, a complete linearization by the action-angle-variable representation [1]. As a consequence of this there are always sufficiently many mastersymmetries to construct the symmetry group in a recursive way. So the observed phenomenon (of nonexistence of further mastersymmetries even for completely integrable systems) can only come up in the infinite dimensional case. If periodic boundary conditions (1.5) are introduced then a reduction to finite dimensional manifolds is performed and the mastersymmetries must come into existence. So the question may come up why we do not consider finite dimensional manifolds in order to strip our computation of its inherent dubiousness.

The reason is simple: Because the algorithm is based on a subgrading and this subgrading totally collapses if a reduction by periodic boundary conditions is carried out. This is so, because the kernel of our reduction contains terms of arbitrary high order. So we are in the following situation: The symmetries (and the action variables) of our system are universal objects whose form does not depend on the reduction whereas the mastersymmetries (angle variables) do depend in their explicit form on the different reductions given by the periodic boundary conditions. Now, the proposed algorithm uses only at the very beginning the existence of mastersymmetries in order to determine the highest order term of the symmetry one is looking for. And this ansatz is correct because by reduction only the order classification of mastersymmetries collapses whereas the order classification of symmetries remains valid. If in the continuation of our algorithmic procedure we would try to compute the missing terms for the symmetry as well as for the mastersymmetry then the computation would not terminate since we would windup in the kernel of reduction which contains terms of arbitrary high order. Fortunately this is not the case since our algorithm, right after the beginning, leaves the range of the somewhat dubious mastersymmetries and performs computations only with respect to symmetries (which are not changed by reduction). Hence, if there are symmetries for the infinite dimensional situation, our algorithm has to find them not withstanding the fact that it is based on nonexistent quantities.

b. The XYh-model

For the Hamiltonian

$$H0:=\sum J_x\sigma^x{}_n\sigma^x{}_{n+1}+J_y\sigma^y{}_n\sigma^y{}_{n+1}+h\sigma^z{}_n \qquad (6.7)$$

$$=J1*H(1,1)+J2*H(2,2)+h*H(3)$$

the call *SYM*(*H*0,*H*0) yields the symmetry

$$SYM(H0,H0):=H1:=(-2*I*h*J1-par1)*H(1,2)+2*I*J1*J2*H(1,3,2)-$$
$$2*I*J1*J2*H(2,3,1)+(par1+2*I*h*J1)*H(2,1) \qquad (6.8)$$

where par1 is an arbitrary parameter (coming from the homogeneous solution of a system of linear equations solved in CS). Hence our call on the symmetry procedure yielded two independent symmetries:

$$H1:=(-2*I*h*J2-2*I*h*J1)*H(1,2)+(2*I*h*J2+2*I*h*J1)*H(2,1)+$$
$$4*I*J2*J1*H(1,3,2)-4*I*J2*J1*H(2,3,1) \qquad (6.10)$$

and

$$T0:=H(2,1)-H(1,2) \tag{6.11}$$

Further symmetries are computed in much less time than for the XYZ-model

Symmetries for the XYh-model on a SUN 3/50 :

Call	words_used (total)	words_used (maximal)	time (sec)
SYM(H0,H0)	180 000	157 000	15
SYM(H0,H1)	400 000	241 000	43
SYM(H0,H2)	580 000	260 000	64
SYM(H0,H3)	1040 000	272 000	118

The free parameter in *SYM(H0,H0)* is essential for finding mastersymmetries, because - as it has to be expected - such a mastersymmetry exists only for special choices of par1.

In order to tell the program that whenever a linear system coming up in CS cannot be solved then it can choose par1 in such a way that a solution can be found, we use a special utility reserveParms( ).

For example reserveParms(7) tells the program that parameters par1,..., par7 are special external parameters which only can be sacrificed when CS runs into an unsolvable linear system.

Now, after giving reserveParms(1) the call GHO(H0, SYM(H0,H0)) gives the mastersymmetry

$$M0:=(-J1*par6+1/2*J1+J1*n)*H(1,1)+ \\ (-J2*par6+1/2*J2+J2*n)*H(2,2)+(-par6*h+h*n)*H(3). \tag{6.9}$$

In this case the parameter par1 was put equal to $-h(J1+J2)$. The part with the free parameter par6 can be dropped since this stands in front of an operator which commutes with H0 anyway. The resulting mastersymmetry is the one already published in [5]. In the same way the higher order mastersymmetries published there can be found. The next one is

$$M1:=I*(J1+J2)*n*((4*I*J2+4*I*J1)*H(3)+4*I*h*H(1,1)+4*I*h*H(2,2)- \\ 4*I*J1*H(1,3,1)-4*I*J2*H(2,3,2))+2*h*(J1-J2)*(H(2,2) \\ -H(1,1))+(-4*h2+4*(J1+J2)2)*H(3). \tag{6.12}$$

So we have now two hierarchies of symmetries:

$$H(n+1)=[M0,Hn] \tag{6.13}$$

$$T(n+1)=[M0,Tn] \tag{6.14}$$

and one hierarchy of mastersymmetries, namely $M0$ and $M1$ as before and

$$M(n+1)=[M0,Mn]. \tag{6.15}$$

Beyond the results of [5] we can find a second hierarchy of mastersymmetries. Taking for $\hat{H}$

$$\hat{H}=SYM(H0,H2)$$

one obtains from $GHO(H0,\hat{H})$ the following mastersymmetry (which is independent of $M0,M1,M2$) as a starting point for a second hierarchy

$N1$=(−8*$J1$**2*h*n+56*$J1$*$J2$*h*n)*H(1,3,1)+ (6.16)

(−16*$J2$*h*$J1$+16*$J2$**2*h+24*$J1$*$J2$*h*n+24*$J2$**2*h*n)*H(2,3,2)+

(−4*$J1$**2*h−8*$J2$*h*$J1$−36*$J2$**2*h+4*h**3+4*1/$J1$*$J2$*h**3)*H(3)−

32*$J2$**2*$J1$*n*H(2,3,3,2)+(8*$J1$*h**2*n−8*$J2$**2*$J1$*n−24*$J2$*h**2*n−

8*$J1$**3*n+16*$J1$**2*$J2$*n)*H(1,1)−32*$J1$**2*$J2$*n*H(1,3,3,1)+

(−4*h**2*$J1$−32*$J2$**3+4*$J2$**2*h**2/$J1$+32*$J1$**2*$J2$+8*$J1$*h**2*n−

24*$J2$*h**2*n+24*$J2$**3*n−40*$J1$**2*$J2$*n+16*$J2$**2*$J1$*n)*H(2,2).

Higher order members of this hierarchy are then obtained by commutations with $M0$.

These four hierarchies of symmetries and mastersymmetries are indeed sufficient in order to write down the complete action-angle representation.

c. New integrable chains

Let us briefly indicate how new completely integrable spin-1/2 chains are found by our program package. We choose a simple example, a complete report will be published elsewhere.

We would like to know which linear combinations of

$$HH1=H(1,2,3)=\sum_{n\in\mathbb{Z}}\sigma^x{}_n\sigma^y{}_{n+1}\sigma^z{}_{n+2} \tag{6.17}$$

$$HH2=H(3,2,1)=\sum_{n\in\mathbb{Z}}\sigma^z{}_n\sigma^y{}_{n+1}\sigma^x{}_{n+2} \tag{6.18}$$

are integrable.

In order to find this out we put external free parameters par1, par2 in front of these parts and call upon *SYM*(*H*0,*H*0) where

$$H0=par1*HH1+par2*HH2. \tag{6.19}$$

Running for the first time through CS the program has no difficulty to solve the linear systems. The second run of CS cannot be completed without giving up one of the external free parameters. It turns out that there are two different choices possible, namely par1=0 or par2 = 0. These choices are run through successively and the program has no difficulty to finish successfully.

For example to par1 = 0 i.e.

$$H0:=par2\sum\sigma^z{}_n\sigma^y{}_{n+1}\sigma^x{}_{n+2} \tag{6.20}$$

it finds the symmetry

$$H1=4i(par2)^2\sum\sigma^x{}_n\sigma^z{}_{n+1}\sigma^z{}_{n+2}\sigma^z{}_{n+3}\sigma^y{}_{n+4}. \tag{6.21}$$

For finding then the hierarchy and the mastersymmetries the same routine as before is needed.

7. Problems for future research

We would like to indicate some directions into which future computer investigations of spin chains should move. These directions also give hints about where efficient applications of our programs can be found.

a. A computer implementation of reduction to finite dimensional manifolds should be given. This would stimulate the following investigations:

a.1. The explanation given in section 6 for the efficiency of nonexistent mastersymmtries could be checked explicitly.

a.2 The explicit diagonalization of the Hamilton operator after reduction to a genuine L^2-Operator could be given by use of its commutant. Probably there arise problems with respect to running time. Therefore the efficiency of the whole package should be streamlined.

a.3. Transfer-matrix and partition function could be found explicitly by computer. Here again running time problems have to be expected.

b. The package computes those symmetry group generators which are given by closed vector fields. This is due to the fact that the operator commutators rather correspond to the POISSON brackets than to the vector field commutators. Therefore the whole package should be rewritten for the vector field Lie algebra instead of the operator algebra. It might then be possible that the mastersymmetries which do not exist on the Poisson-bracket side exist among those vector fields which are not closed. This happens for many classical completely integrable systems (KdV and the like). Furthermore, it certainly is most interesting to see whether there are symmetries which do not correspond to operators, i.e. which do not have a quantum mechanical formulation. A consequence of such a discovery would be the discovery of a hereditary recursion operator for quantum spin chains.

c. A satisfactory treatment of the classical continuous limit should be implemented. Already, the XYZ-model yields the Landau-Lifshitz equation which contains very many of the known classical completely integrable partial differential equations. So, in principle, efficient computer programs to detect and investigate complete integrability for spin chains can be used to deal with completely integrable partial differential equations.

BIBLIOGRAPHY

[1]Arnold, V. I., Mathematical Methods of Classical Mechanics, Graduate Texts in Mathematics, 60, Springer Verlag, Berlin-Heidelberg-New York, 1978.

[2]Ablowitz, M. J. and H. Segur, "Asymptotic Solutions of the Korteweg-de Vries equation," Studies in Appl.Math., vol. 57, pp. 13-44, 1977.

[3]Barouch, E., Statistical Mechanics of the XY-Model, in: Mathematical Methods in Theoretical Physics (W.E. Brittin ed.) Colorado Assoc. U.P, pp. 1-54, 1973.

[4]Barouch, E., "Lax Pair for the Free-Fermion Eight-Vertex Model," Studies in Appl.Math., vol. 70, pp. 151-162, 1984.

[5]Barouch, E. and B. Fuchssteiner, "Mastersymmetries and similarity Equations of the XYh-model," Studies in Appl.Math., vol. 73, pp. 221-237, 1985.

[6]Baxter, R.J., "Partition function of the eight-vertex lattice model," Ann.Physics, vol. 70, pp. 193-228, 1972.

[7]Baxter, R.J., "One-dimensional anisotropic Heisenberg chain," Ann.Physics, vol. 70, pp. 323-337, 1972.

[8]Baxter, R.J., "Eight-vertex model in lattice statistics and one-dimensional anisotropic Heisenberg Chain.II, Equivalence to a generalized ice-type lattice model," Ann.Phys, vol. 76, pp. 25-47, 1973.

[9]Baxter, R.J., "Eight-vertex model in lattice statistics and one-dimensional anisotropic Heisenberg Chain.III, Eigenvectors of the transfer-matrix and Hamiltonian," Ann.Physics, vol. 76, pp. 48-71, 1973.

[10]Baxter, R.J., "Eight-vertex model in lattice statistics and one-dimensional anisotropic Heisenberg chain, I, some fundamental eigenvectors," Ann.Physics, vol. 76, pp. 1-24, 1973.

[11]Bethe, H., "Zur Theorie der Metalle I. Eigenwerte und Eigenfunktionen der Atomkette," Z.Phys., vol. 71, pp. 205-226, 1931.

[12]Capel, H. W., and J. H. H. Perk, "Autocorrelation Functions of the X-component of the magnetization in the one-dimensional XY-model" Physica, vol. 87A, pp. 211-242, 1977

[13]Cardy, J. L., "Conformal Invariance and critical behavior," Physica, vol. 140A, pp. 219-224, 1986

[14]Fokas, A. S. and B. Fuchssteiner, "The Hierarchy of the Benjamin-Ono Equation," Phys. Lett., vol. 86 A, pp. 341-345, 1981.

[15]Fuchssteiner, B., "Application of Hereditary Symmetries to Nonlinear Evolution equations," Nonlinear Analysis TMA, vol. 3, pp. 849-862, 1979.

[16]Fuchssteiner, B. and A. S. Fokas, "Symplectic Structures, Their, B#acklund Transformations and Hereditary Symmetries," Physi-

ca, vol. 4 D, pp. 47-66, 1981.

[17]Fuchssteiner, B., "The Lie algebra structure of Nonlinear Evolution Equations admitting Infinite Dimensional Abelian Symmetry Groups," Progr. Theor. Phys., vol. 65, pp. 861-876, 1981.

[18]Fuchssteiner, B., "Mastersymmetries, Higher-order Time-dependent symmetries and conserved Densities of Nonlinear Evolution Equations," Progr. Theor.Phys., vol. 70, pp. 1508-1522, 1983.

[19]Fuchssteiner, B., "On the Hierarchy of the Landau-Lifshitz equation," Physica, vol. 13D, pp. 387-394, 1984.

[20]Fuchssteiner, B., "Mastersymmetries for completely integrable systems in statistical mechanics," Springer Verlag, Lecture Notes in Physics 216 (L.Garrido ed.) Berlin-Heidelberg-New York, pp. 305-315, 1985.

[21]Fuchssteiner, B., "Some recent results on Solitons, Symmetries and Conservation Laws in Nonlinear Dynamics," Proceedings of the 14th ICGTMP (Y.M. Cho ed.) World Scientific Publ. Singapore, pp. 421-424, 1986.

[22]Fuchssteiner, B., W.Oevel, and W.Wiwianka, "Computer-algebra methods for Investigation of Hereditary Operators of higher order Soliton Equations," Computer Physics Comm., vol. 44, 1987.

[23]Geddes, K. O., G. H. Gonnet, and B. W. Char, MAPLE User's Manual, 3rd edition, Waterloo, 1983.

[24]Hazewinkel, M., "Experimental Mathematics," Math. Modelling, vol. 6, pp. 175-211, 1985.

[25]Hazewinkel, M., "Experimental Mathematics," in: Mathematics and Computer Science (J.W. de Bakker, M. Hazewinkel, J.K. Lenstra, eds.) CWI/North Holland Publ., pp. 193-234, 1986.

[26]Hirota, R., "Direct methods of finding exact solutions of nonlinear evolution equations," in: Bäcklund transformations (R.M. Miura ed.), Lecture Notes in Mathematics 515, Springer-Verlag, Berlin-Heidelberg-New York, pp. 40-68,

1976.

[27]Hirota, R. and J. Satsuma, "A variety of nonlinear network equations generated from the Baecklund transformation for the Toda lattice," Progr.Theoretical Physics, vol. 59, pp. 64-100, 1976.

[28]Hirota, R., "Nonlinear partial differential equations V. Nonlinear equations reducible to linear equations," J.Phys.Soc.Japan, vol. 46, pp. 312-319, 1979.

[29]Kaplansky, I., Lie algebras and locally compact groups, The University of Chicago Press, Chicago-London, 1974.

[30]Kramers, H.A. and G.H. Wannier, "Statistics of the two-dimensional ferromagnetic.I," Phys.Rev., vol. 60, pp. 252-262, 1941.

[31]Landau, L. D. and E. M. Lifshitz, Teoreticheskaya fizika, Nauka, Moscow 1965, 1976. Translation: Course of theoretical physics, 3rd ed. Pergamon, Oxford

[32]Lieb, E.H., "Exact solution of the F-model of an antiferroelectric," Phys.Rev.Lett., vol. 18, pp. 1046-1048, 1967.

[33]Lüscher, M., "Dynamical charges in the quantized renormalized massive Thirring model," Nuclear Physics, vol. B117, pp. 475-492, 1976

[34]Mattis, D. C., The theorie of magnetism I, Springer series in Solid State Sciences 17, Springer Verlag, Berlin-Heidelberg-New York, 1981

[35]Onsager, L., "Crystal statistics. I, A two-dimensional model with an order-disorder transition," Phys.Rev., vol. 65, pp. 117-149, 1944.

[36]Oevel, W. and B. Fuchssteiner, "Explicit Formulas for the Symmetries and Conservation Laws of the Kadomtsev-Petviashvili equation," Phys. Lett., vol. 88 A, pp. 323-237, 1982.

[37]Oevel, W., "Rekursionsmechanismen für Symmetrien und Erhaltungssätze in integrablen Systemen," Dissertation, Paderborn, 1984.

[38]Oevel, W., "A Geometrical approach to Integrable Systems admitting time dependent Invariants," in: Topics in Soliton Theory and Exactly solvable Nonlinear equations (eds: M. Ablowitz, B. Fuchssteiner, M. Kruskal) World Scientific Publ., pp. 108-124, Singapore, 1987.

[39]Perk, J. H. H., "Time-dependent Correlations in the one-dimensional XY-Model" Thesis Amsterdam 1979

[40]Schwarz, F., "An Algorithm for Determining Polynomial First Integrals of Autonomous Systems of Ordinary Differential Equations," J.Symbolic Computation, vol. 1, pp. 229 - 233, 1985.

[41]Schwarz, F., "Automatically Determining Symmetries of Partial Differential Equations," Computing, vol. 34, pp. 91-106, 1985.

[42]Schwarz, F., "A Reduce Package for determining first integrals of Autonomous systems of ordinary differential equations," Computer Physics Communications, vol. 39, pp. 285-296, 1986.

[43]Schwarz, F., "Some experiments in Computer Algebra," in: Topics in Soliton Theory and Exactly solvable Nonlinear equations (eds: M. Ablowitz, B. Fuchssteiner, M. Kruskal) World Scientific Publ., pp. 290-299, Singapore, 1987.

[44]Sklyanin, E. K., On complete integrability of the Landau-Lifshitz equation, USSR Academy of Sciences, Steklov Math. Inst. Leningrad, 1979. Lomi preprint E-3

[45]Sogo, K. and M. Wadati, "Quantum Inverse Scattering Method and Yang-Baxter Relation for Integrable Spin systems," Progress Theoretical Physics, vol. 68, pp. 85-97, 1982.

[46]Suzuki, M., "Equivalence of the two-dimensional Ising Model to the ground state of the linear XY-model," Phys.Lett., vol.

[47]Takhtadzhan, L.A., and L.D. Fadeev, "The Quantum method of the inverse problem and the Heisenberg XYZ model," Russ.Math. Surveys, vol. 34, pp. 11-68, 1979.

HAMILTONIAN STRUCTURE AND STABILITY ANALYSIS

Darryl D. Holm

Center for Nonlinear Studies and Theoretical Division

Los Alamos National Laboratory, MS B284

Los Alamos, NM 87545 USA

Abstract

The Lyapunov method for establishing stability is related to well-known energy principles for nondissipative dynamical systems. A development of the Lyapunov method for Hamiltonian systems due to Arnold establishes sufficient conditions for Lyapunov stability by using the energy plus other conserved quantities, together with second variations and convexity estimates. When treating the stability of ideal fluid dynamics within the Hamiltonian framework, a useful class of these conserved quantities consists of the Casimir functionals, which Poisson-commute with all functionals of the dynamical fluid variables. Such conserved quantities, when added to the energy, help to provide convexity estimates that bound the growth of perturbations. These convexity estimates, in turn, provide norms necessary for establishing Lyapunov stability under the nonlinear evolution. In contrast, the commonly used second variation or spectral stability arguments only prove linearized stability. As ideal fluid examples, in these lectures we discuss planar barotropic compressible fluid dynamics, the three-dimensional hydrostatic Boussinesq model, and a new set of shallow water equations with nonlinear dispersion due to Basdenkov, Morosov, and Pogutse [1985]. Remarkably, all three of these examples possess the *same* Hamiltonian structure and, thus, the same Casimir functionals upon which their stability analyses are based.

1. Introduction

What equilibrium solutions in ideal fluid dynamics are Lyapunov stable? As we shall see, the Hamiltonian structure of ideal fluid dynamics provides the framework for the stability analysis required to answer this question.

In these lectures we recapitulate the Lyapunov stability method discussed in Holm et al. [1985], give the example of its application to barotropic compressible fluids in two dimensions found Holm et al. [1983, 1985], and provide the Hamiltonian framework for discussing Lyapunov stability of equilibrium solutions for the ideal hydrostatic Boussinesq model treated in Holm and Long [1988] and for a new set of shallow water equations with nonlinear dispersion due to Basdenkov, Morosov, and Pogutse [1985] (see also Holm [1988]).

The classical Lyapunov method finds criteria for stability of an equilibrium solution of a conservative dynamical system by seeking a constant of the motion with a local maximum or minimum at the equilibrium. In many examples, the appropriate constant of motion is the energy. An important development for the applicability of the Lyapunov method to fluid dynamics is Arnold's [1965a, 1969a] nonlinear analysis of the stability of planar ideal incompressible fluid motion, providing nonlinear stability results that extend the classical linear theory of Rayleigh [1880]. Arnold adds to the energy H a conserved quantity C which corresponds to the symmetry of Eulerian fluid dynamics under Lagrangian relabeling of fluid particles. (In geometric language, the fluid Hamiltonian in the Lagrangian representation is right invariant on the cotangent bundle of the group of area-preserving diffeomorphisms.) Underlying this method is the fact that the Eulerian equations of motion are Hamiltonian with respect to a certain noncanonical Poisson bracket, called a Lie-Poisson bracket. The added constants of the motion are *kinematic* in the sense that they will be conserved for any system which is Hamiltonian with respect to the Lie-Poisson bracket. In fact, the quantity C Poisson commutes with all functionals of the Eulerian fluid variables; as such, it is called a "Casimir", or a "distinguished functional." The functional C is chosen such that $H+C$ has a critical point at the stationary solution. Arnold [1965a, 1969a] employs convexity properties of $H+C$ to find an explicit norm and *a priori* estimates needed to limit the departure of finite perturbations from equilibrium. In this way, nonlinear Lyapunov stability conditions are established for planar ideal incompressible fluid equilibria.

In Holm et al. [1985] the same method is extended to a number of other conservative systems arising in the physics of fluids and plasmas. The result in each case is that when certain inequalities (the stability criteria) are satisfied by an equilibrium solution (which the method associates to a critical point of a conserved quantity), then *a priori* estimates guarantee Lyapunov stability relative to an explicitly constructed norm, so long as the solutions of interest continue to exist and remain sufficiently smooth.

Holm et al. [1985] distinguish among four interrelated concepts of stability, often encountered in the literature.

(1) *Neutral or spectral stability.* For a dynamical system $du/dt = X(u)$, an equilibrium point u_e satisfying $X(u_e)=0$ is called *spectrally stable*, provided the spectrum of the linearized operator $DX(u_e)$ has no strictly positive real part. A special case is *neutral stability*, for which the spectrum is purely imaginary. This corresponds to the time evolution of normal modes being purely oscillatory. For Hamiltonian systems, spectral stability and neutral stability coincide.

(2) *Linearized stability.* The equilibrium solution u_e is called *linearized stable* or *linearly stable* relative to a norm $\|\delta u\|$ on infinitesimal variations δu, provided for every $\varepsilon > 0$ there is a $\delta > 0$ such that if $\|\delta u\| < \delta$ at $t = 0$, then $\|\delta u\| < \varepsilon$ for $t > 0$, where δu evolves according to $d(\delta u)/dt = DX(u_e)\cdot\delta u$.

Linearized stability implies spectral stability; since, if the spectrum had a strictly positive real part, there would be an unstable eigenspace and in this eigenspace the norm condition on δu would be violated. The converse is not generally true (e.g., the equilibrium solution $(p_e, q_e) = (0,0)$ for the dynamics generated by the Hamiltonian $H = p^2 + q^4$ is neutrally, but not linearized stable). In finite dimensions, a sufficient condition for linearized stability is that $DX(u_e)$ have distinct eigenvalues on the imaginary axis. In infinite dimensions, it is sufficient for $DX(u_e)$ to have a complete set of eigenfunctions with purely imaginary eigenvalues of multiplicity one. In the case of repeated roots on the imaginary axis, instabilities can occur with linear growth rates of a resonance type. There is an extensive theory dealing with this case going back to Krein [1950] (see also Arnold [1978]); this theory gives precise spectral conditions for linearized stability in finite dimensions. See also Levi [1977]. In infinite dimensions, the spectral approach may encounter functional analytic difficulties requiring considerable effort to overcome, and the results in the literature are often only indicative of linearized stability, with no rigorous proof given. See, for example, Penrose [1960], Jackson [1960], Chandrasekhar [1961], Drazin and Reid [1981], and Friedberg [1982]. Another effective method to prove linearized stability is to look for a positive definite conserved quadratic quantity, which serves as the square of a norm. This leads to what Holm et al. [1985] call "formal stability".

(3) *Formal stability.* We say that an equilibrium solution u_e of a system $du/dt = X(u)$ is *formally stable* if a conserved quantity is found whose first variation vanishes at the solution and whose second variation at this solution is positive (or negative) definite. Since the second variation provides a norm preserved by the linearized equations (see Appendix A of Holm et al. [1985]), formal stability implies linearized stability. Again the converse is not generally true (e.g., the equilibrium solution $(p_{1e}, q_{1e}, p_{2e}, q_{2e}) = (0, 0, 0, 0)$ for the dynamics generated by the Hamiltonian $H = (p_1^2 + q_1^2) - (p_2^2 + q_2^2)$ is linearized stable for the Euclidean norm in $\mathbb{R}^4$, but is not formally stable).

Formal stability of fluids and plasmas has been considered by a number of authors, such as Fjortoft [1946, 1950], Eliassen and Kleinschmidt [1957], Bernstein et al. [1958], Kruskal and Oberman [1958], Newcomb (see Appendix I of Bernstein [1958]), Fowler [1963], Gardner [1963], Rosenbluth [1964, p. 137ff.], Dikii [1965a, b], Herlitz [1967], and Davidson and Tsai [1973]. More recently, formal stability has been established by several authors who employ some aspects of Arnold's method (but not the convexity analysis). See for example, Blumen [1968, 1971], Zakharov and Kuznetsov [1974], Sedenko and Iudovitch [1978], Benzi et al. [1982] and Grinfeld [1984].

(4) *Nonlinear stability.* An equilibrium point u_e of a dynamical system is said to be *nonlinearly stable* if for every neighborhood U of u_e there is a neighborhood V of u_e such that trajectories $u(t)$ initially in V never leave U. This definition presupposes well-defined dynamics and a specified topology. In terms of a norm $\|\,\|$, nonlinear stability means that for every $\varepsilon > 0$ there is a $\delta > 0$ such that if $\|u(0) - u_e\| < \delta$, then $\|u(t) - u_e\| < \varepsilon$ for $t > 0$.

Many authors use the term "stability" in one of the weaker senses (1), (2) or (3) above; as in Holm et al. [1985], we use the term stability to mean nonlinear stability, in the sense of (4).

For conservative systems, it is well-known that even in finite dimensions, spectral stability is necessary for nonlinear stability, but is not sufficient (since, if the spectrum had a strictly positive real part, the nonlinear dynamics would have an unstable manifold; see e.g., Marsden and McCracken [1976]). However, neither formal nor linearized stability is necessary for nonlinear stability. (Both counter examples above are also nonlinearly stable.) Linearized stability does not imply nonlinear stability either, as shown by the following counterexample due to T. Cherry [1925] and discussed by Pollard [1966], p. 77] (see also, Siegel and Moser [1971, p. 109]). The dynamics generated by the Hamiltonian

$$H = \frac{1}{2}(q_1^2 + p_1^2) - (q_2^2 + p_2^2) + \frac{1}{2}p_2(p_1^2 - q_1^2) - q_1 q_2 p_1$$

has equilibrium $(p_{1e}, q_{1e}, p_{2e}, q_{2e}) = (0, 0, 0, 0)$ which is linearized stable in the Euclidean norm of $\mathbf{R}^4$. A one-parameter family of solutions for this system is, for any fixed value of the parameter τ,

$$p_1 = \sqrt{2}\,\frac{\sin(t-\tau)}{t-\tau}, \quad p_2 = \frac{\sin 2(t-\tau)}{t-\tau}, \quad q_1 = -\sqrt{2}\,\frac{\cos(t-\tau)}{t-\tau}, \quad q_2 = \frac{\cos 2(t-\tau)}{t-\tau}.$$

The distance at time t from the equilibrium is $\sqrt{3}/(\tau - t)$, which by choosing τ, can be made as small as desired at $t=0$ and which blows up at $t=\tau$. Thus, the equilibrium solution of this system is linearized stable, but nonlinearly unstable in the Euclidean $\mathbf{R}^4$ norm. Also, for a Hamiltonian system with at least three degrees of freedom, an equilibrium solution can be linearly stable but nonlinearly unstable, because of the phenomenon of Arnold diffusion; cf. Arnold [1978, Appendix 8], Chirikov [1979] and Lichtenberg and Lieberman [1982]. Thus, for a Hamiltonian system, spectral analysis can provide sufficient conditions for instability, but it can only give necessary conditions for stability. We are interested in finding sufficient conditions for stability.

In finite dimensions, formal stability implies stability. This is a classical result of Lagrange. (Indeed, if the equilibrium $X_e = 0$ is a nondegenerate minimum of the conserved quantity F, the set $\{x \mid |x| < \varepsilon, F(x) < \mu\}$, where μ is the minimum of F on the sphere of radius ε, is invariant under the flow; ε is chosen such that for all x satisfying $|x| < \varepsilon$ we have $F(0) < F(x)$; see also Siegel and Moser [1971, p. 208]). However, in the infinite dimensional case of concern to us in fluid dynamics, formal stability need not imply stability; indeed, physically realistic examples from elasticity show that an equilibrium solution can have positive second variation of the energy and still have an infinite number of unstable directions (see Ball and Marsden [1984].) Formal stability is a step toward stability, but a further argument is needed. Arnold [1966b] provided a framework for such arguments based on convexity estimates, by using several quantities related to the degeneracy of the Poisson brackets describing the system (or, equivalently, related to the symmetry of the fluid Hamiltonian written in Lagrangian coordinates under relabeling of fluid particles). The papers of which we are aware that actually prove nonlinear stability for conservative fluid and plasma systems are Arnold [1969a], Benjamin [1972], Bona [1975], McKean [1977], Laedke and Spatschek [1980], Holm et al. [1983], Benett et al. [1983], Bona et at. [1983], Wan [1984], Hazeltine et al. [1984], Holm [1984], Holm et al. [1984, 1985], Wan and Pulvirente [1985], Abarbanel et al. [1985, 1986] Lewis et al. [1987], Tang [1987], and Holm and Long [1988].

For dissipative systems, there are several general results that show that linearized stability implies stability; see for example, Marsden and McCracken [1976] and references therein, and for bifurcation results see, e.g., Crawford [1983] and references therein. In the limit of zero dissipation, however, these methods seem to give little, or no information on the stability of the corresponding conservative system. Since conservative systems are the concern of this paper we shall not discuss dissipative systems further.

Holm et al. [1985] give sufficient conditions for stability of equilibria for various two- and three-dimensional models of plasma physics, including magnetohydrodynamics (MHD), multifluid plasmas (MFP), the Poisson-Vlasov, and the Maxwell-Vlasov equations. They explain a general, algorithmic procedure for proving stability and illustrate its application with many examples. Using this procedure, Abarbanel et al. [1985, 1986] treat the stability of two- and three-dimensional incompressible flows, including a Richardson number criterion for the stability of shear flows.

The plan of these lectures is as follows. In Section 2 we recapitulate the general algorithm for proving stability found in Holm et al. [1985]. Section 3 illustrates the application of the stability algorithm for the planar barotropic fluid flow example first treated in Holm et al. [1983] and indicates how the stability criteria for the ideal hydrostatic model can be obtained by analogous calculations. Finally, Section 4 provides the Hamiltonian framework for discussing Lyapunov stability of equilibrium solutions for a new set of shallow water equations with nonlinear dispersion due to Basdenkov, Morosov, and Pogutse [1985] (see also Holm [1988]). A complete discussion of the stability of such shallow water equilibria with nonlinear dispersion has not yet been performed, although Basdenkov, Morosov, and Pogutse [1985] indicate that nonlinear dispersion may, in fact, cause the loss of formal stability.

2. The stability algorithm

We now present the algorithm for proving stability found in Holm et al. [1985]. Some of the steps are facilitated and put into a larger context by the use of a Hamiltonian structure (Poisson brackets); this is explained in remarks following each step.

A. Equations of motion and Hamiltonian

Choose a (Banach) space P of fields u and write the equations of motion on the space P in first-order form as

$$du/dt = X(u) \tag{2.1}$$

for a (nonlinear) operator X mapping a domain in P to P. Find a conserved functional H for Eq. (2.1), usually representing the total energy; that is find a map H:P→**R** such that $dH(u)/dt = 0$ for any C^1 solution u of Eq. (2.1).

Remark A. Often P is a Poisson space, i.e. a linear space (or more generally a manifold) admitting a Poisson bracket operation{ , } on the space of real valued functions on P which makes them into a Lie algebra, and which is a derivation in each variable. There are systematic procedures for obtaining such brackets; these procedures are not reviewed here, although we shall give references relevant to each example.* The equations (2.1) can then be expressed in Hamiltonian form for such a bracket structure:

$$dF/dt = \{F, H\}\ , \tag{2.2}$$

where H is the Hamiltonian, F is any functional of $u \in$ P, and dF/dt is its time derivative through the dependence of u of t.

* As noted in Weinstein [1982], the general notion of a Poisson manifold goes back to Sophus Lie around 1890.

B. Constants of motion

Find a family of constants of the motion for Eq. (2.1). That is, find a collection of functionals C on P such that $dC(u)dt = 0$ for any C^1 solution u of Eq. (2.1).

Remark B. Unless a sufficiently large family is found, the next step may not be possible. A good way to find conserved functionals is to use the Hamiltonian formalism in Remark A to find Casimir* functionals for the Poisson structure, i.e. functionals C such that $\{C, G\} = 0$ for all G. One may also find additional functionals associated with symmetries of the given Hamiltonian.

C. First variation

Relate an equilibrium solution u_e of (EM) to a constant of the motion C by requiring that $H_C := H + C$ have a critical point at u_e.

Note: The constant C may or may not be uniquely determined. Keeping C as general as possible may be useful in step D. Moreover, if C retains some freedom at this stage in terms of unspecified parameters or functions, critical points of H_C will correspond to *classes* of equilibria.

Remark C. If Remarks A and B are followed, then such a C can often be expected to exist. Indeed, level sets of the constants of motion define certain "leaves" in P; if C is a Casimir, they are the "symplectic leaves" of the Poisson structure $\{\,,\}$. Equilibrium solutions are critical points of H restricted to such leaves. If the Casimirs are functionally independent, the Lagrange multiplier theorem implies that H + C has a critical point at u_e for an appropriate Casimir function C. One cannot guarantee that such functions can be found explicitly in all cases: however,

* This term was used in the same context as here by Sudarshan and Mukunda [1974]. Following Sophus Lie, Casimirs are called distinguished functions in Olver [1986]. Also see Peter Olver's lectures in this volume.

they are found in the examples we consider. These points are discussed further in Appendix B of Holm et al. [1985].

D. Convexity estimates

Find quadratic forms Q1 and Q2 on P such that*

$$Q_1(\Delta u) \leq H(u_e + \Delta u) - H(u_e) - DH(u_e)\cdot \Delta u \ , \tag{2.3}$$

$$Q_2(\delta u) \leq C(u_e + \Delta u) - C(u_e) - DC(u_e)\cdot \Delta u \ , \tag{2.4}$$

for all Δu in P. Require that

$$Q_1(\Delta u) + Q_2(\Delta u) > 0 \ \textit{for all } \Delta u \textit{ in } P \ , \ \Delta u \neq 0 \ . \tag{2.5}$$

Remark D. Formal stability – second variation. As a prelude to checking conditions (2.3), (2.4) and (2.5), it is often convenient to see whether the second variation $D^2H_c(u_e)$ is definite, or when feasible, whether $D^2H(u_e)$ restricted to the symplectic leaf through u_e is definite. This property, called *formal stability*, is a prerequisite for Step D to work, but it is not sufficient (see Remark (2) below).

If formal stability is established, then the zero solution of the equation (2.1) linearized at u_e is stable, since in this case $D^2H_c(u_e)$ (which is the Hamiltonian for the linearized motion) provides a conserved norm under the linearized dynamics. (See Appendix A of Holm et al. [1985].)

E. A priori estimates

If steps A through D have been carried out, then for any solution u of (2.1), we have the following estimate on $\Delta u = u - u_e$:

$$Q_1(\Delta u(t)) + Q_2(\Delta u(t)) \leq H_c(u(0)) - H_c(u_e) \ . \tag{2.6}$$

(This is proven below and in Holm et al. [1985].)

* Here $\Delta u = u - u_e$ denotes a *finite* variation of the solution. To avoid confusion, we shall use $\nabla^2 u$ for the Laplacian of u.

F. (Nonlinear) stability

Stability theorem. Suppose that steps A through D have been carried out. Set

$$\|v\|^2 = Q_1(v) + Q_2(v) > 0 \ (for\ v \neq 0) \tag{2.7}$$

so $\|v\|$ defines a norm on P. If H_c is continuous in this norm at u_e, and solutions to (2.1) exist for all time, then u_e is stable. Should solutions to (2.1) not be known to exist for all time, one still has conditional stability: stability for all times during which C^1 solutions exist.

A sufficient condition for continuity of H_c is the existence of positive constants C_1 ana C_2 such that

$$H(u_e + \Delta u) - H(u_e) - DH(u_e) \cdot \Delta u \leq C_1 \|\Delta u\|^2 , \tag{2.8}$$

$$C(u_e + \Delta u) - C(u_e) - DC(u_e) \cdot \Delta u \leq C_2 \|\Delta u\|^2 . \tag{2.9}$$

In this case there follows the stability estimate:

$$\|\Delta u(t)\|^2 = Q_1(\Delta u(t)) + Q_2(\Delta u(t)) \leq (C_1 + C_2) \|\Delta u(0)\|^2 , \tag{2.10}$$

for all Δu in P (these assertions are proved below).

Proof of a priori estimate (2.6). Adding (2.3) and (2.4) gives

$$Q_1(\Delta u) + Q_2(\Delta u) \leq H_c(u_e + \Delta u) - H_c(u_e) - DH_c(u_e) \cdot \Delta u = H_c(u_e + \Delta u) - H_c(u_e) , \tag{2.11}$$

since $DH_c(u_e) = 0$ by step C. Because H_c is a constant of the motion, $H_c(u_e + \Delta u) - H_c(u_e)$ equals its value at $t=0$, which is (2.6). ∎

Proof of the assertions in step F. We prove (Lyapunov) stability of u_e as follows. Given $\varepsilon > 0$, find a δ such that $\|u - u_e\| < \delta$ implies $|H_c(u) - H_c(u_e)| < \varepsilon$. Thus, if $\|u(0) - u_e\| < \delta$, then (2.6) gives

$$\|u(t) - u_e\| \leq |H_c(u(0)) - H_c(u_e)| < \varepsilon . \tag{2.12}$$

Thus, $u(t)$ never leaves the ε-ball about u_e if it starts in the δ ball, so u_e is stable. To see that (2.8) and (2.9) suffice for continuity of H_c at u_e, add them to give, as in the proof of (2.6),

$$H_c(u_e + \Delta u) - H_c(u_e) \leq C_1\|\Delta u\|^2 + C_2\|\Delta u\|^2 = (C_1 + C_2)\|\Delta u\|^2 , \tag{2.13}$$

which implies that H_c is continuous at u_e. This proves the stability estimate (2.10). ∎

<u>Further remarks</u>

(1) In some examples, Q_1 and Q_2 are each positive (so H and C are individually convex). Then (D) is automatic. However, as already noted by Arnold [1969a], there are some interesting examples where Q_1 is positive, Q_2 is negative and yet the sum $Q_1 + Q_2$ is positive and (D) is valid. If the sum $Q_1 + Q_2$ is shown to be negative, then one can replace H_c by $-H_c$ to obtain (D).

(2) It has been presumed that P carries a Banach space topology (although one could merely assume P is a Fréchet space) relative to which the symbols du/dt and $DH(u_e)$ are defined, and steps A, B and C are admissible. The norm $\|\cdot\|$ found in step F is usually not complete; relative to the functions H and C, it need not be differentiable. (This fact is related to the difficulty one encounters when trying to deduce stability from formal stability.) A sufficient condition for the convexity hypothesis (2.3) is that the inequality

$$Q_1(v) \leq D^2H(u) \cdot (v, v) \tag{2.14}$$

holds for all u and v in P. The sufficiency of (2.14) follows from the mean value theorem. There are similar assertions for C and H_c. Note that

$$\|v\|^2 \leq D^2H_c(u)(v, v) \tag{2.15}$$

is considerably stronger than formal stability: $D^2H_C(u_e)(v,v)$ positive definite. Indeed, (2.15) is a *global* convexity condition which reflects the additional hypotheses involved in step D.

(3) As already noted, in systems with a finite number of degrees of freedom, formal stability implies stability. This fact was used by Arnold [1966a] to reproduce the well-known results on stability of rigid body motion. See Marsden and Weinstein [1974] for the relationship of the formal stability ideas to the stability of relative equilibria and reduction. (See also Abraham and Marsden [1978], Sections 4.3 and 4.4 and Arnold [1978], Appendices 2 and 5.)

(4) In many examples, such as compressible flow, one does not have global existence of smooth solutions. The stability algorithm in its present form does not address weak solutions or solutions with shocks. The results will apply only to sufficiently smooth solutions. Moreover, one or more of the steps may require assumptions about some of the variables. For example, in two-dimensional compressible flow, (Section 3), we obtain our estimates only under the assumption that the density satisfies $0<\rho_{min}\leq\rho\leq\rho_{max}\leq\infty$ for constants ρ_{min} and ρ_{max}. (The necessity of such assumptions is revealed by the convexity analysis; formal stability does not reveal this and would tempt one to make unjustified claims in this regard.) This type of stability, which requires one to *monitor* some of the variables is called *conditional stability*.

(5) For Hamiltonian systems with additional symmetries, there will be additional constants of the motion besides Casimirs. These are to be incorporated into the function C in step B. This is needed in fluid examples with a translational symmetry, for example, and in the stability analysis of a heavy top, see Holm et al. [1984].

Table 1: Summary of the energy-Casimir stability method

A.	*Equations of motion and Hamiltonian.* Write the equations of motion (2.1) on P and find the conserved energy H. [Determine the Poisson bracket and Hamiltonian on P.]
B.	*Constants of motion.* Find as many conserved quantities C as possible for (2.1). [Determine the Casimirs of P.]
C.	*First variation.* Let $H_c := H + C$ and u_e be a stationary solution for (2.1). Relate C and u_e by the condition $DH_c(u_e) = 0$. Keep C as general as possible.
D.	*Convexity estimates.* Find quadratic forms Q_1 and Q_2 on P and conditions on u_e such that (2.3), (2.4), and (2.5) hold. [*Formal stability.* Show that $D^2H_c(u_e)$ is definite and conclude linearized stability.]
E.	*A priori estimates.* Write out the estimate (2.6).
F.	*Stability.* Find sufficient conditions on u_e to guarantee that H_c is continuous in the norm (2.7), or prove the estimates (2.8), (2.9), and conclude conditional stability of u_e subject to these conditions. In the presence of a long-time existence theorem, conclude stability.

(6) For two-dimensional incompressible flow, the appropriate Casimir function is the generalized enstropy. This suggests, following Leith (cf. Bretherton and Haidvogel [1976]), that the Casimir functions may play a role in the "selective decay hypothesis" when dissipation is added.

For the convenience of the reader, we summarize schematically the procedure just explained in Table 1, with the optional but useful steps in square brackets. In the example of two-dimensional ideal barotropic flow, we shall follow this procedure and carry out each step explicitly.

For finite dimensional systems, formal stability implies stability. Thus, the energy-Casimir method in this case requires only steps A, B, C, and the formal stability argument in step D.

3. Two-dimensional barotropic flow (Holm et al. [1983], Grinfeld [1984])

A. *Equations of motion and Hamiltonian*

Let D be a domain in $\mathbb{R}^2$ with smooth boundary. The evolution equations for the velocity field $\mathbf{v}(x,y,t)$ and density $\rho(x,y,t)$ are

$$\frac{\partial \mathbf{v}}{\partial t} + (\mathbf{v}\cdot\nabla)\mathbf{v} = -\nabla h(\rho)\ ,\quad \frac{\partial \rho}{\partial t} + div(\rho\mathbf{v}) = 0\ , \tag{3.1}$$

where v is parallel to ∂D and $h(\rho)$ is the specific enthalpy, a given function of $\rho>0$, satisfying $p'(\rho)=\rho h'(\rho)$, where p is the pressure.

We choose P to be a space of v and ρ that are C^1 (say H^s, $s>2$) and tending to a fixed vector field and density at ∞ if D is unbounded, or with v parallel to ∂D. We shall also need to exclude from the beginning of the discussion certain important features over which the present methods have no control. These are as follows, taken as part of our definition of P;

(a) shocks; solutions considered are C^1;

(b) cavitation and extreme compression: the density satisfies $0<\rho_{min}\leq\rho\leq\rho_{max}<\infty$, where ρ_{min} and ρ_{max} are constants (that will shortly be required to satisfy certain inequalities involving other constants in the problem).

The conserved energy is

$$H(\mathbf{v},\rho) = \int_D [\tfrac{1}{2}\rho|\mathbf{v}|^2 + \varepsilon(\rho)]dx\,dy\ , \tag{3.2}$$

where $\varepsilon(\rho)$ is the internal energy per unit area, related to the specific enthalpy by $\varepsilon'(\rho)=h(\rho)$.

Remark A. The equations of motion (3.1) are Hamiltonian. The configuration space of compressible fluid motion is the group of diffeomorphisms of D whose Lie algebra consists of the space V(D) of all vector fields on D. V(D) is represented on the vector space $\Lambda^0(D)$ of functions on D by minus the Lie derivative, i.e.,

$$\mathbf{X}\cdot f := -\mathbf{X}[f] = -df(\mathbf{X}), \; \textit{for } \mathbf{X} \in X(D), \tag{3.3}$$

$f \in \Lambda^0(D)$. On the dual of the semidirect product V(D) s Λ^0(D) with variables $\mathbf{M} = \rho \mathrm{v}$ and ρ, the eq. (3.1) are Hamiltonian (i.e. (2.2) holds) relative to the Lie-Poisson bracket

$$\begin{aligned}\{F, G\} = \int_D \mathbf{M}\cdot\left[\left(\frac{\partial G}{\partial \mathbf{M}}\cdot\nabla\right)\frac{\delta F}{\delta \mathbf{M}} - \left(\frac{\delta F}{\delta \mathbf{M}}\cdot\nabla\right)\frac{\delta G}{\delta \mathbf{M}}\right] dx\,dy \\ + \int_D \rho\left[\frac{\delta G}{\delta \mathbf{M}}\cdot\left(\nabla\frac{\delta F}{\delta \rho}\right) - \frac{\delta F}{\delta \mathbf{M}}\cdot\left(\nabla\frac{\delta G}{\delta \rho}\right)\right] dx\,dy\,.\end{aligned} \tag{3.4}$$

This bracket is found in Iwinski and Turski [1976], Morrison and Greene [1980] and Dzyaloshinsky and Volovick [1980]; see also Dashen and Sharp [1968] and Bialynicki-Birula and Iwinski [1973]. The bracket was derived from Clebsch variables by Enz and Turski [1979], Greene, Holm and Morrison [1980], Morrison [1982] and Holm and Kupershmidt [1983]. This bracket is the Lie-Poisson bracket for a semi-direct product, as noted in Marsden [1982], where it is also pointed out that the bracket could be obtained as an instance of the abstract results concerning the Lagrange to Euler map of Ratiu [1980, 1982] and Guillemin and Sternberg [1980]. Holm and Kupershmidt [1983] also showed that other interesting systems, such as MHD, multifluid plasmas, and nonlinear elasticity are Lie-Poisson for semi-direct products. These and related brackets are derived from canonical (symplectic) brackets in the Lagrangian representation in Marsden, Weinstein et al. [1983], Holm, Kupershmidt and Levermore [1983] and Marsden, Ratiu and Weinstein [1984a,b].

B. Constants of motion

From the equations of motion (3.1) one finds that the quantity (ω/ρ) is advected by the flow, i.e., $\partial(\omega/\rho)/\partial t + \mathbf{v}\cdot\nabla(\omega/\rho) = 0$. Thus, for any function $\Phi:\mathbf{R}\to\mathbf{R}$, the quantity

$$C_{\Phi}(\mathbf{v},\rho) = \int_{D} \rho\Phi(\omega/\rho)\, dx\, dy \tag{3.5}$$

is a constant of the motion, where $\omega = \hat{\mathbf{z}}\cdot(\nabla\times\mathbf{v})$ is the scalar vorticity. Similarly, by Kelvin's circulation theorem the quantities

$$\Gamma_i(\mathbf{v},\rho) = \oint_{(\partial D)_i} \mathbf{v}\cdot d\boldsymbol{\ell}\ ,\ i = 0\ ,\ \cdots\ ,\ g \tag{3.6}$$

are conserved, where $(\partial D)_i$ are the connected components of the boundary.

Remark B. The functions C_{Φ} are Casimirs for the Poisson structure in remark A. This can be checked directly, or it can be proved by noting that C_{Φ}, as a function of $(\mathbf{M},\rho)$, is invariant under the coadjoint action of Diff(D) s F(D) (semidirect product of the group of diffeomorphisms and functions) on P. As discussed in Holm et al. [1985], for $(\eta,f) \in$ Diff(D) s F(D), this action is

$$(\eta,f)\cdot(M,\rho) = (\eta_* M - df \otimes \eta_*\rho, \eta_*\rho) \tag{3.7}$$

where ρ is regarded as a density. Similarly, all Γ_i are also invariant under the coadjoint action, but some subtleties arise in treating their functional derivatives in the usual sense of the formal calculus of variations. Thus, their brackets with arbitrary functionals require care in interpretation; see Lewis et al. [1987].

C. First variation

Let $(\mathbf{v}_e,\rho_e)$ be an equilibrium solution of (3.1). Then $H_C(\mathbf{v},\rho) = H(\mathbf{v},\rho) + C_{\Phi}(\mathbf{v},\rho) + \Sigma_{i=0}^{g} a_i\Gamma_i(\mathbf{v},\rho)$ has a critical point at $(\mathbf{v}_e,\rho_e)$, provided the following holds for all $\delta\mathbf{v}$, $\delta\rho$ (such that $(\mathbf{v}_e+\delta\mathbf{v},\rho_e+\delta\rho)$ lies in P):

$$0 = DH_c(\mathbf{v}_e,\rho_e)\cdot(\delta\mathbf{v},\delta\rho)$$

$$= \int_D [\rho_e\mathbf{v}_e\cdot\delta\mathbf{v} + \Phi'(\omega_e/\rho_e)\mathbf{z}\cdot(\nabla\times\delta\mathbf{v})]\,dx\,dy + \sum_{i=0}^{g} a_i \oint_{(\partial D)_i} \delta\mathbf{v}\cdot d\boldsymbol{\ell} \tag{3.8}$$

$$+ \int_D \left[\frac{|\mathbf{v}_e|^2}{2} + h(\rho_e) + \Phi\left(\frac{\omega_e}{\rho_e}\right) - \frac{\omega_e}{\rho_e}\Phi'\left(\frac{\omega_e}{\rho_e}\right)\right]\delta\rho\,dx\,dy\,.$$

Integrating the second term in the first integral of (3.8) by parts gives

$$0 = \int_D \left\{\left[\frac{|\mathbf{v}_e|^2}{2} + h(\rho_e) + \Phi\left(\frac{\omega_e}{\rho_e}\right) - \frac{\omega_e}{\rho_e}\Phi'\left(\frac{\omega_e}{\rho_e}\right)\right]\delta\rho + \left[\rho_e\mathbf{v}_e - \mathbf{z}\times\nabla\Phi'\left(\frac{\omega_e}{\rho_e}\right)\right]\cdot\delta\mathbf{v}\right\}dx\,dy \tag{3.9}$$

$$+ \sum_{i=0}^{g} \oint_{(\partial D)_i} \Phi'\left(\frac{\omega_e}{\rho_e}\right)\delta\mathbf{v}\cdot d\boldsymbol{\ell} + \sum_{i=0}^{g} a_i \oint_{(\partial D)_i} \delta\mathbf{v}\cdot d\boldsymbol{\ell}\,.$$

For stationary solutions, ω_e/ρ_e is constant along streamlines, so the $2(g+1)$ boundary terms cancel, provided $a_i = -\Phi'((\omega_e/\rho_e)|\partial D)_i)$. From (3.1), stationary flows satisfy

$$\mathbf{v}_e\cdot\nabla(|\mathbf{v}_e|^2/2 + h(\rho_e)) = 0\,,\quad \mathbf{v}_e\cdot\nabla(\omega_e/\rho_e) = 0\,. \tag{3.10}$$

This is consistent with assuming a Bernoulli Law

$$|\mathbf{v}_e|^2/2 + h(\rho_e) = K\left(\frac{\omega_e}{\rho_e}\right), \tag{3.11}$$

for K a smooth function of a real variable. The condition $0 = DH_c(\mathbf{v}_e,\rho_e)\cdot(\delta\mathbf{v},\delta\rho)$ holds if the coefficients of $\delta\rho$ and $\delta\mathbf{v}$ vanish. For $\delta\rho$ this is

$$K(\zeta) + \Phi(\zeta) - \zeta\Phi'(\zeta) = 0\,, \tag{3.12}$$

which determines Φ up to a constant:

$$\Phi(\zeta) = \zeta\left(\int^{\zeta}\frac{K(t)}{t^2}\,dt + const.\right). \tag{3.13}$$

An important point is that the coefficient of $\delta\mathbf{v}$ in (3.9) also vanishes by virtue of the expression (3.13) for Φ. Indeed, from Bernoulli's Law (3.11) we have

$$\nabla(|\mathbf{v}_e|^2/2 + h(\rho_e)) = \nabla K(\omega_e/\rho_e)\,, \tag{3.14}$$

so that for stationary solutions, (3.1) gives

$$0 = \partial \mathbf{v}_e / \partial t = -\nabla(|\mathbf{v}_e|^2/2 + h(\rho_e)) + \mathbf{v}_e \times \omega_e \mathbf{z}\,, \tag{3.15}$$

and hence

$$\mathbf{v} \times \omega_e \mathbf{z} = \nabla K\left(\frac{\omega_e}{\rho_e}\right) \text{ or } \rho_e \mathbf{v}_e = \frac{\rho_e}{\omega_e} \mathbf{z} \times \nabla K\left(\frac{\omega_e}{\rho_e}\right) = \mathbf{z} \times \nabla \Phi'\left(\frac{\omega_e}{\rho_e}\right), \tag{3.16}$$

using the relation $K'(\zeta) = \zeta\Phi''(\zeta)$ between K and Φ. Consequently, we have the following.

Proposition. Stationary solutions $(\mathbf{v}_e, \rho_e)$ of two-dimensional barotropic Euler flow with $\rho_e > 0$ are critical points of $H + C_\Phi + \Sigma_{i=0}^{g} a_i \Gamma_i$, where Φ is given in terms of the Bernoulli function K for the stationary solution by

$$\Phi(\zeta) = \zeta\left(\int^{\zeta} \frac{K(t)}{t^2}\, dt + const.\right), \quad a_i = -\Phi'(\omega_e / \rho_e)\,|\,(\partial D)_i\,. \tag{3.17}$$

Remark D. The second variation of $H_C(\mathbf{v}_e, \rho_e)$ is computed to be

$$D^2 H_c(\mathbf{v}_e, \rho_e)(\delta\mathbf{v}, \delta\rho) = \int_D \left\{ \frac{|\delta(\rho\mathbf{v})|^2}{\rho_e} + \left[\varepsilon''(\rho_e) - \frac{|\mathbf{v}_e|^2}{\rho_e} \right] (\delta\rho)^2 \right.$$
$$\left. + \frac{1}{\omega_e} K'\left(\frac{\omega_e}{\rho_e}\right) \left[\delta\left(\frac{\omega}{\rho}\right)\right]^2 \right\} dx\, dy\,, \tag{3.18}$$

where $\delta(\rho\mathbf{v}) := \mathbf{v}_e \delta\rho + \rho_e \delta\mathbf{v}$ and $\delta(\omega/\rho) := (\rho_e \delta\omega - \omega_e \delta\rho)/\rho_e^2$.

Expression (3.18), suggests that conditions for stability are $\rho_e > 0$ and $\varepsilon''(\rho_e)\rho_e > |\mathbf{v}_e|^2$ (the latter meaning that the stationary flow is subsonic), and $(\rho_e/\omega_e)K'(\omega_e/\rho_e) > 0$. These are the conditions for formal stability, but the nonlinear theory requires more stringent conditions. (The second variation calculation has also recently been done by Grinfeld [1984] using Clebsch variables.)

D. Convexity estimates

We have, after a short computation,

$$H(\mathbf{v}_e + \Delta\mathbf{v}, \rho_e + \Delta\rho) - H(\mathbf{v}_e, \rho_e) - DH(\mathbf{v}_e, \rho_e)\cdot(\Delta\mathbf{v}, \Delta\rho)$$

$$= \int_D \left\{ \frac{|\Delta(\rho\mathbf{v})|^2}{2\rho} - \frac{|\mathbf{v}_e|^2}{2}\frac{(\Delta\rho)^2}{\rho} + [\varepsilon(\rho_e + \Delta\rho) - \varepsilon(\rho_e) - \varepsilon'(\rho_e)\Delta\rho] \right\} dx\, dy\,, \tag{3.19}$$

where $\Delta(\rho\mathbf{v}) := (\rho_e + \Delta\rho)(\mathbf{v}_e + \Delta\mathbf{v}) - \rho\mathbf{v}_e$. Assume $\varepsilon''(\tau) \geq c_{min}{}^2 / \tau$ for all τ and a constant c_{min} (the minimum sound speed). Then we get (2.3) with

$$Q_1(\Delta(\rho\mathbf{v}), \Delta\rho) = \frac{1}{2}\int_D \left\{ \frac{|\Delta(\rho\mathbf{v})|^2}{\rho_{max}} + \left| \frac{c_{min}^2}{\rho_{max}} - \frac{|\mathbf{v}_e|^2}{\rho_{min}} \right| (\Delta\rho)^2 \right\} dx\, dy\,, \tag{3.20}$$

where $0 < \rho_{min} \leq \rho \leq \rho_{max} < \infty$. Note that Q_1 is a quadratic form in the variables $(\rho\mathbf{v}, \rho)$ rather than $(\mathbf{v}, \rho)$.

If the Bernoulli function K satisfies

$$a \leq \frac{1}{\zeta} K'(\zeta) = \Phi''(\zeta)\,, \tag{3.21}$$

then one finds convexity condition (2.4) with a quadratic form in $\Delta(\omega / \rho)$:

$$Q_2(\Delta(\rho\mathbf{v}), \Delta\rho) = \frac{1}{2} a \rho_{min} \int_D [\Delta(\omega / \rho)]^2\, dx\, dy\,, \tag{3.22}$$

where $\Delta(\omega / \rho) := [(\omega_e + \Delta\omega) / (\rho_e + \Delta\rho) - \omega_e / \rho_e]$. Thus, (2.5) holds and the sum $Q_1 + Q_2$ is positive, provided

$$a > 0 \text{ and } c_{min}^2 / \rho_{max} > |\mathbf{v}_e|^2 / \rho_{min} \tag{3.23}$$

E. A priori estimates

The estimate (2.6) holds with Q_1 and Q_2 given as above.

F. Nonlinear stability

If we have

$$\varepsilon''(\iota) \leq c^2_{max} / \rho_{min} \text{ for all } \iota,\ \rho_{min} \leq \iota \leq \rho_{max} \tag{3.24}$$

and

$$\frac{1}{\zeta} K'(\zeta) \leq A < \infty, \tag{3.25}$$

then the continuity conditions (2.8) and (2.9) hold for arguments similar to those given in step D in the algorithm. Thus, with this hypothesis, and for solutions in P satisfying $\rho_{\min} \leq \rho \leq \rho_{\max}$, we have Lyapunov stability in the norm $\| \|^2 = Q_1 + Q_2$ as long as solutions remain in P. (The existence theory for solutions to these equations is not well established, except for short-time solutions, so there is little more that one can expect in the present circumstances.)

We summarize the results for barotropic planar flow as follows.

Barotropic Stability theorem. Stationary solutions $(\mathbf{v}_e, \rho_e)$ of the two-dimensional barotropic Euler flow that satisfy the conditions

$$0 < \rho_{min} \leq \rho_e \leq \rho_{max} < \infty, \tag{3.26}$$

$$0 < a \leq \frac{1}{\zeta} K'(\zeta) \leq A < \infty, \tag{3.27}$$

$$c^2_{min} / \rho_{max} \leq \varepsilon''(\iota) \leq c^2_{max} / \rho_{min}, \tag{3.28}$$

where K is the Bernoulli function for $(\mathbf{v}_e, \rho_e)$, are conditionally stable in the norm on $(\rho\mathbf{v}, \rho)$ given by $Q_1 + Q_2$, that is, perturbations from equilibrium are *a priori* bounded in time in the norm determined by $Q_1 + Q_2$ as long as the solutions satisfy $\rho_{\min} \leq \rho \leq \rho_{\max}$.

Example A. Shear flow. A stationary solution of (3.1) in the strip $\{(x,y)\in\mathbf{R}^2 | Y_1 \le y \le Y_2\}$, is given by the plane parallel flows with arbitrary velocity profile $v_e(x,y)=(u(y),0)$ and constant density $\rho_e = 1$. We can allow x to be unrestricted in $\mathbf{R}$ or to be periodic. In the former case, we require that the allowed perturbations be initially square integrable. Note that $(\omega_e/\rho_e)(x,y) = -u'(y)$. Let c_e denote the sound speed of this stationary solution. By our earlier analysis this flow is formally, hence linearized, stable if and only if $c_e^2 - u(y)^2 > 0$ and $u(y)/u''(y) > 0$.

The hypothesis on the existence of the Bernoulli function K is in this case $u''(y) \neq 0$. In other words, plane parallel flows with constant density and velocity profile with no inflection point are formally, hence linearly, stable. This is analogous to Rayleigh's theorem for the incompressible problem.

We turn now to the study of *a priori* estimates for this shear flow. For this, one must compute the Bernoulli function K from its defining relation (3.11) under the hypothesis $\nabla(\omega_e/\rho_e) = u''(y)\hat{y} \neq 0$. Denote by ϕ the inverse of u; we get $K(\zeta) = u[\phi(\zeta)]^2/2 + h(1)$ and thus $K'(\zeta) = -u(\phi(\zeta))u'(\phi(\zeta))/u''(\zeta)) = \zeta u(\phi(\zeta))/u''(\phi(\zeta))$, so that condition (3.27) becomes $0 < a \le u(y)/u''(y) \le A < \infty$. To get the *a priori* estimate (2.6), one imposes stability condition (3.28), which bounds $\varepsilon''(\tau)$. Condition (3.28), for example, is satisfied for an ideal gas with $\gamma = 2$, i.e., a monatomic gas in two dimensions. The *a priori* estimate (2.6) then results, with $\rho_e = 1$ and velocity profile $u(y)$, satisfying (3.27) but arbitrary otherwise.

For the Mie-Grüneisen equation of state $\varepsilon(\tau) = A\tau + B/\tau + C$, with constants $A = \frac{1}{2}a\rho_e^3$, $B = \varepsilon'(\rho_e) + \frac{1}{2}a\rho_e$, $C = \varepsilon(\rho_e) - \rho_e\varepsilon'(\rho_e) - a\rho_e^2$, where the constant a satisfies $c_{min}^2/\rho_{max} \le a \le c_{max}^2/\rho_{min}$, condition (3.28) is sufficient for the *a priori* estimate for the "elastic fluid", again with $\rho_e = 1$.

Parallel shear flows with one inflection point taking place at $y=0[u''(0)=0]$ can also be considered, under the assumption that the equilibrium velocity profile is antisymmetric about the inflection point: $u(-y)=-u(y)$. For the case in which the ratio $u(y)/u''(y)$ is positive and bounded, as in (3.27), one again obtains a priori bounds. For example, one may take $u(y)=\text{arc tanh } y$, $|y|\leq 1$, as in Herlitz [1967].

Compressible shear flow in the plane can also be stationary if $\mathbf{v}_e(x,y)=(u(y),0)$ and $\rho_e(x,y)=f(y)$, for arbitrary functions $u(y)$, $f(y)$. In this case, $\omega_e(x,y)/\rho_e=u'(y)/f(y)$ and the assumption on the existence of the Bernoulli function K is $[u'(y)/f(y)]'\neq 0$. This flow is formally stable provided $c_e^{\,2}(y)-u(y)^2>0$ and $\zeta^{-1}K'(\zeta)>0$, where $c_e(y)$ is the sound speed. Thus, the stationary flow must be subsonic everywhere, and $K(\zeta)$ must be increasing as a function of $\zeta^2/2$. The *a priori* estimate (2.6) holds, if ε and K satisfy the inequalities in the theorem.

Example B. Circular flows. To illustrate the effect of barotropic compressibility on stability, we consider circular flow in an annular domain, so in polar coordinates (r,θ), $\mathbf{v}_e=\hat{\theta}v_e(r)$ where $\hat{\theta}$ is a unit vector in the aximuthal direction, and $\rho_e=\rho_e(r)$. Because of circular symmetry, there are additional conserved quantities: namely, the angular momentum $\int_D(\rho\mathbf{v}\times\mathbf{r})\cdot\hat{\mathbf{z}}\,dx\,dy$ and moment of inertia $\int_D\rho r^2\,dx\,dy$. Hence, we take

$$H_c=\int_D dx\,dy\,[\frac{1}{2}\rho|\mathbf{v}|^2+\varepsilon(\rho)+\rho\Phi(\omega/\rho)+\frac{1}{2}\Omega\rho\mathbf{v}\times\mathbf{r}\cdot\mathbf{z}+\frac{1}{8}\Omega^2\rho r^2+\Omega]+\oint_{(\partial D)_i}\mathbf{v}\cdot d\boldsymbol{\ell}\,, \tag{3.29}$$

where $\Omega=\text{constant}$ and $(\partial D)_i$ are circularly symmetric. This can be rewritten as

$$H_c=\int_D dx\,dy\,[\frac{1}{2}\rho|\mathbf{v}|^2+\varepsilon(\rho)+\rho\Phi((\omega+\Omega/\rho)]+\oint_{(\partial D)_i}\mathbf{v}\cdot d\boldsymbol{\ell} \tag{3.30}$$

where $\mathbf{v}=\tilde{\mathbf{v}}+\frac{1}{2}\Omega\mathbf{r}\times\hat{\mathbf{z}}$ is the fluid velocity relative to a frame rotating with angular velocity $\Omega/2$, and $\tilde{\omega}=\hat{\mathbf{z}}\cdot\text{curl }\tilde{\mathbf{v}}=\omega-\Omega$. Since H_c in these variables retains its previous form, the stability condition $\Phi''(\omega_e/\rho_e)>0$ can be written as either

$$v_e / \frac{d}{dr}\,[\rho_e^{-1}(\omega_e+\Omega)] > 0\ , \tag{3.31}$$

where $\bar{\omega}_e = r^{-1}d(r\bar{v}_e)/dr$, or, equivalently, using the equilibrium condition $d\rho_e/dr = \rho_e\bar{v}_e(\bar{v}+\Omega r)/rc_e^2$ where $c_e^2=\rho_e h'(\rho_e)$, as

$$\rho_e v_e / \left[\frac{d}{dr}(\omega_e+\Omega) - (v_e+\Omega r)(\omega_e+\Omega)v_e/rc_e^2\right] > 0 \tag{3.32}$$

for stability.* Thus, compressibility can be either stabilizing or not, depending on the relative signs and magnitudes of $\bar{v}_e/r, \bar{\omega}_e$, and Ω, and the magnitude of c_e^2. In the limit that c_e^{-2} tends to zero, the second term in the denominator vanishes in (3.32) and it becomes the counterpart for circular incompressible flow of Rayleigh's inflection point criterion.

For rigidly rotating flows, $\bar{v}_e/=\bar{\omega}_e r, \bar{\omega}_e=\text{const}$, and condition (3.32) becomes

$$(2\Omega+\omega_e)(\omega_e+\Omega) < 0 \tag{3.33}$$

for stability, which is satisfied when $\bar{\omega}_e\Omega<0$ and $2|\Omega|>|\bar{\omega}_e|>|\Omega|$, independently of the (circular) domain considered. For $\Omega=0$, homogeneous flows with $v_e(r)=v_0(r/r_0)^n$ for constants n, v_0, r_0, and $v_e^2/c_e^2 = d\log\rho_e/d\log r = m^2 = \text{const}$, are stable according to (3.32) for either $n>1+m^2$, or $n<-1$, also independently of the domain.

Remarks on analogous problems

1. The barotropic equations in a rotating frame are

$$\partial\mathbf{v}/\partial t = -(\mathbf{v}\cdot\nabla)\mathbf{v} - \nabla h(\rho) + \Omega\mathbf{v}\times\mathbf{z}\ ,\quad \partial\rho/\partial t = -div\,\rho\mathbf{v}\ , \tag{3.34}$$

which imply

* For the homogeneous case, nonlinear stability of circular, elliptical and annular patches of vorticity is studied theoretically by Wan [1984], Wan and Pulvirente [1985] and Tang [1987], and numerically by Dritschel [1986].

$$(\partial/\partial t + \mathbf{v}\cdot\nabla)[(\omega + \Omega)/\rho] = 0 \ , \tag{3.35}$$

and the corresponding stability criteria based on the discussion here have certain meteorological applications and also apply to large-scale topographical planetary waves in the ocean when $h(\rho)=g\rho^2/2$ and ρ is identified with the height of a shallow water surface over a flat bottom (see, e.g., LeBlond and Mysak [1978]).* In the absence of circular symmetry, the nonlinear stability analysis for the barotropic equations in a rotating frame is an obvious modification of what is presented earlier in this section for the case without rotation.

2. The Boussinesq inviscid dynamics of a continuously stratified, rotating, incompressible fluid in three dimensions under hydrostatic balance in the direction of gravity may also be formulated as a Hamiltonian system (Holm and Long [1988]). This Hamiltonian formulation reveals a useful analogy to two-dimensional compressible barotropic fluids by which the Lyapunov stability results of Holm et al. [1983, 1985] for the barotropic fluid equilibria may be transferred to the corresponding hydrostatic Boussinesq equilibria. To see this analogy, we first transform the ideal Boussinesq fluid model to isopycnal (same mass density) coordinates, in which the stratified fluid density ρ becomes an independent variable while the height of an isopycnal surface becomes a dependent variable. Since the mass density moves with the fluid in the ideal Boussinesq model, the velocity in isopycnal coordinates vanishes in the ρ-direction, thereby reducing the motion to isopycnal surfaces and setting the stage for the analogy to two-dimensional barotropic flow. This analogy is revealed when the Lie-Poisson bracket for the Boussinesq fluid model in isopycnal coordinates turns out to be identical in form to that for two-dimensional barotropic fluids. The Lyapunov stability conditions for the

* Stability of group-invariant solutions for shallow water equilibrium in a rotating, cylindrically symmetric container was discussed at the School in Paipa by Holm, Rogers, and Winternitz (private communication). In this regard, see Ripa [1987].

Boussinesq model in isopycnal coordinates then follow via this analogy from the corresponding conditions for barotropic fluid flows in two dimensions, simply by reinterpreting the physical meanings of the variables (and using a different expression for the energy).

Ideal Hydrostatic Boussinesq Model

In the Boussinesq approximation, the ideal stratified fluid equations in hydrostatic equilibrium in three dimensions are given in nondimensional form by

$$\varepsilon \frac{Du}{Dt} - fv = -p_x \tag{3.36a}$$

$$\varepsilon \frac{Dv}{Dt} + fu = -p_y \tag{3.36b}$$

$$0 = -p_z - \rho \tag{3.36c}$$

$$\nabla \cdot u = 0 \tag{3.36d}$$

$$\frac{D\rho}{Dt} = 0 \tag{3.36e}$$

where $D/Dt = \partial_t + \mathbf{u}\cdot\nabla$ is the material derivative in three dimensions. The spatial coordinates (x,y,z) have been scaled by (L,L,H), the velocity components $\mathbf{u} = (u,v,w)$ by (U,U,UH/L), the time t by L/U, the y-varying Coriolis parameter f by f_0, the pressure p by $\rho_0 f_0 LU$, and the density ρ by $\rho_0 f_0 LU/gH$ (with g the gravitational acceleration and ρ_0 the mean density in the Boussinesq approximation). For this scaling, the dimensionless parameter of importance is the Rossby number,

$$\varepsilon = U / f_0 L \, . \tag{3.37}$$

We transform the set of equations (3.36a)–(3.36e) in Cartesian coordinates (x,y,z) into isopycnal coordinates (x,y,ρ) in which density isoclines (isopycnals) are independent vertical coordinates, by inverting the relation for the stratified density $\rho(x,y,z,t)$ (assuming this can be done) and using the hydrostatic condition (3.36c) rewritten as

$$p_\rho = \frac{\partial p}{\partial \rho} = \frac{p_z}{\rho_z}\ [by\,(3.36c)] = \frac{-\rho}{\rho_z} = -\,\rho h_\rho\ , \tag{3.38}$$

in terms of a height function,

$$h(x,y,\rho,t) = \int^{\rho} \frac{d\rho'}{\frac{\partial \rho}{\partial z}(x,y,\rho',t)}\ . \tag{3.39}$$

Thus, the hydrostatic condition (3.36c) is expressible as

$$\psi_\rho = h\ ,\ \psi = p + \rho h\ . \tag{3.40}$$

The fluid velocity $(u,v,\bar{w})$ in isopycnal coordinates (x,y,ρ) is defined as the rate of change of these coordinates along a flow line

$$(u,v,\bar{w}) = \frac{D}{Dt}(x,y,\rho)\ . \tag{3.41}$$

Therefore, by (3.36e) the velocity vanishes in the direction normal to isopycnals, $\bar{w}=0$, since the stratified density is comoving with the fluid. The Eulerian vertical velocity w is given by the material derivative of the height function h in (3.39)

$$w = \frac{Dh}{Dt} = h_t + h_x\frac{Dx}{Dt} + h_y\frac{Dy}{Dt} + h_\rho\frac{d\rho}{dt} = h_t + uh_x + vh_y\ . \tag{3.42}$$

Next, mass conservation is expressed by considering a volume element

$$dx\,dy\,dz = h_\rho\,dx\,dy\,d\rho\ . \tag{3.43}$$

Constancy of the elemental volume along a flow line implies

$$(h_\rho)_t + (uh_\rho)_x + (vh_\rho)_y = 0\ . \tag{3.43$'$}$$

Transforming the remaining motion equations (3.36a,b) and collecting equations (3.40), (3.42), and (3.43′) gives the following set of hydrostatic Boussinesq equations in isopycnal coordinates.

$$\varepsilon u_t - (f + \varepsilon\zeta)v = -B_x \,, \tag{3.44a}$$

$$\varepsilon v_t + (f + \varepsilon\zeta)u = -B_y \,, \tag{3.44b}$$

$$\psi_\rho = h \,, \tag{3.44c}$$

$$h_{\rho t} + (u h_\rho)_x + (v h_\rho)_y = 0 \,, \tag{3.44d}$$

$$w = h_t + u h_x + v h_y \,. \tag{3.44e}$$

Here, the quantity $\zeta = v_x - u_y$ is the "isopycnal relative vorticity" (recall that horizontal gradients are being taken holding ρ fixed, thus ζ is not equivalent to the vertical component of relative vorticity), B is the Bernoulli function

$$B = \frac{\varepsilon}{2}(u^2 + v^2) + \psi \,, \tag{3.45}$$

h(x,y,ρ,t) is the isopycnal height, and $\psi = p + \rho h$ is the Montgomery stream function (Montgomery [1937]).

From Eqs. (3.44a,b,d) it is easy to demonstrate that the Ertel potential vorticity

$$Q = (f + \varepsilon\zeta)/h_\rho \tag{3.46}$$

is conserved along flow lines in isopycnal coordinates, i.e.

$$Q_t + uQ_x + vQ_y = 0 \,, \tag{3.47}$$

and that the energy density

$$E = \frac{1}{2}[\varepsilon h_\rho(u^2 + v^2) - h^2] \tag{3.48}$$

satisfies

$$E_t + (u h_\rho B)_x + (v h_\rho B)_y + (\psi h_t)_\rho = 0 \,. \tag{3.49}$$

Performing the invertible change of variables

$$m_1 = h_\rho(\varepsilon u + R(y)) \,,\; m_2 = h_\rho \varepsilon v \,,\; \sigma = h_\rho \,, \tag{3.50}$$

(where $-dR/dy = f(y)$, the Coriolis parameter) transforms Eqs. (3.44a,b,d) into

$$\partial_t m_i = -(\partial_j m_i + m_j \partial_i) v_j - \sigma \partial_i (\psi - uR - \frac{\varepsilon}{2}(u^2 + v^2)) \tag{3.51a}$$

$$\partial_t \sigma = -\partial_j (\sigma v_j)\ ,\ i,j = 1,2\ , \tag{3.51b}$$

where $\partial_j = \partial/\partial x^j$, j=1,2, and we sum over repeated indices. These equations are Hamiltonian, i.e., are expressible in the form

$$dF/dt = \{F,H\}\ , \tag{3.52}$$

using the Lie-Poisson bracket (cf. Eq. (3.4))

$$\{F,H\} = -\int dx dy d\rho \left[\frac{\delta F}{\delta m_i} (\partial_j m_i + m_j \partial_i) \frac{\delta H}{\delta m_j} + \frac{\delta F}{\delta m_i} \sigma \partial_i \frac{\delta H}{\delta \sigma} + \frac{\delta F}{\delta \sigma} \partial_j \sigma \frac{\delta H}{\delta m_j} \right] \tag{3.53}$$

and Hamiltonian

$$H = \frac{1}{2} \int dx\, dy\, d\rho \left\{ \frac{1}{\varepsilon\sigma} \left| (m_1 - \sigma R)^2 + m_2^2 \right| - h^2 \right\}\ , \tag{3.54}$$

whose variational derivatives are given by

$$\delta H = \int dx\, dy\, d\rho \left\{ [\psi - uR - \frac{\varepsilon}{2}(u^2 + v^2)] \delta\sigma + u \delta m_1 + v \delta m_2 \right\}\ . \tag{3.55}$$

The Lie-Poisson bracket (3.53) is equivalent to that in (3.4) for barotropic fluids in two dimensions up to integrations by parts. That is, the Hamiltonian framework for the ideal hydrostatic Boussinesq model in isopycnal coordinates is identical in form to that for two-dimensional compressible barotropic fluids, discussed earlier. Therefore, the stability criteria for the ideal hydrostatic Boussinesq model can be obtained by calculations that are analogous to those for the barotropic compressible case. The explicit computations that determine sufficient conditions for stability of equilibrium solutions of the ideal hydrostatic Boussinesq model are given in Holm and Long [1988]. Here we merely summarize the results as follows.

Ideal Hydrostatic Boussinesq Stability Theorem. A sufficient criterion for the equilibrium flow to be linearly stable is that the following three conditions all be met everywhere in the domain:

$$h_\rho < 0 \tag{3.56a}$$

$$\kappa^2 < -\frac{1}{\varepsilon h_\rho (u^2 + v^2)} \tag{3.56b}$$

$$\frac{1}{Q}\frac{\partial K}{\partial Q} > 0 \,. \tag{3.56c}$$

Condition (3.56a) demands that the density increase with depth; Condition (3.56b) restricts the vertical scale of variation of the disturbance (i.e., stability cannot be demonstrated with respect to disturbances that vary too rapidly with depth); Condition (3.56c) requires that the Bernoulli function increase with Q^2 along isopycnals.

The second condition (3.56b) can be rewritten in dimensional variables as

$$\kappa^2 < \frac{N^2}{(u^2 + v^2)} \tag{3.57}$$

where N is the buoyancy frequency. This can be compared to the dispersion relation for long (hydrostatic) internal gravity waves of local vertical wavenumber m and phase speed c

$$m^2 = N^2/c^2 \,. \tag{3.58}$$

If we equate κ with m, then (3.57) states that the flow is (conditionally) stable with respect to internal wave disturbances whose intrinsic phase speed is everywhere greater than the mean flow.

4. *Hamiltonian Structure for Two-Dimensional Hydrodynamics with Nonlinear Dispersiōn*

In this section, a Hamiltonian formulation using the barotropic Lie-Poisson bracket (3.4) or (3.64) is presented for a recently proposed model of two-dimensional shallow-water hydrodynamics with *nonlinear* dispersion. As in the barotropic case, nonlinear integral invariants for this model are found to be in the kernel of the Lie-Poisson bracket. A generalized Kelvin theorem is also given for the model.

Bazdenkov, Morozov, and Pogutse [1985] have recently presented a new mathematical model for large-scale flows in planetary atmospheres and oceans. The derivation of this model extends the classical shallow-water equations, by extending the number of terms retained in the expansion of the three-dimensional Euler equations in the small parameter $\varepsilon = h/a$ (where h is the depth of the fluid and a is a characteristic length of the flow). The resulting extended equations account not only for dispersive effects due to the higher order terms in the ratio h/a, but also incorporate topographical forcing due to the spatial variation of the bottom boundary. The extended shallow water equations are expressible in the form of a dynamical system

$$\partial_t \mathbf{u} = -(\mathbf{u}\cdot\nabla)\mathbf{u} - g\nabla(h+b) + h^{-1}\nabla A + h^{-1}B\nabla b\ , \tag{4.1a}$$

$$\partial_t h = -\nabla\cdot(h\mathbf{u})\ . \tag{4.1b}$$

Here, h(x,y,t) is the depth of the fluid above the time-independent bottom boundary at z=b(x,y), in x,y,z Cartesian coordinates; u(x,y,t)=(u,v) denotes the horizontal three-dimensional Euler velocity components v_x and v_y *averaged over the depth*, that is,

$$u = h^{-1}\int_b^{h+b} dz\, v_x \ \ and\ \ v = h^{-1}\int_b^{h+b} dz\, v_y\ . \tag{4.2}$$

The depth-averaged horizontal velocity **u** is taken to be tangential on the boundary of the domain of flow in $\mathbf{R}^2$. In Eq. (4.1a,b), $\nabla = (\partial_x, \partial_y)$ is the horizontal gradient and -g is the constant acceleration due to gravity.

The quantities A and B in (4.1a,b) represent dispersive terms, consisting of hydrostatic pressure and corrections to it arising from averaging over depth the contributions to the momentum flux of the vertical Euler velocity. Namely,

$$A = h^2 \frac{d}{dt}\left[\frac{1}{3}h(\nabla\cdot\mathbf{u}) - \frac{1}{2}(\mathbf{u}\cdot\nabla b)\right], \tag{4.3a}$$

$$B = h\frac{d}{dt}\left[\frac{1}{2}h(\nabla\cdot\mathbf{u}) - (\mathbf{u}\cdot\nabla b)\right], \tag{4.3b}$$

where $d/dt = \partial_t + \mathbf{u}\cdot\nabla$ is the material derivative in the two horizontal dimensions. (Recall that the quantity in the shallow water approximation $\nabla\cdot\mathbf{u}$ characterizes the magnitude of the vertical velocity of the Euler fluid.) When A and B are both absent, Eqs. (4.1a,b) return to the classical shallow-water equations.

The dispersive terms in (4.1a) involving A and B are *nonlinear* in the depth h and the depth-averaged velocity **u** of the fluid.* Hence, dispersion effects in this model can be expected to differ considerably from those in weakly dispersive models such as the Boussinesq and Korteweg-de Vries equations, where the dispersive terms are linear.

Basdenkov, Morozov, and Pogutse [1985] note that the extended shallow-water equations (4.1a,b) conserve an energy functional, H, and advect a generalized scalar vorticity, ξ. That is, the energy functional,

* No notational confusion should arise between the use of h and *u* in this section and in the isopycnal representation of the previous section.

$$H = \int dxdy \left[\frac{1}{2} hu^2 + gh(h+b) + \frac{1}{6} h^3 (\nabla\cdot\mathbf{u})^2 - \frac{1}{2} h^2 (\nabla\cdot\mathbf{u})(\mathbf{u}\cdot\nabla b) + \frac{1}{2} h(\mathbf{u}\cdot\nabla b)^2 \right] , \tag{4.4}$$

is globally conserved and the generalized vorticity ξ is advected by the flow,

$$\frac{d\xi}{dt} = 0 , \tag{4.5}$$

where ξ is defined as follows (with $\hat{\mathbf{z}}$ the unit vector along the z-axis),

$$\xi = h^{-1}\hat{\mathbf{z}}\cdot \Big[curl\,\mathbf{u} + \frac{1}{3} h\nabla h \times \nabla(\nabla\cdot\mathbf{u}) + \frac{1}{2}\nabla b \times \nabla(h\nabla\cdot\mathbf{u}) + \frac{1}{2}\nabla(\mathbf{u}\cdot\nabla b) \times \nabla(h+2b) \Big] . \tag{4.6}$$

The details of the approximations that lead to Eqs. (4.1)-(4.6) and the domain of validity of these equations are discussed further in Basdenko, Morosov, and Pogutse [1985].

In this section, following Holm [1988] we shall cast the dynamical system (4.1a,b) into Hamiltonian form, by defining a Poisson bracket operation $\{ , \}$ and using the Hamiltonian functional H given in (4.4) to express the dynamical system (4.1a,b) in the form

$$dF/dt = \{F,H\} , \tag{4.7}$$

for all functionals F depending on the depth, h, and on the total momentum density m, defined below in Eq. (4.9). Recall that a Poisson bracket operation should be bilinear, skew-symmetric, and satisfy the Jacobi identity

$$\{F , \{G,H\}\} + \{G , \{H,F\}\} + \{H , \{F,G\}\} = 0 . \tag{4.8}$$

Because of the Jacobi identity, whenever two functionals F and G are both constants of motion (i.e., F and G both Poisson-commute with the Hamiltonian, H), the functional given by their Poisson bracket $\{F,G\}$ is also a constant of motion.

The Hamiltonian formalism we present casts the system (4.1a,b) into the framework of a noncanonical (Lie-Poisson) Hamiltonian structure and identifies the integral invariants of this system as Casimirs, i.e., functionals that Poisson-commute with every Hamiltonian expressible in these variables. It also provides a simple geometric derivation of Kelvin's theorem for the extended shallow-water equations, and explains the geometrical meaning of the expression for the generalized vorticity, ξ in Eq. (4.6).

In terms of the following total momentum variable,

$$\mathbf{m} = h\mathbf{u} - \nabla\left(\frac{1}{3}h^3\nabla\cdot\mathbf{u} - \frac{1}{2}h^2\mathbf{u}\cdot\nabla b\right) - \left(\frac{1}{2}h^2\nabla\cdot\mathbf{u} - h\mathbf{u}\cdot\nabla b\right)\nabla b\,, \tag{4.9}$$

the energy in (4.4) becomes

$$\begin{aligned} H = \int dxdy \Big[\, & \mathbf{m}\cdot\mathbf{u} - \frac{1}{2}hu^2 - \frac{1}{6}h^3(\nabla\cdot\mathbf{u})^2 + \frac{1}{2}h^2(\mathbf{u}\cdot\nabla b)(\nabla\cdot\mathbf{u}) \\ & - \frac{1}{2}h(\mathbf{u}\cdot\nabla b)^2 + gh\left(\frac{1}{2}h + b\right)\Big]\,, \end{aligned} \tag{4.10}$$

after integrating by parts and using the boundary condition that u is tangential. The variational derivatives of H in (4.10) are found from

$$\begin{aligned} \delta H = \int dxdy \Big\{ & \mathbf{u}\cdot\delta\mathbf{m} + \Big[\mathbf{m} - h\mathbf{u} - \nabla\left(\frac{1}{3}h^3\nabla\cdot\mathbf{u} - \frac{1}{2}h^2\mathbf{u}\cdot\nabla b\right) \\ & - \left(\frac{1}{2}h^2\nabla\cdot\mathbf{u} - h\mathbf{u}\cdot\nabla b\right)\nabla b\Big]\cdot\delta\mathbf{u} + \Big[\frac{-1}{2}u^2 - \frac{1}{2}h^2(\nabla\cdot\mathbf{u})^2 \\ & + h(\mathbf{u}\cdot\nabla b)(\nabla\cdot\mathbf{u}) - \frac{1}{2}(\mathbf{u}\cdot\nabla b)^2 + g(h+b)\Big]\delta h\Big\}\,. \end{aligned} \tag{4.11}$$

Equations (4.1a,b) are expressible in Hamiltonian form (4.7) by using the Lie-Poisson bracket (cf. Eqs. (3.4) and (3.53)),

$$\{F,H\} = \int dxdy \Big[\frac{\delta F}{\delta m_i} \Big[\Big(m_k \partial_i + \partial_k m_i \Big) \frac{\delta H}{\delta m_k} + h\partial_i \frac{\delta H}{\delta h} \Big] + \frac{\delta F}{\delta h} \partial_k h \frac{\delta H}{\delta m_k} \Big\} , \tag{4.12}$$

(summing over repeated indices) where the derivative operator $\partial_k = \partial/\partial x^k$, $k=1,2$, operates on all terms it multiplies to the right. Equations (4.1a,b) for extended shallow-water theory now result in Hamiltonian form (4.7) by substituting the variational derivatives from (4.11) into the Lie-Poisson bracket (4.12). Namely, with E the integrand in Eq. (4.10),

$$\begin{aligned} \partial_t m_i &= \{m_i, H\} \\ &= -\Big(m_k \partial_i + \partial_k m_i \Big) \frac{\delta H}{\partial m_k} - h\partial_i \frac{\delta H}{\delta h} \\ &= -\partial_k \Big[m_i \frac{\delta H}{\delta m_k} + \delta_i^k \Big(m_j \frac{\delta H}{\delta m_j} + h \frac{\delta H}{\delta h} - E \Big) \Big] \\ &= -\partial_k \Big[m_i u^k + \delta_i^k \Big(-\frac{1}{3} h^3 (\nabla\cdot u)^2 + \frac{1}{2} h^2 (u\cdot\nabla b)(\nabla\cdot u) + \frac{1}{2} g h^2 \Big) \Big] , \end{aligned} \tag{4.13}$$

$$\partial_t h = \{h,H\} = -\partial_k (h u^k) . \tag{4.14}$$

As a bonus, the equations appear in conservative form for both momentum and mass. Rearrangement of the momentum conservation Eq. (4.13) eventually gives (4.1a), while (4.14) reproduces (4.1b). When the terms in (4.9) and (4.13) involving $\nabla\cdot\mathbf{u}$ and $\mathbf{u}\cdot\nabla b$ are absent, the classical shallow-water equations re-emerge, in Hamiltonian form.

The Lie-Poisson bracket (4.12) is the natural Poisson bracket on the dual space associated with the Lie algebra

$$= V \, s \, \Lambda^\circ . \tag{4.15}$$

The symbol s denotes the semidirect product with respect to the natural action of vector fields V on functions Λ°. the commutator for this Lie algebra is given by

$$[(X;f)\,,\,(\overline{X};\overline{f})] = ([X,\overline{X}]\,;\,X(\overline{f}) - \overline{X}(f))\,. \tag{4.16}$$

Dual coordinates on the Lie algebra are: **m** dual to $X \in V$, and h dual to $f \in \Lambda^\circ$. The Lie-Poisson bracket (4.12) can be expressed as

$$\{H,F\} = <\mu, \left[\frac{\delta H}{\delta\mu}, \frac{\delta F}{\delta\mu}\right]> , \tag{4.17}$$

where $< , >$ is the (L^2) pairing between elements of the Lie algebra and its dual , $\mu = (\mathbf{m}, h) \in$ *, and $\delta H/\delta\mu \in$ is determined from

$$DH(\mu)\cdot\delta\mu = <\delta\mu\,, \frac{\delta H}{\delta\mu}> , \tag{4.18}$$

for any $\delta\mu \in$ * (assuming such an element $\delta H/\delta\mu \in$ always exists). See, e.g., Holm and Kupershmidt [1982, 1983], Holm, Kupershmidt, and Levermore [1983], Marsden, Ratiu, and Weinstein [1984], and various articles in Marsden [1984] for further examples, references, and discussions of such Lie-Poisson brackets (i.e., Poisson brackets defined on the dual space of a Lie algebra) as they apply in ideal fluid dynamics.

Using the expression (4.17) for the Lie-Poisson bracket (4.12) enables us to rearrange Eqs. (4.13)-(4.14) into a generalized Kelvin theorem, namely,

$$(\partial_t + L_{\frac{\delta H}{\delta \mathbf{m}}})(h^{-1}\mathbf{m}\cdot d\mathbf{x}) = -d(\frac{\delta H}{\delta h})\,, \tag{4.19}$$

where, $L_{\delta H/\delta \mathbf{m}}$ denotes the Lie derivative with respect to $\delta H/\delta\mathbf{m} = \mathbf{u}$ and, by (4.11),

$$\frac{\delta H}{\delta h} = -\frac{1}{2}u^2 - \frac{1}{2}h^2(\nabla\cdot\mathbf{u})^2 + h(\mathbf{u}\cdot\nabla b)(\nabla\cdot\mathbf{u}) - \frac{1}{2}(\mathbf{u}\cdot\nabla b)^2 + g(h+b)\,. \tag{4.20}$$

Because of (4.19) the circulation loop integral

$$\oint_{\gamma(t)} h^{-1}\mathbf{m}\cdot d\mathbf{x} \tag{4.21}$$

is conserved for every closed curve $\gamma(t)$ moving with the fluid. Taking the exterior derivative of (4.19) and invoking the identities $d^2=0$ and $[d,L_u]=0$ now give

$$\frac{d}{dt}(h^{-1}\mathbf{z}\cdot curl(h^{-1}\mathbf{m})) = 0. \tag{4.22}$$

(after also using $(\partial_t + L_u)(hdxdy)=0$). Equation (4.22) simply restates the advection of the generalized vorticity ξ in (4.6). Because of this advection and the continuity Eq. (4.1b), the quantities

$$C = \int dxdy\, h\, \Phi(\xi) \tag{4.23}$$

are conserved for any function Φ. In fact, these quantities in (4.23) are *Casimirs*, or distinguished functionals for the Lie-Poisson bracket (4.12), i.e.,

$$\{C,H\} = 0 \ \forall H\,, \tag{4.24}$$

as may readily be shown. In other words, the functionals C in (4.23) lie in the kernel of the Lie-Poisson bracket (4.12); therefore, each of them is conserved for any Hamiltonian H, not just for the energy in (4.10). We have seen that the Casimirs play an important role in classifying equilibrium states and in the study of Lyapunov stability conditions for such equilibria. A complete discussion of stability in the present case for shallow water equilibria with nonlinear dispersion has not yet been performed, although Basdenkov, Morosov, and Pogutse [1985] indicate that nonlinear dispersion may, in fact, cause the loss of formal stability.

Acknowledgement

The material in Sections 2 and 3 is excerpted from Holm et al. [1985] with the countenance of my friends and collaborators Jerry Marsden, Tudor Ratiu, and Alan Weinstein to whom I am grateful.

References

H. D. I. Abarbanel, D. D. Holm, J. E. Marsden, and T. Ratiu (1984) Richardson number criterion for the nonlinear stability of three dimensional stratified flow, Phys. Rev. Lett. 52, 2352-2355.

H. D. I. Abarbanel, D. D. Holm, J. E. Marsden, and T. Ratiu (1986) Nonlinear stability analysis of stratified ideal fluid equilibria, Phil. Trans. Roy. Soc. (London) A 318, 349-409.

R. Abraham and J. Marsden (1978) *Foundations of Mechanics* (Addison-Wesley, New York).

V. I. Arnold (1965a) Conditions for nonlinear stability of the stationary plane curvilinear flows of an ideal fluid, Doklady Mat. Nauk, 162 (5) 773-777.

V. I. Arnold (1965b) Variational principle for three dimensional steady-state flows of an ideal fluid, J. Appl. Math. Mech. 29, 1002-1008.

V. I. Arnold (1966a) Sur la géometrie differentielle des groupes de Lie de dimension infinie et ses applications a l'hydrodynamique des fluides parfaits, Ann. Inst. Fourier, Grenoble, 16, 319-361.

V. I. Arnold (1966b) Sur un principe variationnel pour des écoulements stationaires des liquides parfaits et ses applications aux problémes de stabilité non linéaires, J. Mécanique, 5, 29-43.

V. I. Arnold (1969a) On an a priori estimate in the theory of hydrodynamic stability, English Transl: Am. Math. Soc. Transl. 19, 267-269.

V. I. Arnold (1969b) The Hamiltonian nature of the Euler equations in the dynamics of a rigid body and of an ideal fluid, Usp. Mat. Nauk. 24, 225-226.

V. I. Arnold (1978) *Mathematical Methods of Classical Mechanics*, Graduate Texts in Math. No. 60 (Springer, Berlin).

J. M. Ball and J. E. Marsden (1984) Quasiconvexity, second variations and nonlinear stability in elasticity, Arch. Rat. Mech. An. 86, 251-277.

S. V. Bazdenkov, N. N. Morosov and O. P. Pogutse (1985) Dokl. Akad. Nauk SSSR 293, 818 [English transl. Sov. Phys. Dokl. 32 (4), 262 (1987)].

T. B. Benjamin (1972) The stability of solitary waves, Proc. Roy. Soc. London 328A, 153-183.

D. P. Bennett, R. W. Brown, S. E. Stansfield, J. D. Stroughair and J. L. Bona (1983) The stability of internal solitary waves, Math. Proc. Camb. Phil. Soc. 94, 351-379.

R. Benzi, S. Pierini, A. Vulpiani and E. Salusti (1982) On nonlinear hydrodynamic stability of planetary vortices, Geophys. Astrophys. Fluid Dynamics 20, 293-306.

I. B. Bernstein, E. A. Frieman, M. D. Kruskal and R. M. Kulsrud (1958) An energy principle for hydromagnetic stability problems, Proc. Roy. Soc. London, 244A, 17-40.

I. Bialynicki-Birula and Z. Iwinsky (1973) Canonical formulation of relativistic hydrodynamics, Rep. Math. Phys. 4, 139-151.

W. Blumen (1968) On the stability of quasi-geostrophic flow. J. Atmos. Sci. 25, 929-931.

W. Blumen (1971) On the stability of plane flow with horizontal shear to three-dimensional nondivergent disturbances, Geophysical Fluid Dynamics 2, 189-200.

J. Bona (1975) On the stability theory of solitary waves, Proc. Roy. Soc. London, 344A, 363-374.

J. L. Bona, D. K. Bose and R. E. L. Turner (1983) Finite-amplitude steady waves in stratified fluids, J. Math. Pures et Appl. 62, 389-439.

F. P. Bretherton and D. B. Haidvogel (1976) Two-dimensional turbulence above topography, J. Fluid Mech. 78, 129-154.

S. Chandrasekhar (1961) *Hydrodynamic and Hydromagnetic Instabilities* (Oxford Univ Press, London, New York).

T. M. Cherry (1925) Trans. Cambridge Philos. Soc. **23**, 199. (Referred to in E. J. Whittaker, *Analytical Dynamics* (Cambridge, London, 1937), Sec. 182, p. 412 and A. Wintner, *The Analytical Foundations on Celestial Mechanics* (Princeton Univ., 1947), Sec. 136, p. 101.)

B. V. Chirikov (1979) A universal instability of many dimensional oscillator systems, Phys. Rep. 52, 263-379.

J. D. Crawford (1983) The Hopf bifurcation and plasma instabilities, Ph.D. Thesis, Univ. California, Berkeley; see also, Cont. Math. AMS 28, 377-392.

R. Dashen and D. Sharp (1968), Currents as coordinates for hadrons, Phys. Rev. 165, 1857-1878.

R. C. Davidson and S. T. Tsai (1973) Thermodynamic bounds on the magnetic fluctuation energy in unstable anisotropic plasmas. J. Plasma Phys. 9, 101-116.

L. A. Dikii (1965a) On the nonlinear theory of the stability of zonal flows, Izv. Atm. and Oceanic Phys. 1 (11) 1117-1222.

L. A. Dikii (1965b) On the nonlinear theory of hydrodynamic stability, Prikl. Math. Mech. 29, 852-855.

P. G. Drazin and W. H. Reid (1981) *Hydrodynamic Stability* (Cambridge Univ. Press, London).

D. G. Dritschel (1986) The nonlinear evolution of rotating configurations of uniform vorticity, J. Fluid Mech. 172, 157-182.

I. E. Dzyaloshinskii and G. E. Volovick (1980) Poisson brackets in condensed matter physics, Ann. Phys. 125, 67-97.

A. Eliassen and E. Kleinschmidt (1957) *Dynamic Meteorology, Handbuch der Physik, 48, Geophysik II*, J. Bartels, ed. (Springer-Verlag, Berlin) 1-154.

C. P. Enz and L. A. Turski (1979) On the Fokker-Planck description of compressible fluids, Physica 96A, 369-378.

R. Fjortoft (1946) On the frontogenesis and cyclogenesis in the atmosphere, Geofys. Pub. 16, 1-28.

R. Fjortoft (1950) Application of integral theorems in deriving criteria of stability for laminar flows and for the baroclinic circular vortex, Geophys. Publ. 17, 1-52.

T. K. Fowler (1963) Liapunov's stability criteria for plasmas, J. Math. Phys. 4, 559-569.

J. P. Freidberg (1982) Ideal magnetohydrodynamic theory of magnetic fusion systems, Rev. Mod. Phys. 54, 801-902.

C. S. Gardner (1963) Bound on the energy available from a plasma, Phys. Fluids 6, 839-840.

J. M. Greene, D. D. Holm and P. J. Morrison (1980) Canonical and noncanonical Hamiltonian formulations of fluid dynamics and magnetohydrodynamics. Proc. Abstracts of Workshops in Nonlinear Waves and Dynamical Systems, Khania, Crete, Greece (July 1980). See Physica D 2 (1981) 545-548.

M. A. Grinfeld (1984) Variational principles and stability of stationary flows of barotropic ideal fluid, Geophys. Astrophys. Fluid Dynamics 28, 31-54.

V. Guillemin and S. Sternberg (1980) The moment map and collective motion, Ann. Phys. 127, 220-253.

R. D. Hazeltine, D. D. Holm, J. E. Marsden and P. J. Morrison (1984) Generalized Poisson brackets and nonlinear Liapunov stability — Applications to reduced MHD, ICPP Proc. (Lausanne) 2, 204-206.

S. I. Herlitz (1967) Stability of plane flow, Arkiv för Fysik 34, 39-48.

D. D. Holm (1984) Stability of planar multifluid plasma equilibria by Arnold's method, Cont. Math. AMS 28, 25-50.

D. D. Holm (1988) Hamiltonian structure for two-dimensional hydrodynamics with nonlinear dispersion, submitted to Phys. Fluids.

D. D. Holm and B. A. Kupershmidt (1983) Poisson brackets and Clebsch representations for magnetohydrodynamics, multifluid plasmas, and elasticity, Physica D 6, 347-363.

D. D. Holm, B. A. Kupershmidt and C. D. Levermore (1983) Canonical maps between Poisson brackets in Eulerian and Lagrangian descriptions of continuum mechanics, Phys. Lett. 98A, 389-395 (and Hamiltonian differencing for ideal fluids, Adv. in Appl. Math. 6 (1985) 52-84).

D. D. Holm and B. Long (1988), Lyapunov stability of ideal stratified fluid equilibria in hydrostatic balance, submitted to Nonlinearity.

D. D. Holm, J. Marsden and T. Ratiu (1984) Nonlinear stability of the Kelvin-Stuart cat's eyes, Proc. AMS-SIAM Summer Conference, Santa Fe (July 1984); AMS Lecture Series in Applied Mathematics, Vol. 23.

D. D. Holm, J. E. Marsden, T. Ratiu and A. Weinstein (1983) Nonlinear stability conditions and a priori estimates for barotropic hydrodynamics, Phys. Lett. 98A, 15-21.

D. D. Holm, J. E. Marsden, T. Ratiu and A. Weinstein (1984) Stability of rigid body motion using the energy-Casimir method, Cont. Math. AMS, 28, 15-23.

D. D. Holm, J. E. Marsden, T. Ratiu, A. Weinstein (1985) Nonlinear stability of fluid and plasma equilibria, Phys. Rep. 123, 1-116.

Z. R. Iwinski and L. A. Turski [1976] Canonical theories of systems interacting electromagnetically. Lett. Appl. Eng. Sci. 4, 179-191.

J. D. Jackson (1960) Longitudinal plasma oscillations, J. Nucl. Energy, Part C: Plasma Physics 1, 171-189.

M. G. Krein (1950) A generalization of several investigations of A. M. Liapunov on linear differential equations with periodic coefficients. Dokl. Akad. Nauk. SSSR 73, 445-448.

P. H. LeBlond and L. A. Mysak (1978) *Waves in the Ocean* (Elsevier, New York).

E. W. Laedke and K. H. Spatschek (1980) Liapunov stability of generalized Langmuir solutions, Phys. Fluids 23, 44-51.

M. Levi (1977) Stability of linear Hamiltonian systems with periodic coefficients, IBM Research Report RC 6610 (No. 28482).

D. Lewis, J. Marsden, R. Montgomery and T. Ratiu (1987) The Hamiltonian structure of dynamic free boundary problems, Physica D 18, 391-404.

A. J. Lichtenberg and M. A. Lieberman (1982) *Regular and Stochastic Motion*, (Springer-Verlag, New York, Heidelberg, Berlin).

J. E. Marsden (1982) A group theoretic approach to the equations of plasma physics, Can. Math. Bull. 25, 129-142.

J. E. Marsden and M. McCracken (1976) *The Hopf Bifurcation and its Applications*, Applied Mathematical Sciences, Vol. 19 (Springer, Berlin).

J. E. Marsden, A. Weinstein, T. Ratiu, R. Schmid and R. G. Spencer (1983) Hamiltonian systems with symmetry, coadjoint orbits and plasma physics, Proc. IUTAM-ISIMM Symposium on 'Modern Developments' in Analytical Mechanics", Torino, June 7-11, 1982, Atti della Academia della Scienze di Torino, 117, 289-340.

J. E. Marsden, T. Ratiu and A. Weinstein (1984a) Semi-direct products and reduction in mechanics, Trans. Am. Math. Soc. 281, 147-177.

J. E. Marsden, T. Ratiu and A. Weinstein (1984b) Reduction and Hamiltonian structures on duals of semidirect product Lie algebras, Cont. Math. AMS, 28, 55-100.

H. McKean (1977) Stability for the Korteweg-de Vries equation, Comm. Pure Appl. Math. 30, 347-353.

R. B. Montgomery (1937) A suggested method for representing gradient flow in isentropic surfaces, Bull. Am. Meteor. Soc. 18, 21-212.

P. J. Morrison (1982) Poisson brackets for fluids and plasmas in: *Mathematical Methods in Hydrodynamics and Integrability in Related Dynamical Systems*, AIP Conf. Proc. #88 La Jolla, M. Tabor and Y. M. Treve, eds.

P. J. Morrison and J. M. Greene (1980) Noncanonical hamiltonian density formulation of hydrodynamics and ideal magnetohydrodynamics, Phys. Rev. Lett. 45, 790-794, errata, ibid. 48 (1982) 569.

P. J. Olver (1986) *Applications of Lie Groups to Differential Equations*, Graduate Texts in Mathematics, Vol. 107, (Springer, Berlin).

O. Penrose (1960) Electrostatic instabilities of a uniform nonmaxwellian plasma, Phys. Fluids 3, 258-265.

H. Pollard (1966) *Mathematical Introduction to Celestial Mechanics* (Prentice-Hall, Englewood Cliffs, N.J.).

T. Ratiu (1980) Euler-Poisson Equations on Lie algebras, Thesis, Univ. California, Berkeley.

T. Ratiu (1982) Euler-Poisson equations on Lie algebras and the N-dimensional heavy rigid body, Am. J. Math. 104, 409-448, 1337.

Lord Rayleigh (1880) On the stability, or instability, of certain fluid motions, Proc. Lon. Math. Soc. 11, 57-70.

P. Ripa (1987) On the stability of elliptical vortex solution of the shallow-water equations, J. Fluid Mech. **183**, 343-363.

M. N. Rosenbluth (1964) Topics in microinstabilities, in: *Advanced Plasma Physics*, M. Rosenbluth, ed (academic Press, New York) p. 137.

V. I. Sedenko and V. I. Iudovitch (1978) Stability of steady flows of ideal incompresible fluid with free boundary, Prikl. Mat. & Mekh. 42, 1049 (Applied Math. & Mechanics, 42, 1148-1155).

C. L. Siegel and J. K. Moser (1971) *Lectures on Celestial Mechanics* (Springer Verlag, New York).

E. C. G. Sudarshan and N. Mukunda (1974) *Classical Mechanics A Modern Perspective* (Wiley, New York, 1974; 2nd ed., Krieger, Melbourne-Florida, 1983).

Y. Tang (1987) Nonlinear stability of vortex patches, Trans. Am. Math. Soc. 304, 617-637.

Y. H. Wan (1984) On the nonlinear stability of circular vortex patches, Cont. Math. AMS 28, 215-220.

Y. H. Wan and F. Pulvirente (1985) Nonlinear stability of circular vortex patches, Comm. Math. Phys. 99, 435-450.

A. Weinstein (1983a) Sophus Lie and symplectic geometry, Expo. Math. 1, 95-96.

V. E. Zakharov and E. A. Kuznetsov (1974) Three-dimensional solitons, Sov. Phys. JETP 39, 285-286.

An Introduction to the Equivalence Problem of Elie Cartan illustrated by Examples

NIKY KAMRAN*

SCHOOL OF MATHEMATICS
THE INSTITUTE FOR ADVANCED STUDY
PRINCETON, NJ 08540 U.S.A.

1. Introduction.

In a famous paper [1] which appeared eighty years ago in the Annales de l'École Normale, Elie Cartan presented a general construction for determining when two exterior differential systems generated by 1-forms are equivalent under a change of variables belonging to a prescribed group of transformations. This construction is known as Elie Cartan's *Method of Equivalence* and the problem it solves is known as the *Cartan Equivalence Problem.*

There are numerous examples in Cartan's works of equivalence problems for differential equations which were solved using the method of equivalence. Some of these problems were notoriously difficult, like the equivalence problem for an over-determined system of two second-order partial differential equations in two independent and one dependent variables which is in involution under the group of contact transformations, which leads to a non-linear representation of a real form of the exceptional Lie group G_2 as the automorphism group [2].

Until about 1945, several mathematicians inspired by Elie Cartan used the Method of Equivalence with great success in their own research. Prominent amongst these scientists stand Chern [3][4], Debever [5][6] and Vranceanu [7][8]. A long hiatus during which the Method of Equivalence lay almost in oblivion followed. This hiatus lasted until some

*On leave from the Department of Applied Mathematics, University of Waterloo, Waterloo, Ontario N2L3G1, Canada

progress was made towards a better understanding of the Cartan Equivalence Method in modern differential geometrical terms, thanks to the works of Chern [**9**], Ehresmann [**10**] and Singer and Sternberg [**11**].

The procedure to be followed to apply the Method of Equivalence towards the solution of non-trivial problems such as the ones arising in connection with Differential Equations became clear after the publication of a fundamental paper by R.B. Gardner [**12**] which contains an *algorithm* for the practical implementation of the Method of Equivalence. The publication of Gardner's paper led to a lot of new research activity in the subject. Some notable applications to the areas of Ordinary Differential Equations, Partial Differential Equations, the Calculus of Variations, Control Theory and Classical Invariant Theory can be found in References 13 to 26 and References 32 and 33.

Our goal in this paper is to give an elementary account of the solution of the Cartan Equivalence Problem, illustrated by examples that will purposely be chosen so as to enable the reader to reproduce the calculations with no difficulty at all. All the proofs of the theorems quoted are omitted. We therefore strongly encourage the interested reader to consult the basic references on the Method of Equivalence, which are Cartan's papers [**1**][**27**], Sternberg's book [**28**] (in particular Chapter 7) and Gardner's paper [**12**], as well as Cartan's book [**29**] or the review article by Bryant, Chern and Griffiths [**30**] for the fundamental results from the theory of exterior differential systems in involution used in the text. Our paper is thus meant to serve merely as a pedagogical introduction to this beautiful subject, with no claims of originality or exhaustiveness.

2. The Cartan Equivalence Problem.

Let U and V be open subsets of $\mathbf{R}^n$ and $(\omega_U^i)_{1\leq i\leq n}$ and $(\bar{\omega}_V^j)_{1\leq j\leq n}$ be coframes defined on U and V respectively. Let G be a Lie subgroup of $GL(n,\mathbf{R})$. The Cartan Equivalence Problem is to determine all diffeomorphisms $f: U \to V$ such that

$$f^*\bar{\omega}_V^j = \gamma_j^i \omega_U^j \,, \tag{2.1}$$

where (γ_j^i) is a G-valued function on U.

The local equivalence problem for a wide range of differential geometrical structures can be cast in the form of a Cartan equivalence problem. Let us mention a few simple examples, some of which should be familiar from elementary Differential Geometry textbooks.

(a) Equivalence of Riemannian metrics in the plane [27]:

Two Riemannian metrics g_U and $\bar{g}_V$ defined on open subsets U and V of $\mathbf{R}^2$ are equivalent if there exists a diffeomorphism $f: U \to V$ such that $f^*\bar{g}_V = g_U$. Choosing coframes $(\omega_U^i)_{i=1,2}$ and $(\bar{\omega}_V^j)_{j=1,2}$ such that

$$(g_U)_{ij} dx^i \otimes dx^j = \omega_U^1 \otimes \omega_U^1 + \omega_U^2 \otimes \omega_U^2 \,, \tag{2.2a}$$

$$(\bar{g}_V)_{ij} d\bar{x}^i \otimes d\bar{x}^j = \bar{\omega}_V^1 \otimes \bar{\omega}_V^1 + \bar{\omega}_V^2 \otimes \bar{\omega}_V^2 \,, \tag{2.2b}$$

the equivalence condition $f^*\bar{g}_V = g_U$ takes the form of a Cartan equivalence problem

$$f^*\bar{\omega}_V^i = \gamma_j^i \omega_U^j \,, \tag{2.2c}$$

where (γ_j^i) takes values in $SO(2,\mathbf{R})$.

(b) Conformal equivalence of Riemannian metrics in the plane [17]:

The metrics g_U and $\bar{g}_V$ are conformally equivalent if there exists a diffeomorphism $f: U \to V$ and real-valued function λ such that $f^*\bar{g}_V = e^\lambda g_U$. Introducing coframes $(\omega_U^i)_{i=1,2}$ and $(\bar{\omega}_V^j)_{j=1,2}$ as in Eqs. (2.2) and (2.3), the conformal equivalence condition is equivalent to

$$f^*\bar{\omega}_V^i = \gamma_j^i \omega_U^j \,, \tag{2.3}$$

(c) Conformal equivalence of one-parameter families of curves in the plane [27]:

Let $\mathcal{F}$ and $\bar{\mathcal{F}}$ be two one-parameter families of curves in the Euclidean plane. Let $\theta(x, y)$ denote the value of the angle between the X-axis and the tangent to the curve in $\mathcal{F}$ through (x, y). The curves in $\mathcal{F}$ are the integral curves of

$$\omega_U^1 := -\sin\theta\, dx + \cos\theta\, dy \ , \tag{2.4a}$$

while the one-parameter family of orthogonal trajectories given by the integral curves of

$$\omega_U^2 := \cos\theta\, dx + \sin\theta\, dy \ . \tag{2.4b}$$

A diffeomorphism $f : \mathbf{R}^2 \to \mathbf{R}^2$ will map $\mathcal{F}$ to $\bar{\mathcal{F}}$ conformally if and only if

$$f^* \begin{pmatrix} \bar{\omega}_V^1 \\ \bar{\omega}_V^2 \end{pmatrix} = \begin{pmatrix} u & 0 \\ 0 & u \end{pmatrix} \begin{pmatrix} \omega_U^1 \\ \omega_U^1 \end{pmatrix} , \tag{2.4c}$$

where

$$\bar{\omega}_V^1 := -\sin\bar{\theta}\, d\bar{x} + \cos\bar{\theta}\, d\bar{y} \quad , \quad \bar{\omega}_V^2 := \cos\bar{\theta}\, d\bar{x} + \sin\bar{\theta}\, d\bar{y} \ . \tag{2.4d}$$

We see that Eq. (2.4c) has the form of a Cartan equivalence problem (2.1).

(d) Equivalence of first-order particle Lagrangians on the line [27][23][24]:

Consider two variational problems given by

$$\mathcal{L}(u) = \int (j^1 u)^* L\, dx \quad , \quad \bar{\mathcal{L}}(\bar{u}) = \int (j^1 \bar{u})^* \bar{L}\, d\bar{x} \ , \tag{2.5a}$$

where L and $\bar{L}$ are real-valued functions on $J^1(\mathbf{R}, \mathbf{R})$.

The two functionals $\mathcal{L}$ and $\bar{\mathcal{L}}$ will agree on all possible functions u if and only if there exists a contact transformation $\Phi : J^1(\mathbf{R}, \mathbf{R}) \to J^1(\mathbf{R}, \mathbf{R})$ such that

$$\Phi^* \bar{L}\, d\bar{x} = L\, dx + \lambda(du - p\, dx) \ . \tag{2.5a}$$

We thus have a Cartan equivalence problem given by

$$\Phi^* \begin{pmatrix} \bar{\omega}_V^1 := d\bar{u} - \bar{p}\, d\bar{x} \\ \bar{\omega}_V^2 := \bar{L}\, d\bar{x} \\ \bar{\omega}_V^3 := d\bar{p} \end{pmatrix} = \begin{pmatrix} A & 0 & 0 \\ B & 1 & 0 \\ C & D & E \end{pmatrix} \begin{pmatrix} \omega_U^1 := du - p\, dx \\ \omega_U^2 := L\, dx \\ \omega_U^3 := dp \end{pmatrix} . \tag{2.5b}$$

We may also relax the equivalence condition (2.5a) by requiring that the functionals agree only on extremals. By a standard result in the Calculus of Variations, we know that two Lagrangians have the same Euler-Lagrange equations if and only if they differ by a divergence. We thus require that

$$\Phi^* \bar{L} d\bar{x} \equiv (L + D_x f) dx \mod du - p\, dx \ , \tag{2.5c}$$

for some real-valued function f on $J^0(\mathbf{R}, \mathbf{R})$.

The equivalence condition (2.5c) can also be cast in the form of a Cartan equivalence problem by introducing an auxiliary variable w whose transformation rule will incorporate f. More precisely, the existence of a contact transformation $\Phi : J^1(\mathbf{R}, \mathbf{R}) \to J^1(\mathbf{R}, \mathbf{R})$ satisfying Eq. (2.5c) is equivalent to the existence of a diffeomorphism $\tilde{\Phi} : J^1(\mathbf{R}, \mathbf{R}) \times \mathbf{R} \to J^1(\mathbf{R}, \mathbf{R}) \times \mathbf{R}$ such that

$$\tilde{\Phi}^* = \begin{pmatrix} \bar{\omega}^1_V \\ \bar{\omega}^2_V \\ \bar{\omega}^3_V \\ \bar{\omega}^4_V := d\bar{w} \end{pmatrix} = \begin{pmatrix} A & 0 & 0 & 0 \\ B & C & 0 & 0 \\ D & E & F & 0 \\ G & C-1 & 0 & 1 \end{pmatrix} \begin{pmatrix} \omega^1_U \\ \omega^2_U \\ \omega^3_U \\ \omega^4_U := dw \end{pmatrix} , \tag{2.5d}$$

which has the form of a Cartan equivalence problem.

(e) Equivalence of systems of second-order ordinary differential equations [31]:

Consider two systems of second-order ordinary differential equations

$$\frac{d^2 x^i}{dt^2} = F^i \left(\frac{dx^j}{dt}, x^k, t \right) \quad , \quad \frac{d^2 \bar{x}^i}{d\bar{t}^2} = \bar{F} \left(\frac{d\bar{x}^j}{d\bar{t}}, \bar{x}^k, \bar{t} \right) , \tag{2.6a}$$

where $1 \le i, j, k \le n$.

The above systems will be equivalent under a point transformation

$$\bar{x}^i \circ \Phi = f^i(x^j, t) \quad , \quad \bar{t} \circ \Phi = t \ , \tag{2.6b}$$

if and only if

$$\Phi^* d\bar{t} = dt \quad , \quad \Phi^* \bar{\omega}^i_V = A^i_j \omega^j_U \ , \tag{2.6c}$$

$$\Phi^* \bar{\theta}^i_V = C^i_j \theta^j_U + C^i_j B^j_k \omega^k_U \ , \tag{2.6d}$$

where

$$\omega^i_U = dx^i - y^i\,dt \quad , \quad \theta^i_U = dy^i - F^i(y^j, x^k, t)dt \; , \tag{2.6e}$$

$$\bar{\omega}^i_V = d\bar{x}^i - \bar{y}^i\,d\bar{t} \quad , \quad \bar{\theta}^i_V = d\bar{y}^i - \bar{F}^i(\bar{y}^j, \bar{x}^k, \bar{t})d\bar{t} \; . \tag{2.6f}$$

We thus have a Cartan equivalence problem given by Eqs. (2.6c) and (2.6d).

If $U = V$ and $\bar{\omega}^i_V|_q = \omega^i_U|_q$ for all $q \in V$, then the solutions of the Cartan Equivalence Problem given by Eq. (2.1) are called *self-equivalences* or *automorphisms*. We shall see in the next section that the self-equivalence always form a Lie pseudogroup, which finite (i.e. its elements depend on arbitrary constants labelling the elements of a finite-dimensional Lie group of transformations) or infinite (i.e. its elements depend on arbitrary functions labelling the elements of an "infinite-dimensional" Lie group of transformations). If $(\omega^i_U)_{1\leq i\leq n}$ has a (non-trivial) pseudogroup $\mathcal{G}_U$ of self-equivalences, then the general solution of the Equivalence Problem is obtained by composing the most general self-equivalence of $(\omega^i_U)_{1\leq i\leq n}$ with a particular solution of Eq. (2.1). Thus we expect the set of solutions of the Cartan Equivalence Problem given by Eq. (2.1) to depend on a local solvability condition for the existence of a particular solution of Eq. (2.1) and on the structure of $\mathcal{G}_U$. This will be made explicit in the next section.

3. An outline of the solution of the Cartan Equivalence Problem.

We begin by stating a theorem and a corollary which give the solution of the Cartan Equivalence Problem in the special case in which $G = \{e\}$.

The equivalence problem reduces here to the problem of determining all diffeomorphisms $f : U \to V$ such that

$$(3.1) \qquad f^*\bar{\omega}^i_V = \omega^i_U \quad , \quad 1 \leq i \leq n .$$

This is the local equivalence problem for ***parallelisms*** or ***$\{e\}$-structures***. We have

$$(3.2) \qquad d\omega^i_U = \tfrac{1}{2} C^i_{jk}\omega^j_U \wedge \omega^k_U \quad , \quad d\bar{\omega}^i_V = \tfrac{1}{2}\bar{C}^i_{jk}\bar{\omega}^j_V \wedge \bar{\omega}^k_V ,$$

and as a consequence of the commutativity of d and pull-backs, any solution f of the equivalence problem must satisfy the necessary conditions

$$(3.3) \qquad \bar{C}^i_{jk} \circ f = C^i_{jk} \quad , \quad 1 \leq i,j,k \leq n .$$

The functions C^i_{jk} are thus ***invariants*** under any equivalence. Moreover, if we define the covariant derivatives of a function $F : U \to \mathbf{R}$ to be the directional derivatives of F with respect to the vector fields dual to the ω^i_U, i.e.

$$(3.4) \qquad dF = F_{|i}\omega^i_U ,$$

then we obtain as further necessary conditions

$$(3.5) \qquad \bar{C}^i_{jk\,|\,i_1\cdots i_p} \circ f = C^i_{jk\,|\,i_1\cdots i_p} ,$$

for all $1 \leq i,j,k\,, i_1, \ldots, i_p \leq n$ and $p \geq 1$.

We define the set of functions $\mathcal{F}_s$ by

$$(3.6) \qquad \mathcal{F}_s := \left\{C^i_{jk}, C^i_{jk\,|\,i_1}, \ldots, C^i_{jk\,|\,i_1\cdots i_s} \,|\, 1 \leq i,j,k,i_1,\ldots,i_s \leq n\right\} .$$

We have $\mathcal{F}_s \subseteq \mathcal{F}_{s+1}$ and we let for $p \in U$

$$(3.7) \qquad k_s(p) := \dim\langle d\mathcal{F}_s\rangle ,$$

where

$$(3.8)\qquad \langle d\mathcal{F}_s\rangle := \operatorname{span} d\{C^i_{jk}, C^i_{jk\,|\,i_1}, \ldots, C^i_{jk\,|\,i_1\cdots i_s}\,|\,1 \le i,j,k,i_1,\ldots,i_s \le n\} \subseteq T^*_pU \ .$$

The integer $k_s(p)$ thus gives the maximal number of functionally independent elements of $\mathcal{F}_s$ at p. We assume that we are in a non-singular situation in which every $p \in U$ has an open neighborhood U_p such that $k_s(p') = k_s(p)$ for all $p' \in U_p$. It is then straightforward to show using finite induction and the chain rule that if $k_s(p) = k_{s+1}(p) =: r$, then $k_s(p) = k_{s+\ell}(p)$ for all $\ell \ge 1$, and there exists a maximal functionally independent set of functions $\{f^1, \ldots, f^r\}$ such that every $g \in \mathcal{F}_s$ is of the form $g = g(f^1, \ldots, f^r)$.

The following result, due to Elie Cartan [27], shows how the necessary conditions given by Eqs. (3.3) and (3.5) can be used to obtain necessary and sufficient conditions for the existence of an equivalence.

THEOREM 3.1. *If s is the smallest integer such that $k_s = k_{s+1}$, then necessary and sufficient conditions for the existence of a diffeomorphism $f : U \to V$ such that*

$$(3.8)\qquad f^*\bar{\omega}^i_V = \omega^i_U \quad , \quad 1 \le i \le n ,$$

and

$$(3.9)\qquad f(p) = q ,$$

are

(i) $\bar{s} = s$ *and* $k_s = \bar{k}_s =: r$.

(ii) If $\{f^1, \ldots, f^r\}$ *is a maximal functionally independent set of functions such that every* $g \in \mathcal{F}_s$ *is a function of* $f^1, \ldots, f^r$ *and* $\{\bar{f}^1, \ldots, \bar{f}^r\}$ *is the corresponding set for* $\bar{\mathcal{F}}_s$*, then the local solvability condition*

$$(3.10)\qquad f^\alpha(p) = \bar{f}^\alpha(q) \quad , \quad 1 \le \alpha \le r ,$$

is satisfied.

(iii) If g is an element of $\mathcal{F}_{s+1}$ and $\bar{g}$ is the element with same label in $\bar{\mathcal{F}}_{s+1}$, then

$$g = F(f^1, \dots, f^r) \quad , \quad \bar{g} = F(\bar{f}^1, \dots, \bar{f}^r) . \tag{3.11}$$

Theorem 3.1 thus shows that the isomorphism class of an $\{e\}$-structure is characterized locally by the *functional relations* between a complete set of local invariants and their covariant derivatives, and that the equivalence are obtained by solving a completely integrable Pfaffian system.

It is important to note that the conditions (i), (ii) and (iii) in the statement of Theorem 3.1 are always satisfied for self-equivalences, since the identity map is always a solution of the self-equivalence problem. In fact, the set of automorphisms of an $\{e\}$-structure forms a Lie group of transformations under composition, whose dimension is related to the cardinality of a maximal set of functionally independent invariants. This relation is made precise by the following corollary to Theorem 3.1.

COROLLARY. *If $G = \{e\}$, then the set of self-equivalence forms an $(n-r)$-dimensional Lie group of transformations.*

We shall see in the next section that the above corollary can be used to determine the symmetry group of a differential equation.

In view of Theorem 3.1, we are naturally led to look for a construction by means of which the original equivalence problem given by Eq. (2.1) is reduced to a new equivalence problem, for which G is now replaced by a *proper* Lie subgroup $G_{(1)}$ of G, such that the two equivalence problems have the same set of solutions. The result of such a construction is called a *group reduction.*

We begin by deriving some important relations which are the analogues of Eqs. (3.2) and (3.3) in the case $G \neq \{e\}$. On the G-spaces $U \times G$ and $V \times G$ endowed with left translation along the fibres over U and V we define the *lifted* coframes $(\omega^i)_{1 \leq i \leq n}$ and $(\bar{\omega}^j)_{1 \leq j \leq n}$ by

$$\omega^i\big|_{(p,s)} := S^i_j \omega^j_U\big|_p \quad , \quad \bar{\omega}^i\big|_{(q,T)} := T^i_j \bar{\omega}^j_V\big|_q . \tag{3.12}$$

The following result gives a formulation in terms of the lifted coframes of the Cartan Equivalence Problem given by Eq. (2.1).

THEOREM 3.2. *There exists a diffeomorphism* $f : U \to V$ *such that*

$$f^*\bar{\omega}^i_V = \gamma^i_j \omega^j_U \;, \tag{3.13}$$

for some G*-valued function* (γ^i_j) *if and only if there exists a lift* $\tilde{f} : U \times G \to V \times G$ *which satisfies*

$$\tilde{f}^*\bar{\omega}^i = \omega^i \;, \tag{3.14}$$

$$S\tilde{f}(p,T) = \tilde{f}(p,ST) \;. \tag{3.15}$$

The exterior differentials of the elements of the lifted coframe satisfy the following equations, which generalize Eq. (3.1) and are known as Elie Cartan's *first structure equations*

$$d\omega^i = a^i_{j\rho}\pi^\rho \wedge \omega^j + \tfrac{1}{2}\gamma^i_{jk}\omega^j \wedge \omega^k \;, \tag{3.16}$$

where the $a^i_{j\rho}$ are constants such that the $r := \dim G$ $n \times n$ matrices $A_\rho := (a^i_{j\rho})$ form a basis for the Lie algebra $\mathcal{G}$ of G and the r 1-forms π^ρ form a maximal set of linearly independent right-invariant 1-forms on G. The terms $\frac{1}{2}\gamma^i_{jk}\omega^i \wedge \omega^k$ in Eq. (3.16) are called *torsion terms*

Let V be an n-dimensional real vector space with a basis $\{e_i\,,\ 1 \le i \le n\}$ and V^* be the dual of V endowed with the dual basis $\{e^{*j}\,,\ 1 \le j \le n\}$. Also, let $\{\delta_\sigma|_e\,,\ 1 \le \sigma \le r\}$ be the basis of $T_eG \simeq \mathcal{G}$ dual to the basis $\{\pi^\rho|_e\,,\ 1 \le \rho \le r\}$ of $T^*_eG \simeq \mathcal{G}^*$. Define the map $L : \mathcal{G} \otimes V^* \to V \otimes \Lambda^2 V^* : v^\rho_i \delta_\rho|_e \otimes e^{*i} \mapsto (a^i_{j\rho}v^\rho_k - a^i_{k\rho}v^\rho_j)e_i \otimes e^{*j} \wedge e^{*k}$, and consider the spaces

$$\mathcal{G}^{(1)} := \ker L \qquad , \quad \Pi_{\mathcal{G}} := V \otimes \Lambda^2 V^* / \operatorname{im} L \;. \tag{3.17}$$

Let $\mathrm{pr} : V \otimes \Lambda^2 V^* \to \Pi_{\mathcal{G}}$ denote the canonical projection onto the quotient $V \otimes \Lambda^2 V^* / \operatorname{im} L$ and define $\gamma : U \times G \to V \otimes \Lambda^2 V^* : (p,S) \mapsto \frac{1}{2}\gamma^i_{jk}(p,S) e_i \otimes e^{*j} \wedge e^{*k}$. The (first-order) *structure tensor* is by definition the map $\tau_U : U \times G \to \Pi_{\mathcal{G}} : (p,S) \mapsto (\mathrm{pr} \circ \gamma)(p,S)$.

The following theorem gives the generalization to the case $G \neq \{e\}$ of the necessary condition given by Eq. (3.3).

THEOREM 3.3. *If there exists a map* $f : U \to V$ *such that*

$$f^*\bar{\omega}^i_V = \gamma^i_j \omega^j_U , \tag{3.18}$$

for some G-valued function (γ^i_j), *then the lift* $\tilde{f} : U \times G \to V \times G$ *pulls back the structure tensor* τ_V *on* $V \times G$ *to the structure tensor* τ_U *on* $U \times G$, *that is*

$$\tau_V \circ \tilde{f} = \tau_U . \tag{3.19}$$

Effectively, the group reduction construction amounts to normalizing the structure tensors τ_U and τ_V under the G-action $\tilde{\rho}$ in $\Pi_{\mathcal{G}}$. The G-action is well-defined in the quotient $\Pi_{\mathcal{G}} = V \otimes \Lambda^2 V^* / \operatorname{im} L$ since for all $S \in G$ and $w \in \mathcal{G} \otimes V^*$, we have $((\rho \otimes \Lambda^2 \rho^*)(S) \circ L)(w) = (L \circ (Ad \otimes \rho^*)(S))(w)$, where $\rho : G \to GL(V)$ denotes the defining representation of G. We assume for simplicity that $\tau_U(U \times G)$ is a single G-orbit in $\Pi_{\mathcal{G}}$. We choose a fixed vector $\tau_{(1)} = \tau_U(p, S_{(1)})$ and let $G_{(1)} := \{S \in G \mid \tilde{\rho}(S)\tau_{(1)} = \tau_{(1)}\}$ be the stability group of $\tau_{(1)}$. We choose any G-valued map $\overset{(1)}{\delta}_U : U \to G$ whose graph $\Gamma(\overset{(1)}{\delta}_U)$ in $U \times G$ is contained in $\tau_U^{-1}(\tau_{(1)})$ and define the coframe $(\overset{(1)}{\omega}{}^i_U = \overset{(1)}{\delta}{}_U{}^i_j \omega^j_U)_{1 \le i \le n}$. Performing the identical construction in $V \times G$, we have the following result

THEOREM 3.4. *The two Cartan equivalence problems*

$$f^*\bar{\omega}^i_V = \gamma^i_j \omega^j_U , \tag{3.20}$$

where (γ^i_j) *is a G-valued function, and*

$$\overset{(1)}{f}{}^* \overset{(1)}{\bar{\omega}}{}^i_V = \overset{(1)}{\gamma}{}^i_j \overset{(1)}{\omega}{}^j_U , \tag{3.21}$$

where $(\overset{(1)}{\gamma}{}^i_j)$ *is a* $G_{(1)}$*-valued function, have the same set of solutions.*

Theorem 3.4 and 3.1 suggest to iterate the process of group reduction until we obtain a family of nested subgroups of G

$$G_{(k)} \subset G_{(k-1)} \subset \cdots \subset G_{(1)} \subset G , \tag{3.22}$$

such that $G_{(k)} = \{e\}$, or $G_{(k)} \neq \{e\}$ and the $G_{(k)}$-action in $\Pi_{\mathcal{G}(k)}$ is trivial. If $G_{(k)} = \{e\}$, we have reduced the original problem to an equivalent problem of local isomorphism for $\{e\}$-structures which we solve using Theorem 3.1. If $G_{(k)} \neq \{e\}$ and the $G_{(k)}$-action in $\Pi_{\mathcal{G}(k)}$ is trivial, we have to use the Cartan-Kähler Theorem for the existence of integral manifolds of involutive Pfaffian systems and the Cartan-Kuranishi Theorem for the involutivity of a finite prolongation of a Pfaffian system. Thus the lifts $\overset{(k)}{\omega}{}^i$ and $\overset{(k)}{\bar\omega}{}^j$ of $\overset{(k)}{\omega}{}^i_U$ and $\overset{(k)}{\bar\omega}{}^j_V$ satisfy the structure equations

$$d\overset{(k)}{\omega}{}^i = \overset{(k)}{a}{}^i_{j\rho}\overset{(k)}{\pi}{}^\rho \wedge \overset{(k)}{\omega}{}^j + \tfrac{1}{2}\overset{(k)}{\gamma}{}^i_{j\ell}\overset{(k)}{\omega}{}^j \wedge \overset{(k)}{\omega}{}^\ell , \tag{3.23}$$

on $U \times G_{(k)}$, and

$$d\overset{(k)}{\bar\omega}{}^i = \overset{(k)}{a}{}^i_{j\rho}\overset{(k)}{\bar\pi}{}^\rho \wedge \overset{(k)}{\bar\omega}{}^j + \tfrac{1}{2}\overset{(k)}{\bar\gamma}{}^i_{j\ell}\overset{(k)}{\bar\omega}{}^j \wedge \overset{(k)}{\bar\omega}{}^\ell , \tag{3.24}$$

on $V \times G_{(k)}$, and we suppose that $G_{(k)} \neq \{e\}$ has a trivial action $\Pi_{\mathcal{G}(k)}$. We wish to determine the admissible integral manifolds of the Pfaffian system $\mathcal{J}_{(k)}$ on $U \times G_{(k)} \times V \times G_{(k)}$ generated by $p_{1,2}^*\overset{(k)}{\omega}{}^i - p_{3,4}^*\overset{(k)}{\bar\omega}{}^i$, where $p_{1,2} : U \times G_{(k)} \times V \times G_{(k)} \to U \times G_{(k)}$, $p_{3,4} : U \times G_{(k)} \times V \times G_{(k)} \to V \times G_{(k)}$ denote the canonical projections onto the first and second products of factors, with independence condition given by $\overset{(k)}{\omega} = \overset{(k)}{\omega}{}^1 \wedge \cdots \wedge \overset{(k)}{\omega}{}^n$. The answer is obtained by applying Cartan's involutivity test. We have

THEOREM 3.5. *The reduced characters σ'_p, $1 \le p \le n$, of $(\mathcal{I}_{(k)}, \overset{(k)}{\omega})$ are given by*

$$\sigma'_1 + \cdots + \sigma'_p = \max_{v_1,\dots,v_p \in \mathbf{R}^n} \operatorname{rank} \begin{pmatrix} v_1^j \overset{(k)}{a}{}^i_{j\rho} \\ \vdots \\ v_p^j \overset{(k)}{a}{}^i_{j\rho} \end{pmatrix} , \tag{3.25}$$

and $(\mathcal{I}_{(k)}, \overset{(k)}{\omega})$ is in involution if and only if

$$\dim \mathcal{G}^{(1)}_{(k)} = \sum_{p=1}^{n} p\sigma'_p . \tag{3.26}$$

If $(\mathcal{I}_{(k)}, \overset{(k)}{\omega})$ is in involution, then each admissible integral manifold is the graph $\Gamma(\overset{(k)}{\tilde f})$ of the lift $\overset{(k)}{\tilde f} : U \times G_{(k)} \to V \times G_{(k)}$ of an equivalence $\overset{(k)}{f} : U \to V$. In particular, we obtain, using Cartan's theorems or infinite Lie pseudogroups, the following result.

THEOREM 3.6. *If the involutivity condition (3.26) is satisfied and the structure tensors obtained from Eqs. (3.23) and (3.24) have constant components with the same numerical values, then the set of solutions of the equivalence problem forms a transitive infinite Lie pseudogroup under composition.*

If $(\mathcal{I}_{(k)}, \overset{(k)}{\omega})$ is not in involution, that is the condition (3.26) is not satisfied, we know from the Cartan-Kuranishi Theorem that we have to prolong the system $(\mathcal{I}_{(k)}, \overset{(k)}{\omega})$. The equations for the first prolongation of Eq. (3.23) are given by

$$d\overset{(k)}{\omega}{}^i = \overset{(k)}{a}{}^i_{j\rho} \overset{(k)}{\pi}{}^\rho \wedge \overset{(k)}{\omega}{}^j + \tfrac{1}{2} \overset{(k)}{\gamma}{}^i_{j\ell} \overset{(k)}{\omega}{}^j \wedge \overset{(k)}{\omega}{}^\ell , \tag{3.27}$$

$$d\overset{(k)}{\pi}{}^\rho = \overset{(k)}{A}{}^\rho_{si} \overset{(k)}{\theta}{}^s \wedge \overset{(k)}{\omega}{}^i + \overset{(k)}{B}{}^\rho_{\sigma j} \overset{(k)}{\pi}{}^\sigma \wedge \overset{(k)}{\omega}{}^j + \tfrac{1}{2} \overset{(k)}{C}{}^\rho_{\ell m} \overset{(k)}{\omega}{}^\ell \wedge \overset{(k)}{\omega}{}^m . \tag{3.28}$$

The 1-forms θ^s, $1 \le s \le \dim \mathcal{G}^{(1)}_{(k)}$, are congruent modulo the ideal generated by $\overset{(k)}{\omega}{}^i$, $1 \le i \le n$, $1 \le \rho \le \dim G_{(k)}$, to the elements of a basis for the space of right-invariant 1-forms on the first *group prolongation* $G^{(1)}_{(k)}$ of $G_{(k)}$. The group $G^{(1)}_{(k)}$ is by definition the Lie subgroup of $GL(V^* + \mathcal{G}^*_{(k)})$ whose elements t are of the form $t(\overset{(k)}{\omega}{}^i) = \overset{(k)}{\omega}{}^i$, $1 \le i \le n$, $t(\overset{(k)}{\pi}{}^\rho) = \overset{(k)}{\pi}{}^\rho + v^\rho_i \overset{(k)}{\omega}{}^i$, $1 \le \rho \le \dim G_{(k)}$, where $v^\rho_i \delta_\rho \otimes \overset{(k)}{\omega}{}^i \in \mathcal{G}^{(1)}_{(k)}$.

We thus have a new equivalence problem with base space $U \times G_{(k)}$, base coframe $(\overset{(k)}{\omega}{}^A_{U\times G_{(k)}}) = (\overset{(k)}{\omega}{}^1_U, \ldots, \overset{(k)}{\omega}{}^n_U, \overset{(k)}{\pi}{}^1 =: \overset{(k)}{\omega}{}^{n+1}_{U\times G_{(k)}}, \ldots, \overset{(k)}{\pi}{}^{\dim G_{(k)}}_{U\times G_{(k)}} =: \overset{(k)}{\omega}{}^{n+\dim G_{(k)}}_{U\times G_{(k)}})$ and group $G^{(1)}_{(k)}$. The following result asserts that the original equivalence problem given by Eq. (2.1) and this new equivalence problem have the same set of solutions.

THEOREM 3.7. *There is a one-to-one correspondence between diffeomorphisms $f : U \to V$ such that*

$$f^* \bar{\omega}^i_V = \gamma^i_j \omega^j_U , \tag{3.29}$$

for some G-valued function (γ^i_j), and diffeomorphisms $\hat{f} : U \times G_{(k)} \to V \times G_{(k)}$ such that

$$\hat{f}^* \overset{(k)}{\bar{\omega}}{}^A_{V\times G_{(k)}} = \hat{\gamma}^A_B \overset{(k)}{\omega}{}^B_{U\times G_{(k)}} , \tag{3.30}$$

for some $G^{(1)}_{(k)}$-valued function $(\hat{\gamma}^A_B)$.

What Theorem 3.7 tells us is that if the original equivalence problem given by Eq. (2.1) leads after a sequence of group reductions to a family of nested subgroups $G_{(k)} \subset G_{(k-1)} \subset \cdots \subset G_{(1)} \subset G$ such that $G_{(k)} \neq \{e\}$, the $G_{(k)}$-action in $\Pi_{\mathcal{G}(k)}$ is trivial and the involutivity condition (3.26) is not satisfied, then, since the equivalence problem given by Eq. (3.30) has the same set of solutions as the original problem, one can perform the group reduction construction on this new equivalence problem. This will lead to an $\{e\}$-structure on $U \times G_{(k)}$, or a non-trivial reduced group $G^{(1)}_{(k)(\ell)}$ whose action in $\Pi_{\mathcal{G}^{(1)}_{(k)(\ell)}}$ is trivial. In the former case the problem is solved by Theorem 3.1. In the latter case, one checks whether the involutivity condition (3.26) is satisfied. If the system is involutive, then by virtue of Theorem 3.6 we have an infinite Lie pseudogroup of solutions which is transitive if the torsion tensor is constant and otherwise an intransitive infinite Lie pseudogroup of self-equivalences. If the involutivity condition (3.26) is not satisfied, then we prolong and start the procedure all over with the prolonged problem. We know from the Cartan-Kuranishi Theorem that after a finite number of iterations we shall obtain an involutive system corresponding to an $\{e\}$-structure or to an infinite Lie pseudogroup.

We now have concluded our description of the solution of the Cartan Equivalence Problem, except for two points which are crucial in the solution of any concrete example. The first point deals with the determination of the torsion tensor τ_U and the second point has to do with the determination of the G-action in $\Pi_{\mathcal{G}}$.

Rather than compute τ_U directly, Cartan chooses a complement C to $\operatorname{im} L$ in $V \otimes \Lambda^2 V^*$:

$$V \otimes \Lambda^2 V^* = \operatorname{im} L + C \; . \tag{3.31}$$

One procedure for choosing such a complement is to do what Gardner [**12**] calls "Lie algebra-compatible absorption of torsion", that is to solve as many equations of the form

$$\gamma^i_{jk} = -(a^i_{j\rho} v^\rho_k - a^i_{k\rho} v^\rho_j) \; , \tag{3.32}$$

as possible. Suppose that one can solve the $s(:= \operatorname{rank} L)$ equations

$$\gamma^{i_p}_{j_p k_p} = -(a^{i_p}_{j_p \rho} v^\rho_{k_p} - a^{i_p}_{k_p \rho} v^\rho_{j_p}) \; , \tag{3.33}$$

where $1 \leq p \leq s$. Define the projections $\mathrm{pr}_1 : V \otimes \Lambda^2 V^* \to \mathrm{im}\, L : \frac{1}{2}\gamma^i_{jk} e_i \otimes e^{*j} \wedge e^{*k} \to \sum_{p=1}^{s} \frac{1}{2}\gamma^{i_p}_{j_p k_p} e_{i_p} \otimes e^{*j_p} \wedge e^{*k_p}$ and $\mathrm{pr}_2 : V \otimes \Lambda^2 V^* \to C : \frac{1}{2}\gamma^i_{jk} e_i \otimes e^{*j} \wedge e^{*k} \to \sum_{(i,j,k)\neq(i_p,j_p,k_p)} \frac{1}{2}\gamma^i_{jk} e_i \otimes e^{*j} \wedge e^{*k}$. Since any solution of the above inhomogeneous system is defined modulo $\ker L$, it follows that the image of the torsion terms under pr_2 is preserved under the lift $\tilde{f}$ of any solution f of the equivalence problem given by Eq. (2.1). One can now study the G-action in $\Pi_{\mathcal{G}}$ by lifting the G-action to $C \subset V \otimes \Lambda^2 V^*$. Inspired by Cartan's theory of principal components in his method of moving frames, Gardner [**12**] made the fundamental contribution to the application of the Equivalence Method to any non-trivial problem by showing that the most efficient way of computing the G-action in C is to do it first infinitesimally by computing $d\gamma^i_{jk}$, $(i,j,k) \neq (i,j,k)$, modulo the ideal generated by the ω^i from the necessary conditions $d^2\omega^i = 0$ expressed on the first structure equations. This approach has the considerable advantage of only requiring the defining relations of the Lie algebra $\mathcal{G}$ of G as opposed to an explicit parametrization of G, while being sufficient to reconstruct the finite G-action in a neighborhood of the identity. The power of this approach, which Gardner calls the "intrinsic approach of repère mobile" will clearly be illustrated in the next section. We refer the reader to Gardner's paper [**12**] for a description of this intrinsic approach.

4. Examples.

In the present section, we illustrate on some examples the method presented in Section 3 for the solution of the Cartan Equivalence Problem.

The first example we consider is the equivalence problem for Riemannian metrics in the plane. (See example (a) in Section 2). We will sove it by the intrinsic method referred to at the end of Section 3.

The lifts of ω_U^1 and ω_U^2 to $U \times SO(2, \mathbf{R})$ are given by

$$\omega^1 = \omega_U^1 \cos\theta + \omega_U^2 \sin\theta \quad , \quad \omega^2 = -\omega_U^1 \sin\theta + \omega_U^2 \cos\theta \ . \tag{4.1}$$

The first structure equations thus read

$$d\omega^1 = d\theta \wedge \omega^2 + a\omega^1 \wedge \omega^2 \quad , \quad d\omega^2 = -d\theta \wedge \omega^1 + b\omega^1 \wedge \omega^2 \ , \tag{4.2}$$

and the Lie algebra-compatible absorption of torsion is performed by setting

$$\alpha := d\theta + a\omega^1 + b\omega^2 \ , \tag{4.3}$$

so that

$$d\omega^1 = \alpha \wedge \omega^2 \quad , \quad d\omega^2 = -\alpha \wedge \omega^1 \ . \tag{4.4}$$

We see from Eq. (4.4) that $\Pi_{so(1,\mathbf{R})} = 0$ so that the $SO(2, \mathbf{R})$-action in $\Pi_{so(2,\mathbf{R})}$ is of course trivial. (This is the so-called Fundamental Theorem of Riemannian Geometry which states that there is a unique metric and torsion-free connection on the orthonormal frame bundle of a Riemannian manifold, namely the Levi-Civitá connection.)

We thus have to apply Theorem 3.5. The reduced characters are given by $\sigma_1' = 1$, $\sigma_2' = 0$, while $\dim so(2, \mathbf{R})^{(1)} = 0$. The system is therefore *not* involutive and we have to prolong Eq. (4.4).

The structure equations for the prolonged system (which corresponds to Eqs. (3.27) and (3.28)) are given by expressing the integrability conditions $d^2\omega^1 = 0 = d^2\omega^2$ and applying Cartan's Lemma to solve for $d\alpha$. We have

$$d\omega^1 = \alpha \wedge \omega^1 \quad , \quad d\omega^2 = -\alpha \wedge \omega^2 \ , \tag{4.5}$$

(4.6) $$d\alpha = -K\omega^1 \wedge \omega^2 \ ,$$

and we now have an $\{e\}$-structure on $U \times SO(2, \mathbf{R})$, with a torsion coefficient K which is an invariant by Eq. (3.3). Theorem 3.1 now gives the complete solution to the given problem. It should be noted that the invariant K does not depend explicitly on the angle θ since Eqs. (4.5) and (4.6) imply that

(4.7) $$dK \wedge \omega^1 \wedge \omega^2 = 0 \ .$$

(This is an example of a phenomenon called a *transgression.*) The invariant K is nothing but the *Gaussian curvature* of g_U.

The second example we consider is that of the conformal equivalence problem for Riemannian metrics in the plane, which is Example (b), of Section 2. We also use the intrinsic method.

The lifts of ω_U^1 and ω_U^2 to $U \times (CO(2, \mathbf{R}) := \mathbf{R}^* \otimes SO(2, \mathbf{R}))$ are given by

(4.8) $$\omega^1 = \omega_U^1 \rho \cos\theta + \omega_U^2 \rho \sin\theta \quad , \quad \omega^2 = -\omega_U^1 \rho \sin\theta + \omega_U^2 \rho \cos\theta \ .$$

The first structure equations thus read

(4.8) $$d\omega^1 = \frac{d\rho}{\rho} \wedge \omega^1 + d\theta \wedge \omega^2 + a\omega^1 \wedge \omega^2 \quad , \quad d\omega^2 = -d\theta \wedge \omega^1 + \frac{d\rho}{\rho} \wedge \omega^2 + b\omega^1 \wedge \omega^2 \ .$$

The Lie algebra-compatible asbsorption of torsion is performed by setting

(4.10) $$\alpha := d\theta + a\omega^1 + b\omega^2 \ ,$$

so that

(4.11) $$d\omega^1 = \frac{d\rho}{\rho} \wedge \omega^1 + \alpha \wedge \omega^2 \quad , \quad d\omega^2 = -\alpha \wedge \omega^1 + \frac{d\rho}{\rho} \wedge \omega^2 \ .$$

The $CO(2, \mathbf{R})$ action in $\Pi_{co(2,\mathbf{R})}$ is trivial and we have to apply Theorem 3.5. The reduced characters are given by $\sigma_1' = 2\,; \sigma_2' = 0$, while the elements of $co(2, \mathbf{R})^{(1)}$ are given by matrices of the form

(4.12) $$\begin{pmatrix} 1 & 0 & 0 & 0 \\ 0 & 1 & 0 & 0 \\ u & v & 1 & 0 \\ -v & u & 0 & 1 \end{pmatrix} ,$$

where $u, v \in \mathbf{R}$. We thus have $\dim co(2,\mathbf{R})^{(1)} = 2 = \sum_{p=1}^{2} p\sigma'_p$ and the system is in involution. It then follows for Theorem 3.6 that the set of solutions of the equivalence problem considered forms a transitive infinite Lie pseudogroup. What we have recovered is thus the well-known theorem from Complex Analysis which says that every two-dimensional Riemannian metric is conformally equivalent to the flat metric $g_U = \mathrm{diag}(1,1)$.

Next we consider the equivalence problem for systems of second-order ordinary differential equations, which was Example (e) in Section 2. We use the direct parametric approach as opposed to the intrinsic approach and refer the reader to the paper by Chern [**31**] for further details.

Let G be the Lie subgroup of $GL(2n+1,\mathbf{R})$ of matrices of the form

$$\begin{pmatrix} 1 & 0 & \cdots & 0 & 0 & \cdots & 0 \\ 0 & A^1_1 & \cdots & A^1_n & 0 & \cdots & 0 \\ \vdots & \vdots & & \vdots & \vdots & & \vdots \\ 0 & A^n_1 & \cdots & A^n_n & 0 & \cdots & 0 \\ 0 & C^1_j B^j_1 & \cdots & C^1_j B^j_n & C^1_1 & \cdots & C^1_n \\ \vdots & \vdots & & \vdots & \vdots & & \vdots \\ 0 & C^n_j B^j_1 & \cdots & C^n_j B^j_n & C^n_1 & \cdots & C^n_n \end{pmatrix} \tag{4.13}$$

where $\det(A^i_j) \neq 0 \neq \det(C^i_j)$.

The lifts of dt, ω^i_U and θ^i_U to $U \times G$ are given by

$$\omega^0 := dt \quad , \quad \omega^i := A^i_j \omega^j_U \, , \tag{4.14}$$

$$\theta^i := C^i_j \theta^j_U + C^i_k B^k_j \omega^j_U \, . \tag{4.15}$$

From Eqs. (2.6e), (2.6f), (4.14) and (4.15), we obtain

$$d\omega^i \equiv A^i_j (C^{-1})^j_k \omega^0 \wedge \theta^k \mod \omega^i \, , \tag{4.16}$$

from which it follows that the right-hand side of Eq. (4.16) defines torsion terms which cannot be absorbed in a Lie algebra-compatible way. We choose to normalize them under the group action by setting

$$A^i_j = C^i_j \, . \tag{4.17}$$

We thus have

$$dw^i = \theta^i_j \wedge \omega^j + \omega^0 \wedge \theta^i \,, \tag{4.18}$$

where

$$\theta^i_j :\equiv (A^{-1})^i_k dA^k_j - A^i_k B^k_\ell (A^{-1})^\ell_j \omega^0 \quad \mathrm{mod}\ \omega^i \,. \tag{4.19}$$

From Eqs. (2.6e), (2.6f), (4.17) and (4.19) we obtain

$$d\theta^i \equiv \theta^i_j \wedge \theta^j + A^i_j \left(-2B^j_k - \tfrac{\partial F^j}{\partial y^k}\right) A^k_\ell \theta^\ell \wedge \omega^0 \quad \mathrm{mod}\ \omega^i \,. \tag{4.20}$$

From Eqs. (4.18), (4.19) and (4.20) it follows that the $\theta^\ell \wedge \omega^0$ terms in Eq. (4.20) define torsion terms that cannot be absorbed in a Lie algebra-compatible way. They can be normalized under the group action by setting

$$B^j_k = -\frac{1}{2}\frac{\partial F^j}{\partial y^k} \,. \tag{4.21}$$

There is no non-constant torsion which can be absorbed in a Lie algebra-compatible way and the involutivity condition of Theorem 3.5 is not satisfied. We thus have to prolong the first structure equations. An elementary calculation shows that one obtains an $\{e\}$-structure on $U \times GL(n, \mathbf{R})$, given by

$$d\omega^0 = 0 \,, \tag{4.22}$$

$$d\omega^i = \theta^i_j \wedge \omega^j + \omega^0 \wedge \theta^i \,, \tag{4.23}$$

$$d\theta^i = \theta^i_j \wedge \theta^j + P^i_j \omega^0 \wedge \omega^j + T^i_{jk} \omega^j \wedge \omega^k \,, \tag{4.24}$$

$$d\theta^i_j = \theta^i_k \wedge \theta^k_j + (2T^i_{jk} - P^i_{k|j})\omega^0 \wedge \omega^k - T^i_{k\ell|j} \omega^k \wedge \omega^\ell + R^i_{jk\ell} \theta^k \wedge \theta^\ell \,, \tag{4.25}$$

where

$$dP^i_j \equiv P^i_{j|k} \theta^k \quad \mathrm{mod}\ \omega^0 \omega^i \theta^i_j \,, \tag{4.26}$$

$$dT^i_{jk} \equiv T^i_{jk|\ell} \theta^\ell \quad \mathrm{mod}\ \omega^0 \omega^i \theta^i_j \,. \tag{4.27}$$

A further normalization enables us to set

$$A^i_j = \delta^i_j \ , \tag{4.28}$$

in which case

$$\omega^i = dx^i - y^i\, dt \ , \tag{4.29}$$

$$\theta^i = dy^i - F^i\, dt - \frac{1}{2}\frac{\partial F^i}{\partial y^j}(dx^j - y^j\, dt) \ , \tag{4.30}$$

$$\theta^i_j = \frac{1}{2}\frac{\partial^2 F^i}{\partial y^j \partial y^k}\omega^k + \frac{1}{2}\frac{\partial F^i}{\partial y^j}\omega^0 \ , \tag{4.31}$$

$$P^i_j = \frac{\partial F^i}{\partial x^j} + \frac{1}{4}\frac{\partial F^i}{\partial y^k}\frac{\partial F^k}{\partial y^j} - \frac{1}{2}\frac{d}{dt}\frac{\partial F^i}{\partial y^j} \ , \tag{4.32}$$

$$T^i_{jk} = \frac{1}{4}\frac{\partial^2 F^i}{\partial y^j \partial x^k} - \frac{1}{4}\frac{\partial^2 F^i}{\partial y^k \partial x^j} + \frac{1}{8}\frac{\partial^2 F^i}{\partial y^j \partial y^\ell}\frac{\partial F^\ell}{\partial y^k} - \frac{1}{8}\frac{\partial^2 F^i}{\partial y^k \partial y^\ell}\frac{\partial F^\ell}{\partial y^j} \ , \tag{4.33}$$

$$R^i_{jk\ell} = -\frac{1}{2}\frac{\partial^3 F^i}{\partial y^j \partial y^k \partial y^\ell} \ . \tag{4.34}$$

where $\frac{d}{dt}$ denotes the total derivative. As a consequence of Theorem 3.1 and its corollary we obtain:

THEOREM 4.1. *(i) The necessary and sufficient conditions for a system of second-order ordinary differential equations*

$$\frac{d^2 x^i}{dt^2} = F^i\left(\frac{dx^j}{dt}, x^k, t\right) \ , \tag{4.35}$$

to be equivalent to the linear homogeneous system

$$\frac{d^2 \bar{x}^i}{d\bar{t}^2} = a^i_j \bar{x}^j \ , \tag{4.36}$$

where the a^i_j *are constants, under a point transformation*

$$\bar{x}^i \circ \Phi = f^i(x^j, t) \qquad , \qquad \bar{t} \circ \Phi = t \ , \tag{4.37}$$

are given by

$$P^i_j = a^i_j \ , \ T^i_{jk} = 0 = R^i_{jk\ell} \ . \tag{4.38}$$

References.

1. E. Cartan, Ann. Ec. Normale **25** (1908), p. 57, also in *Oeuvres Complètes*, Vol. II, p. 719.
2. E. Cartan, Ann. Ec. Normale **27** (1910), p. 109, also in *Oeuvres Complètes*, Vol. II, p. 927.
3. S.S. Chern, Comptes Rendus Acad. Sci. (Paris) **204** (1937), p. 227.
4. S.S. Chern, Ann. of Math. **43** (1942), p. 545.
5. R. Debever, Bull. Cl. Sci. Acad. r. Belgique **XXVIII** (1942), p. 794.
6. R. Debever, Bull. Cl. Sci. Acad. r. Belgique **XXXI** (1945), p. 262.
7. G. Vranceanu, J. Math. pures et appl. **6** (1937), p. 361.
8. G. Vranceanu, Bull. Math. Soc. Roum. Sci. **40** (1938), p. 361.
9. S.S. Chern, in *Géométrie différentielle*, Colloques Internationaux du CNRS, Strasbourg 1953 (Editions du CNRS, Paris, 1953).
10. C. Ehresmann, in *Géométrie différentielle*, Colloques Internationaux du CNRS, Strasbourg 1953 (Editions du CNRS, Paris, 1953).
11. I. Singer and S. Sternberg, J. Anal. Math. **15** (1965), p. 1.
12. R.B. Gardner, in *Differential Geometric Control Theory* (Birkhäuser, Boston, 1983).
13. N. Kamran, K. Lamb and W.F. Shadwick, J. Diff. Geom. **22** (1985), p. 139.
14. N. Kamran and W.F. Shadwick, in *Lecture Notes in Physics* **246** (Springer-Verlag, New York, 1986).
15. N. Kamran and W.F. Shadwick, Contemporary Mathematics **68** (1987), p. 133.
16. N. Kamran and W.F. Shadwick, Math. Ann. **279** (1987), p. 117.
17. N. Kamran and W.F. Shadwick in *Infinite Dimensional Lie Algebras and their Applications* (World Scientific, Singapore, 1988).
18. L. Hsu and N. Kamran, Proc. London Math. Soc., to appear.
19. L. Hsu and N. Kamran, Lett. Math. Phys., to appear.
20. C. Grissom, G. Thompson and G. Wilkens, J. Diff. Eq., to appear.

21. A. González-López, preprint.

22. R. Gardner and W.F. Shadwick, in *Lecture Notes in Mathematics* **1156** (Springer-Verlag, New York, 1985).

23. R. Bryant, Contemporary Mathematics **68** (1987), p. 65.

24. N. Kamran and P. Olver, preprint.

25 R.B. Gardner and W.F. Shadwick, Contemporary Mathematics **68** (1987), p. 111.

26. R.B. Gardner and W.F. Shadwick, Systems and Control Letters **8** (1987), p. 463.

27. E. Cartan, Séminaire de Math., 11 janvier 1937, also in *Oeuvres Complètes*, Vol. II, p. 1311.

28. S. Sternberg, *Lectures on Differential Geometry* (Chelsea, New York, 1983).

29. E. Cartan, *Les systèmes différentiels extérieurs et leurs applications géométriques* (Hermann, Paris, 1945).

30. R. Bryant, S.S. Chern and P.A. Griffiths, in *Proceedings of the 1980 Beijing Symposium on Differential Geometry and Differential Equations* (Gordon and Breach, New York, 1982).

31. S.S. Chern, Bull. Sci. Math., **LXIII** (1939), p. 206.

32. P. Olver, Bull. Amer. Math. Soc. (N.S.) **18** (1988), p. 21.

33. N. Kamran and W.F. Shadwick, Comptes Rendus Acad. Sci. (Paris) **303** (1986), p. 555.

Acknowledgements.

Financial support from grants awarded by the National Science Foundation, the Natural Sciences and Engineering Research Council of Canada and the Centro Internacional de Fisica are gratefully acknowledged. It is a pleasure to thank the local and non-local organizers of the School for their warm hospitality in Colombia. Finally, I wish to express my gratitude to Bill Shadwick, Robby Gardner and Peter Olver who all contributed to my understanding of the Method of Equivalence, and to Rebecca Davies who expertly handled the typesetting of this paper in record time.

'INTEGRABLE' SYSTEMS AND WATER WAVES

AN EXAMPLE OF SYMMETRY ANALYSIS APPLIED TO A PHYSICAL PROBLEM

D. LEVI
Dipartimento di Fisica, Universita' di Roma 'La Sapienza'
Piazzale A. Moro, 2 - 00185 Roma Italy
I.N.F.N. Sezione di Roma

Abstract

After a physical introduction on the appearance of nonlinear waves in our surrounding world, we construct a model equation, the *Generalized Kadomtsev-Petviashvili equation* (*GKP*), which describe the propagation in shallow water of two dimensional long waves in the presence of a slowly varying bottom, slowly varying or steep side boundaries and vorticity.

The *GKP* is then analyzed by constructing its group of Lie point symmetries thus showing that, for wave solutions, we can reduce it either to a pure *Korteweg-de Vries equation* (*KdV*) for slowly varying boundaries or to a perturbed *KdV* for steep boundaries. An example of 'exact' solution of physical interest is considered in both cases.

I. Introduction

In the study of one dimensional water waves D.J. Korteweg and G. de Vries [1] introduced, at the end of last century, a nonlinear equation, later referred to as *Korteweg-de Vries equation* (*KdV*), which is one of the members of a large class of nonlinear partial differential equations which are integrable, i.e. for which large families of analytic solutions are obtained by essentially linear techniques[2,3]. Later study in this field lead to the discovery of the *Kadomtsev-Petviashvili equation* (*KP*), a two dimensional generalization of the *KdV*, which describes the propagation of two dimensional water waves in a fluid of constant depth in the shallow water - long wave almost one dimensional approximation[3,4].

This is a very idealistic situation which very seldom happens in nature but, however, due to the great stability of this kind of waves (*solitons*), they can be seen in many different physical situations. To confine ourselves just to one physical situation, the geophysical sciences, it has been recognized that such waves occur rather frequently both in the atmosphere and in the oceans and are the result of the cumulative and competing effects of nonlinearity and dispersion. These waves appear on the termocline of lakes, fjords, costal waters(5) or in the straits(6-9), they have also been observed on the nocturnal inversion in the atmosphere(10). Similarly, it is likely that surface long waves associated with solitary internal waves have been observed by satellite observations(11-16).

For the purpose of considering more realistic situations we have, starting from the basic hydrodynamic equations, constructed a model equation, the *Generalized KP equation* (*GKP*) based on the same basic approximations as those considered in the construction of the *KP* equation but for a rotational fluid(17) when the fluid depth is slowly variable,there are side boundaries, either slowly varying or steep.

In Section II we review the derivation of the *GKP*; the *GKP* contains some extra terms with respect to the *KP* and, moreover, due to the presence of the variable functions which describe the bottom topography, the side boundaries and the vorticity, it has variable coefficients. The equation itself is thus in general not integrable. In a preceding article(18) we applied the most general linear transformation to reduce the *GKP* to one of the known integrable cases; the *KP* and the *cylindrical KP*(19) in 2+1 dimensions, the *KdV* and *cylindrical KdV*(2) in 1+1 dimensions. Here in Section III we construct the Lie point symmetry group associated to the *GKP*, whose group invariant travelling wave solutions give us exactly the same results we obtained previously in the case of shallow side boundaries but showing that, in the case of steep boundaries, the most natural reduction of the *GKP* is a perturbed *KdV*.

In Section IV the result obtained are applied to get explicit soliton solutions of the *GKP*.

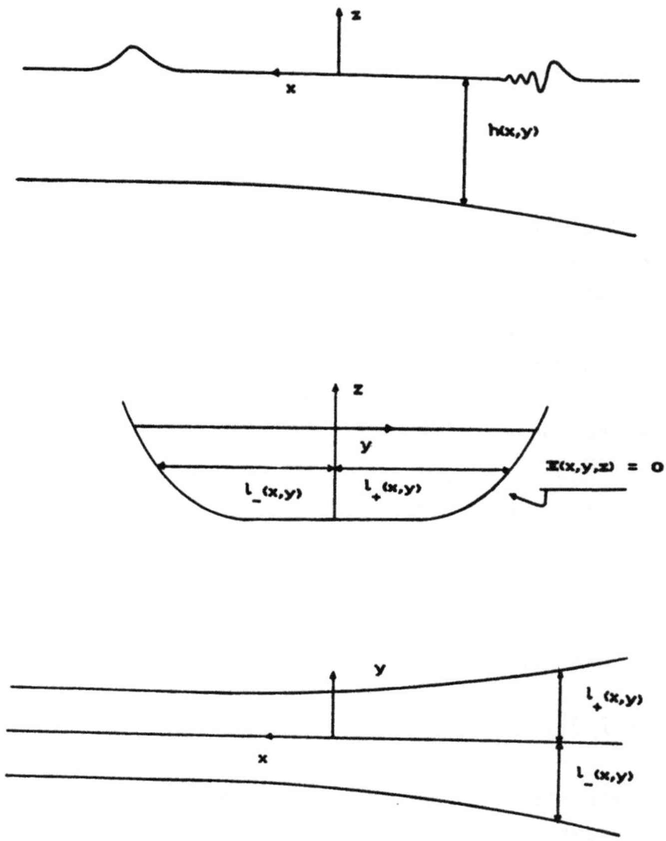

Fig.1 *Three projections of the strait. The quantities* $h(x,y)$ *and* $\ell_{\pm}(x,z)$ *characterize the depth and width, respectively. The cross-section of the strait is given by* $\Xi(x,y,z) = 0$.

II. Derivation of the GKP equation and boundary conditions.

We consider, see Fig.1, a channel of arbitrarily varying depth and width, with bottom and sides described by a given function $\Xi(x,y,z) = 0$. The fluid is assumed to be incompressible and inviscid, subject just to the gravity force, characterized by a constant density ρ, a pressure field p and an Eulerian velocity field $\vec{v} = (v_1,v_2,v_3)$ whose vorticity is given by $\vec{\zeta} = \vec{\nabla}\Lambda\vec{v}$. Under these assumptions the hydrodynamical equations read:

$$(\vec{\nabla},\vec{v}) = 0 \tag{2.1a}$$

$$\rho\vec{v}_{,t} + \rho(\vec{v},\vec{\nabla})v = -\vec{\nabla}p - \rho\,\vec{g} \tag{2.1b}$$

with $\vec{g} = (0,0,g)$ the usual gravity force constant vector. Here and below by a subscript we denote a partial differentiation, by $(.,.)$ the usual scalar product of two vectors and $\vec{\nabla} = (\partial_x,\partial_y,\partial_z)$. For functions of one variable only, a prime will denote a derivative with respect to its argument.

The following boundary conditions must also be satisfied; on the free surface we must have:

$$p\,\big|_{z=\eta(x,y,t)} = 0 \tag{2.2a}$$

$$[\eta_{,t} + v_1\eta_{,x} + v_2\eta_{,y} - v_3]\,\big|_{z=\eta(x,y,t)} = 0 \tag{2.2b}$$

and on the sides and bottom:

$$(\vec{v},\vec{\nabla})\,\Xi\,\big|_{\Xi=0} = 0 \tag{2.3}$$

The approximation that leads to a *KP* is obtained by considering long waves of small amplitudes in shallow water, with wavecrests which vary slowly in the direction perpendicular to the propagation one. More specifically we assume,

$$(H_o / L_x)^2 = \epsilon, \quad (H_o / L_y)^2 = \epsilon^2, \quad N_o / H_o = \epsilon \tag{2.4}$$

where N_o is the characteristic average wave amplitude, L_x is the characteristic wavelength in the propagation direction (width of the soliton), L_y is the characteristic dimension of the perturbations in the y direction, H_o is the average measure of the depth of the fluid and ϵ is a small parameter. Taking this into account we can rewrite the fundamental equations (2.1-2.3) in term of a set of adimensional variables which, for notational simplicity, still are denoted by x, y, z, t, $\vec{v}$, p, ρ, η, Ξ (for more details see ref.18):

$$v_{3,z} + \epsilon\, v_{1,x} + \epsilon^2 v_{2,y} = 0 \qquad (2.5a)$$

$$P_{,x} + \epsilon(v_{1,t} + v_3 v_{1,z}) + \epsilon^2 v_1 v_{1,x} + \epsilon^3 v_2 v_{1,y} = 0 \qquad (2.5b)$$

$$P_{,y} + \epsilon(v_{2,t} + v_3 v_{2,z}) + \epsilon^2 v_1 v_{2,x} + \epsilon^3 v_2 v_{2,y} = 0 \qquad (2.5c)$$

$$P_{,z} + \epsilon(v_{3,t} + v_3 v_{3,z}) + \epsilon^2 v_1 v_{3,x} + \epsilon^3 v_2 v_{3,y} = -1 \qquad (2.5d)$$

$$P|_{z = \epsilon\eta} = 0 \qquad (2.6a)$$

$$[v_3 - \epsilon\eta_{,t} - \epsilon^2 v_1 \eta_{,x} - \epsilon^3 v_2 \eta_{,y}]|_{z = \epsilon\eta} = 0 \qquad (2.6b)$$

$$[\,\Xi_{,z} v_3 + \epsilon\,\Xi_{,x} v_1 + \epsilon^2\,\Xi_{,y} v_2\,]|_{\Xi = 0} = 0 \qquad (2.6c)$$

If we require that the above system should yield a wavelike solution propagating in the x direction, i.e. in the longitudinal direction of the channel then we have to impose constraints on Ξ. To do so we need to solve the equation $\Xi = 0$ in terms of z and this requires that z be a slowly varying function of x and y:

$$z = -\,h(\epsilon x, \epsilon y) \qquad (2.7)$$

If the sides boundaries are not slowly varying functions, as it is the case of a genuine strait or channel, then, near to the boundaries, eq.(2.7) is no more valid and we have to solve $\Xi = 0$ for y in terms of x and z. The existence of wavelike solutions for our system implies that in this case the boundary becomes:

$$y = \ell_{\pm}(\epsilon x, \epsilon^2 z) \qquad (2.8)$$

In conclusion the boundary condition (2.6c) becomes:

$$[v_3 + \epsilon^2 v_1 h_{,\epsilon x} + \epsilon^3 v_2 h_{,\epsilon y}\,]|_{z = -h(\epsilon x, \epsilon y)} = 0 \qquad (2.9)$$

where, in the case of steep boundaries, we have to consider also the following constrain:

$$\left[v_2 - v_1 \ell_{\pm,\epsilon x} - v_3 \ell_{\pm,\epsilon^2 z}\right]\Big|_{y = \ell_\pm(\epsilon x,\epsilon^2 z)} = 0 \tag{2.10}$$

Eq.(2.5) and its boundary conditions (2.6,2.9,2.10) depend on the small parameter ϵ and thus it is obvious to assume that all our fields be ϵ dependent in such a way that we may reduce our system to just one model equation, ϵ independent and valid at a certain order in the ϵ expansion of the fields. Let us express $\vec{v}$, p and η as formal power series in the small parameter ϵ:

$$\vec{v}(x,y,z,t;\epsilon) = \sum_{i=0}^{\infty} \vec{v}^{\,i}(x,y,z,t)\ \epsilon^i \tag{2.11a}$$

$$p(x,y,z,t;\epsilon) = \sum_{i=0}^{\infty} p^i(x,y,z,t)\ \epsilon^i \tag{2.11b}$$

$$\eta(x,y,t;\epsilon) = \sum_{i=0}^{\infty} \eta^i(x,y,t)\ \epsilon^i \tag{2.11c}$$

Then substituting eq.(2.11) into (2.5,2.6,2.9) we find that η^o satisfies the variable coefficient wave equation:

$$\eta^o_{,tt} = h(\epsilon x,\epsilon y)\ \eta^o_{,xx}$$

whose general solution, valid at least up to order ϵ^2, is written in the form:

$$\eta^o(x,y,t) = h^{1/4}(\epsilon x,\epsilon y)\ [\ f^+(X^+) + f^-(X^-)\ ]$$

where $f^\pm$ are arbitrary functions fixed by the initial conditions,

$$X^\pm = R(x,y,\epsilon) \pm t$$

$$R(x,y,\epsilon) = \int_{x_o}^{x} ds\ h(\epsilon s,\epsilon y)^{-1/2}$$

The linear wave equation remain valid as long as higher order corrections, including the nonlinear interaction terms, do not play any role. This happens as long as $t \ll \epsilon^{-1}$. For a long time scale, $t \sim \epsilon^{-1}$, the interactions of the waves with themselves became important. To take them into account we must go over to a reference frame moving with the wave (we choose a wave going to the right):

$$X = R(x,y,\epsilon) - t \tag{2.12a}$$

$$Y = y \tag{2.12b}$$

$$Z = z \tag{2.12c}$$

$$T = \epsilon x \tag{2.12d}$$

Expressing all involved quantities in terms of the new variables and expanding everything in powers of ϵ:

$$h(\epsilon x,\epsilon y) = \sum_{i=0}^{\infty} h_i(T)\ Y^i \epsilon^i$$

$$\ell_\pm(\epsilon x,\epsilon^2 z) = \sum_{i=0}^{\infty} \ell_\pm(T)\ Z^i \epsilon^{2i}$$

we get from the second order terms in the ϵ expansion of the equations (2.5,2.6,2.9) the following nonlinear evolution equation for the zeroth order term of the wave amplitude η, i.e. $\eta^o(X,Y,T)$:

$$\eta^o_{,T} + \frac{3}{2}h_o^{-3/2}(T)\ \eta^o\eta^o_{,X} + \frac{1}{6}h_o^{1/2}(T)\eta^o_{,XXX} + \mathbb{B}(Y,T)\eta^o_{,X} + \mathbb{C}(T)\eta^o$$
$$+\frac{1}{2}h_o^{1/2}(T)\int_{X_o}^{X} ds\ \eta^o_{,YY}(s,Y,T) + h_o^{1/2}(T)\mathbb{A}(T)\eta^o_{,Y} + \mathbb{D}(Y,T) = 0$$

where the functions $\mathbb{A}$, $\mathbb{B}$, $\mathbb{C}$ and $\mathbb{D}$ are given by:

$$\mathbb{A}(T) = -\frac{1}{2}\int_{T_o}^{T} h_1(s)\ h_o^{-3/2}(s)\ ds \tag{2.13a}$$

$$\mathbb{B}(Y,T) = \frac{1}{2}h_o^{1/2}(T)\mathbb{A}^2(T) + h_o^{-5/2}(T)\int_{-h_o}^{0}\Phi_o(Y,Z,T)\,dZ \tag{2.13b}$$

$$\mathbb{C}(T) = \frac{1}{4}(h_o'(T)/h_o(T)) \tag{2.13c}$$

$$\mathbb{D}(Y,T) = \frac{1}{2}\Bigg[\Big(h_o'(T)/h_o(T)\Big)\Phi_o(Y,Z=-h_o(T),T) \tag{2.13d}$$

$$+ h_o^{-1/2}(T)\int_{-h_o}^{0}\Big[[h_o^{-1/2}(T)\Phi_o(Y,Z,T)]_{,T} + \Psi_{o,Y}(Y,Z,T)\quad dZ\Big]\Bigg]$$

where $\Phi_o(Y,Z,T)$ and $\Psi_o(Y,Z,T)$ are two functions related to the zeroth order approximation of the vorticity vector $\vec{\zeta}$:

$$\zeta_1^o = -\Psi_{o,Z} \qquad \zeta_2^o = h_o^{-1/2}\,\Phi_{o,Z} \qquad \zeta_3^o = -h_o^{-1/2}\,\Phi_{o,Y}$$

In the case of steep side boundaries the boundary condition (2.10) is to be added. In this approximation it becomes:

$$\Bigg[\Big(\Psi_o + \mathbb{A}\eta^o + \int_{X_o}^{X}\eta^o{}_{,Y}(s,Y,T)\,ds\Big)-$$

$$-h_o^{-1/2}\Big(\,\eta^o + \Phi_o\,\Big)\,\ell'_{o\pm}\Bigg]\Bigg|_{Y\,=\,\ell_{o\pm}} = 0$$

The physical situation we are interested in is one in which we have a bounded solution at our initial position, $T = T_o$, for instance a soliton. This implies that we can require that

$$\lim_{X\to-\infty}\ \eta^o(X,Y,T) = 0$$

and choose $X_o = -\infty$. In this case we can rewrite the *GKP* equation in local form as the following system:

$$\left[\eta^o_{,T} + \frac{3}{2}h_o^{-3/2}\eta^o\eta^o_{,X} + \frac{1}{6}h_o^{1/2}\eta^o_{,XXX} + h_o^{1/2}\mathbb{A}\eta^o_{,Y} + \mathbb{B}\eta^o_{,X} + \mathbb{C}\eta^o\right]_{,X} + \frac{1}{2}h_o^{1/2}\eta^o_{,YY} = 0 \tag{2.14a}$$

$$\mathbb{D} = 0 \tag{2.14b}$$

together with, in the steep case, the following system of boundary conditions:

$$\left[\mathbb{A}\eta^o_{,X} + \eta^o_{,Y} - h_o^{-1/2}\eta^o_{,X}\ \ell'_{o\pm}\right]\Big|_{Y = \ell_{o\pm}} = 0 \tag{2.15a}$$

$$\left[\Psi_o - h_o^{-1/2}\ \Phi_o\ \ell'_{o\pm}\right]\Big|_{Y = \ell_{o\pm}} = 0 \tag{2.15b}$$

where eqs.(2.14b,2.15b) are just a linear equation for the vorticity functions Φ_o and Ψ_o with its boundary condition.

It is interesting to point out which features of our model influence the propagation of waves at this level of approximation. Let's first of all notice that the depth function was expanded in series and only the first two coefficients h_o and h_1 appear into the *GKP*. If the bottom is symmetric in the y direction, i.e. if $h(x,y) = h(x,-y)$, then $h_1 = 0$ and the equations and boundary conditions are thus insensitive to any variation of the bottom in the y direction. The contribution of h_1, which appears in $\mathbb{A}$ and $\mathbb{B}$, is important if we are interested, for example, to describe the case of waves parallel to a sloping beach.

The functions $\ell_\pm$, which describe the steep boundaries, enter only into the boundary conditions and at the lowest order. This is a consequence of the basic approximations, which include the assumption that the width of the strait is much larger than its depth and thus the variation of the width as function of the depth is immaterial.

The bottom and side boundaries play a different role in our treatment. The shape of the bottom has entered via the depth functions h_o and h_1 into the coefficients of the *GKP* and into the boundary conditions. The side boundaries, on the other hand, figure only into the boundary conditions, so that they can be included or excluded according to the physical situation at study.

III. Symmetry analysis of the GKP equation.

Starting from the *GKP* we would like to find out what we can say of the solution of the physically interesting problem of the propagation of an initial soliton solution into deeper water and into an open sea. To do so in a systematic way we consider the problem of looking for symmetry solutions of *GKP* and among those solutions we shall be mainly interested in the soliton ones, i.e. those which initially can be set in the form of a $\mathrm{sech}^2(X)$. Using a standard algorithm[20], implemented as a MACSYMA package[21], we write down a system of equations depending on the functions $\mathbb{A}$, $\mathbb{B}$, and $\mathbb{C}$ and through them on h_o, h_1 and Φ_o by whose solution we obtain the Lie algebra $\mathcal{L}$ corresponding to the Lie group of local point symmetries of the *GKP*. This solution shall depend on the particular choice made of the functions h_o, h_1 and Φ_o.

If $h_o = 1$, i.e. at first approximation the bottom topography is not varying in the x-direction, the Lie algebra $\mathcal{L}$ of local point symmetries reads:

$$\mathcal{L} = \mathfrak{T}[f(T)] + \mathfrak{Y}[g(T)] + \mathfrak{X}[m(T)] \tag{3.1}$$

where f(T), g(T) and m(T) are arbitrary entire function of their argument, $\mathfrak{T}$, $\mathfrak{Y}$ and $\mathfrak{X}$ are vector fields given by:

$$\begin{aligned}\mathfrak{T}[f] = {} & f\partial_T + \tfrac{2}{3}f'Y\partial_Y + [\tfrac{1}{3}f'X + Y(f\mathbb{A}' + \tfrac{1}{3}\mathbb{A}f') - \tfrac{1}{3}f''Y^2]\partial_X \\ & + \{-\tfrac{2}{9}f'''Y^2 + \tfrac{2}{9}f''X - \tfrac{2}{3}f'\eta^o - \tfrac{2}{3}[fF_o' + \tfrac{2}{3}f'F_o] \\ & + \tfrac{2}{3}Y[f\mathbb{A}'' + \tfrac{4}{3}f'\mathbb{A}' - \tfrac{1}{3}f''\mathbb{A} - fF_1' - \tfrac{4}{3}f'F_1]\}\partial_{\eta^o}\end{aligned} \tag{3.2}$$

$$\mathfrak{Y}[g] = g\partial_Y - g'Y\partial_X - \tfrac{2}{3}[gF_1 + g'\mathbb{A} + g''Y]\partial_{\eta^o} \tag{3.3}$$

$$\mathfrak{X}[m] = m\partial_X + \tfrac{2}{3}m'\partial_{\eta^o} \tag{3.4}$$

where $\mathbb{A}$ now is just an integral of h_1, F_i $(i=0,1)$ are T-dependent functions related to the vorticity function Φ_o by the following formula:

$$\int_{-1}^{0}\Phi_o(Y,Z,T)dZ = F_o(T) + F_1(T)Y \qquad (3.5)$$

This is the only admissible dependence of the integral of the vorticity function Φ_o from T and Y if we want the functions f and g to be completely arbitrary; otherwise, by restricting the symmetry group, we can add to the r.h.s of eq.(3.5) the following expression:

$$f^{-2/3}F_2\left[Y\ f^{-2/3} - \int^T d\tau\ g(\tau)f(\tau)^{-5/3}\right]$$

where F_2 is an arbitrary function of its argument.
For f, g and m arbitrary the Lie point algebra $\mathcal{L}$ is isomorphic to that of the pure *KP* equation, i.e. it has the form of the Kac-Moody-Virasoro algebra studied in ref(22). In fact the *GKP*, which in this case reads:

$$\eta^o_{,TX} + \tfrac{3}{2}(\eta^o\eta^o_{,X})_{,X} + \tfrac{1}{6}\eta^o_{,XXXX} + \tfrac{1}{2}\eta^o_{,YY} + \mathbb{A}(T)\eta^o_{,XY} + \mathbb{B}(T)\eta^o_{,XX} = 0 \qquad (3.6)$$

is just a member of the *KP* hierarchy of integrable equations[23] and it is reducible to the pure *KP* by just a point transformation[19]. So the solutions of eq.(3.6) are obtainable from those of the pure *KP* studied in ref.(22); in particular a wave solution is obtained by reducing the *KP* to a *KdV* equation, i.e. requiring that the wave amplitude η^o be Y-independent. When we require that also the boundary conditions (2.15) be satisfied than the solution we get is equivalent to that of the problem of the propagation of a long wave in a one dimensional flat channel.

When $h_o'(T) \neq 0$ we have to require that $f(T) = k\ h_o(T)/h_o'(T)$, where k is an arbitrary constant. In this case the Lie point algebra $\mathcal{L}$ reads:

$$\mathcal{L} = k\ \mathfrak{T} + \mathfrak{Y}(g) + \mathfrak{X}(m) \qquad (3.7)$$

with

$$\mathfrak{X}(g) = g\partial_Y - \frac{g'Y}{h_o^{1/2}}\partial_X + \frac{1}{3}\Big[-2h_o^{3/2}(gF_1+\mathbb{A}g') \tag{3.8a}$$

$$+Y\,(-4g\alpha h_o'^2/h_o + 3gh_o'' + g'h_o' - 2h_o g'')\Big]\partial_{\eta^o}$$

$$\mathfrak{X}(m) = m\partial_X + \tfrac{2}{3}h_o^{3/2}\; m'\,\partial_{\eta^o} \tag{3.8b}$$

$$\mathfrak{T} = (h_o/h_o')\partial_T + Y\,[1 - \frac{2h_oh_o''}{3h_o'^2}]\partial_Y + \Big\{\tfrac{1}{2}X\,[1 - \frac{2h_oh_o''}{3h_o'^2}]$$

$$+Y\,\Big[\tfrac{1}{2}\,\mathbb{A}\,[1 - \frac{2h_oh_o''}{3h_o'^2}] + \mathbb{A}'\frac{h_o}{h_o'}\Big] + \frac{Y^2}{3h_o^{1/2}}\Big(\frac{h_oh_o''}{h_o'^2}\Big)'\Big\}\partial_X$$

$$+\Big\{\eta^o[1 + \frac{2h_oh_o''}{3h_o'^2}] - \frac{2h_o^{3/2}X}{9}\Big[\frac{h_oh_o''}{h_o'^2}\Big]' - \frac{h_o^{3/2}}{9}\Big[\frac{2h_oh_o''}{h_o'^2}(\mathbb{A}^2h_o^{1/2} - 2F_o)$$

$$+3(F_o - \mathbb{A}^2h_o^{1/2}) - \frac{6h_o}{h_o'}(F_o' - \mathbb{A}\mathbb{A}'h_o^{1/2})\Big] + \frac{h_o^{3/2}Y}{9}\Big[2\mathbb{A}\Big(\frac{h_oh_o''}{h_o'^2}\Big)'$$

$$-6(F_1' - \mathbb{A}'')\frac{h_o}{h_o'} + (F_1 - \mathbb{A}')(\frac{8h_oh_o''}{h_o'^2} - 9)\Big] + \frac{h_oY^2}{18}\Big[\frac{7h_o'}{h_o}\Big(\frac{h_oh_o''}{h_o'^2}\Big)'$$

$$+ 4\Big(\frac{h_oh_o''}{h_o'^2}\Big)''\Big]\Big\}\partial_{\eta^o} \tag{3.8c}$$

where m and g are arbitrary T dependent functions and the coefficient α and the functions F_i (i=0,1) enter into the definition of the vorticity function Φ_o:

$$\int_{-h_o}^{0} \Phi_o(Y,Z,T)dZ = h_o^{5/2}\Big(-\tfrac{1}{2}h_o^{1/2}\mathbb{A}^2 + F_o \tag{3.9}$$

$$+YF_1 + Y^2h_o^{-5/2}(\alpha h_o'^2 - \tfrac{3}{4}h_oh_o'')\Big)$$

Following the analysis carried out when studying the symmetry algebra of the *KP* equation[22] we can deduce that, as in that case, we have three types of group invariant solutions of the *GKP*.

1. Symmetry reduction by the subalgebra $\mathcal{L}_1 = \{ \mathfrak{T} \}$

Invariance under $\exp[\lambda \mathfrak{T}]$ implies that the solution of the *GKP* has the form

$$\eta^{o}(X,Y,T) = W(x,y)h_o h_o'^{2/3} - \frac{2}{9}X\, h_o^{3/2} h_o''/h_o'$$

$$+\frac{Y^2}{27}\left[\frac{4h_o h_o''^{2}}{h_o'^{2}} + 6h_o'\left(\frac{h_o h_o''}{h_o'^{2}} \right)' + \frac{9}{2}h_o'' \right]$$

$$-\frac{2}{3}F_o h_o^{3/2} + \frac{1}{3}A^2 h_o^2 + \frac{2}{9}Y h_o^{3/2}\left(\frac{A h_o''}{h_o'} + 3A' - 3F_1 \right) \qquad (3.10a)$$

$$y = Y h_o'^{2/3}/h_o \qquad (3.10b)$$

$$x = (X - AY)\frac{h_o'^{1/3}}{h_o^{1/2}} - \frac{1}{3}Y^2 \frac{h_o''}{h_o h_o'^{2/3}} \qquad (3.10c)$$

where $W(x,y)$ satisfies the nonlinear partial differential equation:

$$y\, W_{,xy} - \frac{3}{4}W_{,x} - \frac{1}{6}W_{,xxxx} - \frac{3}{2}(WW_{,x})_{,x} + \frac{1}{2}W_{,xx}\{x - 1 - 2y^2\alpha\} = 0 \qquad (3.11)$$

The solutions one can get by symmetry reduction from eq.(3.11) are expressed in terms of polynomial functions of just one variable; moreover the relation between the wave amplitude η^{o} and the solution W of eq.(3.11), see eq.(3.10a), have explicit dependence from the variable X,Y and T which, in no way, if we want to keep a physically meaningful model of our strait, can be killed. So this reduction can't provide us with wavelike solutions for our system.

2. Symmetry reduction by the subalgebra $\mathcal{L}_2 = \{ \mathfrak{Y}[g(T)] \}$

Whenever k = 0 a more general choice for the vorticity

function Φ_o is compatible with the equations determining the Lie algebra. For the choice

$$\int_{-h_o}^{0} \Phi_o(Y,Z,T)dZ = h_o^{5/2}\left(-\frac{1}{2}h_o^{1/2}A^2+F_o+YF_1 + Y^2F_2\right)$$

with F_i (i=0,2) arbitrary T dependent functions, the vector field $\mathfrak{X}[g]$ can be written as

$$\mathfrak{X}(g) = g\partial_Y - \frac{g'Y}{h_o^{1/2}}\partial_X - \frac{2h_o^{3/2}}{3}\left\{gF_1+Ag'+Y\left[2gF_2+\left(\frac{g'}{h_o^{1/2}}\right)'\right]\right\}\partial_{\eta^o}$$

Invariance under exp[$\lambda\mathfrak{X}$] implies that the solutions of *GKP* have the form:

$$\eta^o(X,Y,T) = W(x,y)-\left(\frac{2h_o^{3/2}}{3}\right)\left\{Y\left(F_1+A\frac{g'}{g}\right)+\frac{1}{2}Y^2\left[2F_2+\frac{1}{g}\left(\frac{g'}{h_o^{1/2}}\right)'\right]\right\} \qquad (3.12)$$

with x and y given by:

$$y = T \qquad (3.13)$$

$$x = X + \frac{Y^2g'}{2h_o^{1/2}g} \qquad (3.14)$$

It is to be noticed that we can always reduce eq.(3.12) to η^o=W(x,y) by choosing

$$F_1= - A\frac{g'}{g} \qquad F_2= - \frac{1}{2g}\left(\frac{g'}{h_o^{1/2}}\right)' \qquad (3.15)$$

and this is important because in such a case bounded solutions of the reduced equation shall go over into bounded solutions of the *GKP*. W(x,y) satisfies the following nonlinear partial differential equation:

$$W_{,xy} + \frac{3}{2h_o^{3/2}}\left[W\,W_{,x}\right]_{,x} + \frac{h_o^{1/2}}{6}W_{,xxxx} + W_{,x}\left(\frac{g'}{2g} + \frac{h_o'}{4h_o}\right) \qquad (3.16)$$
$$+ W_{,xx}F_o - \frac{2h_o^2}{3}\left[F_2 + \frac{1}{2g}\left(\frac{g'}{h_o^{1/2}}\right)'\right] = 0$$

were now h_o and g are functions of y. Choosing $F_2(y)$ according to (3.15) and setting

$$g = h_o^{-9/2} \qquad (3.15b)$$

eq.(3.16), written down in terms of the field variable

$$Q(x,y) = -(3/2)h_o^{-2}W(x,y), \qquad (3.12a)$$

reduces just to a pure *KdV* equation,

$$Q_{,y} + F_oQ_{,x} + \frac{h_o^{1/2}}{6}\left[Q_{,xxx} - 6QQ_{,x}\right]=0 \qquad (3.17)$$

whose solutions are travelling waves in the variables x and y (and thus true three dimensional *solitons*). These solutions are the same as those found in ref.(18).

Let us consider also the case of steep side boundaries. To do so we rewrite the boundaries conditions (2.15) in terms of the symmetry variables (3.13,3.14). In such a way we get:

$$\frac{W_{,x}}{h_o^{1/2}}\left[A\,h_o^{1/2} + \frac{g'}{g}\,Y - \ell'_{o\pm}\right]\Bigg|_{Y=\ell_{o\pm}} = 0 \qquad (3.18)$$

By setting, in the whole generality, $\ell_{o\pm} = \frac{1}{2}(\Sigma \pm \Delta)$ the boundary conditions (3.18) are solved by requiring that $g = \Delta$ and $A=\frac{1}{2}h_o^{-1/2}(\Sigma' - \frac{\Delta'}{\Delta}\Sigma)$ and the nonlinear evolution equation (3.16) for the field $Q(x,y)$ becomes:

$$Q_{,y} + \frac{h_o^{1/2}}{6}\left[Q_{,xxx} - 6\,Q\,Q_{,x}\right] + F_o\,Q_{,x} = -\,Q\left(\frac{9h_o'}{4h_o} + \frac{\Delta'}{2\Delta}\right) \qquad (3.19)$$

This is a perturbed *KdV* equation and can be treated following the procedure introduced, for example, by Newell and Kaup[24]. Let us just notice here that, as $\mathbb{A}$ depends on h_1, the solution we get is valid for a channel with meandering sides only if the bottom is not symmetric with respect to the x-axis.

3. Symmetry reduction by the subalgebra $\mathcal{L}_3 = \{ \mathfrak{X}[m(T)] \}$

Invariance under $\exp[\lambda\mathfrak{X}]$ implies that the solutions of *GKP* have the form:

$$\eta^o(X,Y,T) = W(x,y) + \frac{2h_o^{3/2}m'}{3m} X \tag{3.20}$$

with x and y defined as:

$$x = Y \tag{3.21}$$

$$y = T \tag{3.22}$$

In this symmetry variables the *GKP* equation becomes:

$$W_{,xx} + \frac{4h_o m''}{3m} + \frac{7h_o' m'}{3m} = 0 \tag{3.23}$$

i.e. a linear ordinary differential equation for W, whose solution is a polynomial expression of at most second degree in X and Y and thus of no physical interest.

VI Soliton solutions.

In this Section we present the one soliton solutions of the *GKP* equation that we can construct starting from the results presented in the previous Section, i.e. from the one soliton solution of eqs.(3.17,3.19). Following ref.(18) the zeroth order approximation of the wave amplitude in the channel, given by the solution of the *GKP*, is obtained by considering the appropriate point transformation which reduced the *GKP* to a *KdV* equation, i.e. eq.(3.12 - 3.14).

Let us first consider the case of the channel without steep

side boundaries. This corresponds to the reduction of *GKP* to eq.(3.17). In this case (already considered in ref.(18)) a soliton solution of *GKP*, taking into account eq.(3.12a), with F_1, F_2 and g given by eq.(3.15), $F_o = 0$, reads:

$$\eta^o(X,Y,T) = \frac{4\nu^2}{3} h_o^2(T)\ \mathrm{sech}^2\{\nu[\varkappa - \tfrac{2}{3}\chi(y) - \varkappa_o]\} \tag{4.1}$$

with $\varkappa$ given by eq.(3.14), $\varkappa_o$ an initial arbitrary position

$$\chi(y) = \nu^2 \int_o^y ds\ h_o^{1/2}(s) \tag{4.2}$$

and ν an arbitrary positive constant. In Fig.2-3 we shall plot the wavecrest corresponding to the soliton solution (4.1) for $h_1 = A = \varkappa_o = 0$ for a function $h_o(y)$ describing the physical situation of a strait opening out into an ocean. We consider a quadratically increasing depth:

$$h_o(y) = 1 + a^2 y^2 \tag{4.3}$$

with a constant; the function h_o is chosen in such a way that it is smoothly connected to a flat bottom.

Figs.2 and 3 illustrate the evolution in X of a wavecrest which is a strait line at X=0. As X increases the wavecrest begins to get curved and takes a bell shaped form; for longer time scale (see Fig. 3) the wavecrest gets pinched and gradually evolves into a horseshoe curve. For a more detailed analysis of this solution see ref.(18); let us mention just here that, while the wavecrest acquires the proper shape as one observes from space, the amplitude of this wave is increasing as the depth increases.

Let us now consider the case in which steep boundaries are present; in this case the soliton solution of the *GKP* is given by a solution of a perturbed *KdV* (3.19).

A solution is obtained by assuming that the perturbation, the r.h.s. of eq.(3.19) is small. In such a case the soliton solution can be obtained by carrying out a standard perturbation scheme[24] which takes into account the fact that *KdV* is a completely integrable equation and consequently the wave functions of the spectral problem associated to it span the whole space and thus form a complete set in such a way that any solution can be

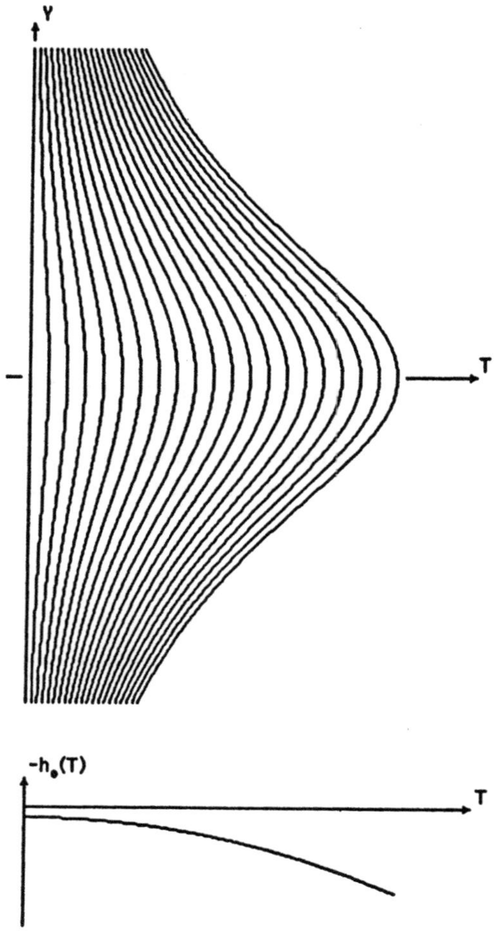

Fig 2. *Solution of the GKP equation for the depth function $h_o(T) = 1 + (T/20)^2$. The curves represent the "time" evolution (see eq. 2.12) of the shape of a wavecrest from X = 0 to X = 0.1. T ∈ [0,0.15] and Y ∈ [-14.5,14.5].*

---

expressed in term of them. Thus the soliton solution is always given by eq.(4.1) but where now ν is a $\nu(y)$ and takes the form:

$$\nu(y) = \frac{\nu_o \Delta_o^{1/3}}{h_o^{3/2}(y)\Delta^{1/3}(y)} \qquad \nu_o = \nu(0) \tag{4.4a}$$

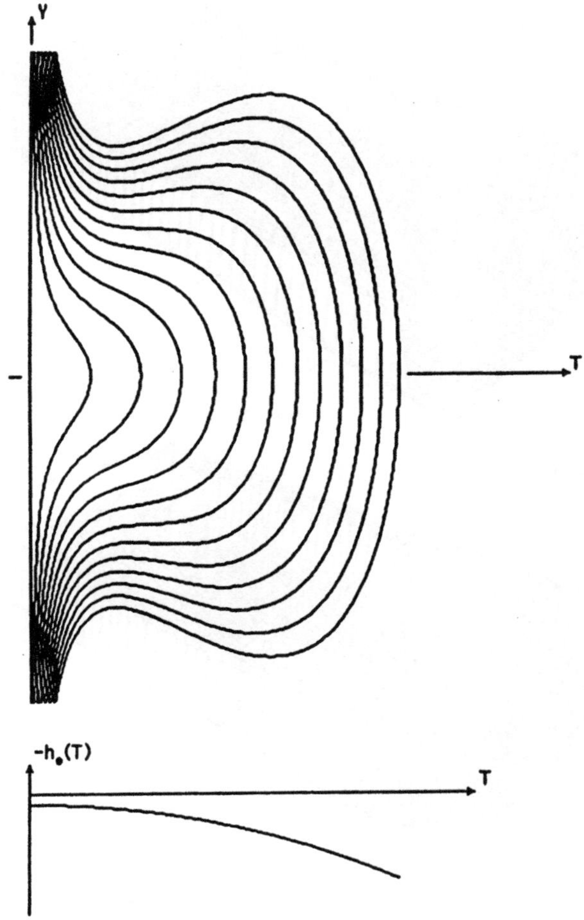

Fig.3. Solution of the GKP equation for the same depth function as Fig. 2. Here the curves represent the "time" evolution (see eq.2.12) of the shape of the wavecrest from X = 0 to X = 120. T ∈ [0,86] and Y ∈ [-54.5,54.5].

---

and

$$\chi(y) = \nu_o^2 \Delta_o^{2/3} \int_o^y \Delta^{-2/3}(s) h_o^{-5/2}(s) ds \qquad (4.4b)$$

In Figs 4-5 we present the three dimensional plotting of the soliton solution at 5 fixed physical times with the steep sides

(4a)

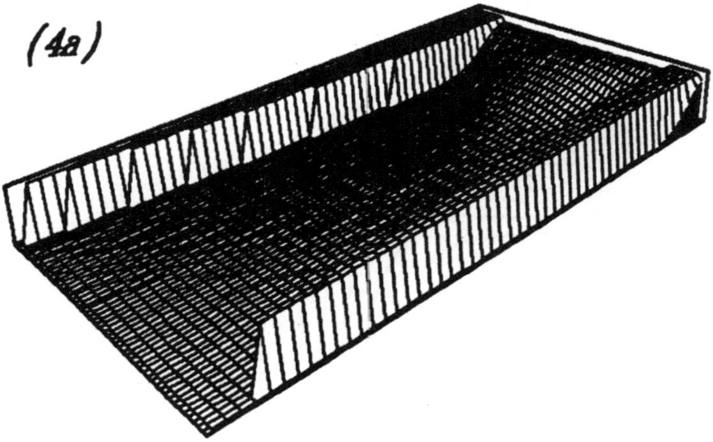

(4b)

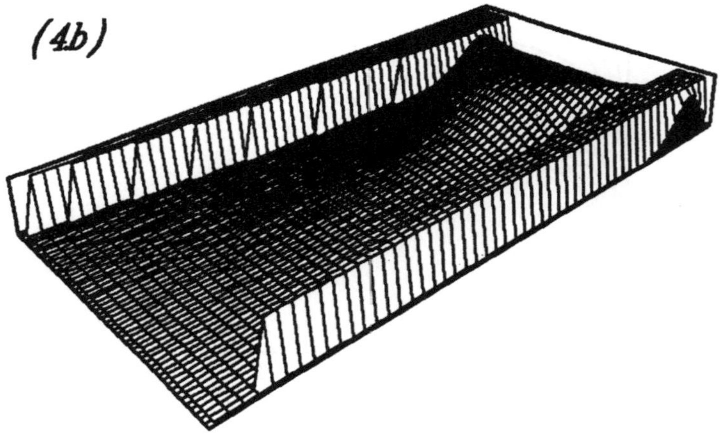

(4c)

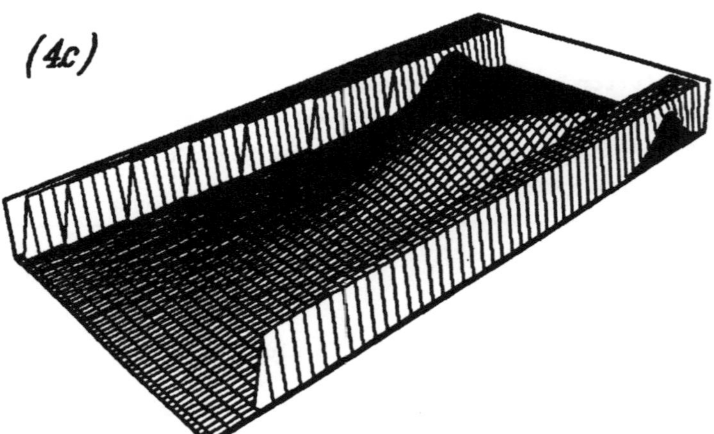

(4d)

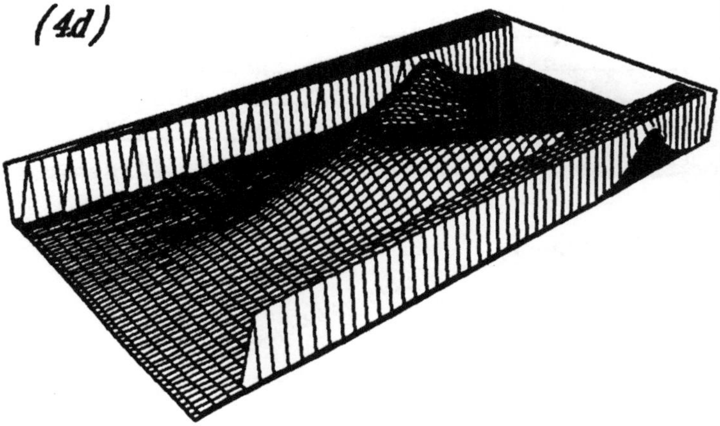

(4e)

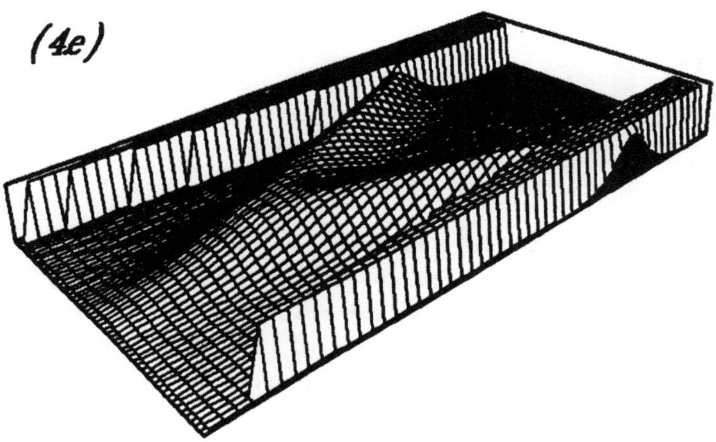

Fig.4. *The soliton solution (4.1,4.4) of the GKP equation with steep boundaries for $\nu_o=1$ is here plotted for 5 different values of the "real" time variable for a channel of quadratically increasing bottom $h_o=1+(T/20)^2$ and enlarging steep sides given by $\Delta=1+T^2$. (4a) represent the solution at $t=0.1$, (4.b) at $t=1.2$, (4.c) at $t=2.1$, (4.d) at $t=3.3$ and (4.e) at $t=4.2$.*

---

described by the functions $\Sigma = 0$ and $\Delta = \Delta_o + \Delta_1 T^2$, i.e. we consider a strait channel with sides quadratically enlarging. In Fig.4 we clearly see the formation of the bell shape of the waves which, however, do not acquire any pinching effect. Moreover the water sticks to the sides of the channel. The bottom topography is the same as in the previous case considered but for the side

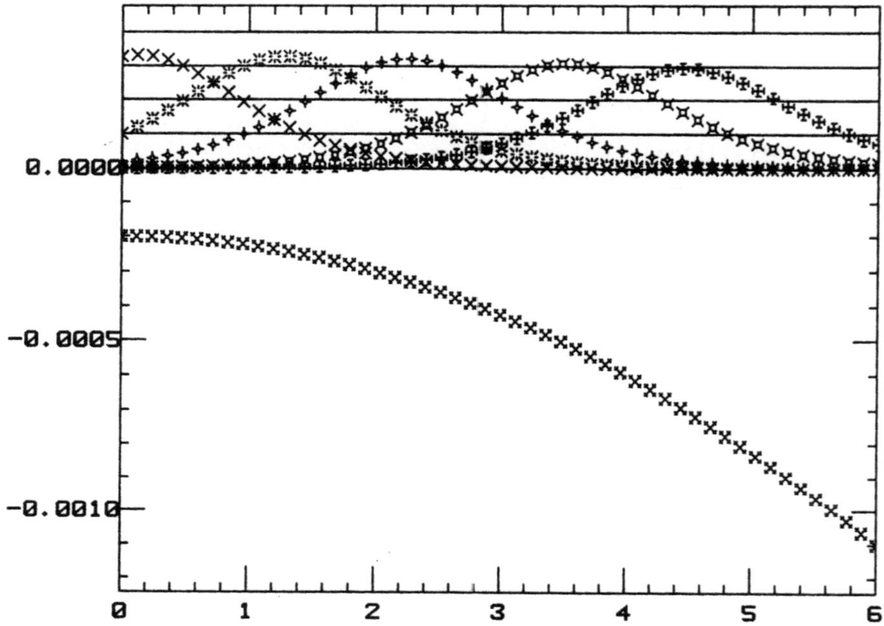

Fig.5. *The bottom topography represented by a cross with hash marks is plotted starting from z=-0.0002 for x varying in the interval [0,6]; the various soliton solutions of eqs.(4.1,4.5) are plotted on the surface z =0 for y =0 at the different times corresponding to the surfaces of Fig.4 with the amplitude reduced 4000 times to make them compatible with the bottom plotting. The curve represented by a cross gives the soliton solution at t=0.1, the star at t=1.2, the plus with a hole at t=2.1, the cross with a hole at t=3.3 and the plus with a hash marks at t=4.2.*

---

boundary we choose $\Delta_o = \Delta_1 = 1$ to increase the effect of the rounding of the wavecrests. Let us notice, moreover, that, at difference with the previous results, the amplitude of the soliton solution is a decreasing function of x, see Fig.5 where the soliton solution is depicted as a function of x for the 5 different values of t considered in Fig.4 at y = 0.

In this case, apart from the adiabatic change of the soliton velocity and amplitude we have an extra contribution coming from the continuum spectrum. This contribution gives rise to a shelf of depression between the rear of the soliton and the point to which the infinitesimal wave would have propagated. This contribution is necessary to satisfy the conservation of the solution's mass and, clearly, it is such that the total motion is not adiabatic.

REFERENCES

1. D.J. Korteweg and G. de Vries, On the change of form of long waves advancing in a rectangular canal, and on a new type of long stationary waves, Phil. Mag. 39(1895)422-443

2. F. Calogero and A. Degasperis, *Spectral Transform and Solitons*, Vol.1 North Holland, Amsterdam 1982

3. M.J. Ablowitz and H. Segur, *Solitons and the Inverse Scattering Transform*, SIAM, Philadelphia 1981

4. B.B. Kadomtsev and V.I. Petviashvili, On the stability of solitary waves in weakly dispersing media, Soviet Phys. Dokl. 15(1970)539-41

5. D. Farmer and J.D. Smith, Nonlinear internal waves in fjord, in *Hydrodynamics of Estuaries and Fjords*, edited by J. Nihoul, Elsevier, New York 1978, p.465-494

6. F.M. Boyce, Internal waves in the Straits of Gibraltar, Deep Sea Res. 22(1975)597-610

7. H. Lacombe and C. Richez, The regime of the Straits of Gibraltar, in *Hydrodynamics of Semi-enclosed Seas*, edited by J. Nihoul, Elsevier, New York 1982, p.13-73

8. V. Artale, D. Levi, E. Salusti and F. Zirilli, On the generation of internal solitary marine waves, Il Nuovo Cimento C7(1984)365-377

9. V. Artale and D. Levi, Nonlinear Internal Waves in a One-dimensional Channel, Il Nuovo Cimento C10(1987)61-76

10. D.R. Cristie, K.J. Muirhead and A.L. Hales, On solitary waves in the atmosphere, J. Atmos. Sci. 35(1978)805-825

11. D. Halpern, Observations on short period internal waves in Massachusetts Bay, J. Mar. Res. 29(1971)116-132

12. J.R. Apel, H.M. Byrne, J.R. Proni and R.L. Charnell, Observations of oceanic internal and surface waves from the Earth Resources Technology Satellite, J. Geophys. Res. 80(1975)865-881

13. L.L. Fu and B. Holt, Internal waves in the Gulf of California: observations from a spaceborne radar, J. Geophys. Res. 89(1984)2053-2060

14. W. Halpers and E. Salusti, Scylla and Charybdis observed from space, J. Geophys. Res. 88(1983)1800-1808

15. P.E. La Violette, The advention of submesoscale thermal features in the Alboran Sea gyres, J. Phys. Oceanogr. 14(1984)550-565

16. J.R. Apel, J.R. Holbrook, A.K. Liu and J.J. Tsai, The Sulu sea internal soliton experiment, J. Phys. Oceanogr. 15(1985)1625-1651

17. D. David, D. Levi and P. Winternitz, Integrable nonlinear equations for water waves in Straits of varying depth and width, Stud. Appl. Math. 76(1987)133-168

18. D. David, D. Levi and P. Winternitz, Soliton in Shallow Seas of variable Depth and in Marine Straits, Stud. Appl. Math. (in press)

19. D. Levi and P. Winternitz, The Cylindrical Kadomtsev-Petviashvili Equation; its Kac-Moody-Virasoro Algebra and relation to the KP equation, Phys. Lett. A129(1988)165-167

20. P.J. Olver, *Application of Lie Groups to Differential Equations*, Springer Verlag, New York 1986

21. B. Champagne and P. Winternitz, A MACSYMA Program for calculating the Symmetry Group of Differential Equations, Preprint CRM-1278, Montreal 1985

22. D. David, N. Kamran, D. Levi and P. Winternitz, Symmetry reduction for the Kadomtsev-Petviashvili equation using a loop algebra, J. Math. Phys. 27(1986)1225-1237

23. P.M. Santini and A.S. Fokas, Recursion Operators and Bi-Hamiltonian Structures in Multidimensions I, Comm. Math. Phys. 115(1988)375-419

24. D.J. Kaup and A.C. Newell, Solitons as particles, oscillators, and in slowly changing media: a singular perturbation theory, Proc. R. Soc. Lond. A361(1978)413-446

THE GEOMETRY OF SOLITON EQUATIONS

F. Magri*, C. Morosi**, .G. Tondo***

*Dipartimento di Matematica, Università di Milano (Italy)
**Dipartimento di Matematica, Università di Perugia (Italy)
***Dottorato in Matematica, Università di Milano (Italy)

ABSTRACT

The ideas emerging from the geometrical study of the soliton equations are used to give a sound explanation of some algorithmic procedures quite popular in the field of the inverse scattering technique. A detailed study of a particular example (the so-called Veselov-Novikov systems) aims to show the feasibility of the approach following from the theory here developed.

1. INTRODUCTION

One of the main issues of the classical theory of integrable Hamiltonian systems has been the complete understanding of the geometry of the phase space where these systems are defined. Many classical results of Liouville and Jacobi have been reformulated, from a global standpoint, in terms of Lagrangian fibrations. The existence of action-angle variables has been shown to be closely related to the topology of such fibrations. As a natural consequence of this change of point of view, the study of integrable systems has progressively shifted from dynamics into differential geometry. The study of particular types of vector fields has been replaced by the study of particular classes of manifolds, such as symplectic manifolds, G-Hamiltonian manifolds, Poisson manifolds and so on.

This paper fits in this line of development. It suggests a new model of phase space, called Poisson-Nijenhuis manifold, emerging from the study of a large class of integrable systems with an infinite-number of degrees of freedom, the so-called equations solvable by the Inverse Scattering Transform[1]. The model is characterized by the introduction of a suitable pair of tensor fields on the phase space. Roughly speaking, the first one (a Poisson tensor field) allows us to introduce the concept of Hamiltonian vector field, the second one (a Nijenhuis tensor field) allows to perform its integration. In this paper we do not enter any detailed study of the geometry of these manifolds (given elsewhere[2]). Our aim is to use the geometrical insight, which it provides, to give a sound theoretical explanation of certain algorithmic computations, often used in the IST theory without a sufficient motivation. The unavoidable drawback of this procedure is that many results have to be checked at the end of the computation (in general, in a very cumbersome way). In our opinion, the geometrical point of view clarifies the computational algorithms, allows one to complete some points of the theory, otherwise not fully understood, provides a more unified point of view and makes the final controls useless.

The paper is organized as follows. The first three sections are devoted to the definition of the manifolds we are interested in: starting from the classical concept of Poisson manifold, we first introduce the idea of Posson-Nijenhuis manifolds and then that of $\mathcal{G}$N manifold, a special class of PN manifolds with a Lie group acting on it in a peculiar way. These last manifolds are the natural framework for the study of the equations solvable by the IST. A hierarchy of $\mathcal{G}$N manifolds, modelled on an associative algebra with unit and general enough to encompass the main examples known from the literature, is constructed in Sect. 5. The next section concerns the main technique to deal with these manifolds, i.e. the reduction technique; one clearly sees that many algorithms of IST are realizations of the projection technique naturally suggested by the geometry of the $\mathcal{G}$N-manifold. This is quite concretely pointed out by the example of Veselov-Novikov systems analyzed in Sect. 7.At last, Sects. 8-9 deal with the construction of the symmetry algebras and their reductions, and Sect. 10 shows how different approaches which have been considered in the literature are simply different realizations (over different associative algebras) of the unique abstract scheme previously defined.

2. POISSON MANIFOLDS

A *Poisson manifold* is usually defined [3)] as a differentiable manifold M endowed with a Poisson bracket, i.e. a composition law for scalar-valued functions $\{,\} : C^\infty(M) \times C^\infty(M) \to C^\infty(M)$ such that :

($\mathbb{R}$-linearity)	$\{c\,f, g\} = c\,\{f,g\}$	$c \in \mathbb{R}$	(2.1)
(skew-symmetry)	$\{f,g\} = -\{g,f\}$		(2.2)
(Jacobi)	$\{\{f,g\},h\} + \{\{g,h\},f\} + \{\{h,f\},g\} = 0$		(2.3)
(Leibnitz)	$\{f\cdot g,h\} = f\{g,h\} + g\{f,h\}$		(2.4)

These brackets endow $C^\infty(M)$ with the structure of Lie algebra, and relate functions to vector fields. This correspondence, which is defined by

$$\varphi_h(f) = \{h,f\}\,, \tag{2.5}$$

is a Lie algebraic homomorphism:

$$[\varphi_f,\varphi_g] = \varphi_{\{f,g\}} \tag{2.6}$$

The vector field φ_h is the *Hamiltonian vector field* associated with the function h.

A different definition of Poisson manifold is obtained by computing, in any local coordinate chart of M, the Poisson brackets of the coordinate function u^k

$$P^{ik}(u^1,u^2,\ldots,u^n):=\{u^i,u^k\} \qquad (2.7)$$

One easily verifies that they behave as the components of a contravariant tensor field of type (2,0)

$$P = P^{jk}(u^1,\ldots u^n)\,\frac{\partial}{\partial u^j}\otimes\frac{\partial}{\partial u^k} \qquad (2.8)$$

This tensor field is skewsymmetric and has a vanishing Schouten torsion[4)] on account of conditions (2.2) and (2.3). This means that P fulfils the conditions

$$\langle\alpha,\ P\beta\rangle = -\ \langle\beta,\ P\alpha\rangle \qquad (2.9)$$

$$\langle L_{P_\alpha}(\beta),\ P\gamma\rangle + \ldots + \ldots = 0 \qquad (2.10)$$

for every triple of one-forms α,β,γ on M. The symbol L_{P_α} denotes the Lie derivative along the vector field

$$P\alpha = P^{kj}\,\alpha_k\,\frac{\partial}{\partial u^j} \qquad (2.11)$$

associated by P with the one-form $\alpha = \alpha_k(u^i)du^k$. In a component-wise form, Eq.s (2.9) and (2.10) are written:

$$P^{ik} + P^{ki} = 0 \qquad (2.12)$$

$$\partial_m\, P^{ij} \cdot P^{mk} + \ldots + \ldots = 0 \qquad (2.13)$$

The tensor field P is called the *Poisson tensor* of the manifold M. Conversely, given P the Poisson bracket is obtained according to

$$\{f,g\} := P^{ik}(u^j)\, \partial_i f \cdot \partial_k g \qquad (2.14)$$

So, a Poisson manifold can be defined also as a manifold M endowed with a Poisson tensor P.

3. POISSON-NIJENHUIS MANIFOLDS

We now assume that a *second* tensor field N, of type (1.1), is defined in M, whose Nijenhuis torsion vanishes[5]:

$$[N\varphi, N\psi] - N[N\varphi, \psi] - N[\varphi, N\psi] + N^2 [\varphi, \psi] = 0 \qquad (3.1)$$

We say that M is a *Poisson-Nijenhuis manifold* if P and N fulfil the compatibility conditions

$$N \cdot P = P \cdot N^* \qquad (3.2)$$

$$N \cdot L_\varphi(P) \cdot N^* = L_{N\varphi}(PN^*) \qquad (3.3)$$

They mean that the tensor field

$$Q := N \cdot P = N_i^j\, P^{ik} \frac{\partial}{\partial u^j} \otimes \frac{\partial}{\partial u^k} \qquad (3.4)$$

is a second Poisson tensor in M. For this reason, Poisson-Nijenhuis manifolds are often called bihamiltonian manifolds[2]. By iteration, one can also show that all the iterated tensors $Q^j = N^j \cdot P$, and all the linear combinations $\sum_{j \in \mathbb{N}} \lambda_j\, Q^j$ of these tensors, are Poisson tensors (a highly non trivial result, since the condition (2.10) for Poisson tensors is quadratic in P). The tensors $\sum_{j\in\mathbb{N}} \lambda_j Q^j$ form the linear hull of Poisson tensors spanned by N and P.

Among several interesting properties enjoied by PN-manifolds, the following three are noteworthy, being closely related with the theory of integrable Hamiltonian systems. The first property concerns the traces of the powers of N

$$I_k(m) := \frac{1}{k} \operatorname{Tr}(N_m)^k \tag{3.5}$$

and it is important for the connection with the theory of finite-dimensional integrable systems. It claims that the previous traces commute in pairs

$$\{I_k, I_j\} = 0 \tag{3.6}$$

The second property concerns the Lie algebra of vector fields *(symmetries)* leaving N and P invariant

$$L_\varphi(N) = 0 \qquad L_\varphi(P) = 0 \tag{3.7}$$

It is useful in connection with the theory of infinite-dimensional Hamiltonian systems. According to it, the iterated vector fields φ_a^j constructed from any algebra of symmetries $\{\varphi_a, a \in \mathcal{G}\}$

$$[\varphi_a, \varphi_b] = \varphi_{[a,b]} \tag{3.8}$$

according to

$$\varphi_a^{j+1} = N\, \varphi_a^j \qquad \varphi_a^o = \varphi_a \tag{3.9}$$

are such that

$$L_{\varphi_a^j}(N) = 0 \qquad L_{\varphi_a^j}(P) = 0 \tag{3.10}$$

$$[\varphi_a^j , \varphi_b^k] = \varphi^{j+k}_{[a,b]} \tag{3.11}$$

So they form a Lie *algebra of Kac-Moody type*.[6] In particular, the vector fields (3.9) commute in pairs. They form the *Lenard complex* engendered by a.

The last property concerns the Lie algebra of vector fields (*master symmetries*) such that

$$L_\tau(N) = \lambda N \qquad \lambda \in \mathbb{R} \tag{3.12}$$

Since the iterated fields $\tau^j := N^j\tau$ fulfil the relations

$$L_{\tau^i}(N) = \lambda N^{i+1} \tag{3.13}$$

$$[\tau^i , \tau^h] = \lambda(h-i)\, \tau^{i+h} \tag{3.14}$$

they make a *Virasoro algebra*[6] . The role of these symmetries in the theory of integrable systems has been emphasized by Oevel, Fuchssteiner and Dorfman [7] . A simple example is given, in $M=S^1$ equipped with the Nijenhuis tensor $N=e^{i\theta}\frac{\partial}{\partial\theta}\otimes d\theta$, by the vector field $\tau=d/d\theta$.

4. GN-MANIFOLDS AND LENARD BICOMPLEXES

We further specialize the PN manifold, by assuming that a Lie group G acts on M and that the action ϕ of G on M is Poissonian and leaves N invariant. The infinitesimal generators φ_a, $a \in \mathcal{G}$, of Φ are consequently symmetries of the PN manifold. Furthermore, we assume that the group G is itself a Nijenhuis manifold, endowed with a left-invariant Nijenhuis tensor $\Delta \in \mathcal{X}^1_1(M)$. We denote by $\Delta_e: \mathcal{G} \to \mathcal{G}$ the value of

Δ at the identity e of G: it is a linear mapping of $\mathcal{G}$ into itself fulfilling the condition

$$[\Delta_e a, \Delta_e b] - \Delta_e[a, \Delta_e b] - \Delta_e[\Delta_e a, b] + \Delta_e^2[a,b] = 0 \qquad (4.1)$$

The existence of two Nijenhuis tensor fields, one on the group G and the other one on the manifold M, allows us to set up an interesting double recursion scheme as follows. Let a_o be any element of $\mathcal{G}$ whose adjoint action commute with Δ_e:

$$\Delta_e \, ad_{a_o} = ad_{a_o} \Delta_e \qquad (4.2)$$

By means of Δ_e, we set up a first recursion scheme in $\mathcal{G}$:

$$a^i = \Delta_e{}^i(a_o) \qquad (4.3)$$

Next, we consider the infinitesimal generators

$$\varphi^i = \varphi_{a^i} \qquad (4.4)$$

associated with the elements a^i, and we set up a second recursion scheme in M by means of N:

$$\varphi^{kj} := N^k \varphi^j \, . \qquad (4.5)$$

We call this double sequence of vector fields of M the *Lenard bicomplex* engendered by $a_o \in \mathcal{G}$ 8). The main statement is that the vector fields φ^{kj} commute in pairs:

$$[\varphi^{kj}, \varphi^{im}] = 0 \qquad (4.6)$$

By suitably choosing the manifold M and the algebra $\mathcal{G}$ (according to the prescriptions given in the next section), one is able to recover in this way the main classes of nonlinear equations solvable by the IST

5. A HIERARCHY OF GN-MANIFOLDS

In this section we exhibit the main examples of $\mathcal{G}$N-manifold we are dealing with. They are affine hyperplanes in an associative algebra with unit. They are constructed as follows. Let A be an associative algebra with unit, and $D:A\to A$ a derivation. We assume that a subalgebra V exists in A such that:

i) V is stable with respect to D: $D(V)\subset V$

ii) the restriction of D to V is invertible

iii) a trace form $Tr: V\to \mathbb{R}$ exists in V allowing to identify V with its dual V^*

iv) D is skew-symmetric with respect to the trace form: $Tr\,\varphi\cdot D\psi = -Tr\,\psi\cdot D\varphi$

We denote by K the Kernel of D, and by K_V the subalgebra of elements of K leaving V invariant with respect to both left and right-multiplication

$$K_V = \{b\in K : b(V)\subset V, \quad (V)b\subset V\} \tag{5.1}$$

We write $\langle\xi,\varphi\rangle = Tr\,\xi\cdot\varphi$ to denote the value of the covector $\xi\in V^*\simeq V$ on the covector $\varphi\in V$. We call the elements of K constants, and the elements of V vectors. Our model of $\mathcal{G}$N-manifold $(M,P,Q,\mathcal{G},\varphi_a,\Delta_e,a_o)$ is defined as follows:

1. *the manifold M* is the affine hyperplane

$$M = V + \{c\} \qquad c\in K_V \tag{5.2}$$

modelled on V. Its points are denoted by $u=v+c$.

2. *the Poisson tensors P and Q* are given by the linear mapping $M\times V^* \to V$ defined by

$$P_u\xi = D\xi + [\xi,u] \tag{5.3}$$

$$Q_u\xi = [\xi,a] \tag{5.4}$$

where $[\ ,\ ]$ is the commutator in A and $a \in K$ is any constant such that $[a,V] \subset V$.

3. *the symmetry algebra* is the subalgebra $\mathcal{G} \subset K_V$ of the constants commuting with both a and c:

$$\mathcal{G} = \{b \in K_V : [a,b] = 0,\ [c,b] = 0\} \tag{5.5}$$

4. *the infinitesimal action* φ_b, *the Nijenhuis tensor* Δ_e *and the starting symmetry* a_o are defined by

$$\varphi_b(u) = -[u,b] \tag{5.6}$$

$$\Delta_e := R_{\bar{b}} \quad : \text{the right-multiplication by any element } \bar{b} \in \mathcal{G} \tag{5.7}$$

$$a_o = 1 \quad : \text{the unit in A} \tag{5.8}$$

We remark that if the derivation P_u is invertible in V, as well as D, M is endowed with the Nijenhuis tensor $N=Q\cdot P^{-1}$. This is the usual case in the applications.

Straightforward generalizations of the previous model are obtained by considering the algebra of matrices of order n (with entries in A) instead of A. In this case the symbol Tr means that one first compute the trace in the matrix sense and then the trace in V, and the derivation is extended by components . Henceforth, we shall consider the case n=2. As we shall see, an extension of the symmetry algebra is coupled with any extension of the ground manifold M. Failing a general theory, we shall limit ourselves to point out this feature in concrete examples.

6. REDUCTION OF POISSON-NIJENHUIS MANIFOLDS

Let M be a bihamiltonian manifold. Without loss of generality,we assume that the first Poisson tensor P is regular. Indeed, we can always reduce ourselves to this situation by a first reduction to the characteristic manifolds of P. We refer to [2)] for more details on this kind of reduction. Let us assume, on the contrary, that N is not kernel-free. At each point $m \in M$, it defines two sequences of subspaces ImN^i and $KerN^i$, i=1,2,... obeying the inclusion relations

$$ImN^{i+1} \subset ImN^i \quad , \quad KerN^{i+1} \supset KerN^i \tag{6.1}$$

If the dimension of M is finite, it is known that a finite index r exists, called the Riezs index of N at the point m, such that both sequences become stationary [9)]:

$$ImN^{r+1} = ImN^r \quad , \quad KerN^{r+1} = KerN^r \; . \tag{6.2}$$

Moreover, they split the tangent space into the direct sum

$$T_mM = ImN^r \oplus KerN^r \; . \tag{6.3}$$

Then let S be any submanifold of M such that:

(A1) The Riezs index of N is constant on S

(A2) the distribution $Ker(N_m)^r$ is transversal to T_mS on S:

$$T_mM = \mathrm{Ker}(N_m)^r \oplus T_mS.$$

Moreover, let us assume that:

(A3) the index r and the rank of the distribution $\mathrm{Ker}N^r$ are constant also in a tubular neighbourhood U containing S.

Since the distribution $\mathrm{Ker}N^r$ is involutive, the assumption on the constancy of the rank means that the distribution $\mathrm{Ker}N^r$ is integrable on U (in the Frobenius sense). The transversality condition (A2) means that S can be identified with the space of the leaves of the foliation induced by N on U.

Under the previous assumptions, one can prove that:

Proposition 6.1 (reduced PN-manifolds)

i) *the manifold S is a Poisson-Nijenhuis manifold*

ii) *the Poisson tensors $\hat{P}$ and $\hat{Q}$ on S are the unique tensors for which the projection $\pi : U \to S$ induced by the foliation is a Poisson morphism*

iii) *these tensors are regular on S*

proof:

1. to show that the first Poisson tensor P is projectable over S, it suffices to show that the Poisson bracket of functions which are invariant along the leaves of the foliation is itself invariant. This follows from the second compatibility condition (3.3). Indeed, if φ is any vector field tangent to the foliation, and f and g are constant functions along the leaves of the foliation, we have:

$$\begin{aligned} L_\varphi\{f,g\}_P &= L_\varphi\langle df, Pdg\rangle \\ &= \langle d\cancel{L_\varphi(f)}, Pdg\rangle + \langle df, L_\varphi(P)\, dg\rangle + \langle df, Pd\cancel{L_\varphi}(g)\rangle \end{aligned} \tag{6.4}$$

Since df and dg annihilate the distribution $\mathrm{Ker}N^r$, there are one-forms α and β such that

$$df = N^{*r}\cdot\alpha \qquad dg = N^{*r}\cdot\beta \tag{6.5}$$

Then since $N^r\varphi=0$, it follows that

$$\begin{aligned} L_\varphi\{f,g\}_P &= \langle N^{*r}\alpha,\ L_\varphi(P)\cdot N^{*r}\beta\rangle \\ &= \langle\alpha,\ N^r\cdot L_\varphi(P)\cdot N^{*r}\beta\rangle \\ &\overset{(3.3)}{=} \langle\alpha,\ L_{N^r_\varphi}(PN^{*r})\cdot\beta\rangle \\ &= 0 \end{aligned} \tag{6.6}$$

2. The Poisson brackets on S are then defined by

$$\{\hat{f},\hat{g}\}_S\cdot\pi = \{\hat{f}\cdot\pi,\ \hat{g}\cdot\pi\}_M \tag{6.7}$$

where f,g are scalar-valued functions on S, and π: $U\to S$ is the projection induced by the foliation. More explicit expressions are obtained by introducing coordinate charts on M and S. Let u^j, $j=1,2,\ldots,m$, and q^a, $a=1,2,\ldots,s$, be local coordinates on M and S respectively. In these coordinates, let

$$u^j = s^j(q^1\ldots q^s) \tag{6.8}$$

and

$$q^a = \pi^a(u^1\ldots u^m) \tag{6.9}$$

be the local representatives of the submanifolds S and of the projection π, and $P^{jk}(u^1\ldots u^m)$ the components of the Poisson tensor. Then the components P^{ab} of the Poisson tensor $\hat{P}$ on S are given by

$$P^{ab}(\pi(u^1 \dots u^m)) = \{\pi^a, \pi^b\}(u^1 \dots u^m) \\ = \frac{\partial \pi^a}{\partial u^j} P^{jk}(u^1 \dots u^m) \frac{\partial \pi^b}{\partial u^k} \tag{6.10}$$

that is

$$P^{ab}(q^1 .. q^s) = \frac{\partial \pi^a}{\partial u^j}(s(q)) P^{jk}(s(q)) \frac{\partial \pi^b}{\partial u^k}(s(q)) \tag{6.11}$$

Thus, in order to compute the components P^{ab} of the reduced tensor it suffices to know the components P^{jk} of P and the differential $\frac{\partial \pi^a}{\partial u^j}$ of the projection evaluated on S. Shortly, we shall see how this differential can be computed without finding the finite equations (6.9) of the foliation.

3. The reduced tensor over S obviously fulfils the Jacobi's condition. Moreover, it is invertible as a consequence of the algebraic compatibility condition (3.2), entailing that

$$P(\mathrm{Im}N^{*r}) \subset \mathrm{Im}N^r \tag{6.12}$$

So, the result follows from the remark that the adjoint of the differential $d\pi$ is an injective mapping having $\mathrm{Im}N^{*r}$ as its image, and that the differential $d\pi$ is Kernel-free when restricted to $\mathrm{Im}N^r$, due to the splitting conditions (6.3) and (A2).

4. Since both the second tensor Q and the linear combination $P+\lambda Q$ obey the same compatibility conditions (3.2) and (3.3) with N as P, we conclude that they also are projectable on S. Thus $\hat{P}+\lambda\hat{Q}$ is still a

Poisson tensor, showing that S is a bihamiltonian manifold. Moreover $\hat{Q}$ is Kernel-free. Indeed, $\text{Ker}Q=\text{Ker}N^{*} \subset \text{Ker}N^{*r}$. Thus Q restricted to $\text{Im}N^{*r}$ is kernel-free and we can repeat the analysis made for P. ■

Having proved the reduction theorem, we wish to exhibit two algorithms to explicitly compute the reduced tensors without having to compute the projection π: they are dual each other.
According to the first one, we compute the differential $d\pi$ of the projection acting on the tangent space;according to the second one, we first compute the adjoint of this differential acting on the cotangent space.

Proposition 6.2 (first reduction algorithm)

In order to compute the components P^{ab} and Q^{ab} of the reduced tensors $\hat{P}$ and $\hat{Q}$ on the immersed submanifold S given by (6.8), the following steps are required:

i) to evaluate the components P^{jk} and Q^{jk} at the points of S
ii) to solve the iterative system

$$Q^{jk}\,\alpha_k^{(p+1)} = P^{jk}\,\alpha_k^{(p)} \tag{6.13}$$

with the initial condition

$$Q^{jk}\,\alpha_k^{(o)} = 0 \tag{6.14}$$

After r steps, the solution of this system becomes stationary by the assumption (A1): $\alpha^{(r)}=\alpha^{(r-1)}$.
iii) to solve, with respect to the vector φ^{a}, tangent to S, the system

$$\varphi^{j} = P^{jk}\,\alpha_k^{(r)} + \frac{\partial s^{j}}{\partial q^{a}}\,\varphi^{a} \tag{6.15}$$

for any tangent vector φ^j to M in S, (where $\alpha_k^{(r)}$ is the general solution of the previous system). The solution

$$\varphi^a = G^a_j(q^1 \ldots q^s)\varphi^j \tag{6.16}$$

exists and is unique by the assumption (A2). It gives directly the differential of the projection π^a evaluated at the points of S:

$$G^a_j\,(q^1 \ldots q^s) = \frac{\partial \pi^a}{\partial u^j}\,(s(q)). \tag{6.17}$$

The remaining steps of the procedure are now trivial:

iv) we compute the adjoint of the differential of the projection by using the duality relation

$$\beta_a\; G^a_j(\varphi^j) = \varphi^j (G^*)^a_j(\beta_a) \tag{6.18}$$

v) we evaluate the components of the reduced tensors by composing, in the given order, the adjoint of the differential of the projection with the Poisson tensor on M and with the differential of the projection, according to (6.11):

$$P^{ab}(q^1 \ldots q^s) = G^a_j(q^1 \ldots q^s) P^{jk}(s(q))(G^*)^b_k(q^1 \ldots q^s) \tag{6.19}$$

$$Q^{ab}(q^1 \ldots q^s) = G^a_j(q^1 \ldots q^s) Q^{jk}(s(q))(G^*)^b_k(q^1 \ldots q^s) \tag{6.20}$$

The proof of these statements is almost obvious since the solution of the first system defines the distribution KerN^{*r} and $P\alpha^{(r)}$ spans the distribution KerN^r. Now let us consider the second procedure, characterizing directly the adjoint of the differential of the projection evaluated at S.

Proposition 6.3 (second reduction algorithm)

In order to compute the components P^{ab} and Q^{ab} of the reduced tensors $\hat{P}$ and $\hat{Q}$ on the immersed submanifold S:

i) we compute the adjoint of the differential of the local immersion s^j

$$\frac{\partial s^j}{\partial q^a}(q^1 \ldots q^s)\,\alpha_j = \beta_a \tag{6.21}$$

by using the duality relation

$$\alpha_j\,(\frac{\partial s^j}{\partial q^a}\ \varphi^a) = \varphi^a(\frac{\partial s^j}{\partial q^a}\,\alpha_j) \tag{6.22}$$

ii) we write the "constrained eigenvalue-problem"

$$P^{jk}(s(\bar{q}))\alpha_k = \lambda Q^{jk}(s(q))\alpha_k \qquad\qquad \beta_a = \frac{\partial s^j}{\partial q^a}\,\alpha_j \tag{6.23}$$

for every covector β_a in S. By the assumptions (A1) and (A2) this problem splits into two parts: the first part (independent of λ) gives the covector α_j as a function of the covector β_a:

$$\alpha_j = (G^*)^a_j\ (q^1 \ldots q^s)\ \beta_a, \tag{6.24}$$

the other one is an eigenvalue problem for β_a:

$$N^{*b}_{\ a}\ \beta_b = \lambda\beta_a\ . \tag{6.25}$$

The functions $N^{*b}_{\ a}(q^1 \ldots q^s)$ are the components of the adjoint of the reduced Nijenhuis tensor on S; the system (6.24) gives directly the adjoint of the differential of the projection π^a, evaluated at the points of S.

In order to compute the components P^{ab} and Q^{ab} of the reduced

Poisson tensor on S, we have now to proceed as in the last part of the previous algorithm:

iii) we compute the differential of the projection by using the duality relation (6.18).

iv) we determine the components we are looking for by using (6.19) and (6.20).

The proof of this statement follows from the remark that any solution of (6.13) has to belong, necessarily, to $\mathrm{Im}N^{*r}$. The other statements are simple consequences of this property, and can be easily proved by the reader. We shall limit ourselves to show, in a concrete example, how these algorithms work in practice.

7. THE VESELOV-NOVIKOV SYSTEMS

In this section we consider the algebraic $\mathcal{G}$N manifold described in Sect. 4, for the case of 2x2 matrices, with entries in an associative algebra A with unit (concrete realizations of this algebra will be considered in Sect. 10).

As a manifold M we take the affine hyperplane given by

$$M = \{u : u = \begin{bmatrix} v_1 & v_2 \\ v_2 & v_4 \end{bmatrix} + \begin{bmatrix} 0 & 1 \\ 0 & c \end{bmatrix}\} \qquad (7.1)$$

where the one-index notation will be used from now on .

The constant c is to be specified below; its choice is related with the "asymptotic boundary conditions" appearing in the usual formulations of IST theory.

The bihamiltonian structure of M is specified by the tensor fields

$$P_u \xi = D\xi + [\xi, u] \qquad Q_u\xi = [\xi, a] \tag{7.2}$$

where ξ is a covector at the point $u \in M$; its value on the vector φ is given by

$$<\xi, \varphi> = Tr(\xi_1\varphi_1 + \xi_2\varphi_3 + \xi_3\varphi_2 + \xi_4\varphi_4) \tag{7.3}$$

according to the notation of Sect. 5. The constant matrix a appearing in (7.2) is given by

$$a_1 = a_2 = a_3 = 0, \qquad a_4 = 1, \tag{7.4}$$

different systems corresponding to different choices of this matrix. In M we consider the submanifold S given by

$$S = \{u : u = \begin{bmatrix} 0 & 0 \\ q & r \end{bmatrix} + \begin{bmatrix} 0 & 1 \\ 0 & c \end{bmatrix}\} \tag{7.5}$$

As the reader can verify, this submanifold fulfils all the assumption of the reduction theorem 6.1. Indeed, one can show that in the tubular neighbourhood U of S given by

$$U = \{u : u = \begin{bmatrix} 0 & u_2 \\ q & r \end{bmatrix} + \begin{bmatrix} 0 & 1 \\ 0 & c \end{bmatrix}\} \tag{7.6}$$

the index of N is r=1 since $KerN^2 = KerN$, $ImN^2 = ImN$ and $T_mM = KerN \oplus ImN$. Moreover, S is transversal to Ker N.

Now we describe the procedure for computing the reduced tensors on S, according to the algorithms shown in the previous section. Firstly, the entries of the matrix u are taken as coordinates in M and the entries (q,r) as coordinates in S. Then the equations of S are

$$u_1 = 0 \quad u_2 = 1 \quad u_3 = q \quad u_4 = r + c \tag{7.7}$$

and their differential is

$$\varphi_1=0 \quad \varphi_2=0 \quad \varphi_3=\psi_1 \quad \varphi_4=\psi_2. \tag{7.8}$$

By means of the identity

$$Tr(\varphi_1\xi_1+\varphi_2\xi_3+\varphi_3\xi_2+\varphi_4\xi_4)=Tr(\psi_1\beta_1+\psi_2\beta_2) \tag{7.9}$$

one obtains that the adjoint of the differential is given by

$$\beta_1=\xi_2 \qquad \beta_2=\xi_4 \tag{7.10}$$

In the coordinates u_j, the components of P and Q on S are explicitly given by

$$\begin{aligned} \varphi=P\xi \quad : \quad & \varphi_1=D\xi_1-\xi_3+\xi_2 q \\ & \varphi_2=D\xi_2-\xi_4+\xi_1+\xi_2(r+c) \\ & \varphi_3=D\xi_3-q\xi_1-(r+c)\xi_3+\xi_4 q \\ & \varphi_4=L\xi_4-q\xi_2+\xi_3 \end{aligned} \tag{7.11}$$

$$\varphi=Q\xi \quad : \quad \varphi_1=\varphi_4=0 \quad \varphi_2=\xi_2 \qquad \varphi_3= -\xi_3 \tag{7.12}$$

where $L\cdot:=D\cdot-[r+c,\cdot]$.

According to the algorithm of Prop. 6.2:

1. The first step is to find the kernel of Q

$$Q\xi=0 : \quad \xi_2=\xi_3=0 \tag{7.13}$$

then to solve the system $Q\eta=P\xi$, i.e.:

$$0=D\xi_1 \qquad \eta_2=\xi_1-\xi_4 \qquad -\eta_3=\xi_4 q-q\xi_1 \qquad 0=L\xi_4 \tag{7.14}$$

entailing that

$$\xi_1=\xi_4=0 \qquad \eta_2=\eta_3=0 \tag{7.15}$$

So, we see that the iterative system (6.13) becomes stationary at the first iteration, according to the fact that ind(N)=1 over S; in particular, we find that KerN is spanned by the vectors $P\xi$, with ξ given by (7.13).

2. The splitting of the tangent space T_mM into the direct sum of KerN_m and T_mS amounts to solve the system

$$\varphi_1=D\xi_1 \qquad \varphi_2=\xi_1-\xi_4 \qquad \varphi_3=\xi_4 q-q\xi_1+\psi_1 \qquad \varphi_4=L\xi_4+\psi_2 \tag{7.16}$$

so that the projection onto S is given by

$$\psi_1=\varphi_3+\varphi_2 q+\left[q,D^{-1}\varphi_1\right] \qquad \psi_2=\varphi_4+L\,(\varphi_2-D^{-1}\varphi_1) \tag{7.17}$$

3. The remaining steps are now trivial. First, we compute the adjoint of the projection (7.17) by means of the duality relation (7.9): the result is

$$\xi_1=D^{-1}\left[q,\beta_1\right]-D^{-1}L\beta_2 \qquad \xi_2=\beta_1 \qquad \xi_3=q\beta_1-L\beta_2 \qquad \xi_4=\beta_2 \tag{7.18}$$

Then the components P^{ab} and Q^{ab} of the reduced tensors are obtained by composing the relations (7.18), (7.2), (7.17) in the given order. The final result is *first reduced Poisson tensor* $\hat{P}$

$$\psi_1=D(q\beta_1)+(D\beta_1)q-\left[q,D^{-1}\left[q,\beta_1\right]\right]-\left[(r+c)q,\beta_1\right] \tag{7.19}$$
$$+\left[q,D^{-1}L\beta_2\right]-DL\beta_2+(r+c)L\beta_2$$

$$\psi_2=L\,D\beta_1+L(\beta_1(r+c))+LD^{-1}\left[q,\beta_1\right]-L\,(\beta_2+D^{-1}L\beta_2)$$

second reduced Poisson tensor $\hat{Q}$

$$\psi_1=-\left[q,\beta_1\right]+L\,\beta_2 \tag{7.20}$$
$$\psi_2=\;L\beta_1$$

According to the second algorithm of Prop. (6.3), we consider the "constrained eigenvalue problem" (6.23), corresponding to the system

$$D\xi_1+\xi_2q-\xi_3=0 \tag{7.21}$$
$$D\xi_2+\xi_1-\xi_4+\xi_2(r+c)=\lambda\xi_2$$
$$D\xi_3+\xi_4q-q\xi_1-(r+c)\xi_3=-\lambda\xi_3$$
$$L\xi_4+\xi_3-q\xi_2=0$$
$$\xi_2=\beta_1$$
$$\xi_4=\beta_2$$

The λ-independent part of this system can be given the form

$$\xi_1=D^{-1}(\left[q,\beta_1\right]-L\beta_2) \tag{7.22}$$
$$\xi_2=\beta_1$$
$$\xi_3=q\beta_1-L\beta_2$$
$$\xi_4=\beta_2$$

corresponding to the adjoint of the differential of the projection onto S (compare with (7.17)); its λ-dependent part can be given the form

$$\lambda\beta_1 = D\beta_1 + D^{-1}[q,\beta_1] + \beta_1(r+c) - \beta_2 - D^{-1}L\beta_2 \tag{7.23}$$

$$\lambda\beta_2 = L^{-1}D(q\beta_1) + L^{-1}(qD\beta_1) + L^{-1}[q\beta_1,(r+c)] + $$
$$-L^{-1}DL\beta_2 - L^{-1}[q,\beta_2] + L^{-1}((r+c)L\beta_2)$$

so that we directly identify the adjoint of the reduced Nijenhuis tensor. By duality, $\hat{N}$ itself is obtained, the final result being

$$\hat{N} = \left[\begin{array}{c|c} -D\cdot + [q,D^{-1}\cdot] + (r+c)\cdot & D((L^{-1}\cdot)q) + \cdot q \\ \hline -2\cdot + [r+c,D^{-1}\cdot] & D\cdot + \cdot(r+c) + (Dr)L^{-1}\cdot - [q,L^{-1}\cdot] \end{array}\right] \tag{7.24}$$

Since the remaining steps for computing the reductions $\hat{P}$, $\hat{Q}$ of P and Q, corresponding to (7.19) and (7.20), are the same as before, we omit them. One can easily verify that $\hat{N} = \hat{P}\cdot\hat{Q}^{-1}$.

Having obtained the reduced bihamiltonian structure on S, we have yet to construct the Veselov-Novikov systems; to this end, we have still to set up the Lenard bicomplexes: this point requires the reduction of the symmetry algebra, which will be considered in the following section.

8. SYMMETRY ALGEBRAS AND LENARD BICOMPLEXES

Let M be a PN manifold and let $\mathcal{G}$ be a Lie algebra, acting on M as a symmetry algebra. Under the assumptions of Prop. (6.1) each infinitesimal generator φ_a, $a \in \mathcal{G}$, can be projected onto S. Indeed, for any vector field φ tangent to the foliation, we have

$$N^r([\varphi_a,\varphi])=N^r L_{\varphi_a}(\varphi)=L_{\varphi_a}(N^r_\varphi) = 0 \tag{8.1}$$

that is the commutator of the generator φ_a with any vector field spanning the distribution belongs to the distribution. As it is known, this condition is necessary and sufficient to assure the projectability of φ_a [10). To compute the projected symmetry, it suffices to evaluate the vector field φ_a on S, and to take its projection by means of the differential of the mapping π: M→S, restricted to S. In components, we have

$$(\varphi_a)^b(q^1\ldots q^s)=G^b_j\ (q^1\ldots q^s)(\varphi_a)^j(s(q))\ . \tag{8.2}$$

Let us now apply this simple results to the algebraic PN manifold associated with the Veselov-Novikov systems. As it has been explained in Sect. 5, the symmetry algebra in the space M is defined by the vector fields

$$\varphi_b(u)=-[u,b] \tag{8.3}$$

where b is any 2x2 matrix such that ($\bar{c}$ given by (7.1))

$$[b,a]=0, \qquad [b,\bar{c}] = 0,\ b_i(V)\subset V,\ (V)b_i\subset V \tag{8.4}$$

The first condition assures that φ_b leaves P and Q invariant; the second condition means that φ_b is tangent to M. For the particular choice of a and c corresponding to the Veselov-Novikov case, we find

$$\mathcal{G} = \{b\ :\ b=-\lambda\ \mathbb{1},\ \lambda\in K_V,\ [c,\lambda] = 0\ \} \tag{8.5}$$

where $\mathbb{1}$ is the 2x2 unit matrix. The fields (8.3) are consequently given by

$$\varphi_1=0 \quad \varphi_2=0 \quad \varphi_3=[q,\lambda] \quad \varphi_4=[r,\lambda] \tag{8.6}$$

Their projections are,by (7.17):

$$\psi_1 = [q,\lambda] \qquad \psi_2 = [r,\lambda] \tag{8.7}$$

This is the first symmetry algebra of the Veselov-Novikov systems. More extended algebras can be found as follows. Let us disregard, for a moment, the second symmetry condition (8.4), that is the tangency condition to M (this means that we consider the vector fields φ_b as defined in $Mat_2(A)$): the single symmetry condition $[b,a] = 0$ gives

$$b = \begin{bmatrix} -\lambda & 0 \\ 0 & -\lambda+\mu \end{bmatrix} . \tag{8.8}$$

We remark that the generators φ_b can be written formally as Hamiltonian vector fields with respect to P:

$$\varphi_b(u) = P_u\, b \tag{8.9}$$

Among these vector fields we look for those that are (formally) Hamiltonian also with respect to Q. This means that we look for those b such that

$$Q\xi = Pb . \tag{8.9'}$$

Explicitly, we must solve the system:

$$\xi_2 = -\mu \qquad \xi_3 = [\lambda,q] - \mu q \qquad 0 = [\lambda-\mu,\ r+c] \tag{8.10}$$

Due to the arbitrariness of r, this entails $\lambda=\mu$ and $\xi_2=-\mu$, $\xi_3=-q\mu$. The remaining two entries ξ_1 and ξ_4 are still arbitrary, and we *assume* that they are constant. Now we iterate the procedure. We try to specify ξ in such a way that $P\xi$ is Hamiltonian also with respect to Q:

$$Q\eta = P\xi \tag{8.11}$$

Of course, these vector fields are good candidates to be symmetries of the bihamiltonian manifold, being Hamiltonian (at least formally) with respect to both Poisson tensors. In our example we find:

$$\begin{aligned} &0 = [\mu, q] \\ &\eta_2 = \xi_1 - \xi_4 - \mu(r+c) \\ &\eta_3 = (Dq)\mu - \xi_4 q + q\xi_1 - (r+c)q\mu \\ &0 = [\xi_4, r+c] \quad . \end{aligned} \tag{8.12}$$

Its solution is:

$$\xi_4 = 0 \qquad \mu = 0 \qquad \eta_2 = \xi_1 \qquad \eta_3 = q\xi_1 \tag{8.13}$$

It exactly coincides with the solution of the previous system, up to a change of name. Consequently, we have no more possibility of iterating the process. We have found, in A, a three-parameters family of bi-hamiltonian vector fields $P\eta$, specified by a matrix η of the form

$$\eta_1 = \lambda \qquad \eta_2 = \mu \qquad \eta_3 = q\mu \qquad \eta_4 = \lambda - \mu, \tag{8.14}$$

which are good candidates to be symmetries of our PN manifold.
At this point, let us reintroduce the second symmetry condition (8.4), by requiring that the vector fields $P\eta$ be tangent to M. This imposes

the further conditions $\nu=\mu c$ and $[\lambda,c] = [\mu,c] = 0$.
So, finally, we are left with a two-parameters family of vector fields. They are all projectable on S, and the corresponding projection is given by

$$\psi_1= [\lambda,q] -Dq\mu+ [q,D^{-1}[q,\mu]] + [(r+c)q,\mu] \tag{8.15}$$
$$\psi_2= [\lambda,r] -\mu Dr+ [\mu,q] + [r,\mu] (r+c)- [r+c,D^{-1} [\mu,q]]$$

One can check directly that these vector fields are symmetries of the reduced PN structure on S. This property is general. Although we cannot give any formal proof of our statement, we have found that the previous algorithm for finding symmetries holds for all the particular examples we have considered, such as KP and AKNS.

Summing up, we have two distinct families of symmetry generators for the Veselov-Novikov systems. The first one is given by

$$\dot{q}= [q,\lambda] \qquad \dot{r} = [r,\lambda] \tag{8.16}$$

for any constant $\lambda \in K_V$ such that $[\lambda,c] = 0$. The second one is

$$\dot{q}= -Dq\mu+ [q,D^{-1}[q,\mu]] + [(r+c)q,\mu] \tag{8.17}$$
$$\dot{r}= -\mu Dr+ [\mu,q] + [r,\mu] (r+c)- [r+c,D^{-1} [\mu,q]]$$

for $\mu \in K_V$ such that $[\mu,c] =0$. Each of them allows to set up a Lenard bicomplex. It suffices to choose $\lambda_o=\mu_o=1$ as the starting symmetry, and to set $\Delta_e=R_c$. The first vector fields obtained according to the standard double recursion scheme

$$\begin{array}{ccccccccc}
1 & \xrightarrow{X} & \varphi^{00} & \xrightarrow{N} & \varphi^{10} & \xrightarrow{N} & \varphi^{20} & \cdots \\
\Delta_e \downarrow & & & & & & & \\
c & \xrightarrow{X} & \varphi^{01} & \xrightarrow{N} & \varphi^{11} & \cdots & & \\
\Delta_e \downarrow & & & & & & & \\
c^2 & \xrightarrow{X} & \varphi^{02} & \cdots & & & & \\
\vdots & & & & & & &
\end{array} \tag{8.19}$$

(where X denotes the action of the symmetry algebra of S) are:

1° *Lenard bicomplex* ($\varphi_\lambda = X^{(1)}\lambda$) :

$$\varphi^{ko} = (0;0) \tag{8.20}$$

$$\varphi^{01} = ([q,c] \; ; \; [r,c])$$

$$\varphi^{11} = (- [Dq,c] + [q,D^{-1} [q,c]] + (r+c) [q,c] + D((L^{-1} [r,c])q) +$$
$$+ [r,c] q; \; -2 [q,c] + [r+c, D^{-1} [q,c]] + [Dr,c] + [r,c] (r+c) +$$
$$+ (Dr) L^{-1} [r,c] - [q,L^{-1} [r,c]])$$

$$\varphi^{02} = ([q,c^2] \; ; \; [r,c^2])$$

2° *Lenard bicomplex* ($\varphi_\mu = X^{(2)}\mu$)

$$\varphi^{00} = - (Dq; \; Dr) \tag{8.21}$$

$$\varphi^{10} = -(-D^2q+(r+c)Dq + D(L^{-1}(Dr)q)+(Dr)q; \; -2Dq+ [r+c,q] +$$
$$+ D^2r+(Dr)(r+c)+(Dr)L^{-1}(Dr)- [q,L^{-1}Dr])$$

$$\varphi^{01} = -(Dqc- [q,D^{-1} [q,c]] - [(r+c)q,c] \; ; \; [q,c] + [r+c, [c,D^{-1}q]] +$$
$$+ cDr- [r,c] (r+c))$$

In practice, these vector fields contain Veselov-Novikov systems; to obtain exactly the VN hierarchy, a suitable combination of these fields and a final *reduction* of the bicomplexes are required, as it will be explained in the next section.

9. THE REDUCTION OF THE LENARD BICOMPLEX

Let us summarize the main steps of the construction which has been developed so far:

i) firstly, we have chosen an associative algebra A with unit and with a derivation $D:A\to A$

ii) secondly, we have selected a subalgebra V in A, which is stable with respect to D and such that D restricted to V is invertible.

iii) thirdly, we have determined the "stability algebra" of V, i.e. the algebra K_V of the constants b leaving V invariant, with respect to both right and left-multiplication:

$$b(V) \subset V \qquad (V)b \subset V \ . \tag{9.1}$$

At this point, we have selected two arbitrary elements a and c in K_V. We have used c to define the manifold M (as the affine hyperplane $M=V+\{c\}$ modelled on V), and the constant a to define the second Poisson tensor $Q\xi = [\xi, a]$ on M. Next, we have investigated the characteristic distributions ImN^r and KerN^r of the bihamiltonian manifold, and we have performed a reduction to the submanifold S fulfilling the assumptions of Prop. 6.1. Finally, we have reduced also the symmetry algebra and we have obtained the Lenard bicomplex.

The difficult point in this construction is the determination of the "stability algebra" K_V. Indeed, in order to solve conditions (9.1), one is in general compelled to work with infinite-dimensional algebras V (see the next section for explicit examples). This entails that each vector field of the Lenard bicomplex defines a dynamical system in an infinite number of field functions. In order to overcome this difficulty a second reduction is required, concerning this time the Lenard bicomplex. The idea is very simple. If we take any linear combination (with constant coefficients) of the vector fields of the Lenard bicomplex

$$\varphi_{(n)} = \sum_{k,j} c^{n}_{jk} \; \varphi^{jk} \tag{9.2}$$

obviously we do not lose the commutation property of the fields, which is the main property we are interested in. Furthermore, we can use the freedom in the choice of the coefficients to require that the vector field $\varphi_{(n)}$ be tangent to some, suitably chosen, submanifold S' in S.

To give a concrete idea of the meaning of this kind of reduction, let us consider the algebra of differential operators in one space-variable y

$$u = \sum_{k\geq 0} u_k(xy) \frac{\partial^k}{\partial y^k} \tag{9.3}$$

whose coefficients are C^∞ scalar-valued functions. Then, in one of the possible approaches to the Veselov-Novikov systems (see the next section) the manifold M is the manifold of 2x2 matrices whose entries are differential operators of arbitrary order; the submanifold S is the affine hyperplane of 2x2 matrices of the forms

$$u = \left[\begin{array}{c|c} 0 & 1 \\ \hline \sum_{k\geq 0} q_k(x,y) \frac{\partial^k}{\partial y^k} & \sum_{k\geq 0} r_k(x,y) \frac{\partial^k}{\partial y^k} \end{array}\right] \tag{9.4}$$

where u_3 and u_4 are still differential operators of arbitrary order, and finally the submanifold S' is the affine hyperplane

$$u = \left[\begin{array}{c|c} 0 & 1 \\ \hline u(x,y) & v(x,y) + \dfrac{\partial}{\partial y} \end{array}\right] \tag{9.5}$$

where u_3 is a zero-order differential operator and u_4 is a monic first-order differential operator. So, by the first reduction we decrease the number of the field functions from ∞^4 to ∞^2, and by the second reduction to just only two functions. One can ask why we do not perform the two reductions simultaneously. The answer is that they concern different objects. In the first (geometric) reduction we consider the whole $\mathcal{G}$N structure, and we can state that S is, in its turn, a bihamiltonian manifold with a symmetry algebra. In the second (algebraic) reduction, that will be explained in this section, we only consider the Lenard bicomplex, and we cannot claim that S' is bihamiltonian. It only possesses a distinguished hierarchy of commuting vector fields.

We are not able to deal with the reduction of the Lenard bicomplex in full generality, but only to give operative criterions for the Lenard bicomplexes arising in the algebraic $\mathcal{G}$N-manifolds shown in Sect. 5. For the sake of clarity, let us use once again the Veselov-Novikov systems as a guiding example. First, we remark that the manifold S on which these equations are defined is modelled on an associative algebra V with unit, that is the coordinates (q,r) on S take their value in V. Moreover, we remark that the components $N^a_b(q,r)$ of $\hat{N}$, given by (7.24), depend in a quite characteristic way on the coordinates (q,r), the derivation D and its inverse D^{-1}, and on the constant c entering parametrically into M. This constant is the last remnant of the reduction process from M to S leading to $\hat{N}$. Indeed, each component depends polynomially on the coordinates q,r, the derivation D and the derivation ad_c associated with c; moreover, only the diagonal entries contain the extra term R_c

$$N^a_b(q,r)=p^a_b(L_q,R_q,L_r,R_r,D,D^{-1},ad_c)+\delta^a_b\ R_c \tag{9.6}$$

This extra-term is the troubling term in the theory. If it were absent, the Nijenhuis tensor would be reducible on any submanifold $S' \subset S$ modelled on a subalgebra V' of V fulfilling the unique condition

$$[c,V'] \subset V' \tag{9.7}$$

instead of the two conditions (9.1). This would give us a great freedom in the choice of V'. Even if this reduction is impossible, we shall now show that it is possible to extract from the Lenard bicomplex a hierarchy of vector fields which are tangent to S', under the only assumption that the components $X^a(q,r)$ of the generators X of the infinitesimal action of $\mathcal{G}$ over S have the following form

$$X^a(q,r) = X^a(L_q,R_q,L_r,R_r,D,ad_c) \tag{9.8}$$

that is they depend on c only through the derivation ad_c. In this case, one can prove that:

Proposition 9.1 (reduction of the Lenard bicomplexes)

Let S be a $\mathcal{G}$N-manifold modelled on an associative algebra V with unit (that is, the coordinates $(q^1,\dots q^s)$ in S take their values in V). Assume that the components $N^a_b(q^1\dots q^s)$ of the Nijenhuis tensor at the point $(q^1,\dots q^s) \in S$ have the form (9.6), and that the generators of the infinitesimal action of $\mathcal{G}$ over S have the form (9.8). Let V' be any subalgebra of V obeying the unique condition (9.7), and let S' be the submanifold of S obtained by restricting the coordinates $(q^1,\dots q^s)$ to take their values in V' instead than in V. Finally, assume that c belongs to the symmetry algebra and construct the Lenard bicomplex according to the double recursion scheme (8.19). Then, the vector fields

$$\varphi_{(n)} = \sum_{j=0}^{n} (-1)^j \binom{n}{j} \varphi^{n-j,j} \qquad (9.9)$$

(obtained by summing up the vector fields φ^{jk} of the bicomplex (8.19) along the diagonals with binomial coefficients) are tangent to S'. Furthermore, let $q^a(\lambda)$ be the flow associated with the adjoint action of c on V:

$$\frac{dq^a(\lambda)}{d\lambda} = \left[q^a(\lambda), c\right] \qquad q^a(0) = q^a \quad (q^1 = q, q^2 = r) \qquad (9.10)$$

and denote by $N^a_b(\lambda)$ and $\varphi^a(\lambda)$ the representatives of N and φ in the "moving coordinates" $q^a(\lambda)$. These representatives are simply obtained by replacing R_q by $R_{q(\lambda)}$ in (9.6) and (9.8) respectively:

$$N^a_b(\lambda) = p^a_b(L_{q^a}, R_{q^a(\lambda)}, D, ad_c) + R_c \ \delta^a_b \qquad (9.11)$$

$$X^a(\lambda) = X^a(L_{q^a}, R_{q^a(\lambda)}, D, ad_c) \qquad (9.12)$$

Then, the vector fields $\varphi_{(n)}$ can also be written as

$$\varphi_{(n)} = (N(\lambda) - R_c + \frac{d}{d\lambda})^n \cdot X(\lambda) \cdot 1 \Big|_{\lambda=0} \quad . \qquad (9.13)$$

The two formulations (9.9),(9.13) will be referred to as the "Schrödinger picture" and the "Heisemberg picture" of the hierarchy extracted from the Lenard bicomplex.

We omit the proof of this Proposition since it is exactly the same of that one given in [8)] (it suffices to remark that $R_c \ \delta^a_b$ gives a representation in V^2 of R_c).

Applications will be shown in the next section. We only add a final remark concerning the symmetry generators. As it has been previously explained, there are two symmetry generators for Veselov-Novikov systems. The first generator $X^{(1)}$ is given by (8.16), the second generator $X^{(2)}$ by (8.17-18). Since $X^{(1)}$ fulfils the condition (9. 8), it allows us to perform a reduction of the Lenard bicomplex; the second one does not fulfil this condition. However, we can deal with symmetry generators as with the vector fields of the Lenard bicomplex. In particular, we can introduce the *third symmetry generator* given by

$$X^{(3)}{}_{\mu} := N\, X^{(1)}{}_{\mu} - X^{(2)}{}_{\mu} \tag{9.14}$$

which is obtained from $X^{(1)}$ and $X^{(2)}$ just according to the rule (9.9). It is easily seen that this new generator fulfils condition (9.8), and thus it allows us to reduce a second Lenard bicomplex. As a matter of fact, this property seems to be quite general, since in all the examples where it is possible to construct extended symmetry generators according to the procedure of Sec.8, and where the first generator (8.16) (the "canonical" one) fulfils condition (9.8), it turns out that it is possible to compose the canonical generator with the remaining ones according to the rule (9.9), so to make them also to fulfil condition (9.8). We cannot, however, prove this statement, failing a formal theory of the extended symmetries.

We end this section by writing the explicit form of the generator $X^{(3)}$ defined by (9.14):

$$X^{(3)}{}_{\mu} = (\mu\, Dq + D(L^{-1}[r,\mu]\cdot q);\ [\mu,q] + (Dr)\mu + (Dr)L^{-1}[r,\mu] - \\ - [q,\, L^{-1}[r,\mu]\,]) \tag{9.15}$$

The first vector fields corresponding to the double recursion scheme (8.19) described at the end of Sect. 8 are :

3° Lenard bicomplex ($\varphi = X^{(3)}{}_{\mu}$)

$$\varphi^{00} = (Dq;\ Dr) \tag{9.16}$$

$$\varphi^{01} = (cDq+D(L^{-1}[r,c]\cdot q);\ [c,q]+(Dr)c+Dr\cdot L^{-1}[r,c]-[q,L^{-1}[r,c]])$$

$$\varphi^{10} = (-D^2q+(r+c)Dq+D(L^{-1}(Dr)\cdot q)+(Dr)q;\ -2Dq+[r+c,q]+D^2r+$$

$$+(Dr)\cdot(r+c)+(Dr)L^{-1}(Dr)-[q,L^{-1}Dr])$$

At this point, we are able to write down the explicit form for the abstract VN equations for both generators $X^{(1)}$ (VN I bicomplex) and $X^{(3)}$ (VN II bicomplex). The first vector fields are:

VNI bicomplex ($X=X^{(1)}$) From the vector fields (8.20) one obtains:

$$\varphi_{(0)} = \varphi^{00} = (0;0)$$

$$\varphi_{(1)} = \varphi^{10}-\varphi^{01} = ([c,q];\ [c,r])$$

$$\varphi_{(2)} = \varphi^{20}-2\varphi^{11}+\varphi^{02} = (-2[c,Dq]+2[q,D^{-1}[c,q]]+2[c,rq]+$$

$$+2D((L^{-1}[c,r])q)+[c,[c,q]];\ -4[c,q]+2[r,D^{-1}[c,q]]+$$

$$+2D^{-1}[c,[c,q]]+2[c,Dr]+2[c,r]r+2(Dr)L^{-1}[c,r]-$$

$$-2[q,L^{-1}[c,r]]-[c,[c,q]])$$

VNII bicomplex ($X=X^{(3)}$) From the vector fields (9.16) one obtains:

$$\varphi_{(0)} = \varphi^{00} = (Dq;Dr)$$

$$\varphi_{(1)} = \varphi^{10}-\varphi^{01} = (-D^2q+2D(rq)+2D(L^{-1}[c,r]q);$$

$$-2Dq+D^2r+2[r,q]+2(Dr)r+2(Dr)L^{-1}[c,r]-$$

$$-2[q,L^{-1}[c,r]])$$

The computation of equations of "higher order" can be made straightforwardly by means of the combination (9.9).

10. VN SYSTEMS

In this section we consider a few realizations of the *unique* abstract scheme previously considered on the algebras of polynomials, differential operators and integral operators with distributional kernels.

Example 1 Let $\mathscr{A}[z]$ be the algebra of polynomials in $z \in \mathbb{C}$, with coefficients in an associative algebra $\mathscr{A}$ with unit (e.g., $\mathscr{A} = \mathbb{C}$ or $\mathscr{A} = \mathrm{Mat}_m(\mathbb{C})$). In this example, the following choices are made:

i) A is the current algebra[11)] of C^∞ functions on $\mathbb{R}$ taking values in $\mathscr{A}[z]$

ii) $D = \partial/\partial x$ is the usual partial derivative with respect to x, so that K is given by the polynomials with constant coefficients. In particular, we set $c=z$.

iii) V is the subalgebra of polynomial functions rapidly vanishing for $|x| \to \infty$. It fulfils the conditions of Sect. 5 with respect to D and $c=z$.

iv) S is the affine hyperplane with equations

$$(q(x)\ ;\ r(x)) = \sum_{k\geq 0} (q_k(x);\ r_k(x))\ z^k + (0;\ z) \tag{10.1}$$

and the reduction submanifold S' is defined by $q_k = u\delta_{ko}$, $r_k = v\delta_{ko}$. Since c commutes with z, the bicomplex VNI is clearly trivial, whereas from the second bicomplex VNII one obtains ($u_x = Du, \ldots$)

$$\begin{aligned} \varphi_{(0)} &= (u_x;\ v_x) \\ \varphi_{(1)} &= (-u_{xx} + 2(vu)_x;\ v_{xx} - 2u_x + 2v_x v + 2\,[v,u]) \end{aligned} \tag{10.2}$$

These are the first two equations of the hierarchy of DLW equation (see [12]for the abelian case).

Example 2. The algebra of $\mathscr{A}$-valued polynomial functions on $\mathbb{R}$ is now replaced by the algebra of differential operators

$$(q(x,y);\ r(x,y)) = \sum_{k\geq 0} (q_k(x,y);\ r_k(x,y))\partial_k \quad (\partial_k := \frac{\partial^k}{\partial y^k}) \tag{10.3}$$

where q_k and r_k are scalar functions (the generalization to the "non-abelian" case of q_k and r_k given by matrices is straightforward). D is still given by $D=\partial/\partial x$ and $c=\partial/\partial y$, V is the subalgebra of differential operators rapidly vanishing for $x^2+y^2\to\infty$ and S is the affine hyperplane modelled on V with equations:

$$(q(x,y);\ r(x,y)) = \sum_{k\geq 0} (q_k(x,y);\ r_k(x,y))\partial_k + (0;\partial\ ) \tag{10.4}$$

The reduction submanifolds S' is modelled on the subalgebra of zero-order differential operators, so that it is: $q_k=u(x,y)\delta_{ko}$, $r_k=v(x;y)\delta_{ko}$ (the choice $c=\partial/\partial y$ is essential in order that S' fulfils the reduction condition (9.7)). Since c does not commute with q and r, non trivial equations are obtained from both bicomplexes VNI and VNII. By writing $u_y := [c,u]$, $v_y := [c,v]$ and so on, and by introducing the operator $I=(\partial_x-\partial_y)^{-1}$, one obtains the following fields

$$\text{VNI:}\quad \varphi_{(0)}=(u_y;\ v_y) \tag{10.5}$$

$$\varphi_{(1)}=(u_{yy}-2u_{xy}+2(uv)_y+2(u\ Iv_y)_x;\ -v_{yy}+2v_{xy}+2vv_y-4u_y+$$

$$+2\ D^{-1}u_{yy}+2v_x Iv_y)$$

$$\text{VNII:}\quad \varphi_{(0)}=(u_x;v_x) \tag{10.6}$$

$$\varphi_{(1)}=(-u_{xx}+2(uv)_x+2(u\ I(v_y))_x;\ v_{xx}+2vv_x-2u_x+2v_x\ Iv_y)$$

The linear combination (VNI-VNII) gives rise to the hierarchy of purely differential equations obtained in[13] by means of the formalism which will be considered in the next Example 3.

Example 3. In this last example, we consider the algebra of integral operators with distributional kernels

$$(uf)(x,y_1) = \int_R u(x,y_1,y_2)\ f(x,y_2)dy_2 \tag{10.7}$$

As it is well-known, it is an associative algebra with unit, with respect to the convolution product

$$(u*v)_{12} = \int_R u_{13}\ v_{32}dy_3 \tag{10.8}$$

and with the unit given by the Dirac's distribution $\delta_{12}=\delta(y_1-y_2)$.(As it is usual, the x-dependence is understood and the two-indices notation is used, so that $u_{12}=u(x,y_1,y_2),\ldots$).

In this example:

i) V is the algebra of the kernels of the form

$$v_{12} = \sum_{k\geq 0} v_{12k}\ \delta_{12}^{(k)} \qquad (\delta_{12}^{(k)} = \partial^k \delta_{12}/\partial y_1^k) \tag{10.9}$$

where v_{12k} are rapidly vanishing functions, $D=\partial/\partial x$, $c=\delta_{12}^{(1)}$.

ii) S is the affine hyperplane with equations

$$(q_{12};r_{12}) = \sum_{k\geq 0} (q_{k12};\ r_{k12})\ \delta_{12}^{(k)} + (0;\ \delta_{12}^{(1)}) \tag{10.10}$$

and the reduction submanifold S' is modelled on the subalgebra corresponding to $q_{012}=u_1$, $r_{012}=v_1$ and $q_{k12}=r_{k12}=0$ for $k>0$ (this subalgebra is isomorphic to the subalgebra of zero-order differential operators, i.e. to the reduction subalgebra of the previous Example 2).

iii) The Nijenhuis tensor on the symmetry algebra is the right-multiplication by $\delta^{(1)}_{12}:(\Delta_e a)_{12}=(a*\delta^{(1)})_{12}$ and the starting symmetry is again the unity of A , i.e. $a_o=\delta_{12}$.

Since all the assumptions of the Reduction Theorem 9.1 are fulfilled by the previous choices, we are guaranteed that the bicomplexes VNI and VNII furnish vector fields which are tangent to S', i.e. we obtain kernels of the form $\varphi_{12(n)}=\varphi_{1(n)}\delta_{12}$. So, the evolution equations are defined by the vector fields $\varphi_{1(n)}$: in this way, the equations previously given in Example 2 are exacly reobtained, as one could naturally expect since the reduction submanifolds in Examples 2 and 3 are isomorphic.

In particular, the Nijenhuis tensor N_{12} and the symmetry generators X_{12} correspond to those obtained by Boiti et al.[13] by using the present distributional approach (which is originally due to Fokas and Santini in relation with the KP equation[14]). As for a more detailed discussion of the relation between the two approaches considered in Example 2 and in this Example, and in particular for the meaning of the so-called "fixed-step" formulation (9.13) of the evolution equations in 2+1 dimensions, see also[8-15].

We limit ourselves to remark that the distributional approach seems to be by no means essential to obtain hierarchies of evolution equations in 2+1 dimensions with a "fixed-step" recursion operator, since it corresponds only to a possible realization of an abstract reduction process defined on an abstract $\mathcal{G}$N-manifold and is clearly equivalent, in this respect, to the differential approach described in Example 2.

11. CONCLUSIVE REMARKS

In this paper we have tried to understand, from a geometrical point of view, some of the algorithmic procedures usually employed in the IST theory to generate integrable systems. We have found helpful to divide the theory into several parts. The first step is to recognize and to critically analyse the starting point of the IST theory, that is, according to a generally accepted opinion, a suitable "spectral problem". In our opinion, the starting object is the bihamiltonian manifold with symmetries, described in Sect. 5. The two points of view are not completely equivalent: the existence of the spectral problem can be naturally deduced from the properties of the bihamiltonian manifold, but not every property of the manifold can be deduced from the spectral problem. One of the main reasons for accepting the enlarged geometric point of view is that it clarifies how many computations connected with the spectral problem are simply realizations of geometric constructions. This justifies a priori the results found at the end of the computations, making unnecessary to check them.
In particular, it allows one to understand that these computations correspond essentially to a *double reduction* procedure. The first one is the reduction of second order tensor fields (described in Sect. 6), allowing one to recover the so-called "recursion operators" and "Hamiltonian operators" of the IST theory: examples of this procedure have been given in Sect. 7. The second one is a reduction of special algebras of vector fields, here called Lenard bicomplexes: it has been explained in Sect. 9 and used in Sect. 10. It allows one to recover the final hierarchies of integrable systems of the IST theory.

In conclusion, the aim of this paper has been to show that the method of the spectral problem is a reduction technique for particular classes of bihamiltonian manifolds with symmetries, and that the method of the so-called compatibility conditions (usually introduced for obtaining the final equations) is a reduction technique for Lenard bi-complexes. This picture is more than a simple reformulation of known techniques. In our opinion, it allows one to better understand the meaning of these techniques and to use them with more awareness. In particular, it avoids a lot of computations, usually required to check the final results.

ACKNOWLEDGMENTS

This work has been partially supported by the Italian M.P.I. (Progetto Geometria e Fisica) and C.N.R. (Gruppo Nazionale per la Fisica Matematica).

REFERECES

1) Calogero, F., De Gasperis, A., "Spectral Transform and Solitons I", North Holland, Amsterdam (1980).
Konopelchenko, B.G., "Non Linear Integrable Equations", Lect. Notes in Phys., 270, Springer Verlag, New York (1987).
2) Magri, F., Morosi, C., Quaderno S/19, Università di Milano (1984).
Magri, F., Morosi, C., Ragnisco, O., Commun. Math. Phys. 99, 115 (1985).
3) Carathéodory, C., "Calculus of Variations and Partial Differential Equations of the First Order", Holden Day, San Francisco (1967).
Weinstein, A., J. Differential Geometry, 18, 523 (1983).
Liberman, P., Marle, C.M., "Symplectic Geometry and Analytical Mechanics", Reidel, Dordrecht (1987).
4) Schouten, J.A., Nederl. Akad. Wetensch. Proc. Sez. A, 43, 449 (1940).
Lichnerowitz, A., "Topics in Differential Geometry", Ed. by H.

Rund and W. Forbes, Acad. Press, New York (1967).

5) Helgason, S., "Differential Geometry, Lie Groups and Symmetric Spaces", Acad. Press, New York (1978).

6) Segal, G., Pressley, A., "Loop Groups",Clarendon Press, Oxford (1986).

7) Oevel, W., "Topics in Soliton Theory", Ed. by M.J. Ablowitz, B. Fuchsteiner and M. Kruskal, p. 108-124, World Scientific, Singapore (1987).
Fuchssteiner, B., Progr. Theor. Phys. 70, 1508 (1983).

8) Magri, F., Morosi, C., "Topics in Soliton Theory", Ed. by M.J. Ablowitz, B. Fuchssteiner and M. Kruskal, p. 78-96, World Scientific Singapore (1987).
Magri, F., Morosi, C., Tondo, G., Commun. Math. Phys., 115 (1988).

9) Heuser, H.G., "Functional Analysis", J. Wiley, New York (1980).

10) Koszul, J.L., Canad. J. Math., 7, 562 (1955).

11) Fuks, D.B., "Cohomology of Infinite Dimensional Lie Algebras", Consultants Bureau, New York (1986).

12) Konopelchenko, B.G., Phys. Letters, 116, 231 (1986).

13) Boiti, M., Leon, J.JP, Pempinelli, F., Stud. Appl. Math. (1988).

14) Fokas, A., Santini, P., Stud. Appl. Math., 75, 179 (1986).

15) Magri, F., Morosi, C., Tondo, G., "Nonlinear Evolutions", Ed. by J.JP. Léon, World Scientific, Singapore (1987).

MECHANISMS FOR VARIABLE SEPARATION IN PARTIAL DIFFERENTIAL EQUATIONS AND THEIR RELATIONSHIP TO GROUP THEORY

WILLARD MILLER, JR.†‡

Abstract. This is a survey of those techniques used to obtain explicit solutions of partial differential equations through the idea of "separation of variables". Frequently, but not always, variable separation is associated with the symmetries of a differential equation. We have a good understanding of variable separation for the scalar equations of mathematical physics. For spinor equations many interesting results are known but there is, as yet, no general theory.

Key words. separation of variables, Dirac equation, spinor equations

AMS(MOS) subject classifications. 33A65, 33A75

1. Introduction. These lectures constitute a presentation of the basic concepts in the theory of separation of variables for partial differential equations and an exploration of the deep relations between variable separation and the generalized Lie symmetries of these equations. Historically, the theory of variable separation has been developed most intensively and proved most useful for two classes of partial differential equations: first order nonlinear equations and linear equations of all orders. However, the concepts are clearly applicable to general nonlinear equations. We will pay particular attention to variable separation for systems of equations, the part of the field which is of most active current interest.

The primary use of variable separation is for computation of explicit solutions of partial differential equations. The solutions can be calculated by solving ordinary differential equations (the separation equations). Many of the solutions obtained by this method prove so important that these functions are studied and tabulated in their own right: the special functions of mathematical physics. For Hamilton-Jacobi equations variable separation is used to obtain complete integrals which in turn lead to explicit solutions of the associated Hamiltonian system. For linear equations the Fourier method can be used to solve boundary value problems and represent a wide variety of functions as sums or integrals of separated solutions.

Basically, a partial differential equation is (additively) separable in the independent variables $x_1, \ldots, x_n$ if the equation admits a nontrivial solution of the form

$$u = \sum_{i=1}^{n} S^{(i)}(x_i).$$

†Supported in part by the National Science Foundation under grant DMS 86-00372

‡School of Mathematics, and Institute for Mathematics and its Applications, University of Minnesota, Minneapolis, MN 55455

One can also talk about product separation $v = \prod_{i=1}^{n} T^{(i)}(x_i)$ or more complicated types of separation such as $w = \tan[\sum_{i=1}^{n} S^{(i)}(x_i)]$. However a change of dependent variable reduces these other types to additive separation, e.g., $u = \ln v$ or $u = \arctan w$. In §2 we give a precise and useful definition of variable separation.

In §3 we shall review briefly the theory of generalized Lie symmetries of partial differential equations and show that the standard procedure for computing *symmetry adapted* solutions of these equations from a knowledge of the Lie symmetries is an example of variable separation of a particularly simple type.

In §4 we shall apply the theory of variable separation to two particularly simple and physically important problems: orthogonal variable separation for Hamilton-Jacobi and Helmholtz equations. This will lead us to the mathematics of Stäckel form. Then in §5 we will provide an intrinsic characterization of variable separation for these scalar equations in terms of Lie symmetries. In §6 we introduce the basics of variable separation for systems of equations of Dirac type. Finally, in §7 we shall make some general comments about variable separation for systems of equations. The basic problems in this area remain unsolved.

These lectures are concerned with the method of variable separation itself; space does not permit a study of the properties of the separable solutions and the relationship between these properties and Lie symmetries. This relationship is explored in the monograph [1]. Excellent expositions of applications of separable functions to the solution of initial and boundary value problems can be found in [2] and [3].

2. Additive separation. We start with a definition of additive separability for a system of partial differential equations of the form

$$H^{(K)}(x_i, u^{(L)}, u_i^{(L)}, u_{ij}^{(L)}, \dots) = E^{(K)} \tag{2.1}$$

in coordinates $x_1, \dots, x_n$. Here, $u^{(L)}$ are the dependent variables, $u_i^{(L)} = \partial_{x_i} u^{(L)}, u_{ij}^{(L)} = \partial_{x_i}\partial_{x_{ij}} u^{(L)}$, etc., where $1 \le i, j, \dots \le n, 1 \le K, L, \dots \le N$ and the $E^{(K)}$ are parameters. Note that the number of equations, N, equals the number of dependent variables. We assume that each $H^{(K)}$ is a polynomial in a finite number of the variables $u^{(L)}, u_i^{(L)}, u_{ij}^{(L)}, \dots$ with coefficients which are real analytic functions of the variables $x_i, u^{(L)}$, all defined in a common domain $D \times J^N, D \subseteq R^n$ with $(0, \dots, 0) \in D$, and J^N an open set in R^N.

A *solution* of (2.1) is a vector valued function $\mathbf{u} = \mathbf{S}(\mathbf{x}, \mathbf{E})$ where $\mathbf{u} = (u^{(1)}, \dots, u^{(N)})$, defined and analytic for $\mathbf{x} = (x_1, \dots, x_n)$ in a nonzero domain $D' \subseteq D$ and $\mathbf{E}$ in an open interval $R' \subseteq R^N$, such that the substitution of this function into (2.1) renders (2.1) an identity for all $(\mathbf{x}, \mathbf{E}) \in D' \times R'$. An *additive separable solution* is a solution of the form $\mathbf{u} = \sum_{j=1}^{n} \mathbf{S}^{(j)}(x_j, \mathbf{E})$.

Since a separable solution in the coordinates $\mathbf{x}$ satisfies $u_{ij}^{(L)} = 0$ for $i \neq j$, without loss of generality we can set all mixed partial derivatives identically equal to zero and write (2.1) in the form

$$H^{(K)}(x_i, u^{(L)}, u_i^{(L)}, u_{ii}^{(L)}, \dots) = E^{(K)}. \tag{2.2}$$

For convenience we set $u^{(L)}_{i,1} \equiv u^{(L)}_i, u^{(L)}_{i,j+1} = \partial_{x_i} u^{(L)}_{i,j}, j = 1, 2, \ldots$ and define m_i to be the largest number j such that $\partial_{u^{(L)}_{i,j}} \mathbf{H} \not\equiv 0$ for some L. To avoid discussion of degenerate cases we require $m_i > 0$ for $i = 1, 2, \ldots, n$. Let D_i denote the total differentiation operators

$$D_i = \partial_{x_i} + \sum_L u^{(L)}_{i,1} \partial_{u^{(L)}} + \sum_{j=1}^{\infty} \sum_L u^{(L)}_{i,j+1} \partial_{u^{(L)}_{i,j}}. \tag{2.3}$$

Our final nondegeneracy assumption is that $\mathbf{u}$ is a separable solution of (2.2) such that the $N \times N$ matrix $(\mathcal{H}(i)^K_L) = (H^{(K)}_{u^{(L)}_{i,m_i}})$ is nonsingular for all i, i.e.,

$$\det \mathcal{H}(i) \neq 0, \quad i = 1, \ldots, n. \tag{2.4}$$

Then the evident equations $D_j H^{(K)}(\mathbf{x}, \mathbf{u}) = 0$ can be solved uniquely for $u^{(L)}_{i,m_i+1}$:

$$u^{(L)}_{i,m_i+1} = -\sum_{K=1}^{N} \tilde{D}_i H^{(K)} \mathcal{A}(i)^L_K, \quad i = 1, \ldots, n, \tag{2.5}$$

where

$$\tilde{D}_i = \partial_{x_i} + \sum_L u^{(L)}_{i,1} \partial_{u^{(L)}} + \sum_{j=1}^{m_i - 1} \sum_L u^{(L)}_{i,j+1} \partial_{u^{(L)}_{i,j}} \tag{2.6}$$

and

$$\sum_{K=1}^{N} \mathcal{H}(i)^K_L \mathcal{A}(i)^M_K = \delta^M_L, \quad 1 \le L, M \le N. \tag{2.7}$$

Here, δ^M_L is the Kronecker delta. Clearly, $\mathbf{u}$ satisfies integrability conditions $D_j \mathbf{u}_{i,m_i+1} = 0, \quad j \neq i$, or

$$\begin{aligned} &\tilde{D}_j \tilde{D}_i H^{(L)} - \sum_{J,K=1}^{N} [\tilde{D}_i H^{(K)} \mathcal{A}(i)^J_K \tilde{D}_j \mathcal{H}(i)^L_J + \tilde{D}_j H^{(K)} \mathcal{A}(j)^J_K \tilde{D}_i \mathcal{H}(j)^L_J] \\ &+ \sum_{J,K,P,Q=1}^{N} \tilde{D}_j H^{(P)} \mathcal{A}(j)^J_P \tilde{D}_i H^{(K)} \mathcal{A}(i)^Q_K H^{(L)}_{u^{(Q)}_{i,m_i}, u^{(J)}_{j,m_j}} = 0. \end{aligned} \tag{2.8}$$

Multiplying both sides of (2.8) by $\det \mathcal{H}(i) \det \mathcal{H}(j)$ we see that the resulting expression is a polynomial in the derivatives of $\mathbf{u}$. In general, (2.8) is a restriction both on the coefficients of $\mathbf{H}$ and the form of the separable solution $\mathbf{u}$. However, there is an important special case where (2.8) is an identity in the variables $\mathbf{u}, \mathbf{u}_{k,\ell}$. [Indeed, this case will occur if (2.2) admits so many separable solutions that for each $\mathbf{x}^0 \in D$ and each set of real constants

$\mathbf{u}^0, \mathbf{u}_i^0, \mathbf{u}_{ii}^0, \ldots, \quad i = 1, 2, \ldots, n$ satisfying $\mathbf{H}(\mathbf{x}^0, \mathbf{u}^0, \mathbf{u}_i^0, \mathbf{u}_{ii}^0, \ldots) = \mathbf{E}$, there is a separable solution $\mathbf{u}(\mathbf{x})$ such that $\mathbf{u}(\mathbf{x}^0) = \mathbf{u}^0, \mathbf{u}_i(\mathbf{x}^0) = \mathbf{u}_i^0, \ldots$.] Then conditions (2.8) reduce to restrictions on the coefficients of $\mathbf{H}$ which are independent of the choice of separable solution. If the polynomial version of (2.8) is an identity we say that $\{x_i\}$ is a *regular separable coordinate system* (for the equation (2.2)).

Suppose $\{x_i\}$ is a regular separable coordinate system and consider the equations

$$
\begin{aligned}
D_i \mathbf{v} &= \mathbf{v}_{i,1}, \\
D_i \mathbf{v}_{j,1} &= \delta_{i,j} \mathbf{v}_{j,2}, \\
&\vdots \\
D_i \mathbf{v}_{j,m_j - 1} &= \delta_{i,j} \mathbf{v}_{j,m_j}, \\
D_i \mathbf{v}_{j,m_j} &= -\delta_{i,j} \tilde{D}_i \mathbf{H} \mathcal{A}(i), \quad 1 \le i, j \le n,
\end{aligned}
\tag{2.9}
$$

where $\mathbf{v}$ is an N component vector. The integrability condition for this system of equations is $D_k D_i \mathbf{v}_{j,m_j} = D_i D_k \mathbf{v}_{j,m_j}$, equivalent to (2.8). Since (2.8) is satisfied identically, it follows that for each $\mathbf{x}^0 \in D$ and each set of constants $\mathbf{v}_{i,j}^0, 1 \le i \le n, 0 \le j \le m_i$, such that $\det \mathcal{H}(\mathbf{x}^0, \mathbf{v}^0) \ne 0$, there is a unique solution $\mathbf{v}$ of the system (2.9) such that $\mathbf{v}(\mathbf{x}^0) = \mathbf{v}^0, D_i \mathbf{v}(\mathbf{x}^0) = \mathbf{v}_{i,1}^0, D_i \mathbf{v}_{i,j}(\mathbf{x}^0) = \mathbf{v}_{i,j+1}^0$ (see [10, Chapter 1]). Choose $\mathbf{E}$ such that $\mathbf{E} = \mathbf{H}(\mathbf{x}^0, \mathbf{v}^0)$. Then $\mathbf{u} = \mathbf{v}(\mathbf{x})$ is a separable solution of (2.2) such that $\mathbf{u}(\mathbf{x}^0) = \mathbf{v}^0$ and $\mathbf{u}_{i,j}(\mathbf{x}^0) = \mathbf{v}_{i,j}^0, 1 \le i \le n, 1 \le j \le m_i$. Indeed $D_{x_i} \mathbf{H}(\mathbf{x}, \mathbf{v}) = 0$, so $\mathbf{H}(\mathbf{x}, \mathbf{v}) = \mathbf{E}$.

THEOREM 1. *If $\{x_i\}$ is a regular separable system for the equation $\mathbf{H} = \mathbf{E}$, i.e., if equations (2.8) are satisfied identically, then for every set of $N(m_1 + m_2 + \cdots + m_n + 1)$ constants $\{\mathbf{v}^0, \mathbf{v}_{i,j}^0\}$ with $\mathbf{H}(\mathbf{x}^0, \mathbf{v}^0) = \mathbf{E}$ and $\det \mathcal{H}(i)(\mathbf{x}^0, \mathbf{v}^0) \ne 0, 1 \le i \le n$, there is a unique separable solution $\mathbf{u}$ of $\mathbf{H}(\mathbf{x}, \mathbf{u}) = \mathbf{E}$ such that $\mathbf{u}(\mathbf{x}^0) = \mathbf{v}^0, \mathbf{u}_{i,j}(\mathbf{x}^0) = \mathbf{v}_{i,j}^0, 1 \le i \le n, 1 \le j \le m_i$.*

If equations (2.8) are not satisfied identically, separable solutions still may exist, but they will depend on fewer than $N(\sum_{i=1}^n m_i + 1)$ parameters. This type of separation is *nonregular*.

An interesting special case of the preceding considerations occurs when $E^{(k)} = E$ in (2.2), i.e., when all of the constants $E^{(k)}$ are equal. (We shall see that this corresponds to systems of equations analogous to the Dirac equations in relativistic quantum theory.) It is easy to see that the following result holds [4,5].

COROLLARY 1. *If $\{x_i\}$ is a regular separable system for the equation $\mathbf{H} = \mathbf{E}$ with $E^{(k)} = E$, i.e., if equations (2.8) are satisfied identically, then for every set of $N(m_1 + m_2 + \cdots + m_n + 1)$ constants $\{\mathbf{v}^0, \mathbf{v}_{i,j}^0\}$ with $\mathbf{H}(\mathbf{x}^0, \mathbf{v}^0) = \mathbf{E}$ and $\det \mathcal{H}(i)(\mathbf{x}^0, \mathbf{v}^0) \ne 0$, $1 \le i \le n$, there is a unique separable solution $\mathbf{u}$ of $\mathbf{H}(\mathbf{x}, \mathbf{u}) = \mathbf{E}$ such that $\mathbf{u}(\mathbf{x}^0) = \mathbf{v}$, $\mathbf{u}_{i,j}(\mathbf{x}^0) = \mathbf{v}_{i,j}^0$, $1 \le i \le n$, $1 \le j \le m_i$.*

Another important special case occurs for $N = 1$. Then we have a single partial differential equation

$$H(x_i, u, u_i, u_{ii}, \dots) = E, \tag{2.10}$$

rather than a system. Then conditions (2.8) take the simpler form ($i \neq j$)

$$\begin{aligned} &H_{u_{i,m_i}} H_{u_{j,m_j}} (\tilde{D}_i \tilde{D}_j H) + H_{u_{i,m_i} u_{j,m_j}} (\tilde{D}_i H)(\tilde{D}_j H) \\ &\quad = H_{u_{j,m_j}} (\tilde{D}_i H)(\tilde{D}_j H_{u_{i,m_i}}) + H_{u_{i,m_i}} (\tilde{D}_j H)(\tilde{D}_i H_{u_{j,m_j}}) \end{aligned} \tag{2.11}$$

COROLLARY 2. *If $\{x_i\}$ is a regular separable system for the equation $H = E$, i.e., if equations (2.11) are satisfied identically, then for every set of $m_1 + m_2 + \cdots + m_n + 1$ constants $\{v^0, v^0_{i,j}\}$ with $H(\mathbf{x}^0, v^0) = E$ and $H_{u_{j,m_j}}(\mathbf{x}^0, v^0) \neq 0$, there is a unique separable solution u of $H(\mathbf{x}, u) = E$ such that $u(\mathbf{x}^0) = v^0$, $u_{i,j}(\mathbf{x}^0) = v^0_{i,j}$, $1 \leq i \leq n$, $1 \leq j \leq m_i$.*

We shall postpone the analysis of separation of variables for systems of equations to §6 and restrict our attention here to a single partial differential equation. The following examples exhibit regular and nonregular separation for scalar equations.

Example 1. $H = (x_1+x_2)(u_{11}+u_{22})-2(u_1+u_2)$. Equations (2.11) are satisfied identically so $\{x_1, x_2\}$ is a regular separable system. The general separable solution depends on five parameters and is given by

$$u = (\alpha x_1^3 + \beta x_1^2 + \gamma x_1 - \frac{1}{2} E x_1) + (-\alpha x_2^3 + \beta x_2^2 - \gamma x_2 + \delta). \tag{2.12}$$

Example 2. $H = u_{11}^2 + u_1 + u_{22}$. Here we have $u_{111} = -\frac{1}{2}$ (provided $u_{11} \neq 0$) and $u_{222} = 0$ so equations (2.11) are satisfied identically and $\{x_1, x_2\}$, is a regular separable system. The general separable solution depends on five parameters:

$$u = (-\frac{1}{12} x_1^3 + \alpha x_1^2 + \beta x_1) + \left(\frac{1}{2}(E - 4\alpha^2 - \beta) x_2^2 + \gamma x_2 + \delta\right). \tag{2.13}$$

Example 3. $H = x_2 u_{11} + x_1 u_{22} + u_1 + u_2$. Equations (2.11) reduce to the requirement $u_{11} + u_{22} = 0$. The general separable solution depends on four parameters:

$$u = (\alpha x_1^2 + \beta x_1) + (-\alpha x_2^2 + (E - \beta) x_2 + \gamma). \tag{2.14}$$

This is a nonregular separable system.

Example 4. $H = (u_{11} + u_{22})/u$. Equations (2.11) are satisfied identically for $u \neq 0$. The general separable solution depends on five parameters:

$$u = \alpha \exp(x_1 \sqrt{E}) + \beta \exp(-x_1 \sqrt{E}) + \gamma \exp(x_2 \sqrt{E}) + \delta \exp(-x_2 \sqrt{E}) \tag{2.15}$$

for $E > 0$, with obvious modifications for $E \leq 0$.

There is a similar theory of additive separation for partial differential equations of the form (2.1) with $E^{(k)} = 0$, i.e., equations not depending on a parameter. For simplicity we consider only the case of a scalar equation; the spinor case is similar. We make the same assumptions on H as before and take the equation in the form (2.10) with $E = 0$:

$$H(x_i, u, u_i, u_{ii}, \dots) = 0 \tag{2.16}$$

Then a separable solution u of (2.16) must satisfy the integrability conditions (2.11). In case the integrability conditions are identities in the sense that there exist functions $P_{i,j}(x_k, u, u_{k,\ell})$, polynomials in $u_{k,\ell}$ such that

$$\begin{aligned} \mathcal{F}_{ij} \equiv\ & H_{u_{i,m_i}} H_{u_{j,m_j}} (\tilde{D}_i \tilde{D}_j H) + H_{u_{i,m_i}, u_{j,m_j}} (\tilde{D}_i H)(\tilde{D}_j H) \\ & - H_{u_{j,m_j}} (\tilde{D}_i H)(\tilde{D}_j H_{u_{i,m_i}}) - H_{u_{i,m_i}} (\tilde{D}_j H)(\tilde{D}_i H_{u_{j,m_j}}) \\ =\ & P_{i,j} H, \qquad i \neq j, \end{aligned} \tag{2.17}$$

we say that $\{x_k\}$ is a *regular separable coordinate* system for the equation $H = 0$.

THEOREM 2. *If $\{x_k\}$ is a regular separable system for $H = 0$ then for every set of $m_1 + m_2 + \cdots + m_n + 1$ constants $\{v^0, v^0_{i,j}\}$ with $H(x^0, v^0) = 0$ and $H_{u_{j,m_j}}(x^0, v^0) \neq 0$, there is a unique separable solution u of $H(x, u) = 0$ such that $u(x^0) = v^0$, $u_{i,j}(x^0) = v^0$, $u_{i,j}(x^0) = v^0_{i,j}$, $1 \leq i \leq n$, $1 \leq j \leq m_i$.*

Again we observe that if equations (2.17) are not satisfied identically, separable solutions still may exist but will depend on fewer that $\sum_{i=0}^{n} m_i + 1$ independent parameters. This is *nonregular* separation. Examples 1-4 above for $E = 0$ are instances of regular and nonregular separation. Less trivial is

Example 5. $H = (x_2 - x_3)u_{11} + (x_3 - x_1)u_{22} + (x_1 - x_2)u_{33}$. Equations (2.17) are satisfied with $P_{i,j} \not\equiv 0$, so $\{x_k\}$ is a regular separable system for $H = 0$, though not for $H = E$. The general separable solution depends on six parameters and is given by

$$u = \frac{1}{6}\alpha(x_1^3 + x_2^3 + x_3^3) + \frac{1}{2}\beta(x_1^2 + x_2^2 + x_3^2) + \gamma_1 x_1 + \gamma_2 x_2 + \gamma_3 x_3 + \delta. \tag{2.18}$$

For other definitions of separability see [6].

3. Generalized Lie symmetries.

In this section we list some concepts and formulas from the Lie symmetry theory of partial differential equations that are essential for understanding the significance of variable separation. We limit ourselves here to an enumeration of basic results. For detailed proofs and background material see Olver's recent book[7]. See also [8,9].

In the following all functions are assumed to be locally real analytic. We consider local independent coordinates $x_1, \ldots, x_n$ and a dependent coordinate u. Given an assignment $u = f(x)$ we introduce the notation

$$u^J = \partial^J f(x), \qquad \partial^J = \frac{\partial^{|J|}}{\partial_{x_1}^{j_1} \partial_{x_2}^{j_2} \ldots \partial_{x_n}^{j_n}}, \tag{3.1}$$

to represent the kth order derivatives of f. Here $J = (j_1, \ldots, j_n)$ with each j_i a nonnegative integer and $|J| = j_i + \cdots + j_n = k$. A *generalized vector field* on the $(n+1)$-dimensional space of independent and dependent variables is an expression of the form

$$\hat{Z} = \sum_{i=1}^{n} \xi_i(x, u^J) \frac{\partial}{\partial x_i} + \varphi(x, u^J) \frac{\partial}{\partial u} \tag{3.2}$$

where ξ_i and φ depend on x, u and finitely many derivates of $u, (|J| \leq \ell < \infty)$. Locally this field generates new coordinates $X^*(\alpha), u^*(\alpha)$ obtained by solving

$$\frac{\partial x_i^*(\alpha)}{\partial \alpha} = \xi_i(x^*, u^{*J}), \quad x_i^*(0) = x_i,$$
$$\frac{\partial u^*(\alpha)}{\partial \alpha} = \varphi(x^*, u^{*J}), \quad u^*(0) = u = f(x).$$

If ξ_i and φ depend only on $\mathbf{x}$ and u, then $\hat{Z}$ is the generator of a Lie point transformation. In an obvious way $\hat{Z}$ can be prolonged to a vector field on the jet space with coordinates $(\mathbf{x}, u, u^J)$:

$$Z = \hat{Z} + \sum_{|K|>0} \varphi^K(x, u^J) \frac{\partial}{\partial u^K}, \tag{3.3}$$

where

$$\begin{aligned} \varphi^K &= D^K(\varphi - \sum_{i=1}^{n} u^i \xi_i) + \sum_{i=1}^{n} u^{K,i} \xi_i, \\ K &= (k_1, \ldots, k_n), \qquad J = (j_1, \ldots, j_n), \\ J, i &= (j_1, \ldots, j_{i-1}, j_i + 1, j_{i+1}, \ldots, j_n), \\ D^K &= D_1^{k_1} D_2^{k_2} \cdots D_n^{k_n}, \end{aligned} \tag{3.4}$$

And D_i is the *total derivative*

$$D_i = \frac{\partial}{\partial x_i} + \sum_{|K| \geq 0} u^{K,i} \frac{\partial}{\partial u^K}. \tag{3.5}$$

[Although the sums in (3.3) and (3.5) are formally infinite, in practice we will apply these operators only to functions that depend on finitely many derivatives u^K.]

Let $H(\mathbf{x}, u^K) = 0$ be a partial differential equation for u. We say that the generalized vector field $\hat{Z}$ is a *generalized symmetry operator* for this equation provided

$$ZH(\mathbf{x}, u^K) = 0, \tag{3.6}$$

whenever $H(\mathbf{x}, u^K) = 0$ i.e., $Z\,H$ vanishes for all solutions u of $H = 0$. In the following we will make the technical hypothesis, almost always satisfied by partial differential equations of physical interest, that for each generalized symmetry operator $\hat{Z}$ there are a finite number of functions $\chi_J(\mathbf{x}, u^K)$ such that

$$ZH = \sum_J \chi_J D^J H. \tag{3.7}$$

See [7,page 135], for a discussion of this specialization of (3.6).

The *characteristic* ϕ of a generalized vector field $\hat{Z}$ is defined by

$$\phi = \varphi - \sum_{i=1}^{n} u^i \xi_i. \tag{3.8}$$

Note that the generalized vector field $\hat{X}(\phi) = \phi\partial_u$ has the prolongation

$$X(\phi) = \phi\partial_u + \sum_{|K|>0} D^K \phi \frac{\partial}{\partial_{u^K}}. \tag{3.9}$$

We call $\hat{X}(\phi)$ the *standard representation* of $\hat{Z}$. It follows from (3.3) and (3.4) that

$$Z = X(\phi) + \sum_{i=1}^{n} \xi_i D_i, \tag{3.10}$$

and from (3.7) that $\hat{Z}$ is a generalized symmetry for the equation $H = 0$ if and only if its standard representation $\hat{X}(\phi) = \phi\partial_u$ is a generalized symmetry. Thus, from the viewpoint of Lie symmetries there is no loss of generality in restricting to operators of the form $\phi\partial_u$.

The *commutator* of two prolongations of standard representation operators is an operator of the same form:

$$[X(\phi_1), X(\phi_2)] \equiv X(\phi_1)X(\phi_2) - X(\phi_2)X(\phi_1) = X(\{\phi_1, \phi_2\}), \tag{3.11}$$

where

$$\begin{gathered} \{\phi_1, \phi_2\} = -\{\phi_2, \phi_1\} \\ \{a_1\phi_1 + a_2\phi_2, \phi_3\} = a_1\{\phi_1, \phi_3\} + a_2\{\phi_2, \phi_3\}, \quad a_i \in R, \\ \{\{\phi_1, \phi_2\}, \phi_3\} + \{\{\phi_2, \phi_3\}, \phi_1\} + \{\{\phi_3, \phi_1\}, \phi_2\} = 0. \end{gathered} \tag{3.13}$$

In terms of this bracket the symmetry condition (3.6)–(3.7) becomes

$$\{\phi, H\} = 0 \text{ whenever } H = 0. \tag{3.14}$$

If ϕ_1 and ϕ_2 depend only on the variables x_i and u^i , then (3.13) reduces to the Poisson bracket:

$$\{\phi_1, \phi_2\} = \sum_{i=1}^{n} (\partial_{x_i}\phi_1 \partial_{u^i}\phi_2 - \partial_{x_i}\phi_2 \partial_{u^i}\phi_1). \tag{3.15}$$

If ϕ_1 and ϕ_2 are linear in the u^K,

$$\phi_i = \sum_K a_K^{(i)}(\mathbf{x}) u^K, \quad i = 1, 2,$$

then the bracket agrees with the usual operator commutator bracket:

$$\begin{aligned} \{\phi_1, \phi_2\} &= -[L_1, L_2]u = -(L_1 L_2 - L_2 L_1)u, \\ L_i &= \sum_K a_K^{(i)}(x) D^K, \qquad i = 1, 2, \end{aligned} \tag{3.16}$$

For simplicity we have restricted the above considerations to a scalar partial differential equation with a single dependent variable. However, the extension to spinor equations with many dependent variables is completely straightforward [7].

We are now ready to highlight the simplest point of contact between separation of variables methods and the symmetries of a partial differential equation

$$H(\mathbf{x}, u^K) = 0. \tag{3.17}$$

Let $\hat{Z} \neq 0$ be an infinitesimal point symmetry for (3.17) which is *projectable*, i.e. of the form

$$\hat{Z} = \sum_{i=1}^{n} \xi_i(x) \frac{\partial}{\partial x_i} + \varphi(x, u) \frac{\partial}{\partial u}, \tag{3.18}$$

and, in addition, suppose not all ξ_i are zero. Then by Lie's theorem (see [10, pages 34,49, and 50]) we can find new coordinates $s, y_2, \ldots, y_n, v$ such that

$$\mathbf{x} = A(s, \mathbf{y}), \qquad \mathbf{u} = B(s, \mathbf{y}, v), \tag{3.19}$$

and, in the new coordinates

$$\hat{Z} = \frac{\partial}{\partial s}. \tag{3.20}$$

Since $\hat{Z}$ is a point symmetry, it follows from (3.7) (and very mild technical assumptions on H) that H must be of the form

$$H = H_1(s, \mathbf{y})H_2(\mathbf{y}, v^L), \tag{3.21}$$

where H_2 has nontrivial dependence on the v^L. Thus, each solution $v(s, \mathbf{y})$ of

$$H_2(y, v^L) = 0 \tag{3.22}$$

determines a solution u of $H(\mathbf{x}, u^K) = 0$. The equations $H = 0$ and $H_2 = 0$ are essentially equivalent. However, the second equation depends on s only through derivatives terms in v. Again, under mild technical assumptions on H, we can find solutions $v(\mathbf{y})$ of (3.22), i.e. solutions such that $\partial_s v \equiv 0$. [If $\partial_v H_2 \equiv 0$ then we can find solutions $v(s, \mathbf{y}) = \lambda s + v'(\mathbf{y})$ where λ is a constant.] For $n = 2$, Eq (3.22) reduces to an ordinary differential equation and the solutions of this equation yield *symmetry adapted* solutions of (3.17). Furthermore $(s, \mathbf{y})$ is a separable coordinate system for (3.17), though not necessarily a regular separable system. In general we have reduced a partial differential equation in n variables to one in $n-1$ variables, and this provides an instance of partial separation.

As a simple example, consider the symmetry

$$\hat{Z} = t\partial_x + c\partial_t - \partial_u, \qquad c \neq 0 \tag{3.23}$$

for the Korteweg-de Vries equation

$$\partial_t u - \partial_{xxx} u - u\partial_x u = 0. \tag{3.24}$$

(The symmetry algebra for this equation can be found in many references, e.g., [7]). The requirement $\hat{Z} = \partial_s$ leads to new coordinates (s, y, v) such that

$$t = cs, \quad y = \frac{1}{2}cs^2 + x, \quad v = -s + u.$$

In terms of the new coordinates, (3.24) becomes

$$\partial_s v - c\partial_{yyy} v - cv\partial_y v - 1 = 0. \tag{3.25}$$

The ordinary differential equation obtained by setting $\partial_s v \equiv 0$ is related to the first Painlevé transcendent, [7, page 107].

For a more general approach to symmetry adapted solutions see [9,64].

4. Separability for Hamilton-Jacobi, Helmholtz, and Laplace equations.

We now apply the results of §2 to determine the possible regular separable coordinate systems for the Hamilton-Jacobi equation on a pseudo-Riemannian manifold V^n

$$H(x^i, u_i) \equiv \sum_{i,j=1}^{n} g^{ij} u_i u_j = E, \quad g^{ij} = g^{ji}, \tag{4.1}$$

where $u_i = \partial_i u$. In the local coordinates $\{x^i\}$ the metric on V^n is $ds^2 = \sum_{i,j=1}^{n} g_{ij} dx^i dx^j$ and $\sum_{j=1}^{n} g_{ij} g^{jk} = \delta_i^k$, $g = \det(g_{ij}) \neq 0$. (Here we change notation again to that of Eisenhart's book [11]. Recall that a solution $u(\mathbf{x}, \lambda_1, \ldots, \lambda_n)$, $\lambda_1 = E$ is a *complete integral* of (4.1) provided $\det(\partial_{x^i \lambda_j} u) \neq 0$, and that the knowledge of a complete integral enables us to solve the associated Hamiltonian system $p_i = -\partial_{x^i} H(\mathbf{x}, \mathbf{p})$, $x^i = \partial_{p_i} H(\mathbf{x}, \mathbf{p})$. Separation of variables is a powerful technique for obtaining explicit complete integrals for many Hamilton-Jacobi equations [12]). Initially we limit ourselves to *orthogonal* coordinates $\{x^i\}$, i.e., coordinates for which $ds^2 = \sum_{i=1}^{n} H_i^2 (dx^i)^2$, so that $g_{ij} = 0$ if $i \neq j$. Thus (4.1) becomes

$$\sum_{i=1}^{n} H_i^{-2} u_i^2 = E, \tag{4.2}$$

and from the integrability conditions (2.11) we see that $\{x^i\}$ is a regular separable system if and only if

$$\partial_{jk} H_i^{-2} = \partial_j H_i^{-2} \partial_k \ln H_j^{-2} + \partial_k H_i^{-2} \partial_j \ln H_k^{-2}, \quad j \neq k. \tag{4.3}$$

These are the standard Levi-Civita separability conditions and are well known to be equivalent to the requirement that the metric coefficients be in Stäckel form with respect to the coordinates $\{x^i\}$ (see [11,13,14]). That is, there exists an $n \times n$ matrix $s_{ji}(x^j)$, whose j^{th} row depends only on x^j, such that $S = \det(s_{ji}) \neq 0$, (a *Stäckel matrix*), and

$$H_j^{-2} = \frac{s^{j1}}{S}, \tag{4.4}$$

where s^{j1} is the $(j,1)$ minor of (s_{ij}).

We will work out the proof of this fact, indeed we will generalize it, and study the theory of Stäckel matrices in §6. Here we only note that if the H_i^{-2} are in Stäckel form and u satisfies the set of ordinary differential equations (the separation equations)

$$u_i^2 + \sum_{j=1}^{n} \lambda_j s_{ij}(x^i) = 0, \quad i = 1, \ldots, n, \tag{4.5}$$

where $\lambda_1 = -E, \lambda_2, \ldots, \lambda_n$ are parameters and $u_i = u_i(x^i)$, then u is a separable solution of (4.2).

For the Hamilton-Jacobi equation with potential

$$\sum_{i=1}^{n} H_i^{-2} u_i^2 + V(x) = E, \tag{4.6}$$

the results are similar. The integrability conditions reduce to (4.3) and

$$\partial_{ik} V - \partial_k \ln H_j^{-2} \partial_j V - \partial_j \ln H_k^{-2} \partial_k V = 0, \quad j \neq k. \tag{4.7}$$

As shown in reference[15], this last condition means precisely that the potential function can be expressed in the form

$$V = \sum_{i=1}^{n} f^{(i)}(x^i) H_i^{-2}. \tag{4.8}$$

We will prove a generalization of this result in §6.

A regular orthogonal separable system for the Hamilton-Jacobi equation (4.2) with $E = 0$ is characterized by the integrability conditions

$$\partial_{jk} H_i^{-2} - \partial_j H_i^{-2} \partial_k \ln H_j^{-2} - \partial_k H_i^{-2} \partial_j \ln H_k^{-2} = \rho_{jk}(x) H_i^{-2} H_j^2 H_k^2, \qquad j \neq k, \tag{4.9}$$

for some functions ρ_{jk}. These equations are equivalent to

$$\partial_{jk} \ln K_i^{-2} + \partial_j \ln K_i^{-2} \partial_k \ln K_i^{-2} - \partial_j \ln K_i^{-2} \partial_k \ln K_j^{-2} - \partial_j \ln K_k^{-2} \partial_k \ln K_i^{-2} = 0, \tag{4.10}$$

for $K_i^{-2} = H_i^{-2}/H_1^{-2}$. Furthermore, as shown in reference [16], the equations are equivalent to the requirement that $H_i^{-2} = Q(\mathbf{x}) \mathcal{H}_i^{-2}$, where the metric $ds^2 = \sum_{j=1}^{n} \mathcal{H}_j^2 (dx^j)^2$ is in Stäckel form. The separation equations have the appearance (4.5) with $E = 0$.

Next we study the problem of (multiplicative) separation of variables for the Helmholtz (or Schrödinger) equation

$$(\Delta + V(\mathbf{x}))\Psi(\mathbf{x}) = E\Psi(\mathbf{x}), \tag{4.11}$$

on the pseudo-Riemannian manifold V^n. Here

$$\Delta = \sum_{i,j=1}^{n} \frac{1}{\sqrt{g}} \partial_i (\sqrt{g} g^{ij} \partial_j) \tag{4.12}$$

is the Laplace-Beltrami operator on V^n, defined independent of local coordinates [11]. To convert this product separation problem, $\Psi = \Pi_{i=1}^{n} \Psi^{(i)}(x^i)$, to the standard additive

separation form, we introduce the new dependent variable $u = \ln \Psi$. Further, we restrict ourselves to orthogonal separable systems

$$ds^2 = \sum_{i=1}^{n} H_i^2 (dx^i)^2.$$

Then (4.11) becomes

$$H \equiv \sum_{i=1}^{n} [H_i^{-2}(u_{ii} + u_i^2) + s_i u_i] + V = E, \tag{4.13}$$

where

$$s_i = \frac{1}{\sqrt{g}} \partial_i (\sqrt{g} H_i^{-2}), \quad \sqrt{g} = H_1 H_2 \dots H_n. \tag{4.14}$$

The integrability conditions (2.11) for regular separation lead to (4.3), (Stäckel form), upon comparison of the coefficients of u_i^2. Comparison of the coefficients of u_{ii} in (2.11) yields the *Robertson condition*[13]

$$\partial_{ij} \ln(\sqrt{g} H_i^{-2}) = 0, \qquad i \neq j. \tag{4.15}$$

Comparison of the constant terms in (2.11) yields the conditions (4.7) on the potential $V(\mathbf{x})$, i.e., the potential must be expressable in the form (4.8) to permit separation. There are no additional consequences of the integrability conditions. The separation equations take the form

$$u_{ii} + u_i^2 + g_i(x^i) u_i + f_i(x^i) + \sum_{j=1}^{n} \lambda_j s_{ij}(x^i) = 0, \tag{4.16}$$

where $\lambda_1 = -E$.

It follows that every orthogonal coordinate system permitting product separation of the Helmholtz equation (4.11) corresponds to a Stäckel form; hence it permits additive separation of the Hamilton-Jacobi equation (4.1). Eisenhart has shown, [11], that the additional Robertson condition for product separation is equivalent to the requirement $R_{ij} = 0$ for $i \neq j$, where R_{ij} is the Ricci tensor of V^n expressed in the Stäckel coordinates $\{x^i\}$. It follows that the Robertson condition is automatically satisfied in Euclidean space, a space of constant curvature or any Einstein space.

More generally we can introduce the notion of R-separation for the Helmholtz equation (4.13) in orthogonal coordinates $\{x^i\}$. Here, R-separable solutions take the form $\Psi = \exp(R(\mathbf{x}))\Pi_{i=1}^{n} \Psi^{(i)}(x^i) = e^R \Theta$, where $R(\mathbf{x})$ is a fixed function, independent of parameters. If $R \equiv 0$ we have *separation*, and if $R(x) = \sum_{i=1}^{n} R^{(i)}(x^i)$ we have *trivial* R-separation.

Otherwise the R-separation is *nontrivial.* Writing $u = \ln\Theta = \ln\Psi - R$, we have the following generalization of (4.13):

(4.17)

$$H \equiv \sum_{i=1}^{n}\left[H_i^{-2}(u_{ii}+u_i^2)+(2H_i^{-2}\partial_i R+s_i)u_i+H_i^{-2}(\partial_{ii}R+(\partial_i R)^2)+s_i\partial_i R\right]+V=E.$$

Comparing the coefficients of u_i^2 in the integrability conditions (2.11) we again find the metric $ds^2 = \sum_{i=1}^{n} H_i^2(dx^i)^2$ must be in Stäckel form. Comparison of the coefficients of u_{ii} yields

$$\partial_{ij}[2R+\ln(\sqrt{g}H_i^{-2})]=0, \quad i\neq j. \tag{4.18}$$

Finally, comparison of the constant terms in (1.6) and use of (4.18) leads to requirement (4.7) for the *modified potential*,

$$\tilde{V} = V - \frac{1}{2}\sum_{i=1}^{n} H_i^{-2}(\partial_i \ell_i + \frac{1}{2}\ell_i^2), \tag{4.19}$$

where

$$\ell_i = \partial_i \ln(\sqrt{g}H_i^{-2}) = \partial_i \ln\frac{\sqrt{g}}{S}. \tag{4.20}$$

We see that whenever $\tilde{V}$ satisfies (4.7), hence (4.8), equation (4.11) permits orthogonal R-separation with

$$R = -\frac{1}{2}\ln\frac{g}{S} + \sum_{i=1}^{n} L^{(i)}(x^i), \tag{4.21}$$

where the functions $L^{(i)}$ are arbitrary. Thus through appropriate choice of V, every additively separable coordinate system $\{x^i\}$ for the zero-potential Hamilton-Jacobi equation can occur as a multiplicatively separable system for the Helmholtz equation. Details are given in reference [17].

The question arises whether nontrivial R-separation occurs for $V = 0$. From (4.15), (4.18) and Eisenhart's formulation of Robertson's condition as $R_{ij} = 0$, $i \neq j$, we see that only trivial orthogonal R-separation can occur in an Einstein space. However, as shown in reference [17], nontrivial R-separation can occur for $V = 0$, even in conformally flat spaces. An example is

$$\begin{aligned} &ds^2 = (x+y+z)[(x-y)(x-z)dx^2+(y-z)(y-x)dy^2+(z-x)(z-y)dz^2],\\ &e^R = (x+y+z)^{-\frac{1}{4}}. \end{aligned} \tag{4.22}$$

Orthogonal R-separation for the Laplace equation on V^n, $\Delta\Psi(\mathbf{x}) = 0$, can be treated by analogy to the foregoing [18,19], but will not be presented here.

Also, because of space and time limitations we shall not treat here the general case of (possibly) nonorthogonal separation for the Hamilton-Jacobi, Helmholtz, heat and Laplace equations. The details of the general case can be found in [18,20-25].

For explicit computations of the possible separable coordinate systems for an equation on a fixed manifold (e.g., Euclidean space or a space of constant curvature), see [1,15,26-30].

5. Intrinsic characterization of variable separation for scalar equations.

For the Hamilton-Jacobi, Helmholtz, and Laplace equations in V^n, introduced in the previous section, ($R-$)separable coordinate systems can always be characterized intrinsically, i.e., in a coordinate-free manner. Consider first the problem of additive orthogonal separation for the Hamilton-Jacobi equation (4.1). Let H be the quadratic form

$$H = \sum_{i,j=1}^{n} g^{ij} u_i u_j, \quad u_i = \partial_{y_i} u, \tag{5.1}$$

where $(g^{ij}(y))$ is the metric expressed in terms of the general coordinates $\{y^k\}$ and u is the dependent variable. Expression (5.1) is invariant under a change of independent coordinates. Now assume that (4.1) is separable in the orthogonal coordinates $\{x^i\}$. From the separation equations (4.5), we are led to the quadratic forms $(\sum_{j=1}^n A^{\ell j} s_{jk}(x^j) = \delta_k^\ell)$

$$\mathcal{A}^\ell = \sum_{j=1}^{n} A^{\ell j} u_j^2 \quad \ell = 1, 2, \ldots, n, \tag{5.2}$$

which can easily be shown to have the following properties:

(1) $\mathcal{A}^1 = H$.

(2) The n element set $(\mathcal{A}^\ell)$ is linearly independent (as a set of quadratic forms).

(3) The differentials of the separable coordinates, $\omega^j = dx^j$, constitute a simultaneous eigenbasis for the $(\mathcal{A}^\ell)$. [Here, ρ is a root of a quadratic form $\mathcal{A} = (a^{ij})$ with respect to the metric g^{ij} if $\det(a^{ij} - \rho g^{ij}) = 0$, and $\omega = \sum_\ell \lambda_\ell dx^\ell$ is an eigenform corresponding to ρ if $\omega \neq 0$ and $\sum_{j=1}^n (a^{ij} - \rho g^{ij})\lambda_j = 0$.]

(4) $\mathcal{A}^\ell(\mathbf{x}, u_x) = -\lambda_\ell$ for each additively separable solution $\mathbf{u} = \sum_{i=1}^n u^{(i)}(x^i, \lambda)$ of (4.1).

The following converse of these statements holds [24,31]:

THEOREM 3. *Let $(\mathcal{A}^\ell)$, $\mathcal{A}^1 = H$, be a linearly independent set of n second order symmetric quadratic forms such that*

(1) $\{\mathcal{A}^\ell, \mathcal{A}^m\} = 0$, $1 \le \ell, m \le n$,

(2) *The $(\mathcal{A}^\ell)$ have a common eigenbasis $\{\omega^{(j)}\}$.*

Then there is a separable coordinate system $\{x^j\}$ for the Hamilton-Jacobi equation $H(\mathbf{y}, v_y) = E$ on V^n such that $\omega^{(j)} = f^{(j)}(x)dx^j$ for some functions $f^{(j)}$. The separable solutions u are determined by $\mathcal{A}^\ell(\mathbf{x}, u_\mathbf{x}) = -\lambda_\ell$.

The main point of this theorem is that, under the required hypotheses the eigenforms ω^ℓ of the quadratic forms a^{ij} are normalizable, i.e., that up to multiplication by a nonzero function, ω^ℓ is the differential of a coordinate. This fact, which is proved through use of the commutation relations $\{\mathcal{A}^\ell, \mathcal{A}^m\} = 0$ to verify appropriate integrability conditions, permits us to compute the coordinates directly from a knowledge of the symmetry operators.

As an example of the use of Theorem 3 we consider the Hamilton-Jacobi equation for two dimensional Minkowski space. In cartesian coordinates this equation is

$$H \equiv u_x^2 - u_t^2 = E.$$

The vector space of all symmetries of the form $\mathcal{L} = a(x,t)u_x + b(x,t)u_y$ (Killing vectors) is easily shown to be closed under the bracket $\{\cdot,\cdot\}$; hence the symmetries form a Lie algebra. Furthermore we have $\{H, \mathcal{L}\} \equiv 0$ for each linear symmetry. The Lie algebra is three dimensional over the field of real scalars, with basis

$$\begin{aligned} &\mathcal{L}_1 = u_x, \qquad \mathcal{L}_2 = u_t, \qquad \mathcal{L}_3 = tu_x + xu_t, \\ &\{\mathcal{L}_1, \mathcal{L}_2\} = 0, \quad \{\mathcal{L}_3, \mathcal{L}_1\} = \mathcal{L}_2, \quad \{\mathcal{L}_3, \mathcal{L}_2\} = \mathcal{L}_1. \end{aligned} \tag{5.3}$$

It can be shown that every symmetry which is quadratic in the first derivatives of u (second order Killing tensor) is a polynomial in the linear symmetries $\mathcal{L}_i$. (This is true for all spaces of constant curvature, e.g., [24].) Thus all candidates for variable separation can be built from the basis symmetries (5.3).

Consider, for example, the quadratic symmetry $\mathcal{A}^2 = 2\mathcal{L}_3\mathcal{L}_1$. With respect to cartesian coordinates, the corresponding symmetric quadratic forms are

$$\mathcal{A}^2 \sim \begin{pmatrix} 2t & x \\ x & 0 \end{pmatrix}, \quad \mathcal{A}^1 = \mathcal{H} \sim \begin{pmatrix} 1 & 0 \\ 0 & -1 \end{pmatrix}.$$

Clearly, $\mathcal{A}^2$ has roots $\rho = t \pm \sqrt{t^2 - x^2}$ (assuming $t > |x|$) with a basis of eigenforms $\omega_1 = (t + \sqrt{t^2 - x^2})dx - xdt$, $\omega_2 = (t - \sqrt{t^2 - x^2})dx - xdt$. By Theorem 3, $\mathcal{A}^2$ does define a regular separable coordinate system $\{\xi^1, \xi^2\}$ for (5.2) and there exist functions f_i such that $d\xi^i = f_i\omega_i$, $i = 1, 2$. We find $f_1 = [\xi^2((\xi^2)^2 - (\xi^1)^2)]^{-1}$, $f_2 = -[\xi^1((\xi^2)^2 - \xi^1)^2)]^{-1}$,

$$t = \frac{1}{2}((\xi^1)^2 + (\xi^2)^2), \qquad x = \xi^1\xi^2. \tag{5.4}$$

On the other hand the symmetry

$$\mathcal{A} = 2\mathcal{L}_3(\mathcal{L}_1 - \mathcal{L}_2) \tag{5.5}$$

has two equal roots and only one eigenform. Thus $\mathcal{A}$ cannot determine a separable coordinate system.

For manifolds of dimension $n \geq 3$ there is a second way that a system of $n-1$ commuting symmetries may fail to determine separable coordinates: although each quadratic symmetry determines a basis of eigenforms, there is no basis of eigenforms for all symmetries simultaneously. See [24] for a simple example. Reference [31] considers *partial* separation for Hamilton-Jacobi equations.

These results extend to R-separation of the Helmholtz equation on V^n,

$$\Delta\Psi(\mathbf{x}) = E\Psi(\mathbf{x}), \tag{5.6}$$

where in local coordinates

$$\Delta = \frac{1}{\sqrt{g}} \sum_{i,j=1}^{n} \partial_i(\sqrt{g} g^{ij} \partial_j). \tag{5.7}$$

Here a linear differential operator A on V^n is a *symmetry operator* for Δ if

$$[\Delta, A] \equiv \Delta A - A\Delta = 0. \tag{5.8}$$

[This agrees with the bracket (3.16).] Note that uniquely associated with every second order symmetry operator

$$A = \sum_{\substack{i=1\\j=1}}^{n} a^{ij}\partial_{ij} + \sum_{i=1}^{n} b^i \partial_i + c$$

in local coordinates $\{y^\ell\}$, is the second order Killing tensor

$$\mathcal{A} = \sum_{i,j=1}^{n} a^{ij} u_i u_j.$$

Indeed, $[\Delta, A] = 0$ implies $\{\mathcal{H}, \mathcal{A}\} = 0$, though the converse is false.

THEOREM 4. *Necessary and sufficient conditions for the existence of an orthogonal R-separable coordinate system $\{x^j\}$ for the Helmholtz equation $\Delta\Psi = E\Psi$ are that there exist n second order differential operators $A^1 = \Delta, A^2, \ldots, A^n$ such that*

(1) $[A^i, A^j] = 0, \quad 1 \leq i, j \leq n.$

(2) *The associated set of Killing tensors $(\mathcal{A}^i)$ is linearly independent.*

(3) *There is a basis $\{\omega^{(j)} : 1 \leq j \leq n\}$ of simultaneous eigenforms for the $\mathcal{A}$.*

If these conditions are satisfied then there exist functions $f^j(\mathbf{x})$ such that $\omega^{(j)} = f^j dx^j$, $1 \leq j \leq n$. Theorem 4 follows from Theorem 3 through exploitation of the commutation relations (1). Indeed, these relations can be used to show that the separation conditions for $\tilde{V}$ are valid.

COROLLARY 3. *Suppose the second order differential operators $A^1 = \Delta, A^2, \ldots, A^n$ satisfy conditions 1-3 of Theorem 4 and in addition that they are self-adjoint form with respect to the measure $dV = \sqrt{g}dy^1 \ldots dy^n$:*

$$A^\ell = \frac{1}{\sqrt{g}} \sum_{\substack{i=1 \\ j=1}}^{n} \partial_{y^i}(\sqrt{g} a^{ij}_{(\ell)} \partial_{y^j}) + c_\ell(\mathbf{y}), \quad \ell = 1, 2, \ldots, n, \tag{5.9}$$

(a form which is independent of the choice of local coordinates $\{y^j\}$). Then the R-separable solutions $\Psi = e^R \prod_{i=1}^n \Psi^{(i)}(x^i)$ of $\Delta\Psi = E\Psi$ are characterized as the eigenfunctions of the A^ℓ:

$$A^\ell \Psi = -\lambda_\ell \Psi, \quad \ell = 1, 2, \ldots, n, \tag{5.10}$$

where $\lambda_1 = -E$ and $\lambda_2, \ldots, \lambda_n$ are separation constants.

This result is implied by theorems stated without published proof by Shapovalov, e.g. [18,33,34]. The results were obtained independently and proved in [17].

For the Hamilton-Jacobi equation

$$H(x, \partial_i u) = 0 \tag{5.11}$$

i.e., $E = 0$, there is an analogous characterization of additive separation by conformal symmetries. A function $\mathcal{L}(x, \partial_i u)$ is said to be a *conformal symmetry* provided there is a function $q(x, \partial_i u)$ such that

$$\{\mathcal{L}, H\} = qH. \tag{5.12}$$

THEOREM 5 [16,31],. *Necessary and sufficient conditions for the existence of an orthogonal separable coordinate system $\{x_j\}$ for the Hamilton-Jacobi equation (5.11) are that there exist $n-1$ symmetric quadratic functions $\mathcal{B}^k = \sum_{i,j=1}^n b^{ij}_{(k)}(\mathbf{x}) u_i u_j$, $2 \le k \le n$, such that*

(1) *Each $\mathcal{B}^k$ is a conformal symmetry.*
(2) *$\{\mathcal{B}^i, \mathcal{B}^j\} = 0$, $2 \le i, j \le n$.*
(3) *The set $(H, \mathcal{B}^k)$ is linearly independent (as n quadratic forms).*
(4) *There is a basis $\{\omega^{(j)} : 1 \le j \le n\}$ of simultaneous eigenforms for the $\{\mathcal{B}^k\}$.*

If conditions **1-4** *are satisfied then there exist functions $f^j(\mathbf{x})$ such that $\omega^{(j)} = f^j dx^j$, $1 \le j \le n$.*

Again, these results extend to R-separation of the Laplace equation on V^n:

$$\Delta\Psi(\mathbf{x}) = 0, \tag{5.13}$$

see [18,19,34] .

The preceding theorems characterize orthogonal R-separation for Hamilton-Jacobi, Helmholtz, and Laplace equations in terms of symmetries. There are similar results for nonorthogonal separation of these equations, though the results are more complicated to state and prove: see [20,25,31,34].

We can reach several important conclusions concerning variable separation and R-separation for Hamilton-Jacobi, Helmholtz, and Laplace (or wave) equations. First, one must recognize the intrinsic geometric nature of R-separation. The apparently technical conditions for R-separation are equivalent to the existence of an n-dimensional family of commuting symmetry operators which can be simultaneously diagonalized. In spaces, such as those of constant curvature, for which all symmetry operators can be constructed from the Lie symmetry algebra, all R-separation questions become problems in algebra [1]. These problems have been solved explicitly for the Hamilton-Jacobi and Helmholtz equations on n-dimensional spheres and n-dimensional Euclidean space, [26,27,35].

Second, comparing Theorems 3 and 4, it is obvious that R-separation, not ordinary separation, for the Helmholtz equation is the natural analogy of additive separation for the Hamilton-Jacobi equation. Finally, we note the close relationship between variable separation and quantization theory. Corresponding to a separable system $\{x^j\}$ for the Hamilton-Jacobi equation we have an involutive family $\{\mathcal{A}^\ell\}$ of quadratic constants of the motion. The Helmholtz equation R-separates in these same coordinates if and only if second order operators $\{A^\ell\}$ can be found (with the pure second order terms in A^ℓ agreeing with those of $\mathcal{A}^\ell$) such that the A^ℓ pairwise commute.

There are similar results relating symmetry and R-separation for heat and time-dependent Schrödinger equations, [1,23,33,29,34].

For early references to variable separation see [14,36,37].

6. Spinor equations.

In this section our aim will be to study multiplicative separation for spinor equations that take the form of the Dirac equation, i.e.,

$$H\psi(x) = E\psi(x) \tag{6.1}$$

where H is an $N \times N$ first order matrix differential operator in the variables $x^1, \ldots, x^n$, ψ is an N-component spinor and E is an eigenvalue. However, it is instructive to begin by studying additive separation for the system

$$\sum_{i=1}^{n} H^i \partial_i \varphi(x) = D \tag{6.2}$$

where $\partial_i = \partial_{x^i}$,

$$D = \begin{pmatrix} E^{(1)} & & 0 \\ & \ddots & \\ 0 & & E^{(N)} \end{pmatrix},$$

φ is an $N \times N$ matrix, and the $N \times N$ matrices $H^i(x)$ are nonsingular. Applying Theorem 1, we see that necessary and sufficient conditions that the coordinates $\{x^i\}$ are a regular separable system for (6.2) are

$$\partial_{jk} H^i = \partial_j H^k (H^k)^{-1} \partial_k H^i + \partial_k H^j (H^j)^{-1} \partial_j H^i, \quad j \neq k. \tag{6.3}$$

(Note that for $N = 1$ equations (6.3) agree with the Levi-Civita conditions (4.3) where $H^i \equiv H_i^{-2}$.)

We will determine the significance of these conditions. Suppose $S_{ij}(x)$, $1 \leq i, j \leq n$ are $N \times N$ matrices such that $\partial_k S_{ij}(x) = 0$ for $k \neq i$ and $\det S = \det(S_{ij}) \neq 0$ where S is a (bordered) $Nn \times Nn$ matrix. Let $A = S^{-1} = (A^{ij})$ where each A^{ij} is an $N \times N$ matrix. Then

$$\sum_{i=1}^{n} A^{ki} S_{ij} = \sum_{i=1}^{n} S_{ji} A^{ik} = \delta_j^k, \quad 1 \leq j, k \leq n. \tag{6.4}$$

Now consider the separated equations

$$\varphi_i = \sum_{j=1}^{n} S_{ij}(x^i) D^{(j)}, i = 1, \ldots, n \tag{6.5}$$

where each $D^{(j)}$ is an $N \times N$ diagonal matrix with independent parameters $d_I^{(j)}$, $1 \leq I \leq N$, as diagonal elements. Suppose $D^{(1)} = D$ and set $A^{1i} = H^i$. Then it follows from (6.4) that

$$\sum_{i=1}^{n} H^i \varphi_i = D, \tag{6.6}$$

and, under the assumption that each H^i is nonsingular, equation (6.6) has a family of separated solutions depending on $N(n+1)$ constants. Hence $\{x^i\}$ is a regular separable system and the matrices H^i must satisfy conditions (6.3).

To prove the converse, we first suppose that $(S_{ij}(x^i))$ is a nonsingular bordered $nN \times nN$ matrix such that conditions (6.5) hold with $A^{1\ell} \equiv H^\ell$ nonsingular for $1 \leq \ell \leq n$. It follows from the second equation (6.4) that

$$\sum_{i=1}^{n} S_{ji} \partial_\ell A^{ik} = 0, \quad j \neq \ell.$$

Hence

$$\partial_\ell A^{ik} = A^{i\ell} f(\ell, k, x) \tag{6.7}$$

where f is an $N \times N$ matrix, independent of i. Setting $i = 1$ we find

$$f(\ell, k, x) = (H^\ell)^{-1}\partial_\ell H^k.$$

Hence,

$$\partial_\ell A^{ik} = A^{i\ell}(H^\ell)^{-1}\partial_\ell H^k. \tag{6.8}$$

As is easily checked, the integrability conditions $\partial_j(\partial_\ell A^{ik}) = \partial_\ell(\partial_j A^{ik})$, $j \neq \ell$, for this system of equations are precisely (6.3). Conversely, if conditions (6.3) hold for nonsingular H^k, we can find solutions (A^{ij}) of (6.8) such that $A^{1k} = H^k$ and (locally) $\det A = \det(A^{ik}) \neq 0$. Then, differentiating the second equation (6.4) with respect to x^ℓ and making use of (6.8), we find

$$\sum_{i=1}^{n} \partial_\ell S_{ji} A^{ik} = 0, \qquad j \neq \ell,$$

so that $\partial_\ell S_{ji}(x) = 0$ for $j \neq \ell$. This proves that the separation equations (6.5) are the mechanism for regular variable separation in equations (6.2) and that the conditions (6.3) imply that the matrices H^i are in Stäckel form: $H^i = (S^{-1})^{1i}$ where $S_{ij} = S_{ij}(x^i)$. (The scalar case $N = 1$ corresponds to the classical Stäckel form [14].)

THEOREM 6. *Suppose $\{H^{(i)}\}$ satisfies equations (6.3), i.e., it is in Stäckel form. Let $\{C_{ij}(x^i)\}$ be a particular Stäckel form matrix:*

$$\sum_{i=1}^{n} H^{(i)} C_{ij}(x^i) = \delta_j^1 I, \quad 1 \leq j \leq n,$$
$$\det(C_{ij}) \neq 0.$$

Then the most general Stäckel form matrix $\{D_{ij}(x^i)\}$ associated with $\{H^{(i)}\}$ can be expressed as $D_{ij}(x^i) = \sum_{\ell=1} C_{i\ell}(x^i) F_j^\ell$ where $\{F_j^\ell\}$ is a constant nonsingular (bordered) $Nn \times Nn$ matrix such That $F_j^1 = \delta_j^1 I$

If we require regular separation of the system

$$\sum_{i=1}^{n} H^i \partial_i \varphi(x) + V(x) = D \tag{6.9}$$

where assumptions (6.2) hold and $V(x)$ is an $N \times N$ matrix, the separation conditions are (6.3) again (which imply that the H^i are in Stäckel form) and

$$\partial_{jk} V = \partial_j H^k (H^k)^{-1} \partial_k V + \partial_k H^j (H^j)^{-1} \partial_j V, \qquad j \neq k. \tag{6.10}$$

Condition (6.10) simply says that

$$V = \sum_{i=1}^{n} H^i S_i(x^i) \tag{6.11}$$

for some $N \times N$ matrices S_i with $\partial_j S_i = 0$, $i \neq j$. Indeed it is straightforward to verify (using (6.3)) that this expression satisfies (6.10). On the other hand, suppose that $V'(x)$ is an invertible matrix such that the matrices $H'^i = (V')^{-1}H^i$ are also in Stäckel form. Then both the $\{H'^i\}$ and the $\{H^i\}$ satisfy (6.3). A direct computation shows that the necessary and sufficient condition that V' maps the Stäckel form H^i to another Stäckel form is that it satisfies (6.10). Thus if V' satisfies (6.10) and is invertible then $(V')^{-1}H^i = A'^{1i}$ where $\sum_i A'^{ji}C'_{i\ell}(xi) = \delta^j_\ell$. Thus $\sum_i (V')^{-1}H^i C'_{i1}(x^i) = 1$ or $V' = \sum_i H^i C'_{i1}(x^i)$ which is of the form (6.11). In particular the identity matrix I satisfies (6.10) and is nonsingular so there exist matrices $C_i(x^i)$ such that $I = \sum_i H^i C_i(x^i)$. Now let V be any solution of (6.10). Clearly we can always find a constant μ such that $V(x)+\mu I \equiv V'(x)$ is locally nonsingular. Since V' is also a solution of (6.10) we can find matrices $C'_{i1}(x^i)$ such that

$$V'(x) = \sum_i H^i C'_{i1}(x^i) = V(x) + \mu \sum_i H^i C_i(x^i).$$

Setting, $S_i(x^i) = C'_{i1}(x^i) - \mu C_i(x^i)$ we obtain (6.11).

The appropriate separation equations for this case are

$$\varphi_i = \sum_{j=1}^{n} S_{ij}(x^i)D^{(j)} + S_i(x^i), \qquad i = 1, \ldots, n.$$

It follows easily that the separated solutions φ satisfy the "eigenvalue" equations

$$\sum_{i=1}^{n} A^{ki}\varphi_i + V^k = D^{(k)}, \qquad 1 \leq k \leq n. \tag{6.12}$$

where $V^k = \sum_{i=1}^{n} A^{ki}S_i$. However the expressions on the left-hand side of (6.12) have no apparent connection with symmetries of (6.9).

We *can* obtain a relation between symmetry and separation for systems of linear equations. To motivate this connection let us consider the system (6.9) again where the diagonal matrix $D = EI$, a multiple of the identity matrix I. We write $\varphi(x) = \sum_{i=1}^{n} \varphi^{(i)}(x^i)$. Then we can obtain (multiplicative) separable solutions of the linear equation

$$\sum_{i=1}^{n} H^i \partial_i \psi + V\psi = E\psi$$

for the N-component spinor ψ if $\psi = \Pi_{i=1}^{n}\Theta^{(i)}(x^i)\xi$ where $\Theta^{(i)} = \exp\varphi^{(i)}$ and ξ is a constant N-spinor. This characterization of ψ will hold provided $\Theta^{(i)}\Theta^{(j)} = \Theta^{(j)}\Theta^{(i)}$ for $1 \le i, j \le n$.

With these motivational items out of the way, we proceed to an examination of multiplicative separation for equations of Dirac type:

$$\mathbf{H}\psi = E\psi, \qquad E \text{ constant} \tag{6.13}$$

where

$$\mathbf{H} = \sum_{i=1}^{n} H^i(x)\partial_i + V(x),$$

H^i, V are $N \times N$ matrices and ψ is an N-component spinor. We require that the H^i, $1 \le i \le n$, are nonsingular matrices. We define a *factorizable system* for (6.13) as a set of equations

$$\partial_i\psi = (\sum_{j=1}^{n} C_{ij}(x)\lambda^j - C_i(x))\psi \qquad i = 1, \ldots, n. \tag{6.14}$$

where the $C_{ij}(x)$, $C_i(x)$ are $N \times N$ matrices such that $\det(C_{ij}) \neq 0$, the λ^j are independent parameters with $\lambda^1 = E$, and for every initial point x^0 and N-spinor ξ there is a solution $\psi(x)$ of (6.14) which also satisfies (6.13) and has the property $\psi(x^0) = \xi$. The integrability conditions for (6.14): $\partial_i(\partial_j\psi) = \partial_j(\partial_i\psi)$, $i \neq j$, imply

$$\begin{aligned} &a) \quad C_{jk}C_{i\ell} + C_{j\ell}C_{ik} = C_{ik}C_{j\ell} + C_{i\ell}C_{jk}, \\ &b) \quad \partial_j C_{ik} - C_{ik}C_j - C_iC_{jk} = \partial_i C_{jk} - C_{jk}C_i - C_jC_{ik}, \\ &c) \quad \partial_i C_j + C_iC_j = \partial_j C_i + C_jC_i. \end{aligned} \tag{6.15}$$

Let the $nN \times nN$ bordered matrix $A = (A^{ij})$ be the inverse of (C_{jk}):

$$\sum_{j=1}^{n} A^{ij}C_{jk} = \sum_{j=1}^{n} C_{kj}A^{ji} = \delta^i_k I. \tag{6.16}$$

It follows that the solutions ψ of (6.14) satisfy the eigenvalue equations

$$\mathcal{A}^k\psi \equiv \sum_{i=1}^{n} A^{ki}(x)\partial_i\psi + B^k(x)\psi = \lambda^k\psi, \quad 1 \le k \le n.$$

Clearly,

$$\begin{aligned} A^{1i}(x) = H^i(x), B^1(x) = V(x), \\ B^i(x) = \sum_{j=1}^{n} A^{ij}(x)C_j(x). \end{aligned} \tag{6.17}$$

THEOREM 7. *The integrability conditions (6.15) for a factorizable system are satisfied if and only if there exists $N \times N$ matrices $A^{ki}(x)$, $B^k(x)$, $1 \le k, i \le n$ such that*

(1) *The operators*

$$\mathbf{A}^k = \sum_{i=1}^{n} A^{ki}\partial_i + B^k, \qquad k = 1, \ldots, n$$

commute, i.e.,

$$\mathbf{A}^k\mathbf{A}^\ell = \mathbf{A}^\ell\mathbf{A}^k.$$

(2) $\mathbf{H} = \mathbf{A}^1$

(3) $\sum_{i=1}^{n} A^{ki}C_{ij} = I\delta_j^k$

where I is the $N \times N$ identity matrix.

Proof. Condition (1) is equivalent to the requirements

$$\text{(6.18)}\quad \begin{aligned} &i) && A^{ji}A^{k\ell} + A^{j\ell}A^{ki} = A^{ki}A^{j\ell} + A^{k\ell}A^{ji}, \\ &ii) && \sum_i (A^{ji}\partial_i A^{k\ell} - A^{ki}\partial_i A^{j\ell}) = A^{k\ell}B^j - A^{j\ell}B^k \\ &&& \qquad + B^k A^{j\ell} - B^j A^{k\ell}, \\ &iii) && \sum_i (A^{ji}\partial_i B^k - A^{ki}\partial_i B^j) = B^k B^j - B^j B^k. \end{aligned}$$

Let us assume that conditions (6.15) hold and define

$$B_{i\ell;hm} = C_{ih}C_{\ell m} + C_{im}C_{\ell h} = C_{\ell h}C_{im} + C_{\ell m}C_{ih}$$

(by (6.15a)). It follows that

$$\text{(6.19)}\qquad \sum_{i,\ell}(A^{ki}A^{j\ell} + A^{k\ell}A^{ji} - A^{j\ell}A^{ki} - A^{ji}A^{k\ell})B_{i\ell;hm} = 0,$$

for all k, j, h, m. But

$$\sum_{i,\ell} A^{k\ell}A^{ji}B_{i\ell;hm} = (\delta_m^k\delta_h^j + \delta_h^k\delta_m^j)I,$$

so $(B_{i\ell,hm})$ is invertable. It follows that (6.18i) holds. A straightforward computation shows that (6.18iii) is a direct consequence of (6.15c), (6.16) and (6.17). Multiplying (6.15b) on the right by $A^{k\ell}$, summing on k and using the identity $\sum_k \partial_j C_{ik}A^{k\ell} = -\sum_k C_{ik}\partial_j A^{k\ell}$, we obtain

$$\begin{aligned} &\sum_k (C_{ik}\partial_j A^{k\ell} + C_{ik}C_j A^{k\ell}) + C_i\delta_j^\ell \\ &\qquad = \sum_k (C_{jk}\partial_i A^{k\ell} + C_{jk}C_i A^{k\ell}) + C_j\delta_i^\ell. \end{aligned}$$

Multiplying this equation on the left by

$$A^{tj}A^{si} - A^{sj}A^{ti},$$

summing on j, i and using (6.18i) we obtain (6.18ii). The proof of the converse is similar.

This result implies that a factorizable system is equivalent to the existence of a family of $n-1$ commuting matrix symmetry operators $\mathbf{A}^k$ of $\mathbf{H}\psi = E\psi$ such that $\det(A^{ki}) \neq 0$. Note that conditions (6.15) and (6.18) take the same form in all coordinate systems. Moreover, by solving each of the ordinary differential equations successively in a given coordinate system we can "factor" a solution as a product of n $N \times N$ matrices. For example, if $n = 2$ a solution might take the form

$$\psi(x, y) = \Theta^{(1)}(x, y, y^0)\Theta^{(2)}(x, x^0, y^0)\xi$$

where $\Theta^{(1)}(x^0, y^0, y^0) = \Theta^{(2)}(x^0, x^0, y^0) = I$.

Factorizable systems also retain their form under changes of frame $\psi = R(x)\psi'$ where $R(x)$ is a given nonsingular $N \times N$ matrix (independent of the parameters λ^j). The symmetry operators $\mathbf{A}^k$ acting on ψ induce symmetry operators

$$(6.20)\qquad \begin{aligned} \mathbf{A}'^k = R^{-1}\mathbf{A}^k R = \sum_i (R^{-1}A^{ki}R\partial_i + \\ R^{-1}A^{ki}\partial_i R) + R^{-1}B^k R \\ = \sum_i A'^{ki}\partial i + B'^k \end{aligned}$$

acting on ψ'. By choosing R appropriately it is always possible to transform the factorizable system in a form where $B'^k \equiv C'_k \equiv 0$. Indeed, the condition that $B'^k \equiv 0$ corresponds to the equations

$$(6.21)\qquad \partial_i R = -C_i R.$$

The integrability conditions for these equations are identical to (6.15c). Thus a solution R exists and is determined uniquely (up to multiplication on the right by a constant nonsingular matrix). It follows from (6.17) that if $B'^k \equiv 0$ then $C'_k \equiv 0$.

We say that a factorizable system is *separable* if (by a change of frame $\psi = R\psi'$ if necessary) the equations (6.14) take the form

$$(6.14)'\qquad \partial_i\psi' = \left(\sum_{j=1}^n C_{ij}(x^i)\lambda^j - C_i(x^i)\right)\psi' \qquad i = 1, \ldots, n.$$

in a *particular* coordinate system $\{x^1, \ldots, x^n\}$, i.e., $\partial_\ell C_{ij} = \partial_\ell C_i = 0$ if $\ell \neq i$.

Let $\mathcal{C}_i(x^i, \lambda) = \sum_{j=1}^n C_{ij}(x^i)\lambda^j - C_i(x^i)$ and let ξ be a constant N-spinor.

THEOREM 8. *Suppose (6.14)′ is a separable system for (6.13) in the coordinates $\{x^\ell\}$ and let $\mathbf{x}_0$ be a fixed vector. Then (6.1) admits solutions of the form*

$$\psi(\mathbf{x}) = R(\mathbf{x})\Theta^{(1)}(x^1)\dots\Theta^{(n)}(x^n)\xi \tag{6.22}$$

where the Θ^ℓ are $N \times N$ matrices such that $\Theta^{(i)}(x^i)\Theta^{(j)}(x^j) = \Theta^{(j)}(x^j)\Theta^{(i)}(x^i)$ for all i, j and $\Theta^{(i)}(x_0^i) = I$.

Proof. Here R is the frame change that leads to (6.14)′. Consider the equation

$$\partial_\ell\Theta^{(\ell)}(x^\ell) = \mathcal{C}_\ell(x^\ell, \lambda)\Theta^{(\ell)}(x^\ell),$$
$$\Theta^{(\ell)}(x_0^\ell) = I.$$

This equation with initial condition has a unique solution for each ℓ. It follows from eqns. (6.15) that $\mathcal{C}_i(x^i, \lambda)\mathcal{C}_j(x^j, \lambda) = \mathcal{C}_j(x^j, \lambda)\mathcal{C}_i(x^i, \lambda)$ for $i \neq j$. We claim that

$$\Theta^{(i)^{-1}}(x^i)\mathcal{C}_j\Theta^{(i)}(x^i) = \mathcal{C}_j$$

for $i \neq j$. Indeed this is clearly true for $x^i = x_0^i$ and

$$\begin{aligned}\partial_i(\Theta^{(i)-1}\mathcal{C}_j\Theta^{(i)}) &= -\Theta^{(i)-1}\mathcal{C}_i\Theta^{(i)}\Theta^{(i)-1}\mathcal{C}_j\Theta^{(i)} \\ &\quad + \Theta^{(i)-1}\mathcal{C}_j\mathcal{C}_i\Theta^{(i)} \\ &= \Theta^{(i)-1}(\mathcal{C}_j\mathcal{C}_i - \mathcal{C}_i\mathcal{C}_j)\Theta^{(i)} = 0\end{aligned}$$

by (6.15c).

It follows that $\Theta^{(j)-1}\Theta^{(i)}\Theta^{(j)} \equiv \Theta^{(i)}$ for all x^i, x^j near x_0^i, x_0^j. Indeed this is true for $x^j = x_0^j$ and

$$\begin{aligned}\partial_j(\Theta^{(j)-1}\Theta^{(i)}\Theta^{(j)}) &= -\Theta^{(j)-1}\mathcal{C}_j\Theta^{(i)}\Theta^j \\ &\quad + \Theta^{(j)-1}\Theta^{(i)}\mathcal{C}_j\Theta^j \\ &= \Theta^{(j)-1}(\Theta^{(i)}\mathcal{C}_j - \mathcal{C}_j\Theta^{(i)})\Theta^{(j)} \\ &= 0.\end{aligned}$$

Setting $\psi' = \Theta^{(1)}\dots\Theta^{(n)}\xi$ we see that ψ' satisfies (6.14)′ with $\psi'(\mathbf{x}_0) = \xi$. □

The "Dirac" equation corresponding to (6.14)′ is of the form $\sum H^i\partial_i\psi' + V\psi' = E\psi'$ with $E = \lambda^1$, $H^i = A^{1i}$, $V = \sum_{j=1}^n H^jC_j(x^j)$. It follows immediately that H^i is in Stäckel form, i.e., it satisfies equations (6.4) and that V is a Stäckel multiplier, i.e., it satisfies equations (6.10). (On the other hand, a set of Stäckel form matrices H^i and a Stäckel multiplier V in a given coordinate system do not necessarily correspond to a multiplicative separable system.)

Theorem 9. *Suppose $\{x^i\}$ is an $(R-)$separable coordinate system for the equation*

$$(\sum_i H^i(x)\partial_i + (V + \mu W))\psi = E\psi$$

where μ is a parameter, $H^i, V, W, \{x^i\}$ are independent of μ and W is nonsingular. Then $\{x^i\}$ is also an $(R-)$separable coordinate system for the equation

$$(\sum_i W^{-1}H^i\partial_i + W^{-1}V)\psi' = E\psi'.$$

This result generalizes a result of [38] in the case $N = 1$.

It is easy to find examples of factorizable systems for which there is no associated separable or R-separable coordinate system. The author is not aware of a general (coordinate independent) criterion for determining when a factorizable system is separable. Moreover, a characterization of the separable coordinates in terms of properties of the commuting symmetry operators $\mathbf{A}^k$ in Theorem 7 is still open.

In the case of the Dirac equation itself somewhat more is known, but many problems remain. Corresponding to arbitrary local coordinates $\{x^1, x^2, x^3, x^4\}$ on a four-manifold with contravariant metric tensor (g^{ij}) of signature $(1, 1, 1, -1)$ the Dirac equation takes the form

$$(\sum_{i=1}^{4} \gamma^i(\partial_i - \Gamma_i) + m)\psi = 0. \tag{6.23}$$

Here γ^i, Γ_i are 4×4 matrices, $m \neq 0$ is the rest mass of the electron and ψ is a 4-spinor. The γ^i are determined up to similarity by the requirement

$$\gamma^i\gamma^j + \gamma^j\gamma^i = 2g^{ij}I, \quad 1 \leq i, j \leq 4. \tag{6.24}$$

Expressions for the Christoffel matrices $\Gamma_i(x)$ can be found in [39-41], for example. (In Minkowski space with cartesian coordinates $\{x, y, z, t\}$ and diagonal metric $(1, 1, 1, -1)$ we have $\Gamma_i \equiv 0$.) For a fixed orthogonal coordinate system $\{g^{ij}\}$ is diagonal and $g^{ii} \neq 0$. It follows that each of the matrices $\gamma^i(x)$ is nonsingular so that the general theory developed in this section applies. All of the examples of variable separation for the Dirac equation known to the author are separable systems in the sense of $(6.14)'$. Most of the known examples occur for Minkowski space and correspond to symmetry operations associated with the evident Poincaré symmetries of (6.23) and or to Casimir operators associated with subalgebras of the Poincaré group. A Poincaré symmetry takes the form $\mathbf{A} = Id + B$ where $d(x)$ is a (scalar) first order partial differential operator, I is the identity matrix and $B(x)$ is a 4×4 matrix. For such a symmetry, standard local Lie theory can be used to show that a coordinate system $\{y^i\}$ exists such that $\mathbf{A} = I\partial_{y^1} + B$. For independent commuting

Poincaré symmetry operators $\mathbf{A}^1, \ldots, \mathbf{A}^\ell$, $1 \le \ell \le 3$, we can construct a coordinate system $\{y^i\}$ such that $\mathbf{A}^i = I\partial_{y^i} + B^i$. See [42-86] for examples. New interest in variable separation for the Dirac equation was ignited when Chandrasekhar [49] found a separable system for this equation in a Kerr metric gravitational background, i.e., in the geometry of a spinning black hole [50]. In the limit that the Kerr metric degenerates to a Minkowski space metric, Chandrasekhar's separable system led to a new Minkowski separable system corresponding to oblate spheroidal coordinates in an unexpected frame. Moreover, one of the associated symmetry operators $\mathbf{A}$ was not associated with the Poincaré symmetries. In [44] it was pointed out that $\mathbf{A}$ is associated with a Killing-Yano tensor and in [51] that the appropriate frame for separation is defined by the eigenvectors of this tensor. Also in [51] it was shown that the tensor $A^{ij} = \frac{1}{2}(A^iA^j + A^jA^i)$ where $\mathbf{A} = \sum_i A^i\partial_i + B$ can be expressed as $A^{ij} = a^{ij}I$ and the tensor a^{ij} determines the separable coordinates in a manner analogous to Theorem 4. However, these are observations, not proofs, and the problem of intrinsic characterization of separable systems for the Dirac equation remains open even for the case of Minkowski space. The Minkowski case should be settled very soon because Kamran et. al. [52] have now classified (under the action of the Poincaré group) all triplets of commuting symmetry operators for the Dirac equation. They find that at least two of the symmetries always correspond to Poincaré symmetries. Thus separable coordinate systems will always have solutions such that dependence on at least two of the variables is exponential. (In particular the general separable coordinates of ellipsoidal or paraboloidal type for the Hamilton-Jacobi equation in Minkowski space cannot separate the Dirac equation.) The availability of this classification should enable us to characterize the separable systems in an invariant manner.

As an alternate procedure to that of Kamran et. al. I propose an approach to separation of the Dirac equation in the spirit of Eisenhart. Given a Dirac equation with potential in a four-dimensional space time, not necessarily Minkowski, we can always write it in the form

$$\sum_{i=1}^{4} \gamma^i \partial_i \psi + V\psi = -m\psi$$

where the true potential and the Christoffel matrices have been incorporated into V. We assume that a fixed frame has been chosen, as have the orthonormal coordinates $\{x^i\}$:

$$ds^2 = \sum_{i=1}^{4} h_i^2 (dx^i)^2.$$

Here

$$\gamma^i\gamma^j + \gamma^j\gamma^i = 2h_i^{-2}\delta^{ij} I, \qquad \mathbf{A}^1 = \sum_{i=1}^{4} \gamma^i \partial_i + V.$$

Since we chose a fixed frame, the γ^i and V are functions of h^i. Our problem is to determine a possible frame change R , ($\psi = R\Phi$), such that the Dirac equation for Φ is separable in the coodinates x^i. The conditions for separation are:

(1) factorization conditions

$$[\mathbf{A}^i, \mathbf{A}^j] = 0, \quad 1 \leq i, j \leq 4$$

(2) Stäckel form conditions

$$\partial_k A^{i\ell} + [\partial_k RR^{-1}, A^{i\ell}] = A^{ik}(\gamma^k)^{-1}(\partial_k \gamma^\ell + [\gamma^\ell, \partial_k RR^{-1}])$$

(3) Stäckel multiplier conditions

$$\begin{aligned}\partial_\ell B^i + [B^i, \partial_\ell RR^{-1}] =& A^{i\ell}(H^\ell)^{-1}\{\partial_\ell V + [V, \partial_\ell RR^{-1}] + \sum_s H^s \partial_{\ell s} RR^{-1} \\ & - \sum_s \partial_\ell RR^{-1} H^s \partial_s RR^{-1}\} \\ & + \sum_s (\partial_\ell RR^{-1} A^{is} \partial_s RR^{-1} - A^{is} \partial_{s\ell} RR^{-1})\end{aligned}$$

(4) conditions that the metric belong to the desired spacetime. In particular, the metric should be of signature $(+,+,+,-)$ and for Minkowski space the Riemann curvature tensor should vanish

$$R_{ijk\ell} = 0.$$

When two of the symmetry operators $\mathbf{A}^i$ can be taken in Poincaré form it appears feasable to solve these equations explicitly.

7. Comments and problems.

Among the unsolved problems in this subject is that of finding all solutions of the separation equations (2.8), or (more simply) of the scalar separation equations (2.11). In other words the most general mechanism for variable separation is still not determined. In all cases known to the author this mechanism is a variant of Stäckel form. To understand some of the variations consider Example 1 again:

$$(x_1 + x_2)(\partial_{11} u + \partial_{22} u) - 2(\partial_1 u + \partial_2 u) = E.$$

This equation admits a five-parameter separable solution in the coordinates x_1, x_2:

$$\begin{aligned}u = (\alpha x_1^3 + \beta x_1^2 + \gamma x_1 - \frac{1}{2} E x_1) \\ + (-\alpha x_2^3 + \beta_2^2 - \gamma x_2 + \delta).\end{aligned}$$

The mechanism of separation is puzzling until one realizes that the appropriate separation equations are

$$\begin{array}{llllll}\partial_1 u & +E/2 & -\gamma & -2\beta x_1 & -3\alpha x_1^2 = 0, \\ \partial_{11} u & & & -2\beta & -6\alpha x_1 = 0, \\ \partial_2 u & & +\gamma & -2\beta x_2 & +3\alpha x_2^2 = 0, \\ \partial_{22} u & & & -2\beta & +6\alpha x_2 = 0. \end{array}$$

The associated "Stäckel matrix" responsible for the separation is

$$\begin{bmatrix} \frac{1}{2} & -1 & -2x_1 & -3x_1^2 \\ 0 & 0 & -2 & -6x_1 \\ 0 & 1 & -2x_2 & 3x_2^2 \\ 0 & 0 & -2 & 6x_2 \end{bmatrix}.$$

This is not a true Stäckel matrix since more than one row depends on a given variable x_i. Moreover, the second and fourth rows are the derivatives of the first and third rows, respectively. It is a nontrivial example of a differential-Stäckel matrix, [53].

In [53] it was shown that *all* additive separable systems for n^{th} order linear differential equations are associated with differential Stäckel matrices. However, there is no apparent connection between separation and symmetry for the equations.

In [54] differential Stäckel matrices were further generalized by allowing each row of the matrices to depend not only on a single coordinate x^i but also on a finite number of derivatives of the dependent variable u with respect to x^i. These matrices correspond to additive separation in classes of nonlinear equations.

Other interesting unsolved problems concern multiplicative separation for spinor equations of the form

$$H\psi = D\psi$$

where H is an $N \times N$ first order matrix differential operator in the variables $x^1, \dots, x^n$, ψ is an N-component spinor and D is an $N \times N$ diagonal constant matrix. In particular, some or all of the diagonal elements of D may be zero. Maxwell's equations in an arbitrary space-time (and more generally, equations for massless particles with spin) are of this form with $D = 0$. Here even the definition of separation is not yet clear, nor is the relationship between symmetry and separation. For some partial results see [49,55]. We mention some of the results known for the Kerr metric:

$$\begin{aligned} ds^2 =& (1 - \frac{2Mr}{\rho^2})dt^2 - (\frac{\rho^2}{\Delta})dr^2 - \rho^2 d\theta^2 \\ &+ (\frac{2aMr\sin^2\theta}{\rho^2})dtd\phi - [(r^2+a^2) + \frac{2a^2Mr\sin^2\theta}{\rho^2}]\sin^2\theta d\phi^2, \end{aligned}$$

where

$$\Delta = r^2 + a^2 - 2Mr, \qquad \rho^2 = r^2 + a^2\cos^2\theta.$$

M is mass of black hole.
a is angular momentum per unit mass of black hole.

When $a = 0$ the Kerr metric reduces to the Schwarzschild metric.

When $M = 0$ this is the Minkowski (flat) space metric in oblate spheroidal coordinates.

The following equations are known to "separate" in these coordinates:

(1) The Hamilton-Jacobi equation. This follows from the standard theory of variable separation.

(2) The Dirac equation (spin $\frac{1}{2}$ particles with mass). See [49,56]. This can be understood from the theory of the preceding section.

(3) Neutrino equation (spin $\frac{1}{2}$ particles with zero mass). See [49]. This is essentially the $m = 0$ case of the Dirac equation.

(4) Maxwell's equations (spin 1 particles with zero mass). See [49,57,58]. The separation in this case is not well understood.

(5) Rarita-Schwinger equations (spin $\frac{3}{2}$ particles with zero mass). See [59]. Poorly understood.

(6) Gravitational perturbation equations (spin 2 particles, zero mass). See [58,60]. Poorly understood.

One reasonably successful approach to obtaining explicit solutions of mass zero, spin s equations is the method of Hertz potentials. To explain this we adopt the Newman-Penrose spinor formalism in Minkowski space, [61,62].In this formalism the covariant zero-mass field equation for a free spin-s field is

$$\nabla^{AX'}\phi_{AB\cdots K} = 0,$$

where $\phi_{AB\cdots K}$ is a totally symmetric spinor with $2s$ indices.

We define the spinor d'Alembertian operator (a scalar) as

$$\Box \equiv \nabla_{AX'}\nabla^{AX'}.$$

A *Hertz potential* is a totally symmetric $2s$-spinor $P_{AB\cdots K}$ satisfying

$$\Box\bar{P}^{X'N'\cdots W'} = 2\nabla^{A(X'}G_A^{N'\cdots W')}$$

where $G_{AN'\cdots W'}$ is an arbitrary gauge spinor with one unprimed and $2s-1$ symmetrized primed indices. The basic result is [60]:

THEOREM (KEGELES AND COHEN). *The spinor*

$$\phi_{AB\cdots K} = \nabla_{AM'}\nabla_{BN'}\cdots\nabla_{KW'}\bar{P}^{M'N'\cdots W'} - \nabla_{(BN'}\ldots\nabla_{KW'}G^{N'\cdots W'}_{A)}$$

is a solution of the zero-mass field equation

$$\nabla^{AX'}\phi_{AB\cdots K} = 0.$$

(See also [63].)

This result has been extended to curved spacetimes for $s = \frac{1}{2}, 1, \frac{3}{2}, 2$. For the Kerr metric, G can be chosen such that the equation for the Hertz potential has only one nontrivial component and variables separate in a simple fashion. This separable Hertz potential then yields a "separable" solution of the field equations. (However, the field equation solution is not strictly separable according to the definition provided in this paper.) This method is not general; it works only for algebraically special spacetimes.

REFERENCES

[1] W. MILLER, JR., *Symmetry and Separation of Variables*, Addison-Wesley, Reading, Massachusetts, 1977.

[2] H. BATEMAN, *Partial Differential Equations of Mathematical Physics, (1st ed., 1932)*, Cambridge Univ. Press, London and New York, 1969.

[3] P. MORSE AND H. FESHBACH, *Methods of Theoretical Physics, Part I*, McGraw-Hill, New York, 1953.

[4] E. G. KALNINS AND W. MILLER, *Intrinsic characterization of variable separation for the partial differential equations of mechanics, Proceedings of Symposium on Modern Developments in Analytical Mechanics, Torino, 1982*, Modern Developments in Analytical Mechanics, Acta Academiae Scientiarum Taurinensis, Torino, 1983.

[5] W. MILLER, *The technique of variable separation for partial differential equations, Proceedings of School and Workshop on Nonlinear Phenomena, Oaxtepec, Mexico, November 29- December 17, 1982, Lecture Notes in Physics*, Springer-Verlag, New York, 1983.

[6] T.W. KOORNWINDER, *A precise definition of separation of variables, Proceedings of the Scheveningen Conferences of Differential Equations. Lecture Notes in Mathematics*, Springer-Verlag, 1980.

[7] P.J. OLVER, *Applications of Lie Groups to Differential Equations*, Graduate Texts in Mathematics, 107, Springer-Verlag, New York (1986).

[8] G.W. BLUMAN AND J.D. COLE, *Similarity Methods for Differential Equations, Applied Mathematical Series # 13*, Springer-Verlag, 1974.

[9] L.V. OVSJANNIKOW, *Group Properties of Differential Equations*, Academic Press, 1982.

[10] L. P. EISENHART, *Continuous Groups of Transformations*, Dover Reprint, Dover, Delaware, 1961.

[11] L.P. EISENHART, *Riemannian Geometry*, Princeton University Press, 2nd printing, 1949.

[12] V.I. ARNOLD, *Mathematical Methods of Classical Mechanics (translated by K. Vogtmann and A. Weinstein), Graduate Texts in Mathematics, 60*, Springer-Verlag, New York, 1978.

[13] L.P. EISENHART, *Separable systems of Stäckel*, Ann. Math., 35 (1934), pp. 284–305.

[14] P. STÄCKEL, *Über die integration der Hamilton-Jacobischen differentialgeichung mittels separation der variabeln*, Habilitationschrift, Halle (1891).

[15] E.G. KALNINS AND W. MILLER, JR., *R-separation of variables for the four-dimensional flat Laplace and Hamilton-Jacobi equations*, Trans. A.M.S., 244 (1978), pp. 241–261.

[16] E.G. Kalnins and W. Miller, Jr., *Conformal Killing tensors and variable separation for the Hamilton-Jacobi equation*, SIAM J. Math. Anal, 14 (1983), pp. 126-137.

[17] E.G. Kalnins and W. Miller, Jr., *The theory of orthogonal R-separation for Helmholtz equations*, Advances in Mathematics, 51 (1984), pp. 91–106.

[18] V.N. Shapovalov, Differential Equations, 16 (1980), p. 1864.

[19] E.G. Kalnins and W. Miller, Jr., *Intrinsic characterization of orthogonal R-separation for Laplace equations*, J. Phys. A: Math. Gen, 15 (1982), pp. 2699–2709.

[20] E.G. Kalnins and W. Miller, Jr., *The general theory of R-separation for Helmholtz equations*, J. Math. Phys.

[21] S. Benenti and M. Francaviglia, *The theory of separability of the Hamilton-Jacobi equation and its application to general relativity*, in *General Relativity and gravitation, Vol. 1, A. Held, ed.*, Plenum, New York, 1980.

[22] S. Benenti, *Separability structures on Riemannian manifolds*, in *Proceedings of the Conference on Differential Geometric Methods in Mathematical Physics, Lecture Notes in Mathematics, Vol. 836*, Springer-Verlag, Berlin, 1980.

[23] E.G. Kalnins and W. Miller, Jr., *R-separation of variables for the time-dependent Hamilton -Jacobi and Schrödinger equations*, J. Math. Phys, 28 (1987), pp. 1005–1015.

[24] E.G. Kalnins and W. Miller, Jr., *Killing tensors and variable separation for Hamilton-Jacobi and Helmholtz equations*, SIAM J. Math. Anal, 11 (1980), pp. 1011-1026.

[25] E.G. Kalnins and W. Miller, Jr., *Killing tensors and nonorthogonal variable separation for the Hamilton-Jacobi equation*, SIAM J. Math. Anal, 12 (1981), pp. 617–638.

[26] E.G. Kalnins and W. Miller, Jr., *Separation of variables on n-dimensional Riemannian manifolds I. The n-sphere and Euclidean n-space*, J. Math. Phys., 27 (1986), pp. 1721–1736.

[27] E.G. Kalnins, *Separation of Variables for Riemannian Spaces of Constant Curvature*, Pitman, Monographs and Surveys in Pure and Applied Mathematics 28, Longman, Essex, England, 1986.

[28] M. Bôcher, *Die Reihentwickelungen der Potentialtheorie*, B.G. Teubner, Leipzig, 1894.

[29] G.J. Reid, *R-separation for heat and Schrödinger equations I.*, SIAM J. Math. Anal., 17 (1986), p. 646.

[30] C.P. Boyer, E.G. Kalnins and P. Winternitz, *Separation of variables for the Hamilton-Jacobi equation on complex projective spaces*, SIAM J. Math. Anal, 16 (1985), pp. 93–109.

[31] V.N. Shapovalov, *Stäckel spaces*, Siberian Math. J., 20 (1980), pp. 790-800.

[32] N.M.J. Woodhouse, *Killing tensors and separation of the Hamilton-Jacobi equation*, Commun. Math. Phys, 44 (1975), pp. 9–38.

[33] V.N. Shapovalov, *Separation of variables in the nonstationary Schrödinger equation*, Sov. Phys. J., 17 (1976), p. 1718.

[34] V.N. Shapovalov, *Separation of variables in second-order linear differential equations*, Differential Equations, 10 (1981), p. 1212.

[35] P. Winternitz, I. Lukac and Ya A. Smorodinskii, *Quantum numbers of the little group of the Poincaré group*, Soviet J. Nucl. Phys, 7 (1968), pp. 139–145.

[36] C. Neumann, *De problemate quodan mechanico, quod ad priman integralium ultraellipticorum classen revocatur*, J. Reine Angew. Math, 56 (1859), pp. 46-63.

[37] E. Jacobi, *Vorlesungen über Dynamik*, Supplement band, Berlin, 1884.

[38] C.P. Boyer, E.G. Kalnins and W. Miller, Jr., *Stäckel-equivalent integrable Hamiltonian systems*, SIAM J. Math. Anal, 17 (1986), pp. 778–797.

[39] D.R. Brill and J.A. Wheeler, *Interaction of neutrinos and gravitational fields*, Revs. Mod. Phys., 29 (1957), pp. 465–479.

[40] V. Fock, *Geometrisierung der Diracschen theorie des elektrons*, Zeitschrift für Physik, 57 (1936), pp. 261–277.

[41] V. Bargmann, *Bemerkungen zur allgemein-relativistischen fassung der quantentheorie*, Stzber. Preuss, Akad. Wiss., Physick-Math., 28 (1932), pp. 346–354.

[42] R.G. McLenaghan and Ph. Spindel, *Quantum numbers for Dirac spinor fields in a curved space time*, Phys. Rev. D, 20 (1979), pp. 409–413.

[43] N. Kamran and R.G. McLenaghan, *Symmetry operators for neutrino and Dirac fields on curved spacetime*, Phys. Rev., D 30 (1984), pp. 357–362.

[44] B. Carter and R.G. McLenaghan, *Generalized total angular momentum operator for the Dirac equation in curved space-time*, Phys. Rev., D 19 (1979), pp. 1093–1097.

[45] A.H. Cook, *On separable solutions of Dirac's equation for the electron*, Proc. R. Soc. Lond. A, 383 (1982), pp. 247–278.

[46] V.G. Bagrov, I.L. Buckbinder and D.M. Gittman, J. Phys. A, 9 (1976), pp. 1955–1965.

[47] V.G. Bagrov, D.M. Gittman, A.V. Shapovalov and V.N. Shapovalov, Izv. viyssh. ucheb. Zared., Fiz, no.6 (1977), pp. 105–114; 7, 46–51.

[48] V.G. Bagrov, D.M. Gittman, V.N. Zhadorzhnyi, N.B. Sukhomlin and V.N. Shapo8alov, Izv. vyssh. ucheb. zaved, Fiz, no.2 (1978), pp. 13–18.

[49] S. Chandrasekhar, *The Mathematical Theory of Black Holes*, Oxford University Press, 1984.

[50] R.P. Kerr, *Gravitational field of spinning mass as an example of algebraically special metrics*, Phys. Rev. Lett, 11 (1963), pp. 237–239.

[51] E.G. Kalnins, G.C. Williams and W. Miller, *Matrix operator symmetries of the Dirac equation and separation of variables*, J. Math. Phys., 27 (1986), pp. 1893-1900.

[52] N. Kamran, M. Légaré, R.G. McLenaghan and P. Winternitz, *The classification of complete sets of operators commuting with the Dirac operator in Minkowski space-time*, CRM-1483 (1987).

[53] E.G. Kalnins & W. Miller, *Differential-Stäckel matrices*, J. Math. Phys., 26 (1985), pp. 1560-1565.

[54] E.G. Kalnins and W. Miller, *Generalized Stäckel matrices*, J. Math. Phys., 26 (1985), pp. 2168-2173.

[55] E.G. Kalnins and W. Miller, *Electromagnetic waves in Kerr geometry*, Proc. R. Soc. Lond., A 408 (1986), pp. 23-30.

[56] R. Güven, *The solution of Dirac's equation in a class of type D backgrounds*, Proc. R. Soc. Lond A, 356 (1977), pp. 465–470.

[57] S. Teukolsky, *Rotating black holes: separable wave equations for gravitational and electromagnetic perturbation*, Phys. Rev. Lett, 29 (1972), pp. 1114–1118.

[58] N. Kamran and R.G. McLenaghan, *Separation of variables for higher spin, zero rest-mass field equations on type D vacuum backgrounds with cosmological constant*, in *Gravitation and Geometry, a Festschift in honor of Ivor Robinson, W. Rindler and A. Trautman, eds.*, Bibliopolis Publishers, 1986.

[59] N. Kamran, *Separation of variables for the Rarita-Schwinger equation on all type D vacuum backgrounds*, J. Math. Phys, 26 (1985), pp. 1740–1742.

[60] L.S. Kegeles and J.M. Cohen, *Constructive procedure for perturbations of spacetimes*, Phys. Rev. D., 19 (1979), pp. 1641–1664.

[61] E.T. Newman and R. Penrose, *An approach to gravitational radiation by a method of spin coefficients*, J. Math. Phys, 3 (1962), pp. 566–573.

[62] R. Penrose and W. Rindler, *Spinors and Space-Time, Vol. 1, Cambridge Monographs on Mathematical Physics*, Cambridge University Press, 1984.

[63] G.F. Torres Del Castillo, *Killing spinors and massless spinor fields*, Proc. R. Soc. Lond., A 400 (1985), pp. 119–126.

[64] P.J. Olver and P. Rosenau, *Group-invariant solutions of differential equations*, SIAM J. Appl. Math, 47 (1987), pp. 263–278.

Recursion Operators
and
Hamiltonian Systems

Peter J. Olver †
School of Mathematics
University of Minnesota
Minneapolis, MN
USA 55455

Abstract

This paper reviews the basic concepts in the theory of recursion operators and their applications to infinite dimensional Hamiltonian systems. Connections with the Poisson complex are explained in detail. The theory is illustrated by new results for first order hyperbolic systems, including the equations of gas dynamics.

† *Research Supported in Part by NSF Grant DMS 86-02004 and NATO Collaborative Research Grant RG 86/0055.*

1. Introduction.

Generalized symmetries first make their appearance the original paper of E. Noether on the correspondence between symmetries of variational problems and conservation laws of the associated Euler-Lagrange equations. The terminology *generalized* refers to the fact that the infinitesimal generators are allowed to depend on derivatives of the dependent variables, which makes the corresponding group transformations nonlocal. Recurson operators were first introduced in [9] to provide a mechanism for generating infinite families of generalized symmetries. A fundamental advance in the subject was the work of Magri, [7], who showed how recursion operators could be constructed for systems with two compatible Hamiltonian structures. In this paper, we develop the general theory of recursion operators and biHamiltonian systems, based on the important constuct of the Poisson complex. New Hamiltonian structures and recursion operators for systems of hyperbolic conservation laws, including the equations of gas dynamics and one-dimensional elasticity are found. Many of the topics in this article are more extensively developed in [11], [12], [14] to which we refer the interested reader for a more complete exposition of the theory, applications and history.

2. Generalized Symmetries, Recursion Operators and Conservation Laws.

Consider a system of partial differential equations

$$\Delta_\nu(x, u^{(n)}) = 0, \qquad \nu = 1, \dots, m, \tag{2.1}$$

defined on some open subset $M^{(n)}$ of the jet space, whose coordinates $(x,u^{(n)})$ consist of the independent variables $x = (x^1,\dots,x^p)$, the dependent variables $u = (u^1,\dots,u^q)$, and their partial derivatives $u^\alpha_J = \partial^J u^\alpha/\partial x^J$ up to order n. A *generalized vector field* is a partial differential operator of the form

$$v_Q = \sum_{\alpha=1}^{q} Q_\alpha(x,u^{(k)}) \frac{\partial}{\partial u^\alpha}, \tag{2.2}$$

in which the *characteristic* $Q[u] = (Q_1,\dots,Q_q)$ is a q-tuple of *differential functions*, meaning smooth functions of x, u, and derivatives of u. The vector field v_Q generates a one-parameter group of transformations on a suitable space of functions. Specifically, if $u = f(x)$ is a prescribed function, then the transformed function $\tilde{f}_\varepsilon(x) = g_\varepsilon \cdot f(x) = f(x,\varepsilon)$ is found by evaluating the solution $u = f(x,\varepsilon)$ to the Cauchy problem

$$\frac{\partial u^\alpha}{\partial \varepsilon} = Q_\alpha(x,u^{(k)}), \qquad \alpha = 1, \ldots, q, \qquad u(x,\varepsilon = 0) = f(x). \tag{2.3}$$

at time ε. (We ignore complications involving the existence and uniqueness of solutions of the Cauchy problem (2.3).) The vector field v_Q is called an (infinitesimal) *generalized symmetry* of the system (2.1) if it takes solutions to solutions (at least formally), i.e. if $u = f(x)$ is a solution, so is $u = g_\varepsilon \cdot f(x)$. In particular, if the characteristic takes the special form

$$Q_\alpha(x, u^{(1)}) = \varphi_\alpha(x,u) - \sum_{i=1}^{p} \xi^i(x,u) \frac{\partial u^\alpha}{\partial x^i}, \qquad \alpha = 1, \ldots, q,$$

then the group corresponds to the geometrical group of transformations generated by the vector field

$$\mathbf{v} = \sum_{i=1}^{p} \xi^i(x,u) \frac{\partial}{\partial x^i} + \sum_{\alpha=1}^{q} \varphi_\alpha(x,u) \frac{\partial}{\partial u^\alpha}.$$

Other generalized symmetries act non-locally on functions.

To obtain the infinitesimal invariance criterion for the system of differential equation (2.1), we prolong v_Q to the infinite jet space $M^{(\infty)}$, which we realize as the direct limit of the finite jet spaces $M^{(n)}$ as $n \to \infty$, leading to the partial differential operator

$$\mathrm{pr}\, v_Q = \sum_{\alpha=1}^{q} \sum_{J} D_J Q_\alpha \frac{\partial}{\partial u_J^\alpha}.$$

Here $D_J = D_{j_1} \cdot \ldots \cdot D_{j_k}$ denotes the kth order total derivative corresponding to the multi-index $J = (j_1,\ldots,j_k)$.

Theorem 1. Suppose the system of partial differential equations (2.1) is totally nondegenerate in the sense of [11; Definition 2.83]. Then v_Q is a generalized symmetry of the system if and only if

$$\text{pr}\, \mathbf{v}_Q(\Delta_\nu) = 0, \qquad \nu = 1, \ldots ,m, \tag{2.4}$$

for all solutions $u = f(x)$ to the system (2.1).

The nondegeneracy condition required for the validity of the theorem is very mild, and is satisfied by well-nigh every system of partial differential equations which arises in physical applications. The symmetry conditions (2.4) constitute a large, over-determined system of elementary partial differential equations, called the *determining equations*, for the characteristic Q of $\mathbf{v}_Q$. In practice, these can be systematically solved to determine all the generalized symmetries of the system (2.1).

Define the Fréchet derivative of $\Delta[u] = (\Delta_1,\ldots,\Delta_m)$ to be the differential operator

$$\mathbf{D}_\Delta(v) = \frac{d}{d\varepsilon}\bigg|_{\varepsilon = 0} \Delta[u + \varepsilon v].$$

Note the elementary identity

$$\text{pr}\, \mathbf{v}_Q(\Delta) = \mathbf{D}_\Delta(Q),$$

so that the symmetry condition (2.4) can be rewritten as

$$\mathbf{D}_\Delta(Q) = 0 \qquad \text{whenever } u \text{ is a solution to } \Delta = 0. \tag{2.5}$$

There is a natural Lie bracket operation between generalized vector fields. Specifically, if $\mathbf{v}_Q$ and $\mathbf{v}_R$ are generalized vector fields, their Lie bracket $\mathbf{v}_S = [\mathbf{v}_Q, \mathbf{v}_R]$ is the generalized vector field with characteristic

$$S = \text{pr}\, \mathbf{v}_Q(R) - \text{pr}\, \mathbf{v}_R(Q). \tag{2.6}$$

The reader can verify the usual binlinearity, skew-symmetry and Jacobi identity for this bracket. In particular, if $\mathbf{v}_Q$ and $\mathbf{v}_R$ are generalized symmetries of the system (2.1), so is the vector field $\mathbf{v}_S = [\mathbf{v}_Q, \mathbf{v}_R]$.

By definition, a *recursion operator* $\mathcal{R}$ for a system of differential equations is a linear operator which maps symmetries to symmetries; in other words if $\mathbf{v}_Q$ is a generalized symmetry, and $\tilde{Q} = \mathcal{R} \cdot Q$, then $\mathbf{v}_{\tilde{Q}}$ is also a generalized symmetry. There is a simple criterion for detemining when a given operator is a recursion operator. The proof is an elementary application of formula (2.5).

Theorem 2. A linear operator $\mathfrak{R}$ is a recursion operator for the system of differential equations $\Delta[u] = 0$ if there is a second linear operator $\tilde{\mathfrak{R}}$ such that the identity

$$\mathbf{D}_\Delta \cdot \mathfrak{R} = \tilde{\mathfrak{R}} \cdot \mathbf{D}_\Delta \tag{2.7}$$

holds on solutions to Δ.

Example 3. We shall illustrate our concepts using the elementary example of the *Riemann equation*

$$u_t = u\, u_x, \tag{2.8}$$

which is the simplest possible scalar nonlinear conservation law. Later we will see how many of the results for this very simple equation have direct counterparts in the case of two-dimensional hyperbolic systems, including the equations for polytropic gas dynamics and one-dimensional elasticity .

We begin by determining all the generalized symmetries of the Riemann equation. Since (2.5) is required to hold only on solutions to (2.8), we can always replace t-derivatives of u by equivalent expressions involving only x-derivatives. If $\Delta = u_t - uu_x$, then the Fréchet derivative operator is $\mathbf{D}_\Delta = D_t - u\cdot D_x - u_x$. Therefore, a differential function $Q[u] = Q(x,t,u,u_x,\ldots,u_k)$, where $u_k \equiv d^k u/dx^k$, is the characteristic of a generalized symmetry if and only if Q is a solution to the first order partial differential equation

$$D_t Q = u\cdot D_x Q + u_x \cdot Q$$

whenever u solves (2.8). Expanding the total derivatives, and replacing t-derivatives of u by x-derivatives, we see that Q must be a solution to the first order partial differential equation

$$w(Q) \equiv Q_t + u\cdot Q_x + u_x^2 \cdot Q_{u_x} + 3u_x \cdot u_{xx} \cdot Q_{u_{xx}} + \ldots + \{D_x^k(u\cdot u_x) - u\cdot u_{k+1}\}\cdot Q_{u_k} = u_x \cdot Q, \tag{2.9}$$

subscripts on Q denoting partial derivatives. By the method of characteristics for a first order linear partial differential equation, it is easy to determine the general solution:

Theorem 4. Define the rational differential functions

$$I_0 = u, \quad I_1 = x - t\cdot u, \quad I_2 = \frac{u}{u_x} - x, \quad I_3 = \frac{u_{xx}}{u_x^3}, \quad I_{j+1} = \frac{1}{u_x} D_x I_j, \quad j \geq 3.$$

A differential function Q is the characteristic of a generalized symmetry $\mathbf{v}_Q = Q\cdot\partial_u$ of (2.8) if and only if

$$Q = u_x \cdot G(I_0, I_1, I_2, \ldots, I_{k+1}), \tag{2.10}$$

where G is an arbitrary smooth function of its arguments.

It turns out that there are several recursion operators for the Riemann equation. Two zero^th order ones are given by

$$\mathcal{R}_1 = 2u + u_x \cdot D_x^{-1}, \qquad \mathcal{R}_2 = u^2 + u\, u_x \cdot D_x^{-1}. \tag{2.11}$$

For instance, to prove (2.7) for $\mathcal{R}_1$, we note that since $u\cdot D_x + u_x = D_x \cdot u$, the commutator

$$\begin{aligned}[D_\Delta, \mathcal{R}_1] &= [D_t - u\cdot D_x - u_x,\ 2u + u_x\cdot D_x^{-1}] \\ &= 2u_t + u_{xt}\cdot D_x^{-1} - 2u\cdot u_x - (u\cdot u_{xx} + u_x^2)\cdot D_x^{-1}\end{aligned}$$

vanishes on solutions, which proves (2.7), with $\tilde{\mathcal{R}} = \mathcal{R}_1$. The verification for $\mathcal{R}_2$ is similar. Therefore, starting with the translational symmetry, with characteristic $Q_0 = u_x$, we generate a hierarchy of higher order symmetries with characteristics $Q_n = u^n \cdot u_x$; explicitly

$$\mathcal{R}_1(Q_n) = \frac{2n+3}{n+1} Q_{n+1}, \qquad \mathcal{R}_2(Q_n) = \frac{n+2}{n+1} Q_{n+2}.$$

(Interestingly, even though there are two independent recursion operators, the two hierarchies happen to coincide. However, this is special to the polynomial symmetries; on other symmetries, these recursion operators will act differently.)

The Riemann equation admits an additional first order recursion operator

$$\mathfrak{R} = D_x \cdot \frac{1}{u_x} . \tag{2.12}$$

This latter operator acts on the hierarchy Q_n according to

$$\mathfrak{R}(Q_n) = n\, Q_{n-1},$$

and so, up to multiple, "inverts" the first order recursion operator $\mathfrak{R}_1$. Again, this is special to the polynomial hierarchy. For instance, starting with the rational second order generalized symmetry with characteristic $\hat{Q}_2 = u_x \cdot I_3 = u_x^{-2} \cdot u_{xx}$, the recursion operator $\mathfrak{R}$ generates the additional hierarchy of higher order symmetries $\hat{Q}_k = u_x \cdot I_{k+1}$, $k = 2, 3, \dots$, whereas $\mathfrak{R}_1$ and $\mathfrak{R}_2$ lead to yet other second order symmetries.

Given a system of partial differential equations (2.1), a *conservation law* is a p-tuple of differential functions $P[u] = (P_1, \dots, P_p)$ whose divergence

$$\mathrm{Div}\, P = \sum_{i=1}^{p} D_i P_i = 0,$$

vanishes on all solutions to (2.1). For dynamic problems, the conservation law takes the form

$$D_t T + \mathrm{Div}\, X = 0.$$

(Div here refers to the spatial variables.) The t-component of such a conservation law is referred to as the conserved density, and, for suitable solutions, (in particular those for which the flux X vanishes on the boundary) the integral $\int T[u]\, dx$ provides a constant of the motion. A conserved density is called *trivial* if it is a (spatial) divergence $T = \mathrm{Div}\, Y$ on solutions. In the Lagrangian framework, Noether's Theorem, cf. [11; Theorem 5.42], provides a complete correspondence between generalized symmetries of a variational problem and conservation laws of the associated Euler-Lagrange equations. In the next section, we shall see how this extends to the Hamiltonian framework.

Example 5. For the Riemann equation (2.8), any conserved density T, which, without loss of generality, we can take to depend only on $x, t, u, \dots, u_n$, must satisfy

$$D_t T + D_x X = 0$$

on solutions to the equation for some flux X. Writing this out, and replacing t-derivatives by the corresponding expressions in terms of x-derivatives, we find

$$u \cdot D_x T + w(T) + D_x X = 0,$$

where **w** is the vector field given in (2.9). Let $X = Y - u \cdot T$, so this becomes

$$w(T) - u_x \cdot T + D_x Y = 0.$$

If we rewrite $T = u_x \cdot F(t, I_0, I_1, I_2, \ldots, I_{k+1})$, in terms of the invariants I_j of **w**, and the single parametric variable t, then $w(T) - u_x \cdot T = u_x \cdot G_t$, hence the functions

$$u_x \cdot F(t, I_0, I_1, I_2, \ldots, I_{k+1}) - u_x \cdot F(0, I_0, I_1, I_2, \ldots, I_{k+1}) = D_x\{ \int_0^t Y \, ds \}$$

differ by a trivial conserved density. Setting $t = 0$ in the formula for T, we conclude that it is equivalent to a conserved density of the form

$$T = u_x \cdot G(I_0, I_1, I_2, \ldots, I_k).$$

Thus, surprisingly, for the Riemann equation the expressions for symmetries and conserved densities are the same! In particular, we note the infinite sequence of zeroth order conserved densities

$$H_n(u) = u^n, \qquad n = 1,2,3, \ldots . \tag{2.13}$$

3. The Poisson Complex and Hamiltonian Systems.

We now present an approach to the theory of Hamiltonian systems based on the important Poisson complex, which plays as fundamental a role here as the deRham complex does in the theory of differential forms. The Poisson complex, though, involves the dual objects to differential forms, which are known as multi–vectors, or, in the infinite–dimensional case, functional multi–vectors. This complex, in the finite–dimensional case, is due to Lichnerowicz, [6], and was generalized to infinite dimensions in Olver, [10]. We begin by recalling the basic definitions; see [11] for many of the details.

On the infinite jet space, $M^{(\infty)}$, the space $\bigwedge_*^0$ of *functionals* is defined as the cokernel of the total divergence operator, so that two differential functions $L[u]$ and $\bar{L}[u]$ define the same functional $\mathcal{L}[u] = \int L[u]\, dx$ if and only if they differ by a total divergence: $L = \bar{L} + \operatorname{Div} P$. More generally, define a *vertical k-form* to be a finite sum

$$\hat{\omega} = \sum P_J^\alpha[u]\, du_{J_1}^{\alpha_1} \wedge \ldots \wedge du_{J_k}^{\alpha_k},$$

where the coefficients P_J^α are arbitrary differential functions. The total derivatives D_i act as Lie derivatives on the vertical forms, and the space Λ_*^k of *functional k-forms* is analogously defined as the cokernel of the total divergence. In particular, an easy integration by parts argument shows that any functional one-form is uniquely equivalent to one in the form

$$\omega = \int \Big\{ \sum_{\alpha=1}^{q} P_\alpha[u]\, du^\alpha \Big\}\, dx = \int \{P \cdot du\}\, dx. \qquad (3.1)$$

Similarly, it can be shown that any functional 2-form can be placed into canonical form

$$\Omega = \tfrac{1}{2} \int \Big\{ \sum_{\alpha,\beta} du^\alpha \wedge \mathcal{D}_{\alpha\beta} \cdot du^\beta \Big\}\, dx = \tfrac{1}{2} \int \{du \wedge \mathcal{D} \cdot du\}\, dx, \qquad (3.2)$$

uniquely determined by the skew-adjoint matrix differential operator $\mathcal{D} = (\mathcal{D}_{\alpha\beta})$.

The *vertical differential* $\hat{d}$ takes a vertical k-form to a vertical (k+1)-form, and is induced by its action

$$\hat{d} P = \sum_{\alpha, J} \frac{\partial P}{\partial u_J^\alpha}\, du_J^\alpha$$

on differential functions. It can be shown that $\hat{d}$ commutes with each total derivative D_i, and hence induces a well-defined map

$$\delta : \Lambda_*^k \longrightarrow \Lambda_*^{k+1}$$

on the spaces of functional forms, called the *variational differential*. An easy argument based on the finite-dimensional Poincaré lemma shows that the *variational complex*

$$0 \longrightarrow \Lambda_*^0 \xrightarrow{\ \delta\ } \Lambda_*^1 \xrightarrow{\ \delta\ } \Lambda_*^2 \xrightarrow{\ \delta\ } \Lambda_*^3 \xrightarrow{\ \delta\ } \ldots .$$

is locally exact, meaning that, over suitable (star-shaped) subdomains $\delta\omega = 0$ if and only if $\omega = \delta\zeta$ for some functional form ζ. In particular, if $\mathcal{L}[u] = \int L[u]\,dx$ is a functional, its variational differential is the one-form

$$\delta\mathcal{L} = \int \{E(L) \cdot du\}\,dx \tag{3.3}$$

determined by the Euler-Lagrange expression $E(L)$ or variational derivative of $\mathcal{L}$. The exactness of the varational complex at the $\wedge^1_*$ - stage leads to the well-known Helmholtz conditions for a differential equation to be the Euler-Lagrange equation for some variational problem, [11; Theorem 5.68], namely $\Delta = E(L)$ for some Lagrangian L if and only if its Fréchet derivative is self-adjoint: $\mathbf{D}_\Delta = \mathbf{D}^*_\Delta$.

Our main interest here is not in the functional forms, but rather in the dual objects - the functional multi-vectors. By definition, a *functional k-vector* is an alternating, k-linear map from the space $\wedge^1_*$ of functional one-forms to the space $\wedge^0_*$ of functionals. It can be shown that each functional k-vector can be written in the form

$$\Theta = \int \Big\{ \sum R^\alpha_J[u]\, \theta^{\alpha_1}_{J_1} \wedge \ldots \wedge \theta^{\alpha_k}_{J_k} \Big\}\,dx,$$

where the θ^α_J form a basis for the *vertical vectors*, dual to the basis du^α_J of vertical forms. We find

$$\langle\Theta; \omega_1 \wedge \ldots \wedge \omega_k\rangle = \int \Big\{ \sum R^\alpha_J[u] \cdot \det(D_{J_i} P^j_{\alpha_i}) \Big\}\,dx,$$

where $\omega_j = \int \{ \sum P^j_\alpha[u]\, du^\alpha \}\,dx$ are functional one-forms written in canonical form (3.1). The total derivatives act as Lie derivatives on vertical multi-vectors, and so the space $\wedge^*_k$ of functional k-vectors is again the cokernel of the total divergence operator. In particular, integation by parts can be used to place any functional uni-vector (i.e. $k = 1$) in the canonical form

$$v_Q = \int \{ \sum_{\alpha=1}^{q} Q_\alpha[u]\, \theta^\alpha \}\,dx = \int \{Q \cdot \theta\}\,dx, \tag{3.4}$$

and the space $\wedge^*_1$ can be identified with the space of generalized vector fields. Similarly, any functional bi-vector, i.e. element of $\wedge^*_2$, can be placed in canonical form

$$\Theta = \Theta_{\mathcal{B}} = \frac{1}{2}\int \{\theta \wedge \mathcal{B}\theta\}\, dx, \tag{3.5}$$

determined by a unique $q \times q$ skew-adjoint matrix differential operator $\mathcal{B}$.

Warning: The space $\wedge_k^*$ of functional multi-vectors is *not* the dual space to the space $\wedge_*^k$ of functional k-forms. This is because the wedge product of two functional forms is not a well-defined functional form!

If $\mathbf{v}_Q$ is a generalized vector field or uni-vector, we can define its prolongation to act as a Lie derivative on the space of functional multi-vectors. The key formula is

$$\text{pr}\, \mathbf{v}_Q(\theta) = \mathbf{D}_Q^* \cdot \theta,$$

where $\mathbf{D}_Q^*$ is the adjoint of the Fréchet derivative of Q, and θ denotes the column vector of basis uni-vectors θ^α, $\alpha = 1,\ldots,q$. (See [10], [12] for a justification of this formula.) The prolongation $\text{pr}\, \mathbf{v}_Q$ acts on differential functions as before, and the action is extended to the entire space by the usual rules of derivation and commutation with the total derivatives. In particular, as the reader can check, this definition of the Lie derivative recovers the correct form of the Lie bracket (2.6) between generalized vector fields.

If Θ is a functional k-vector, and $\omega_I \equiv \omega_{i_1} \wedge \ldots \wedge \omega_{i_m}$, $m \le k$, a wedge product of functional one-forms, we define the *interior product* to be the functional $(k - m)$-vector $\omega_I \lrcorner \Theta$ determined by the formula

$$\langle \omega_I \lrcorner \Theta; \eta_1 \wedge \ldots \wedge \eta_{k-m} \rangle = \langle \Theta; \omega_I \wedge \eta_1 \wedge \ldots \wedge \eta_{k-m} \rangle . \tag{3.6}$$

In particular, if $m = k - 1$, then $\omega_I \lrcorner \Theta \in \wedge_1^*$, and hence can be viewed as an generalized vector field, as in (3.4).

The most important operation on functional multi-vectors is the Schouten bracket, which generalizes the Lie bracket between vector fields. If Φ is a k–vector and Ψ an ℓ–vector, then their Schouten bracket $[\Phi, \Psi]$, is a $(k+\ell-1)$–vector. It is uniquely defined by the following formula:

$$\langle [\Phi,\Psi];\, \delta\mathcal{L}_1 \wedge \ldots \wedge \delta\mathcal{L}_{k+\ell-1} \rangle =$$

$$\frac{(-1)^{k\cdot\ell+\ell}}{\ell} \sum_I (\operatorname{sgn} I)\, \langle \Phi;\, \{\delta\mathcal{L}_I \lrcorner \Psi\}\delta\mathcal{L}_{I'} \rangle + \frac{(-1)^k}{k} \sum_J (\operatorname{sgn} J)\, \langle \Psi;\, \{\delta\mathcal{L}_J \lrcorner \Phi\}\delta\mathcal{L}_{J'} \rangle. \tag{3.7}$$

which must hold for every set of functionals $\mathcal{L}_1, \ldots, \mathcal{L}_{k+\ell-1}$, with variational derivatives $\delta\mathcal{L}_i$ given by (3.3). In (3.7), the first sum is over all multi-indices $I = (i_1, \ldots, i_{\ell-1})$, $1 \le i_1 < \ldots < i_{\ell-1} \le k+\ell-1$, with $I' = (i'_1, \ldots, i'_k)$ being the complementary multi-index, so $I \cup I' = \pi(1, \ldots, k+\ell-1)$ for some permutation π, and $\operatorname{sgn} I$ denoting the sign of the permutation π. Similarly, the second sum is over all multi-indices $J = (j_1, \ldots, j_{k-1})$, $1 \le j_1 < \ldots < j_{k-1} \le k+\ell-1$, with J' and $\operatorname{sgn} J$ being defined analogously. Note also that, according to the remark in the previous paragraph, the terms $\delta\mathcal{L}_I \lrcorner \Psi$ and $\delta\mathcal{L}_J \lrcorner \Phi$ are in $\wedge^*_1$, and hence determine generalized vector fields, which act on the remaining wedge products $\delta\mathcal{L}_{I'}$ and $\delta\mathcal{L}_{J'}$ as Lie derivatives. (This definition, first proposed in [10], has the advantage of being the only one I know of which works equally well in both finite and infinite dimensions.)

The Schouten bracket satisfies the following properties. Let $\Phi \in \wedge^*_k$, $\Psi \in \wedge^*_\ell$, $\Theta \in \wedge^*_m$ be functional multi-vectors.

i) *Bilinearity*: $[\Phi, \Psi]$ is an $\mathbb{R}$–bilinear function of Φ and Ψ.

ii) *Super–symmetry*: $$[\Phi, \Psi] = (-1)^{k\cdot\ell} [\Psi, \Phi].$$

iii) *Super–Jacobi Identity*:

$$(-1)^{k\cdot m} [[\Phi,\Psi],\Theta] + (-1)^{\ell\cdot m} [[\Theta, \Phi],\Psi] + (-1)^{k\cdot\ell} [[\Psi,\Theta], \Phi] = 0.$$

iv) *Lie Derivative*: If $\mathbf{v}_Q$ is a generalized vector field or functional uni-vector, then the Schouten bracket $[\mathbf{v}_Q, \Phi]$ is also a functional k–vector, and coincides with the Lie derivative of Φ with respect to $\operatorname{pr} \mathbf{v}_Q$. In particular, the Schouten bracket of two generalized vector fields is the same as their Lie bracket (2.6).

Each functional bi–vector $\Theta_{\mathcal{D}}$ determines an alternating, bilinear map on the space of one–forms, and hence a bilinear, skew–symmetric "bracket" on the space of real–valued function(al)s:

$$\{\mathcal{F}, \mathcal{H}\} \equiv \langle \Theta_{\mathcal{D}}; \delta\mathcal{F}, \delta\mathcal{H} \rangle.$$

Explicitly, using (3.5), we see that this bracket is given by the standard formula

$$\{\mathcal{F}, \mathcal{H}\} = \int E(F) \cdot \mathcal{D} \cdot E(H) \; dx, \tag{3.8}$$

where F and H are the integrands of the functionals $\mathcal{F}$, $\mathcal{H}$. The bracket automatically satisfies the Leibniz rule, and hence to be a genuine Poisson bracket must only satisfy the additional restriction imposed by the Jacobi identity. This can be easily expressed in invariant form using the tri–vector $[\Theta_{\mathcal{D}}, \Theta_{\mathcal{D}}]$ obtained by bracketing $\Theta_{\mathcal{D}}$ with itself:

$$\{\{F, H\}, P\} + \{\{P, F\}, H\} + \{\{H, P\}, F\} = \tfrac{2}{3} \langle [\Theta_{\mathcal{D}}, \Theta_{\mathcal{D}}]; dF, dH, dP \rangle.$$

Therefore a functional bi-vector $\Theta_{\mathcal{D}}$ determines a Poisson bracket if and only if it satisfies the extra condition

$$[\Theta_{\mathcal{D}}, \Theta_{\mathcal{D}}] = 0. \tag{3.9}$$

This condition is a nonlinear condition on the underlying differential operator $\mathcal{D}$. Any functional bi-vector satisfying (3.9) is called a Hamiltonian bi-vector; similarly, any skew-adjoint differential operator coming from a Hamiltonian bi-vector is called a *Hamiltonian operator.*

Given a Hamiltonian bi-vector, let $\vartheta = \vartheta_{\Theta}$ be the map taking functional k–vectors to functional (k+1)–vectors defined by bracketing with the Poisson bivector Θ:

$$\vartheta(\Psi) = [\Theta, \Psi]. \tag{3.10}$$

The determining property (3.9) along with the super Jacobi identity for the Schouten bracket immediately implies that

$$\vartheta(\vartheta(\Psi)) = [\Theta, [\Theta, \Psi]] = 0$$

for any multi–vector Ψ. Therefore the maps ϑ determine a complex, called the *Poisson complex* corresponding to the Poisson bivector Θ:

$$0 \longrightarrow \Lambda_0^* \xrightarrow{\vartheta} \Lambda_1^* \xrightarrow{\vartheta} \Lambda_2^* \xrightarrow{\vartheta} \Lambda_3^* \xrightarrow{\vartheta} \cdots .$$

The composition of two successive maps is always trivial: $\vartheta \circ \vartheta = 0$.

The first stage of the Poisson complex, $\vartheta : \wedge_0^* \to \wedge_1^*$, maps functionals to generalized vector fields. Specifically, if $\mathcal{H}[u] = \int H[u]\, dx$ is a (Hamiltonian) functional, the corresponding generalized vector field

$$\hat{v}_{\mathcal{H}} = \vartheta(\mathcal{H}) = [\Theta, \mathcal{H}] \tag{3.11}$$

is called the *Hamiltonian vector field* determined by $\mathcal{H}$. Explicitly, it is readily seen that $\hat{v}_{\mathcal{H}}$ has characteristic $Q = \mathcal{D}\cdot E(H)$, where $\mathcal{D}$ is the Hamiltonian operator determined by Θ. The corresponding Hamiltonian flow, cf. (2.3), is governed by the Hamiltonian system of evolution equations

$$u_t = \mathcal{D}\cdot E(H). \tag{3.12}$$

We note the standard formula

$$\{\mathcal{H}, \mathcal{F}\} = -\operatorname{pr} \hat{v}_{\mathcal{H}}(\mathcal{F}) = \operatorname{pr} \hat{v}_{\mathcal{F}}(\mathcal{H}) \tag{3.13}$$

for any pair of functionals $\mathcal{H}, \mathcal{F}$, which proves that a functional $\mathcal{F}$ determines a conserved density for the Hamiltonian system (3.12) if and only if $\{\mathcal{H}, \mathcal{F}\} = 0$. Therefore, every conserved density $\mathcal{F}$ determines a generalized (Hamiltonian) symmetry $\hat{v}_{\mathcal{F}}$ of the Hamiltonian system (3.12).

Conversely, if an generalized vector field v_Q is a Hamiltonian vector field, so $Q = \mathcal{D}\cdot E(H)$ for some differential function H, then closure of the Poisson complex at the $\wedge_1^*$-stage implies that

$$\vartheta(v_Q) = [\Theta_{\mathcal{D}}, v_Q] = \operatorname{pr} v_Q(\Theta_{\mathcal{D}}) = 0. \tag{3.14}$$

If the Poisson complex is *exact* at the $\wedge_1^*$-stage, then (3.14) is both necessary and sufficient for v_Q to be a Hamiltonian vector field. In this case, (3.14) provides a simple and readily verifiable condition that will tell whether or not a given vector field is Hamiltonian with respect to the given Poisson bracket. Writing out (3.14) explicitly leads to the following characterization of Hamiltonian vector fields, [12]; see also [5] for a similar result.

Proposition 6. Let $\Theta_{\mathcal{D}}$ be a Hamiltonian bivector, with $\mathcal{D}$ the corresponding Hamiltonian differential operator. If the evolutionary vector field $\mathbf{v}_Q$ is Hamiltonian, meaning that $Q = \mathcal{D}\cdot E(H)$ for some differential function H, then

$$D_Q\cdot\mathcal{D} + \mathcal{D}\cdot D_Q^* = \mathbf{v}_Q(\mathcal{D}). \tag{3.15}$$

Conversely, if the Poisson complex corresponding to $\Theta_{\mathcal{D}}$ is exact at the $\wedge_1^*$–stage, then (3.15) is both necessary and sufficient for Q to be of the Hamiltonian form (3.12).

Explicit conditions on the Hamiltonian bi-vector that imply exactness of the corresponding Poisson complex are not known in general; this is one of the main open problems in the subject. However, in the case of constant coefficient skew-adjoint differential operators (which are always automatically Hamiltonian), one can prove the following theorem on exactness of the Poisson complex at the initial stage:

Theorem 7. Let $\mathcal{D}$ be a nondegenerate, skew–adjoint, constant coefficient $q \times q$ matrix differential operator. Then, except for a finite dimensional space of linear differential functions, the Poisson complex for $\Theta_{\mathcal{D}}$ is exact at the $\wedge_1^*$–stage. More specifically, there exist linear differential functions $Q_1[u],\ldots,Q_n[u]$ such that an generalized vector field $\mathbf{v}_Q$ satisfies the condition

$$D_Q\cdot D_x + D_x\cdot D_Q^* = 0,$$

if and only if

$$Q = \mathcal{D}\cdot E(H) + c_1Q_1 + \ldots + c_nQ_n$$

for some Hamiltonian $H[u]$, and some constants $c_1, \ldots, c_n$. (In fact, even the linear functions Q_j are Hamiltonian provided one admits nonlocal Hamiltonian functionals depending on the potential w.)

Example 8. For the Hamiltonian operator D_x, the Poisson complex is exact at the $\wedge_1^*$–stage. This is equivalent to the statement that $Q[u] = D_xE(H)$ for some differential function H if and only if $D_Q\cdot D_x + D_x\cdot D_Q^* = 0$. Exactness fails for the third order operator D_x^3, but there are just two linear counterexamples: $Q_1 = u_x$, and $Q_2 = xu_x + u$. Thus we have the analogous result that $D_Q\cdot D_x^3 + D_x^3\cdot D_Q^* = 0$ if and only if $Q[u] = D_x^3\,E(H) + c_1Q_1 + c_2Q_2$, for some constants c_1, c_2.

A similar exactness result should hold for the later $\wedge_k^*$–stages, $k > 1$, of the Poisson complex, but I have not tried to construct a proof. The only field dependent Hamiltonian operator for which exactness at the $\wedge_1^*$–stage is known is the second Korteweg–deVries Hamiltonian operator $\mathcal{D} = D_x^3 + 2uD_x + u_x$, cf. [12], although the Darboux Theorems of [2], [13] greatly extend the range of application of the constant coefficient results. I believe a detainled investigation into the Poisson complex corresponding to a Hamiltonian operator will lead to significant results in infinite-dimensional geometry and differential algebra.

4. Bi-Hamiltonian Systems.

The most productive way to derive recursion operators is through the theory of biHamiltonian systems, first proposed by Magri, [7].

Definition. Two functional bi-vectors $\Theta_{\mathcal{D}}$ and $\Theta_{\mathcal{E}}$ are said to form a *Hamiltonian pair* if every linear combination $c\cdot\Theta_{\mathcal{D}} + d\cdot\Theta_{\mathcal{E}}$, $c, d \in \mathbb{R}$, is a Hamiltonian bi-vector.

Since the condition (3.9) is quadratic, to check whether $\Theta_{\mathcal{D}}$ and $\Theta_{\mathcal{E}}$ form a Hamiltonian pair it suffices to prove that the three functional bi-vectors $\Theta_{\mathcal{D}}$, $\Theta_{\mathcal{E}}$, and $\Theta_{\mathcal{D}} + \Theta_{\mathcal{E}}$ are Hamiltonian. Equivalently, we need check

$$[\Theta_{\mathcal{D}}, \Theta_{\mathcal{D}}] = [\Theta_{\mathcal{E}}, \Theta_{\mathcal{E}}] = 0$$

and the additional compatibility condition

$$[\Theta_{\mathcal{D}}, \Theta_{\mathcal{E}}] = 0. \tag{4.1}$$

A system of differential equations is said to be *biHamiltonian* with respect to a Hamiltonian pair $\Theta_{\mathcal{D}}$ and $\Theta_{\mathcal{E}}$ if it can be written in the two Hamiltonian forms

$$u_t = \mathcal{D}\cdot\delta\mathcal{H}_1 = \mathcal{E}\cdot\delta\mathcal{H}_0.$$

for some Hamiltonian functionals $\mathcal{H}_0$ and $\mathcal{H}_1$. We now state a version of Magri's theorem on the "complete integrability" of biHamiltonian systems, and give an elementary proof based on the exactness of the Poisson complex.

Theorem 9. Let $\Theta_{\mathcal{D}}$ and $\Theta_{\mathcal{E}}$ form a Hamiltonian pair, and assume that the Poisson complex for the bi-vector $\Theta_{\mathcal{D}}$ is exact at the Λ_1^*–stage. Let

$$u_t = Q_1[u] = \mathcal{D}\cdot\delta\mathcal{H}_1 = \mathcal{E}\cdot\delta\mathcal{H}_0 \tag{4.2}$$

be an associated biHamiltonian system. Then the operator $\mathcal{R} = \mathcal{E}\cdot\mathcal{D}^{-1}$ is a recursion operator, and leads to a hierarchy of differential functions

$$Q_{n+1}[u] = \mathcal{R}\cdot Q_n[u].$$

Each of the corresponding evolutionary vector fields $v_n \equiv v_{Q_n}$ is also biHamiltonian

$$u_t = Q_n[u] = \mathcal{D}\cdot\delta\mathcal{H}_n = \mathcal{E}\cdot\delta\mathcal{H}_{n-1}. \tag{4.3}$$

The functionals $\mathcal{H}_0, \mathcal{H}_1, \mathcal{H}_2,\ldots$ are in involution with respect to either Poisson bracket:

$$\{\mathcal{H}_n,\mathcal{H}_m\}_{\mathcal{D}} = 0 = \{\mathcal{H}_n,\mathcal{H}_m\}_{\mathcal{E}}.$$

Thus, for any m, n, $\mathcal{H}_m$ is a conservation law for the evolutionary system governed by the vector field v_n, and v_m is the corresponding generalized symmetry.

Note that the theorem automatically implies the invertibility of the Hamiltonian operator $\mathcal{D}$ on the hierarchy $Q_n[u]$, which shows the advantage of the Poisson complex approach.

Proof.

The proof that $\mathcal{R}$ is recursion operator in general can be found in [11; Theorem 7.27]; interestingly, this does not require the compatibility of the two bivectors. Here we just prove the properties about the hierarchy Q_n. Proceeding by induction on n, according to (3.11), the condition that the evolution equation (4.3) be biHamiltonian is equivalent to the fact that the corresponding generalized vector field $v_n \equiv v_{Q_n}$ can be written in the two forms

$$v_n = [\Theta_{\mathcal{D}}, \mathcal{H}_n] = [\Theta_{\mathcal{E}}, \mathcal{H}_{n-1}].$$

Let

$$\mathbf{v}_{n+1} = [\Theta_{\mathcal{E}}, \mathcal{H}_n]$$

be the next evolutionary vector field in the presumed hierarchy. The main task is to prove that $\mathbf{v}_{n+1}$ is a Hamiltonian vector field for the operator $\mathcal{D}$, i.e.

$$\mathbf{v}_{n+1} = [\Theta_{\mathcal{D}}, \mathcal{H}_{n+1}]$$

for some functional $\mathcal{H}_{n+1}$. By exactness of the Poisson complex for $\Theta_{\mathcal{D}}$, we need only verify that $\mathbf{v}_{n+1}$ is closed, i.e.

$$[\Theta_{\mathcal{D}}, \mathbf{v}_{n+1}] = 0.$$

To verify this latter condition, we use the super–Jacobi identity and the compatibility condition (4.1) for the Hamiltonian pair:

$$[\Theta_{\mathcal{D}}, \mathbf{v}_{n+1}] = [\Theta_{\mathcal{D}}, [\Theta_{\mathcal{E}}, \mathcal{H}_n]] = -[\Theta_{\mathcal{E}}, [\Theta_{\mathcal{D}}, \mathcal{H}_n]] = -[\Theta_{\mathcal{E}}, [\Theta_{\mathcal{E}}, \mathcal{H}_{n-1}]] = 0,$$

the last equality being a consequence of the closure of the $\mathcal{E}$–Poisson complex. (Note that we do not require exactness of the $\mathcal{E}$–Poisson complex.)

To prove the involutiveness of the resulting sequence of functionals note that according to (3.12),

$$\{\mathcal{H}_n, \mathcal{H}_m\}_{\mathcal{D}} = \mathrm{pr}\,\mathbf{v}_n(\mathcal{H}_m), \qquad \{\mathcal{H}_n, \mathcal{H}_m\}_{\mathcal{E}} = \mathrm{pr}\,\mathbf{v}_{n-1}(\mathcal{H}_m),$$

hence

$$\{\mathcal{H}_n, \mathcal{H}_m\}_{\mathcal{D}} = \{\mathcal{H}_n, \mathcal{H}_{m-1}\}_{\mathcal{E}}.$$

We now employ the skew-symmetry of the Poisson bracket to work our way down:

$$\{\mathcal{H}_n, \mathcal{H}_m\}_{\mathcal{D}} = \{\mathcal{H}_n, \mathcal{H}_{m-1}\}_{\mathcal{E}} = \{\mathcal{H}_{n+1}, \mathcal{H}_{m-1}\}_{\mathcal{D}} = \dots = \{\mathcal{H}_k, \mathcal{H}_k\} = 0,$$

where k is the integer part of $\frac{m-n}{2}$ and the final Poisson bracket is the $\mathcal{D}$-Poisson bracket if $m-n$ is even, the $\mathcal{E}$-Poisson bracket if $m-n$ is odd. This completes the proof.

We also note that, for a biHamiltonian system, there is a hierarchy of higher order

Hamiltonian operators, cf. [5]. The only problem with this result is that these operators are usually, but not always, integro-differential operators.

Proposition 10. If $\mathcal{D}$, $\mathcal{E}$ are the Hamiltonian operators for a Hamiltonian pair, and $\mathcal{R} = \mathcal{E}\cdot\mathcal{D}^{-1}$ the associated recursion operator, then the operators

$$\mathcal{R}^{k}\cdot\mathcal{E} = \mathcal{E}\cdot\mathcal{D}^{-1}\cdot\mathcal{E}\cdot\mathcal{D}^{-1}\cdot \ldots \cdot\mathcal{D}^{-1}\cdot\mathcal{E} \tag{4.4}$$

are Hamiltonian operators.

Example 11. The Riemann equation is a "quadri-Hamiltonian system", meaning that it can be written in Hamiltonian form in four distinct ways; however not all pairs of Hamiltonian operators are compatible. The three first order Hamiltonian operators

$$\mathcal{D}_0 = D_x, \qquad \mathcal{D}_1 = 2u\cdot D_x + u_x, \qquad \mathcal{D}_2 = u^2\cdot D_x + u\, u_x, \tag{4.5}$$

are all compatible, i.e. any two of them form a Hamiltonian pair. We find that the Riemann equation (2.8) can be written in the three Hamiltonian forms

$$u_t = \mathcal{D}_0\cdot E(\tfrac{1}{6}H_3) = \mathcal{D}_1\cdot E(\tfrac{1}{6}H_2) = \mathcal{D}_2\cdot E(H_1),$$

using the hierarchy of zeroth order conserved densities (2.13). Moreover, these Hamiltonian operators are not trivially related by (4.4). The resulting recursion operators $\mathcal{R}_1 = \mathcal{D}_1\cdot\mathcal{D}_0^{-1}$, $\mathcal{R}_2 = \mathcal{D}_2\cdot\mathcal{D}_0^{-1}$ are as given in (2.11), while $\mathcal{R}_3 = \mathcal{D}_2\cdot\mathcal{D}_1^{-1}$ is trivially related by the equation $\mathcal{R}_2 = \mathcal{R}_3\cdot\mathcal{R}_1$. The infinite hierarchy of commuting Hamiltonian flows generated by Theorem 9 consists of the generalized vector fields $v_n = Q_n[u]\cdot\partial_u$, with characteristics $Q_n[u] = u^n\cdot u_x$. The nth flow is also tri-Hamiltonian:

$$\begin{aligned} u_t &= Q_n[u] = u^n\cdot u_x \\ &= \mathcal{D}_0\cdot E\left(\frac{1}{(n+2)(n+1)}H_{n+2}\right) = \mathcal{D}_1\cdot E\left(\frac{1}{(2n+1)(n+1)}H_{n+1}\right) = \mathcal{D}_2\cdot E\left(\frac{1}{n^2}H_n\right). \end{aligned}$$

Each of these Hamiltonian systems admits an additional third order Hamiltonian operator

$$\mathcal{E} = D_x\cdot\frac{1}{u_x}\cdot D_x\cdot\frac{1}{u_x}\cdot D_x, \tag{4.6}$$

and can be written in yet another Hamiltonian form

$$u_t = u^n u_x = \mathfrak{E}\cdot E\left(\frac{1}{(n+1)(n+2)(n+3)(n+4)} H_{n+4}\right). \tag{4.7}$$

The Hamiltonian operators $\mathcal{D}_0$ and $\mathfrak{E}$ are compatible; however, $\mathfrak{E}$ is not compatible with either $\mathcal{D}_1$ or $\mathcal{D}_2$. The consequent recursion operator $\hat{\mathfrak{R}} = \mathfrak{E}\cdot\mathcal{D}_0^{-1} = \mathfrak{R}^2$ is the square of the simpler first order recursion operator (2.12). Note also that by proposition 10, there is an entire hierarchy of Hamiltonian differential operators $\left(D_x \cdot \frac{1}{u_x}\right)^{2k} \cdot D_x$ and each of the equations (4.7) can be written in Hamiltonian form using any one of these higher order Hamiltonian differential operators!

The third order generalized symmetry $\hat{v}_3$, with characteristic $\hat{Q}_3 = u_x \cdot I_4$, cf. (2.10) is bi-Hamiltonian

$$u_t = \mathcal{D}_0\cdot E(\hat{H}_1) = \mathfrak{E}\cdot E(\hat{H}_0),$$

using the conserved densities $\hat{H}_0 = u_x\cdot I_1$, $\hat{H}_1 = u_x\cdot I_3$. Consequently, the odd order vector fields $\hat{v}_{2n+1}$ are also bi-Hamiltonian, corresponding to the higher order rational conserved densities $\hat{H}_n = u_x\cdot I_{n+2}$. For any solution to the general first order flow (4.7) corresponding to the vector field v_n, each of the higher order quantities $\int \hat{H}_m[u]\,dx$ is a linear function of t, (provided that the integral converges). In fact, if $H(u)$ is any zero^th order Hamiltonian, with flow $u_t = D_x E(H)$, then $\hat{H}_m + t\,\frac{\partial^{2m+1} H}{\partial u^{2m+1}}$ is a conserved density. In particular, $\hat{H}_m$ is a conserved density for v_n whenever $2m \geq n-1$.

5. Gas Dynamics

In the final section, I will describe some recent joint work with Y. Nutku on Hamiltonian structure, symmetries and conservation laws for quasi-linear hyperbolic systems, including the classical equations of gas dynamics, [14]. The general form of a two-component hyperbolic system of conservation laws of Hamiltonian type is

$$u_t = D_x\cdot H_v, \qquad v_t = D_x\cdot H_u,$$

where the Hamiltonian density $H(u,v)$ depends only on u, v. The Hamiltonian operator is the matrix differential operator $\mathcal{D}_0 = \sigma_1\cdot D_x$, where $\sigma_1 = \begin{pmatrix} 0 & 1 \\ 1 & 0 \end{pmatrix}$. The system can be written in the convenient matrix form

$$u_t = \mathcal{D}_0 E[H] = \mathbf{H} \cdot \mathbf{u}_x, \qquad (5.1)$$

where

$$\mathbf{H} = \sigma_1 \cdot D^2 H = \begin{pmatrix} H_{uv} & H_{vv} \\ H_{uu} & H_{uv} \end{pmatrix}.$$

In the terminology of Dubrovin and Novikov, [4], [15] these systems are of "hydrodynamic type".

There are three important examples. The equations of gas dynamics are of this form, where u represents the velocity, and v the density of the fluid, and the Hamiltonian is $H(u,v) = -\left(\frac{1}{2}u^2 v + F(v)\right)$; the function F is related to the physical pressure P according to the equation $P'(v) = v \cdot F''(v)$, cf. [19; Chapter 6]. The equations of polytropic gas dynamics correspond to the choice $F(v) = \frac{v^\gamma}{\gamma(\gamma-1)}$ for some $\gamma \neq 0,1$; the case $\gamma = 2$ also arises in shallow water theory, [8], [19; p.84]. A second important example is provided by the Hamiltonian $H = \frac{u}{v} + \frac{v}{u}$, in which case (1.2) is equivalent to the Born-Infeld equation from nonlinear electrodynamics, [1], [19; p. 579]. Finally, the Hamiltonian density $H(u,v) = \frac{1}{2}u^2 + F(v)$ gives many simple models for a one-dimensional nonlinear elastic media; the derivative F′ being a monotone function of v corresponds to an ideal fluid or elastic solid; nonmonotone functions provide simple models of phase transitions, [16], [19; p. 123]. The case $F(v) = (1+v)^{-\gamma}$ corresponds to the Euler equation arising in nonlinear acoustics, cf. [8].

We begin by characterizing all the zeroth order conserved densities for the Hamiltonian system (5.1), cf. [15].

Proposition 12. A functional $\mathcal{F}[u] = \int F(u,v)\, dx$ is a conservation law for the hyperbolic system (5.1) if and only if F is a solution to the second order linear partial differential equation

$$A(u,v)\, F_{uu} = B(u,v)\, F_{vv}, \qquad (5.2)$$

with $A = H_{vv}$, $B = H_{uu}$.

The most important class of Hamiltonian systems (5.1) are the *separable* systems, for which the corresponding partial differential equation (5.2) admits a separation of variables in the rectangular (u,v)-coordinates, meaning that $\dfrac{H_{vv}}{H_{uu}} = \dfrac{\mu(v)}{\lambda(u)}$. The special case when $\lambda \equiv 1$, which includes gas dynamics and the elastic models, has added importance; such systems are said to be of *generalized gas dynamics type*. For simplicity, we will restrict our attention to generalized gas dynamics systems here, although extensions to more general separable systems can be found in [14].

For a gas dynamics system, there are two fundamental hierarchies of conserved densities, each of the form

$$H_{2m+\varepsilon}(u,v) = \sum_{k=0}^{m+\varepsilon} \frac{u^{2k+\varepsilon}}{(2k+\varepsilon)!} \cdot G_k(v) \,,$$

where ε is 0 or 1, depending on whether $n = 2m + \varepsilon$ is even or odd, and the functions G_k are generated recursively by

$$G_{k+1}(v) = \int_0^v (v - w)\cdot\mu(w)\cdot G_k(w)\, dw,$$

where $\mu(v) = H_{vv}/H_{uu}$. The two hierarchies depend on the initial selection of G_0; the first takes $G_0 = 1$, while the second has $G_0 = v$. Thus, we have the explicit conserved densities

$$\begin{aligned}
H_0 &= 1, & \tilde{H}_0 &= v, \\
H_1 &= u, & \tilde{H}_1 &= uv, \\
H_2 &= \tfrac{1}{2}u^2 + G_1(v), & \tilde{H}_2 &= \tfrac{1}{2}u^2 v + \tilde{G}_1(v), \qquad (5.3)\\
H_3 &= \tfrac{1}{6}u^3 + u\, G_1(v), & \tilde{H}_3 &= \tfrac{1}{6}u^3 v + u\, \tilde{G}_1(v), \\
H_4 &= \tfrac{1}{24}u^4 + \tfrac{1}{2}u^2 G_1(v) + G_2(v), & \tilde{H}_4 &= \tfrac{1}{24}u^4 v + u^2\, \tilde{G}_1(v) + \tilde{G}_2(v),
\end{aligned}$$

etc. Note that the elastic Hamiltonian appears in the first hierarchy as H_2, whereas the gas dynamics Hamiltonian appears in the alternative hierarchy as $-\tilde{H}_2$.

Each of these Hamiltonian functions generates a Hamiltonian flow, governed by the corresponding evolutionary system. We let

$$Q_n = \mathcal{D}_0 E[H_n] = H_n \cdot u_x, \tag{5.4}$$

cf. (5.1), denote the right hand side of this equation, which is also the characteristic for the symmetry vector field $v_n = Q_n \cdot \partial_u$. We define $\tilde{H}_n$, $\tilde{Q}_n$ and $\tilde{v}_n$ for the alternative hierarchy $\tilde{H}_n$ similarly. All the Hamiltonians H_n and $\tilde{H}_n$ are in involution with respect to the Poisson bracket determined by the Hamiltonian operator $\mathcal{D}_0$.

In the case of polytropic gas dynamics, there are two additional first order Hamiltonian structures. Using the Hamiltonian hierarchies, we find that we can write the polytropic gas dynamics equations in the alternative Hamiltonian forms

$$u_t = \mathcal{D}_1 \cdot E(\tfrac{1}{\gamma}\tilde{H}_1) = \mathcal{D}_2 \cdot E(\tilde{H}_0),$$

with the two Hamiltonian operators, [8],

$$\mathcal{D}_1 = \begin{pmatrix} v^{\gamma-2} \cdot D_x + D_x \cdot v^{\gamma-2} & (\gamma-1)u \cdot D_x + u_x \\ (\gamma-1)u \cdot D_x + (\gamma-2)u_x & v \cdot D_x + D_x \cdot v \end{pmatrix},$$

$$\mathcal{D}_2 = \begin{pmatrix} uv^{\gamma-2} \cdot D_x + D_x \cdot uv^{\gamma-2} & \left\{\frac{1}{2}(\gamma-1)u^2 + 2\frac{v^{\gamma-1}}{\gamma-1}\right\} \cdot D_x + uu_x + v^{\gamma-2}v_x \\ \left\{\frac{1}{2}(\gamma-1)u^2 + 2\frac{v^{\gamma-1}}{\gamma-1}\right\} \cdot D_x + (\gamma-2)uu_x + v^{\gamma-2}v_x & uv \cdot D_x + D_x \cdot uv \end{pmatrix}.$$

The Hamiltonian operators $\mathcal{D}_0$, $\mathcal{D}_1$, $\mathcal{D}_2$ are mutually compatible, leading to three distinct Hamiltonian pairs. The corresponding recursion operators

$$\mathcal{R}_1 = \mathcal{D}_1 \cdot \mathcal{D}_0^{-1}, \qquad \mathcal{R}_2 = \mathcal{D}_2 \cdot \mathcal{D}_0^{-1}, \qquad \mathcal{R}_3 = \mathcal{D}_2 \cdot \mathcal{D}_1^{-1},$$

are trivially related by the identity $\mathcal{R}_2 = \mathcal{R}_3 \cdot \mathcal{R}_1$, but are otherwise distinct. Nevertheless, they both give rise to the same series of gas dynamics Hamiltonians, since

$$\mathcal{R}_1(Q_n) = Q_{n+1}, \qquad \mathcal{R}_2(Q_n) = Q_{n+2},$$

and similarly for the alternative hierarchy $\tilde{v}_n$. Strangely, there does not appear to be a counterpart of these two recursion operators in the general non-polytropic case, i.e. when the pressure is not proportional to a power of the density.

There is, in addition, a third order Hamiltonian operator, analogous to (4.6), for any generalized gas dynamics system. Let $M(v) = \int_0^v \mu(s)\, ds$, and define the matrix variable

$$U(u,v) = \begin{pmatrix} u & M(v) \\ v & u \end{pmatrix}.$$

We use the notation

$$U_x = \begin{pmatrix} u_x & \mu(v)v_x \\ v_x & u_x \end{pmatrix} \quad \text{and} \quad U_x^{-1} = \frac{1}{\delta}\begin{pmatrix} u_x & -\mu(v)v_x \\ -v_x & u_x \end{pmatrix},$$

where $\delta = u_x^2 - \mu(v)\cdot v_x^2$, for the total x derivative of the matrix U and its matrix inverse. A nontrivial calculation proves that the operator

$$\mathcal{E} = D_x\cdot U_x^{-1}\cdot D_x\cdot U_x^{-1}\cdot \sigma_1\cdot D_x = D_x\cdot U_x^{-1}\cdot D_x\cdot \sigma_1\cdot U_x^{-T}\cdot D_x$$

is Hamiltonian. The operators $\mathcal{E}$ and $\mathcal{D}_0$ form a Hamiltonian pair; however, for polytropic gas dynamics, the Hamiltonian operators $\mathcal{E}$ and $\mathcal{D}_1$ are *not* compatible, nor are the Hamiltonian operators $\mathcal{E}$ and $\mathcal{D}_2$.

Theorem 13. Let $H(u,v)$ be any generalized gas dynamics Hamiltonian density. Then there exists a second zeroth order conserved density H^* such that the corresponding Hamiltonian system (5.1) can be written in biHamiltonian form

$$u_t = \mathcal{D}_0 E[H] = \mathcal{E}\, E[H^*].$$

If the Hamiltonian density H in Theorem 13 is one of the densities H_n in the hierarchy (5.3), then it is not hard to see that the corresponding density $H^*(u,v)$ can be taken to be the density H_{n+2}; similarly, if $H = \tilde{H}_n$, then $H^* = \tilde{H}_{n+2}$.

According to Theorem 9, the operator $\hat{\Re} = \mathcal{E}\cdot\mathcal{D}_0^{-1} = D_x\cdot U_x^{-1}\cdot D_x\cdot U_x^{-1}$ is a recursion operator which, just as in the one-dimensional case, is the square of a simpler recursion operator $\Re = D_x\cdot U_x^{-1}$. On the zero[th] order symmetries, $\Re(Q_n) = Q_{n-1}$, $\Re(\tilde{Q}_n) = \tilde{Q}_{n-1}$. In the polytropic case, $\Re$ is the "inverse" to the recursion operator $\Re_1$ on the hierarchies (5.3), although as always, this is special to these particular hierarchies. There is also an additional hierarchy of higher order symmetries.

Theorem 14. Let $H = H_n$ be one of the nth order generalized gas dynamics Hamiltonians, and let v_n be the corresponding first order Hamiltonian flow. Let $\hat{v}_m$ denote the generalized vector field of order m with characteristic

$$\hat{Q}_m = \Re^m(x\mathbf{u}_x) = \Re^{m-1}\cdot\begin{pmatrix}1\\0\end{pmatrix}.$$

Then $\hat{v}_m$ is a symmetry for the flow generated by v_n provided $m \geq n-1$. Similarly, $\hat{v}_m$ is a symmetry for the flow generated by $\tilde{v}_n$ corresponding to the Hamiltonian $H = \tilde{H}_n$ provided $m \geq n$.

In polytropic gas dynamics, we can construct additional recursion operators by combining the Hamiltonian operator $\mathcal{E}$ with the operators $\mathcal{D}_1$, $\mathcal{D}_2$, even though they are not compatible. However, the resulting higher order symmetries appear to always be non-local since we cannot explicitly invert $\mathcal{D}_1$ or $\mathcal{D}_2$.

Finally, we indicate how to construct higher order conservation laws for any generalized gas dynamics system. First note that, if (5.1) is of gas dynamics type, it is equivalent to the matrix evolution equation $U_t = \mathbf{H}\cdot U_x$. This, and the fact that the matrices $\mathbf{H}$ and U_x commute, immediately leads to the important matrix identity

$$D_t\cdot(U_x^{-1}) - D_x(\mathbf{H}\cdot U_x^{-1}) = -(\mathbf{H}_x\cdot U_x^{-1} + U_x^{-1}\cdot\mathbf{H}_x), \tag{5.5}$$

which holds on solutions to the system (5.1). In particular, the (2,1)-entry of (5.5) reads

$$D_t\left(\frac{v_x}{\delta}\right) + D_x\left(\frac{H_{uu}\cdot v_x - H_{uv}\cdot u_x}{\delta}\right) = -2\,H_{uuu}.$$

For classical gas dynamics, $H_{uuu} \equiv 0$, and we recover the conserved density

$$\hat{H}_1[u,v] = \frac{v_x}{\delta} = \frac{v_x}{u_x^2 - \mu(v)\cdot v_x^2} . \tag{5.6}$$

due to Verosky, [17]. For more general gas dynamics Hamiltonians, H_{uuu} will no longer be 0, and $\delta^{-1}\cdot v_x$ will no longer be a conserved density; however, we can simply modify it to get a time-dependent conservation law with density $\hat{H}_1^* = \delta^{-1}\cdot v_x + 2t\, H_{uuu}$. Equivalently, the integral $\hat{\mathcal{H}}_1 = \int \delta^{-1}\cdot v_x \, dx$, when it converges, is a linear function of t.

The first order conserved density $\hat{H}_1 = \delta^{-1}\cdot v_x$ leads to a Hamiltonian flow using the basic Hamiltonian operator $\mathcal{D}_0$. This will allow us to apply Theorem 9 to the Hamiltonian pair $\mathcal{E}$ and $\mathcal{D}_0$, and thereby generate a new hierarchy of higher order conservation laws in gas dynamics. We find that, as with the Riemann equation, the symmetry $\hat{v}_3$ is Hamiltonian with respect to the Hamiltonian pair $\mathcal{D}_0$, $\mathcal{E}$ and the corresponding conserved density is -2 times Verosky's density (5.6). Therefore, there is a hierarchy of mth order Hamiltonian densities $\hat{H}_m$, $m = 1,2,\ldots,$ and corresponding commuting biHamiltonian systems

$$u_t = \hat{Q}_{2m+1} = \mathcal{D}_0 E[\hat{H}_m] = \mathcal{E}\, E[\hat{H}_{m-1}], \qquad m \geq 1.$$

These Hamiltonians are in involution with respect to both the $\mathcal{D}_0$ and $\mathcal{E}$ Poisson brackets. More generally, if $H = H_n$ is a Hamiltonian density in the first generalized gas dynamics hierarchy, then the higher order density $\hat{H}_m$ is conserved for the Hamiltonian system (5.4) provided $n \leq 2m+1$. If $H = \tilde{H}_n$ is in the second generalized gas dynamics hierarchy, then $\hat{H}_m$ is conserved provided $n \leq 2m+2$.

Extensions to higher dimensional hyperbolic systems, cf. [18], and applications to discontinuous solutions and shock waves, as in [3], are under investigation.

References

[1] Arik, M., Neyzi, F., Nutku, Y., Olver, P.J. and Verosky, J.M., Hamiltonian structures for the Born-Infeld equation, in preparation.

[2] Astashov, A.M. and Vinogradov, A.M., On the structure of Hamiltonian operator in field theory, *J. Geom. Phys.* **2** (1986), 263-287.

[3] Benjamin, T.B. and Bowman, S., Discontinuous solutions of one-dimensional Hamiltonian systems, *Proc. Roy. Soc. London* **B413** (1987), 263-295.

[4] Dubrovin, B.A. and Novikov, S.A., Hamiltonian formalism of one-dimensional systems of hydrodynamic type and the Bogolyubov-Whitham averaging method, *Sov. Math. Dokl.* **27** (1983), 665-669.

[5] Fuchssteiner, B., and Fokas, A.S., Symplectic structures, their Bäcklund transformations and hereditary symmetries, *Physica* **4D** (1981), 47-66.

[6] Lichnerowicz, A., Les variétés de Poisson et leurs algèbres de Lie associées, *J. Diff. Geom.* **12** (1977), 253-300.

[7] Magri, F., A simple model of the integrable Hamiltonian equation, *J. Math. Phys.* **19** (1978) 1156-1162.

[8] Nutku, Y., On a new class of completely integrable systems. II. Multi-Hamiltonian structure, *J. Math. Phys.*, **28** (1987), 2579-2585.

[9] Olver, P.J., Evolution equations possessing infinitely many symmetries, *J. Math. Phys.* **18** (1977), 1212-1215.

[10] Olver, P.J., Hamiltonian perturbation theory and water waves, *Contemp. Math.* **28** (1984), 231-249.

[11] Olver, P.J., *Applications of Lie Groups to Differential Equations*, Graduate Texts in Mathematics, vol. 107, Springer-Verlag, New York, 1986.

[12] Olver, P.J., BiHamiltonian systems, in: *Ordinary and Partial Differential Equations*, B.D. Sleeman and R.J. Jarvis, eds., Pitman Research Notes in Mathematics Series, No. 157, Longman Scientific and Technical, New York, 1987, pp. 176-193.

[13] Olver, P.J., Darboux' theorem for Hamiltonian differential operators, *J. Diff. Eq.* **71** (1988), 10-33.

[14] Olver, P.J. and Nutku, Y., Hamiltonian structures for systems of hyperbolic conservation laws, *J. Math. Phys.*, **29** (1988), to appear.

[15] Sheftel', M.B., Integration of Hamiltonian systems of hydrodynamic type with two dependent variables with the aid of the Lie-Bäcklund group, *Func. Anal. Appl.* **20** (1986), 227-235.

[16] Slemrod, M., Dynamics of first order phase transitions, in: *Phase Transformations and Material Instabilities in Solids*, M.E. Gurtin, Ed., Academic Press, New York, 1984, pp. 163-203.

[17] Verosky, J.M., Higher order symmetries of the compressible one-dimensional isentropic fluid equations, *J. Math. Phys.* **25** (1984), 884-888.

[18] Verosky, J.M., First order conserved densities for gas dynamics, *J. Math. Phys.* **27** (1986), 3061-3063.

[19] Whitham, G.B., *Linear and Nonlinear Waves*, John Wiley & Sons, New York, 1974.

POSITIVE PERIODIC SOLUTIONS OF A NON LINEAR PARABOLIC EIGENVALUE PROBLEM

by
L.A.Ortega and J.F.Caycedo
Department of Mathematics, National University of Colombia

Abstract.
Sufficient conditions are obtained for the existence and stability of a unique positive periodic solution of the nonlinear eigenvalue problem:

$$L[u]\,(x,t) = \lambda f(x,t,u(x,t)) \quad \text{on} \quad \Omega x \mathbb{R}\,,$$
$$u \equiv 0 \quad \text{on} \quad \partial\Omega x \mathbb{R}$$

where L is a second order uniformly parabolic differential operator.

Introduction

We consider the nonlinear parabolic eigenvalue problem:

$$\left.\begin{aligned} &L[u]\,(x,t) = \lambda f(x,t,u(x,t)) \text{ on } \Omega x\mathbb{R}, \\ &u \equiv 0 \qquad\qquad\qquad\qquad \text{on } \partial\Omega x\mathbb{R}, \\ &u(x,t+T) \equiv u(x,t). \end{aligned}\right\} \tag{0.1}$$

where, $L[u] = \frac{\partial u}{\partial t} - (\sum_{j=1}^{N}\sum_{i=1}^{N} a_{ij}\frac{\partial^2 u}{\partial x_i \partial x_j} + \sum_{i=1}^{N} b_i \frac{\partial u}{\partial x_i} + C\cdot u)$ is uniformly parabolic operator, $T>0$,

$\Omega \subset \mathbb{R}^N$ $(N>1)$ is a bounded domain with boundary $\partial\Omega$ of class $C^{2+\alpha}$ $(0 < \alpha < 1)$, $f{:}\bar{\Omega}x\mathbb{R}x\mathbb{R} \to \mathbb{R}$, is a function sufficiently smooth T-periodic in t and the coefficients a_{ij}, b_i, C belong to the real Banach space $\mathbb{F} = \{u \in C^\alpha(\bar{\Omega}x\mathbb{R}) | \, u \text{ is T-periodic in t}\}$, with $a_{ji} \equiv a_{ji}$ and $C \leq 0$ on $\Omega x\mathbb{R}$.

The existence of positive solutions of parabolic eigenvalue problems of the form (0.1) has been investigated by Beltramo and Hess in [1]. We do not generalize the Beltramo-Hess results. We will be interested to prove existence and uniqueness of the solution of the problem (0.1) in the case of periodic solutions which are zero on the boundary under certain hypotheses (0.4). These hypotheses are derived from those assumes by R.W.Dickey in [2], for ordinary differential equations.

In this paper we adopt the following hypothesis on f:

$$\text{If } u \in \mathbb{F} \text{ then } f(\cdot, u) \in \mathbb{F}. \tag{0.2}$$

$$f(x,t,0) \geq 0, \text{ and } f(x,t,0) \not\equiv 0 \text{ on } \overline{\Omega} \times \mathbb{R} \tag{0.3}$$

$$\frac{\partial f}{\partial u}(x,t,u) > 0 \qquad \text{if } u \geq 0$$

$$\text{and} \tag{0.4}$$

$$\frac{\partial f}{\partial u}(x,t,u_1) < \frac{\partial f}{\partial u}(x,t,u_2), \qquad \text{if } u_1 > u_2 \geq 0, \text{ for all } (x,t) \in \overline{\Omega} x \mathbb{F}.$$

We are interested to prove the existence of positive solutions of (0.1), which belong to the Banach space:

$$\mathbb{E} = \left\{ u \in C^{2+\alpha}(\overline{\Omega} \times \mathbb{R}) \mid u \text{ is periodic in } t \text{ and } u \equiv 0 \text{ on } \partial\Omega x \mathbb{R} \right\}$$

To do so we follow the scheme given by R.W.Dickey in [2] for ordinary differential equations.

Section I.

The main purpose of this section is to prove the following result.

Theorem 1.1. *If* f *satisfies* (0.2), (0.3), (0.4) *then for all* $\lambda \in (0, \lambda_1)$, *the problem* (0.1) *has a* ***unique solution*** $u(\cdot, \lambda) \in E$, $u(\cdot, \lambda) > 0$ *everywhere on* $\Omega x \mathbb{R}$, *where* λ_1 *is the unique eigenvalue with positive eigenfunction* ϕ *of the linear parabolic eigenvalue problem*:

$$\left\{ L[u] \equiv \lambda \frac{\partial f}{\partial u}(x,t,0)u, \ \text{on } \Omega x \mathbb{R}, u \in \mathbb{E} \right\}.$$

(For the existence of λ_1 see [1]. Theorem 1).

For simplicity, we will denote by A the set of λ for which (0.1) has positive solution.

In this section we refer repeatedly to the following well-known result.

Proposition 1.2. *Given* $f \in \mathbb{F}$ *there exists a unique* $u \in \mathbb{E}$ *such that* $L[u] \equiv f$. *Moreover there exist two constants,* $\bar{k}_1 > 0$ *depending on* $\beta \in (\alpha, 1)$ *and* $\bar{k}_2 > 0$ *such that*

$\| u \|_{\mathbb{E}} \le \bar{k}_1 \| L[u] \|_{\mathbb{F}}$, *for all* $u \in \mathbb{E}$. (I)

$\| u \|_{1+\beta} \le \bar{k}_2 \| L[u] \|_{\infty}$, *for all* $u \in \mathbb{E}$. (II)

(See references [3], [4]).

For each $\lambda \in A$, a positive solution $\underline{u}(\cdot, \lambda) \in \mathbb{E}$ of (0.1), is called a minimal positive solution of (0.1), if $\underline{u}(\cdot, \lambda) \le u(\cdot, \lambda)$ on $\bar{\Omega} x \mathbb{R}$, for any positive solution $u(\cdot, \lambda) \in \mathbb{E}$ of (0.1).

For all positive number λ, we define the sequence $\{u_n(\cdot, \lambda)\}^{\infty}{}_{n=1}$ in $\mathbb{E}$, by

$$L[u_1(\cdot, \lambda)] \equiv \lambda f(\cdot, 0) \quad \text{on } \Omega x \mathbb{R},$$
$$L[u_n(\cdot, \lambda)] \equiv \lambda f(\cdot, u_{n-1}(\cdot, \lambda)) \quad \text{on } \Omega x \mathbb{R}, \tag{1.3}$$

for each positive integer n.

Lemma 1.4. *The sequence* $\{u_n(\cdot, \lambda)\}^{\infty}{}_{n=1}$ *is uniformly bounded iff* $\lambda \in A$.

If $\lambda \in A$ *then* $\{u_n(\cdot, \lambda)\}^{\infty}{}_{n=1}$ *converges to a function* $\underline{u}(., \lambda)$ *in* $\mathbb{F}$*-norm, where* $\underline{u}(., \lambda)$ *is minimal solution of* (0.1).

Proof. First we will prove that the sequence $\{u_n(\cdot, \lambda)\}^{\infty}{}_{n=1}$ is increasing on $\bar{\Omega} x \mathbb{R}$.

Since

$$L[u_1(., \lambda)] \equiv \lambda f(., 0) > 0 \quad \text{on } \Omega x \mathbb{R},$$
$$L[u_2(., \lambda) - u_1(., \lambda)] = \lambda[f(., u_1(., \lambda)) - f(., 0)] > 0 \quad \text{on } \Omega x \mathbb{R},$$

it follows by the maximum principle for parabolic equations, (see [5]) that:

$u_2(\cdot, \lambda) \ge u_1(\cdot, \lambda) > 0$ on $\Omega x \mathbb{R}$.

If for some positive integer n,

$$u_n(., \lambda) \ge u_{n-1}(., \lambda) \ge \ldots \ge u_1(., \lambda) > 0 \text{ on } \Omega x \mathbb{R},$$

then

$$L\left[u_{n+1}(\cdot, \lambda) - u_n(\cdot, \lambda)\right] = \lambda\left[f(\cdot, u_n(\cdot, \lambda)) - f(\cdot, u_{n-1}(\cdot, \lambda))\right] > 0 \quad \text{on } \Omega x \mathbb{R}.$$

It follows by the maximum principle, that

$$u_{n+1}(\cdot, \lambda) \ge u_n(\cdot, \lambda) \quad \text{on } \Omega x \mathbb{R}.$$

To prove the sufficient condition, suppose that $\lambda \in A$. Then there exists a function $u \in \mathbb{E}$ such that $u > 0$ everywhere on $\Omega x \mathbb{R}$, and $L[u] \equiv \lambda f(\cdot, u)$ on $\Omega x \mathbb{R}$. Using the above argument, one can prove that $u_n(\cdot, \lambda) \leq u$ on $\Omega x \mathbb{R}$, for every positive integer n and so,

$$\| u_n(\cdot, \lambda)\|_{\infty}, \leq \| u \|_{\infty}, \text{ for each positive integer n.}$$

To prove the necessary condition, suppose the existence of a positive real number M such that $\|u_n(\cdot, \lambda)\|_{\infty}, \leq M$, for all positive integer n.

Since $\{u_n(\cdot, \lambda)\}^{\infty}{}_{n=1}$ is an increasing sequence by the last inequality, there exists a function $\underline{u}(\cdot, \lambda)$ defined on $\Omega x \mathbb{R}$ such that:

$$u_n(x,t,\lambda) \rightarrow \underline{u}(x,t,\lambda), \text{ as } n \rightarrow \infty \text{ on } \Omega x \mathbb{R}. \tag{1.5}$$

Since the sequence $\{u_n(\cdot, \lambda)\}^{\infty}{}_{n=1}$ is uniformly bounded, by proposition (1.2) there exists a real number $r_1 > 0$ such that:

$$\| u_n(., \lambda)\|_{1+\beta} \leq \bar{k} \| \lambda f(., u_n(., \lambda))\|_{\infty} \leq r_1$$

for all positive integer n, ($\beta \in (\alpha, 1)$).

Since the injection:

$$C^{1+\beta}(\bar{\Omega}x[0,T]) \xrightarrow{i} C^{\alpha}[\bar{\Omega}x[0,T]] ,$$

is a compact operator, it follows by the last inequality that there exists a subsequence of $\{u_n(\cdot, \lambda)\}^{\infty}{}_{n=1}$ which converges in $\mathbb{F}$-norm. Since every subsequence of $\{u_n(\cdot, \lambda)\}^{\infty}{}_{n=1}$ also has a convergent subsequence in $\mathbb{F}$-norm, then by (1.5):

$$\| u_n(\cdot, \lambda) - \underline{u}(\cdot, \lambda)\|_{\mathbb{F}} \rightarrow 0 \quad \text{as } n \rightarrow \infty .$$

To prove that $\underline{u}(\cdot, \lambda)$ is a solution of (0,1), we define an operator T, from $\mathbb{F}$ to $\mathbb{F}$ by:

$$T(u) = J \cdot L^{-1}(\lambda f(\cdot, u)), \quad \text{for all } u \in \mathbb{F},$$

where $E \xrightarrow{J} F$, is the injection.

By the proposition 1.2, T is a continuous operator.

Since,

$$u_n(\cdot,\lambda) \to \underline{u}(\cdot,\lambda), \quad \text{as } n \to \infty,$$

in $\mathbb{F}$-norm, then

$$\| T(u_n(\cdot,\lambda)) - T(\underline{u}(\cdot,\lambda)) \|_{\mathbf{F}} \to 0, \text{ as } n \to \infty$$

and

$$\| T(u_n(\cdot,\lambda)) - T(\underline{u}(\cdot,\lambda) \|_{\mathbf{F}} = \| u_{n+1}(\cdot,\lambda) - T(\underline{u}(\cdot,\lambda)) \|_{\mathbf{F}}$$

$$\to \| \underline{u}(\cdot,\lambda) - T(\underline{u}(\cdot,\lambda)) \|_{\mathbf{F}}, \text{ as } n \to \infty$$

Therefore,

$$\underline{u}(\cdot,\lambda) \equiv T(\underline{u}(\cdot,\lambda)) \equiv L^{-1}(\lambda f(\cdot, \underline{u}(\cdot,\lambda))),$$

and so,

$$L[\underline{u}(\cdot,\lambda)] \equiv \lambda f(\cdot, \underline{u}(\cdot,\lambda)) \text{ on } \Omega x \mathbb{R}.$$

Lemma 1.6. *Suppose,* $F:\bar{\Omega}x\mathbb{R}\,x\mathbb{R} \to \mathbb{R}$, *is another function, with the following properties: if* $\theta \geq \psi \geq 0$, *then* $F(x,t,\theta) \geq f(x,t,\psi)$, $F(x,t+T,\theta) \equiv F(x,t,\theta)$, *for all* $(x,t) \in \bar{\Omega}x\mathbb{R}$.
If λ *is a positive real number and* $V \in \mathbb{E}$ *is solution of the problem*:

$$\{L[u] \equiv \lambda F(\cdot, u) \text{ on } \Omega x\mathbb{R}, V > 0 \text{ on } \Omega x \mathbb{R}\},$$

then, $\lambda \in A$ *and* $\underline{u}(\cdot,\lambda) \leq V$ *on* $\Omega x\mathbb{R}$, *where* $\underline{u}(\cdot,\lambda)$ *is a minimal positive solution of* (0,1).

Proof. With the same argument used in the first part of the proof of Lemma 1.4, we can prove that $V \geq u_n(\cdot,\lambda)$ on $\Omega x\mathbb{R}$, for every positive integer n, where $\{u_n(\cdot,\lambda)\}^{\infty}{}_{n=1}$ is the sequence defined in (1.3).

It follows by Lemma 1.4 that there exists a function $\underline{u}(\cdot,\lambda) \in \mathbb{E}$ such that $\underline{u}(\cdot,\lambda)$ is a minimal positive solution of (0.1),

$$\| u_n(\cdot,\lambda) - \underline{u}(\cdot,\lambda) \|_{\mathbf{F}} \to 0 \text{ as } n \to \infty, \text{ and so } \underline{u}(\cdot,\lambda) \leq V \text{ on } \Omega x\mathbb{R}.$$

Lemma 1.7. *If* $\lambda^* \in A$ *then* $\lambda \in A$ *for every,* $0 < \lambda < \lambda^*$.

Proof. We fix λ, $0 < \lambda < \lambda^*$ and define

$$F(x,t,\theta) \equiv \frac{\lambda^*}{\lambda} f(x,t,\theta), \text{ on } \overline{\Omega} x \mathbb{R} x \mathbb{R}.$$

If $\theta > \psi \geq 0$, then

$$F(x,t,\theta) - f(x,t,\psi) = \frac{\lambda^*}{\lambda} f(x,t,\theta) - f(x,t,\psi) > f(x,t,\theta)$$
$$- f(x,t,\psi) \geq 0 \text{ on } \overline{\Omega} x \mathbb{R}.$$

Since $\lambda^* \in A$, there exists $u(\cdot, \lambda^*) \in \mathbb{E}$ such that, $u(\cdot, \lambda^*) > 0$ on $\Omega x \mathbb{R}$ and

$$L[u(\cdot, \lambda^*)] = \lambda^* f(\cdot, u(\cdot, \lambda^*)) = \lambda F(\cdot, u(\cdot, \lambda^*)) \text{ on } \overline{\Omega} x \mathbb{R}.$$

If follows by Lemma 1.6, and the above inequalities, that $\lambda \in A$ and $\underline{u}(\cdot, \lambda) \leq u(\cdot, \lambda^*)$.

Lemma 1.8. *Assume*, $\rho, F \in \mathbb{F}$, $\rho > 0$, $F > 0$ *on* $\Omega x \mathbb{R}$. *If* $f(x,t,\theta) \leq F(x,t) + \rho(x,t)\theta$, *for every* $(x,t) \in \Omega x \mathbb{R}$, $\theta \geq 0$ *and* $\lambda_1 > 0$ *is the principal eigenvalue of the problem*:

$$\{L[V] \equiv \lambda \rho V \text{ on } \Omega x \mathbb{R}, \ V > 0 \text{ on } \Omega x \mathbb{R}, \quad V \in \mathbb{E}\}$$

then $\lambda \in A$, *for every* $0 < \lambda < \lambda_1$.

Proof. If $0 < \lambda < \lambda_1$, then there exists $V \in \mathbb{F}$ such that, $V > 0$ on $\Omega x \mathbb{R}$ and $L[V] \equiv \lambda \rho V + h$ on $\Omega x \mathbb{R}$, where $h(x, t) \equiv \lambda F(x. t)$, (see [1], Theorem II). Then

$$L[V](x, t) = \lambda[\rho(x, t) V(x, t) + F(x, t)] = \lambda \overline{F}(x, t, V),$$

where,

$$\overline{F}(x, t, \theta) \equiv \rho(x, t)\theta + F(x, t)$$

It follows by Lemma 1.6 that $\lambda \in A$.

Proof of Theorem 1.1. If $u \geq 0$ then for every $(x, t) \in \overline{\Omega} x \mathbb{R}$ there exists $s \in (0,1)$ such that

$$f(x, t, u) - f(x, t, 0) = \frac{\partial f}{\partial u}(x, t, su)u \leq \frac{\partial f}{\partial u}(x, t, 0)u,$$

therefore,

$$f(x, t, u) \leq f(x, t, 0) + \frac{\partial f}{\partial u}(x, t, 0)u, \quad \text{for all } u > 0.$$

It follows that $f(x, t, u) \leq F(x, t) + \rho(x, t)u$, for every $(x, t) \in \Omega x \mathbb{R}$, and $u \geq 0$, where $F(x, t) \equiv f(x, t, 0)$, $\rho(x, t) \equiv (\partial f/\partial u)(x, t, 0)$. Finally from Lemma 1.8 we get that $\lambda \in A$ for all $0 < \lambda < \lambda_1$, where $\lambda_1 > 0$ is the principal eigenvalue of the problem:

$$\left\{L[V] \equiv \lambda \frac{\partial f}{\partial u}(x, t, 0)V \text{ on } \Omega x \mathbb{R},\ V \in \mathbb{E}, V > 0 \text{ on } \Omega x \mathbb{R}\right\}$$

To prove the uniqueness of the solution of (0.1) for $\lambda \in (0, \lambda_1)$, we assume that $0 < \lambda < \lambda_1$, $\underline{u}(\cdot, \lambda) \in \mathbb{E}$ is the minimal positive solution of (0.1) and $u((\cdot, \lambda) \in \mathbb{E}$, some other positive solution of (0.1) for this λ. Then, if $u(\cdot, \lambda) \neq \underline{u}(\cdot, \lambda)$, by (0.4) and the maximum principle we have:

$$u(\cdot, \lambda) - \underline{u}(\cdot, \lambda) > 0 \text{ on } \Omega x \mathbb{R},$$

and

$$\begin{aligned} L[u(x, t, \lambda) - \underline{u}(x, t, \lambda)] &= \lambda[f(x, t, u(x, t, \lambda)) - f(x, t, \underline{u}(x, t, \lambda))] \\ &= \lambda\left[\frac{\partial f}{\partial u}\left(x, t, \underline{u}(x, t, \lambda) + s[u(x, t, \lambda) - \underline{u}(x, t, \lambda)]\right)\left(u(x, t, \lambda) - \underline{u}(x, t, \lambda)\right)\right] \\ &< \lambda \frac{\partial f}{\partial u}\left(x, t, \underline{u}(x, t, \lambda)\right)\left(u(x, t, \lambda) - \underline{u}(x, t, \lambda)\right) \\ &< \lambda \frac{\partial f}{\partial u}(x, t, 0)\left(u(x, t, \lambda) - \underline{u}(x, t, \lambda)\right), \text{ on } \Omega x \mathbb{R}, \end{aligned}$$

Thus

$$L[\underline{u}(., \lambda) - u(., \lambda)] > \lambda\left[\frac{\partial f}{\partial u}(., 0)\left(\underline{u}(., \lambda) - u(., \lambda)\right)\right] \text{ on } \Omega x \mathbb{R}.$$

It follows that

$$L[\underline{u}(\cdot,\lambda)-u(\cdot,\lambda)] \equiv \lambda\frac{\partial f}{\partial u}(\cdot,0)\,[\underline{u}(\cdot,\lambda)-u(\cdot,\lambda)]+h, \text{ on } \Omega x\mathbb{R}, \tag{1.9}$$

where

$$h \equiv L[\underline{u}(\cdot,\lambda)-u(\cdot,\lambda)] - \lambda\frac{\partial f}{\partial u}(\cdot,0)(\underline{u}(\cdot,\lambda)-u(\cdot,\lambda))>0 \text{ on } \Omega x\mathbb{R}.$$

Since $h > 0$ on $\Omega x\mathbb{R}$, the above identity (1.9) implies (see [1] Theorem II), that $\underline{u}(\cdot,\lambda) > u(\cdot,\lambda)$ on $\Omega x\mathbb{R}$. This contradiction proves that $\underline{u}(\cdot,\lambda) \equiv u(\cdot,\lambda)$.

Section II

Theorem 2.1. *Let the hypothesis of theorem* 1.1 *be satisfied. If the map:*
$(x, t) \rightarrow \partial f/\partial u(x, t, 0)$ *is continuously differentiable and bounded on* $\Omega x\mathbb{R}$ *and* $\lambda \in (0, \lambda_1)$, *then the solution* $\underline{u}(\cdot\lambda)$ *of the problem* (0.1) *is globally asymptotically stable: i.e. if*

$$z \in C(\overline{\Omega}\, x[0,\infty)) \cap C^{2+\alpha}(\Omega x(0,\infty))$$

is a solution of the initial value boundary-value problem

$$\left\{\begin{array}{ll} L[z] \equiv \lambda f(x,t,z) & \text{on } \Omega x(0,\infty), \\ z(x,t) \equiv 0 & \text{on } \partial\Omega x[0,\infty),\ z(x,0) \in C^{2+\alpha}(\Omega) \end{array}\right\}.$$

then $\lim_{t\to\infty} \| z(x,t) - \underline{u}(x,\lambda)\| = 0$, *uniformly in the variable* x.

Proof. Since $f \geq 0$, $f \not\equiv 0$ and $\lambda > 0$, it follows by the maximum principle that $z > 0$ on $\Omega x(0, \infty)$. Moreover, there exists a constant $\beta > 0$ large enough such that:

$$\underline{u}(x,0,\lambda) - \beta\phi(x,0) \leq z(x,0) \leq \underline{u}(x,0,\lambda) + \beta\phi(x,0), \text{ on } \overline{\Omega}. \tag{2.1}$$

Such a choice is possible since $z(x, 0) = u(x, 0, \lambda) = \phi(x, 0) = 0$, for $x \in \Omega$, $\phi(x, 0) > 0$ for

$x \in \Omega$ and $(\partial\phi/\partial\vec{n})(x, 0) < 0$ on $\partial\Omega$, where ϕ is the eigenfunction corresponding to λ_1, mentioned in Theorem 1.1.

Let us introduce a new function $\tilde{f}$ defined on $\bar{\Omega}x\mathbb{R}x\mathbb{R}$ by:

$$\tilde{f}(x, t, u) = \begin{Bmatrix} f(x, t, u) & \text{if } u \geq 0 \\ f(x, t, 0) + \frac{\partial f}{\partial u}(x, t, 0)u, & \text{if } u < 0 \end{Bmatrix}.$$

Since the map $(x, t) \to (\partial f/\partial u)(x, t, 0)$ is continuously differentiable and bounded on $\Omega x\mathbb{R}$, it is easy to see that $\tilde{f}$ belongs to $C^1(\Omega x\mathbb{R}x\mathbb{R})$ and $(\partial f/\partial u)(x, t, u) \equiv (\partial f/\partial u)(x, t, 0)$, for all $(x, t, u) \in \Omega x\mathbb{R}x\mathbb{R}(-\infty, 0)$.

Let $0 < \gamma < (\lambda_1 - \lambda) \inf\limits_{\Omega x\mathbb{R}} \frac{\partial f}{\partial u}(x, t, 0)$ and define,

$$V(x, t) \equiv \underline{u}(x, t, \lambda) + \beta\phi(x, t)\, e^{-\gamma t} \text{ on } \bar{\Omega}x\mathbb{R},$$
$$W(x, t) \equiv \underline{u}(x, t, \lambda) - \beta\phi(x, t)\, e^{-\gamma t} \text{ on } \bar{\Omega}x\mathbb{R}.$$

Since $L[\phi](x, t) \equiv \lambda_1(\partial f/\partial u)(x, t, 0)\, \phi(x, t)$ on $\Omega x\mathbb{R}$, it follows by the mean value theorem that,

$$\begin{aligned} &L[W](x, t) - \lambda\tilde{f}(x, t, W(x, t)) = \lambda f\big(x, t, \underline{u}(x, t, \lambda)\big) \\ &-\beta e^{-\gamma t}\lambda_1 \frac{\partial f}{\partial u}(x, t, 0)\, \phi(x, t) + \beta\gamma\phi(x, t)e^{-\gamma t} - \lambda\tilde{f}(x, t, W(x, t)) \\ &= \lambda\Big[\tilde{f}\big(x, t, \underline{u}(x, t, \lambda)\big) - \tilde{f}(x, t, W(x, t))\Big] - \beta\lambda_1 e^{-\gamma t} \frac{\partial f}{\partial u}(x, t, 0)\, \phi(x, t) \\ &+ \beta\gamma\phi(x, t)\, e^{-\gamma t} \leq \lambda\frac{\partial f}{\partial u}(x, t, 0)\, \beta\phi(x, t)\, e^{-\gamma t} - \beta\lambda_1 e^{-\gamma t}\frac{\partial f}{\partial u}(x, t, 0)\, \phi(x, t) \\ &+ \beta\gamma\phi(x, t)e^{-\gamma t} \leq \Big[-(\lambda_1 - \lambda)\inf_{\Omega x\mathbb{R}} \frac{\partial f}{\partial u}(x, t, 0) + \gamma\Big]\beta\phi(x, t)e^{-\gamma t} \leq 0 \quad \text{on } \Omega x(0, \infty), \end{aligned} \tag{2.2}$$

and

$$\begin{aligned} &L[V](x, t) - \lambda\tilde{f}(x, t, V(x, t)) = L[V](x, t) - \lambda f(x, t, V(x, t)) \\ &= \lambda f\big(x, t, \underline{u}(x, t, \lambda)\big) + \beta\lambda_1 \frac{\partial f}{\partial u}(x, t, 0)\, \phi(x, t)\, e^{-\gamma t} \\ &- \beta\gamma\phi(x, t)\, e^{-\gamma t} - \lambda f(x, t, V(x, t)) = \lambda\Big[f(x, t, \underline{u}(x, t, \lambda)) - f(x, t, V(x, t))\Big] \end{aligned}$$

$$+\beta\lambda_1 \frac{\partial f}{\partial u}(x, t, 0)\,\phi(x, t)\,e^{-\gamma t} - \beta\gamma\phi(x, t)\,e^{-\gamma t} \geq -\lambda \frac{\partial f}{\partial u}(x, t, 0)\,\beta\phi(x, t)e^{-\gamma t} \tag{2.3}$$

$$+\beta\lambda_1 \frac{\partial f}{\partial u}(x, t, 0)\,\phi(x, t)\,e^{-\gamma t} - \beta\gamma\phi(x, t)e^{-\gamma t} \geq \beta\phi(x, t)\left[(\lambda_1 - \lambda)\inf_{\Omega\times\mathbb{R}} \frac{\partial f}{\partial u}(x, t, 0) - \gamma\right]e^{-\gamma t} \geq 0,$$

for all $(x, t) \in \Omega x(0, \infty)$. It follows from the inequalities (2.1), (2.2), (2.3) and a standard result on parabolic differential inequalities, (see [5], p.187, Theorem 12), that $W(x, t) \leq z(x, t) \leq V(x, t)$ on $\Omega x(0, \infty)$, that is

$$\| z(x, t) - \underline{u}(x, t, \lambda) \| \leq \beta\phi(x, t)\,e^{-\gamma t} \leq ke^{-\gamma t},$$

for all $(x, t) \in \Omega x(0, \infty)$, where $k = \beta \|\phi\|_\infty$.

References

(1) A.Beltramo and P.Hess: "On Principal Eigenvalue of a Periodic-Parabolic Operator", Comm. In. Partial Differential Equations, 9 (9) 919 (1984).

(2) R.W.Dickey: "*Bifurcation Problem in Non-Linear Elasticity*", Univ. of Wisconsin, Madison, Pilman-Publishing.

(3) A.Lazer: "*Some Remarks on Periodic Solutions of Parabolic Differential Equations, Dynamical Systems II*", edited by A.R.Bednarek, L.Cesari, Academic Press, 1982.

(4) I.I.Smulev: "Periodic Solutions of the First Boundary Problem for Parabolic Equations", Trans. Amer. Math. Sc. 79 (1969) 216-229.

(5) M.Protter and H.Weinberger: "*Maximum Principle in Differential Equations*", Prentice-Hall, Englewood Cliffs, N.J., 1967.

Physical Applications of Reciprocal Transformations

Colin Rogers
University of Waterloo
Canada

Abstract

A review is presented of the diversity of applications of reciprocal transformations to physical models. The provenance of such transformations is traced to early work on invariant transformations in gasdynamics. A reciprocal invariance theorem is presented and an extension of an involutory gasdynamic invariance property to an orthogonal magneto-gasdynamic system is noted. The application of reciprocal transformations to the solution of boundary value problems is illustrated both by a Stefan problem in nonlinear heat conduction and a wave propagation problem in nonlinear visco-elasticity. The article concludes with a summary of other physical applications of reciprocal transformations.

1. Origin of Reciprocal Transformations

Bäcklund transformations of reciprocal-type originated in an analysis of aerodynamic lift and drag by Batemen [1,2]. In that work, a remarkable invariance property of plane isentropic gasdynamics was derived. The preservation of shock structure under such transformations was subsequently established under an additional assumption by Power and Smith [3]. The application of reciprocal transformations to the approximation of subsonic gas flows via the Kármán-Tsien approximation was noted in [4].

The steady, two dimensional flow of an inviscid thermally non-conducting gas, subject to no external force, is governed by the following system of equations:

$$\mathrm{div}(\rho \underline{q}) = 0, \tag{1.1}$$

$$\rho \underline{q} \cdot \nabla \underline{q} + \nabla p = \underline{0}, \tag{1.2}$$

$$\underline{q} \cdot \nabla s = 0, \tag{1.3}$$

to which must be adjoined an appropriate equation of state

$$\rho = \Phi(p,s), \qquad \left.\frac{\partial p}{\partial \rho}\right|_s > 0 \tag{1.4}$$

Here, $\underline{q} = (u,v)$, p, ρ and s denote, in turn, the velocity vector, gas pressure, density and specific entropy.

Scalar multiplication of the continuity equation (1.1) by $\underline{q}$ and addition to the equation of motion (1.2) produces a pair of conservation laws

$$(\rho uv)_x + (p + \rho v^2)_y = 0, \tag{1.5}$$

$$(p + \rho u^2)_x + (\rho uv)_y = 0, \tag{1.6}$$

which allow the introduction of new independent variables x^*, y^* according to

$$\left.\begin{aligned} dx^* &= \beta_1^{-1}[-(p + \beta_2 + \rho v^2)dx + \rho uvdy], \\ dy^* &= \beta_1^{-1}[\rho uvdx - (p + \beta_2 + \rho u^2)dy]. \end{aligned}\right\} \tag{1.7}$$

$$\beta_1 \neq 0,\ \beta_2 \in \mathbb{R},$$

subject to the requirement

$$0 < |J(x^*,y^*;x,y)| < \infty \tag{1.8}$$

so that

$$0 < |(p + \beta_2)(p + \beta_2 + \rho q^2)| < \infty. \tag{1.9}$$

It may be shown [5] that if new dependent variables are now introduced according to

$$\left.\begin{aligned} \underline{q}^* &= -\frac{\beta_1 \underline{q}}{p + \beta_2}, \\ p^* &= \beta_4 - \frac{\beta_1^2 \beta_3}{p + \beta_2}, \qquad \beta_3 \neq 0 \\ \rho^* &= \frac{\beta_3 \rho (p + \beta_2)}{p + \beta_2 + \rho q^2}, \\ s^* &= s \end{aligned}\right\} \tag{1.10}$$

then the transformations (1.10) augmented by (1.7) leave the original gasdynamic system (1.1)–(1.3) invariant.

In general, the gas law (1.4) is *not* invariant under the class of transformations determined by (1.7) and (1.10). This fact may be exploited to link solutions of the gasdynamic system (1.1)–(1.3) with different constitutive laws. In particular, if the original flow is incompressible with $\rho = \rho_0$ then the associated new gas law is of the Kármán-Tsien type

$$p^* = A - B/\rho^*, \tag{1.11}$$

where

$$A = \beta_4 - \beta_1^2\beta_3/(2(p_0 + \beta_2), \qquad b = \beta_1^2\beta_3^2\rho_0/2(p_0 + \beta_2) \tag{1.12}$$

and p_0 corresponds to stagnation pressure [5]. The importance of the Kármán-Tsien law (1.11) in the approximation of the subsonic flow of an adiabatic gas

$$p = C\rho^{\gamma}, \tag{1.13}$$

is well established (von Mises [6], Shapiro [7]). The above link allows the application of classical analytic methods of hydrodynamics to subsonic gasdynamics [5]. It is noted that circumstances under which invariance under the transformations (1.7), (1.10) extend to the equation of state (1.4) have been investigated in detail by Rogers and Baker [8-9].

On specialisation of the scaling parameters β_i $i = 1,2$ and translation parameters β_i $i = 3,4$ according to

$$\beta_1 = -\beta_3 = 1; \qquad \beta_2 = -\beta_4 \tag{1.14}$$

it is seen that

$$\underline{q}^{**} = \frac{-\underline{q}^*}{p^* + \beta_2} = \underline{q},$$

$$p^{**} = -\beta_2 + \frac{1}{p^* + \beta_2} = p,$$

$$\rho^{**} = \frac{-\rho^*(p^* + \beta_2)}{p^* + \beta_2 + \rho^* q^{*2}} = \rho,$$

$$\begin{aligned} dx^{**} &= -(p^* + \beta_2 + \rho v^{*2})dx^* + \rho^* u^* v^* dy^* \\ &= -\frac{1}{(p+\beta_2)(p+\beta_2+\rho q^2)}[(p+\beta_2+\rho u^2)dx^* + \rho uvdy^*] \\ &= -\frac{1}{(p+\beta_2)(p+\beta_2+\rho q^2)}[-(p+\beta_2+\rho u^2)(p+\beta_2+\rho v^2) + (\rho uv)^2]dx \\ &= dx, \end{aligned}$$

$$dy^{**} = \rho^* u^* v^* dx^* - (p^* + \beta_2 + \rho^* u^{*2}) dy^*$$
$$= -\frac{1}{(p+\beta_2)(p+\beta_2+\rho q^2)} [(\rho uv dx^* + (p+\beta_2+\rho v^2) dy^*]$$
$$= [-(\rho uv)^2 + (p+\beta_2+\rho u^2)(p+\beta_2+\rho v^2)] dy$$
$$= dy.$$

Accordingly, in this case an involutary transformation is obtained.

To summarize, it is seen that the gasdynamic system (1.1)—(1.3) is invariant under the transformation R_1 given by

$$R_1 \qquad \left.\begin{aligned} \underline{q}^* &= \frac{\underline{q}}{p + \beta_2}, \\ p^* &= -\beta_2 + \frac{1}{p + \beta_2}, \\ \rho^* &= -\frac{\rho(p + \beta_2)}{p + \beta_2 + \rho q^2}, \\ dx^* &= -(p + \beta_2 + \rho v^2) dx + \rho uv dy \\ dy^* &= \rho uv dx - (p + \beta_2 + \rho u^2) dy \end{aligned}\right\}$$

subject to the Jacobian condition (1.9).

In view of the involutory property $R_1^2 = I$, the invariant transformation R_1 is termed *reciprocal*. It may be conjugated with other known invariances of the system (1.1)—(1.3) such as substitution principles [10]. In general, the result of such conjugations will be non-involutory transformations.

That R_1 is a Bäcklund transformation is readily demonstrated in terms of the Martin formulation of the system (1.1)—(1.3) as a Monge-Ampère system. The details are given in [5].

2. Reciprocal Transformations for Systems of Conservation Laws

Here, a more general result is derived on reciprocal transformations for systems of conservation laws. It not only includes the result of the previous section as a special case, but also delivers a number of other reciprocal properties with diverse physical applications [11].

We consider the system of conservation laws

$$(P_{i1})_t - (P_{i2})_x = 0, \qquad i = 1,\ldots,n \tag{2.1}$$

where the P_{ij}, $i = 1,\ldots,n$; $j = 1,2$ involve the independent variables x,t together with dependent variables $u_1,u_2,\ldots,u_m$ and their partial derivatives. The system (2.1) is a consequence of integrability of the associated system

$$d\underline{Z} = P(\frac{\partial}{\partial x}, \frac{\partial}{\partial t}; \underline{\xi})d\underline{X} \tag{2.2}$$

where

$$\left.\begin{array}{c} \underline{X} = (x,t)^T \\ \underline{Z} = (z_1,z_2,\ldots,z_n)^T, \qquad \underline{\xi} = (u_1,u_2,\ldots,u_m)^T \end{array}\right\} \tag{2.3}$$

and P is the n $\times$ 2 matrix with entries P_{ij}.

In the manner of the preceding section, new independent variables x^*, t^* are now introduced according to

$$\begin{aligned} dx^* &= (A_{11} + B_{11}P_{11} + \ldots + B_{1n}P_{n1})dx + (A_{12} + B_{11}P_{12} + \ldots + B_{1n}P_{n2})dt \\ dt^* &= (A_{21} + B_{21}P_{11} + \ldots + B_{2n}P_{n1})dx + (A_{22} + B_{21}P_{12} + \ldots + B_{2n}P_{n2})dt \end{aligned} \tag{2.4}$$

that is, in matrix form

$$d\underline{X}^* = (A + BP)d\underline{X} = Ad\underline{X} + Bd\underline{Z}, \tag{2.5}$$

where A and B are 2 $\times$ 2 and 2 $\times$ n constant matrices respectively. In a similar manner, we introduce

$$d\underline{Z}^* = (C + DP)d\underline{X} = Cd\underline{X} + Dd\underline{Z}, \tag{2.6}$$

where C and D are, in turn, n $\times$ 2 and n $\times$ n constant matrices.

If reciprocity is imposed then

$$d\underline{X}^{**} = d\underline{X}, \qquad d\underline{Z}^{**} = d\underline{Z}, \tag{2.7), (2.8}$$

so that, since

$$\begin{aligned} d\underline{X}^{**} &= (A + BP^*)d\underline{X}^* \\ &= (A + BP^*)(A + BP)d\underline{X}, \end{aligned}$$

and

$$\begin{aligned} d\underline{Z}^{**} &= (C + DP^*)d\underline{X}^* \\ &= (C + DP^*)(A + BP)P^{-1}d\underline{Z}, \end{aligned}$$

it follows that

$$(A + BP^*)(A + BP) = I_2, \tag{2.9}$$

and

$$(C + DP^*)(A + BP)P^{-1} = I_n. \tag{2.10}$$

In the above, P^* is given by

$$d\underline{Z}^* = P^*d\underline{X}^* = P^*(A + BP)d\underline{X} = (C + DP)d\underline{X},$$

so that

$$P^* = (C + DP)(A + BP)^{-1} \tag{2.11}$$

provided $|A + BP| \neq 0$.

Elimination of P^* in (2.9) and (2.10) yields

$$A^2 + BC + (AB + BD)P = I_2 \tag{2.12}$$

provided $|A + BP| \neq 0$.

Elimination of P^* in (2.9) and (2.10) yields

$$A^2 + BC + (AB + BD)P = I_2 \tag{2.12}$$

and

$$(CA + DC)P^{-1} + (CB + D^2) = I_n \tag{2.13}$$

respectively. Thus, since P is non-constant, the conditions (2.12) and (2.13) require that

$$\left.\begin{array}{ll} A^2 + BC = I_2, & AB + BD = 0_{2\times n}, \\ CA + DC = 0_{n\times 2}, & CB + D^2 = I_n. \end{array}\right\} \tag{2.14}$$

Moreover, one use of (2.11) together with (2.9) and (2.10), it is seen that

$$P^{**} = (C + DP^*)(A + BP^*)^{-1} = (C + DP^*)(A + BP) = P. \tag{2.15}$$

To summarise, the following result has been established:

<u>Reciprocal Theorem</u>

The system

$$d\underline{Z} = Pd\underline{X},$$

is mapped to the system

$$d\underline{Z}^* = P^*d\underline{X}^*,$$

under the reciprocal transformation

$$\left.\begin{array}{c} d\underline{X}^* = Ad\underline{X} + Bd\underline{Z} \\ d\underline{Z}^* = Cd\underline{X} + Dd\underline{Z} \\ P^* = (C + DP)(A + BP)^{-1} \\ |A + BP| \neq 0 \end{array}\right\} R_2 \tag{2.16}$$

where A,B,C,D are constant matrices subject to the constraints (2.14).

In particular, if we set

$$P = \begin{pmatrix} -(p+\rho v^2) & \rho uv \\ \rho uv & -(p+\rho u^2) \end{pmatrix} \tag{2.17}$$

then the specialisations

$$\left.\begin{aligned} A &= -\beta_2 I_2, & B &= I_2, \\ C &= (1-\beta_2^2) I_2, & D &= \beta_2 I_2, \end{aligned}\right\} \tag{2.18}$$

satisfy the constraints (2.14) while the relation (2.11) yields

$$\begin{aligned} P^* &= \begin{pmatrix} -(p^*+\rho^* v^{*2}) & \rho^* u^* v^* \\ \rho^* u^* v^* & -(p^*+\rho^* u^{*2}) \end{pmatrix} \\ &= ((1-\beta_2^2) I_2 + \beta_2 P)(-\rho_2 I_2 + P)^{-1}. \end{aligned} \tag{2.19}$$

Insertion of (2.17) and (2.18) into (2.5) and (2.6) together with solution of (2.19) for p^*, ρ^*, u^*, v^* produces gasdynamic reciprocal relations of the type (1.14).

Invariant transformations may likewise be constructed in both homentropic and anisentropic non-steady gasdynamics (Rogers [12,13]). In [61], such invariances were used to generate solutions to moving boundary-value problems involving piston-drive shocks. It is noted that invariant transformations in non-steady anisentropic gasdynamics have been the subject to renewed interest lately due to a series of papers by Gaffet [14-16]. Thus, in particular, if a certain power-law constitutive relation is adopted then the corresponding anisentropic gasdynamics system may be reduced to the Mikhailov-Shabat equation

$$z_{xy} = e^z - e^{-2z}, \tag{2.20}$$

amenable to the inverse scattering method [17-18].

It is remarked that reciprocal-type invariant transformations may also be constructed in magneto-gasdynamics [19, 20]. Thus, in particular, the system of equations governing the steady motion of a gas with infinite electrical conductivity moving under the action of a magnetic field $\underline{H}$ is, in the absence of dissipative mechanisms:

$$\begin{aligned}
&\mathrm{div}(\rho \underline{q}) = 0, \\
&\rho \underline{q} \cdot \nabla \underline{q} + \underline{H} \times \mathrm{curl}\, \underline{H} + \underline{\nabla} p = \underline{0}, \\
&\mathrm{curl}(\underline{q} \times \underline{H}) = \underline{0}, \qquad (2.21) \\
&\mathrm{div}\, \underline{H} = 0, \\
&\underline{q} \cdot \underline{\nabla} s = 0,
\end{aligned}$$

together with an appropriate constitutive law.

Reciprocal-type transformations were introduced for certain two-dimensional magneto-gasdynamic systems of the type (2.21) by Rogers, Kingston and Shadwick [20]. In particular, it may be shown that if

$$\underline{v} \cdot \underline{H} = 0, \qquad (2.22)$$

so that the velocity and magnetic fields are mutually orthogonal then as originally shown in [21], the system (2.21), in two-dimensions, may be linked to a *non-conducting* system via the reciprocal-type transformations

$$\begin{aligned}
&\underline{q}^* = \frac{-\underline{q}(1 + H^2/\rho q^2)}{\lambda^2(p - H^2/2 + \varepsilon)}, \\
&p^* = \nu - \frac{1}{\lambda^2 \mu (p - H^2/2 + \varepsilon)}, \\
&\rho^* = \frac{\mu\rho(p - H^2/2 + \varepsilon)}{(1 + H^2/\rho q^2)(p + H^2/2 + \rho q^2 + \varepsilon)}, \\
&s^* = s, \qquad R_3 \qquad (2.23) \\
&dx^* = -(p + \frac{H_x^2 - H_y^2}{2} + \rho v^2 + \varepsilon)dx + (\rho uv - H_x H_y)dy, \\
&dy^* = (\rho uv - H_x H_y)dy - (p + \frac{H_y^2 - H_x^2}{2} + \rho u^2 + \varepsilon)dy, \\
&0 < |J(x^*, y^*; x, y)| < \infty
\end{aligned}$$

$$\lambda,\mu,\nu,\xi \in \mathbb{R}$$

$$\lambda,\mu \neq 0.$$

In the non-conducting case $\underline{H} = \underline{0}$ an invariant transformation of the type given by (1.7) and (1.10) for the gasdynamic system (1.1)—(1.3) is retrieved.

3. Reciprocal Transformations and Single Conservation Laws: Application to a Stefan Problem is Nonlinear Heat Conduction

If we set n = 1 together with

$$A = \begin{pmatrix} 0 & 0 \\ 0 & 1 \end{pmatrix}, \quad B = \begin{pmatrix} 1 \\ 0 \end{pmatrix}, \quad C = \begin{pmatrix} 1 \\ 0 \end{pmatrix}^T, \quad D = 0, \tag{3.1}$$

and

$$P = (T(\frac{\partial}{\partial x}; \frac{\partial}{\partial t}; u), -F(\frac{\partial}{\partial x}; \frac{\partial}{\partial t}; u)) \tag{3.2}$$

then the Reciprocal Theorem of the preceding section shows that the single conservation law

$$\frac{\partial T}{\partial t} + \frac{\partial F}{\partial x} = 0, \tag{3.3}$$

is mapped to the reciprocally associated conservation law

$$\frac{\partial T^*}{\partial t^*} + \frac{\partial F^*}{\partial x^*} = 0 \tag{3.4}$$

via the reciprocal transformation

$$\left.\begin{array}{c} dx^* = Tdx - Fdt \\ t^* = t \\ 0 < |T| < \infty \end{array}\right\} R_4 \tag{3.5}$$

where

$$P^* = (T^*, -F^*) = (C + DP)(A + BP)^{-1} = \frac{C}{T}\begin{pmatrix} 1 & F \\ 0 & T \end{pmatrix} = (\tfrac{1}{T}, \tfrac{F}{T}). \qquad (3.6)$$

Accordingly [22],

$$T^* = \frac{1}{T(D^*;\partial^*;u)}, \qquad F^* = -\frac{F(D^*;\partial^*;u)}{T(D^*;\partial^*;u)} \qquad (3.7), (3.8)$$

where

$$\left.\begin{aligned} D^* &:= \frac{\partial}{\partial x} = \frac{1}{T^*}\frac{\partial}{\partial x^*}, \\ \partial^* &:= \frac{\partial}{\partial t} = \frac{F^*}{T^*}\frac{\partial}{\partial x^*} + \frac{\partial}{\partial t^*}. \end{aligned}\right\} \qquad (3.9)$$

It is noted that R_5 may be written in integral operator-notation as

$$\left.\begin{aligned} x^* = \partial_x^{-1}T, \qquad t^* = t \\ 0 < |T| < \infty \\ (R^2 = I). \end{aligned}\right\} R_5 \qquad (3.10)$$

In the case

$$T(\sigma,t) = u(\sigma,t) \qquad (3.11)$$

the ***extended hodograph transformation*** of Clarkson, Fokas and Ablowitz [23] is retrieved. In the latter work, such transformations were shown to play an important role in the construction of canonical sets of integrable equations. In the present work, we retain the differential notation in which the reciprocal transformations were originally derived since this proves more convenient for application to boundary-value problems. In the next section, in the context of nonlinear visco-elasticity, it will be necessary to extend reciprocal transformations of the type (3.10) to incorporate involutory transformations on t.

Here, a reciprocal transformation of the type (3.5) is used to reduce a single phase Stefan problem descriptive of solidification in simple metals to a form which admits a class of exact solutions that extend the classical Neumann solution.

Thus, we consider the solidification problem (Rogers [24])

$$\rho c_p(T)\frac{\partial T}{\partial t} = \frac{\partial}{\partial x}\left[\kappa(T)\frac{\partial T}{\partial x}\right], \qquad 0 < x < X(t) \tag{3.12}$$

$$\kappa(T)\frac{\partial T}{\partial x} = U(t) \qquad \text{on } x = 0,\ t > 0, \tag{3.13}$$

$$\left.\begin{aligned} \kappa(T)\frac{\partial T}{\partial x} &= L\rho\dot{X}(t), && (3.14)\\ T &= T_f, && (3.15)\end{aligned}\right\} \text{ on } x = X(t)$$

$$X(0) = 0, \tag{3.16}$$

where the liquid phase occupies the region $x > X(t)$ and $T(x,t)$ denotes the temperature distribution in the solid phase region $0 < x < X(t)$. The specific heat at constant pressure c_p and the thermal conductivity κ are allowed to be temperature dependent. The density ρ of the material is taken to be constant.

If we set

$$\Phi(T) = \int_{T_0}^{T} S(\sigma)d\sigma, \qquad S = \rho c_p T, \qquad (3.17), (3.18)$$

then (3.12) may be re-written as

$$\frac{\partial}{\partial t}[\Phi(T)] - \frac{\partial}{\partial x}[\kappa(T)\frac{\partial T}{\partial x}] = 0. \tag{3.19}$$

On introduction of the reciprocal transformation

$$\left.\begin{aligned} dx^* &= \Phi(T)dx + \kappa(T)T_x dt, \qquad t^* = t,\\ T^* &= \frac{1}{\Phi(T)}, \end{aligned}\right\} R_6 \tag{3.20}$$

then it is readily shown that the *nonlinear* heat conduction equation (3.19) is reduced to the *linear* canonical heat equation

$$\frac{\partial T^*}{\partial t^*} = \kappa^* \frac{\partial^2 T^*}{\partial x^{*2}}, \qquad (\kappa^* > 0, \text{ a constant}) \tag{3.21}$$

subject only to the condition

$$\frac{\kappa^* \Phi'}{\Phi^2} = \kappa(T) \tag{3.22}$$

Such a reduction was originally obtained by Storm [25] in an investigation of heat conduction in simple monatomic metals. There, the validity of the approximation was examined for aluminum, silver, sodium, cadmium, zinc, copper and lead as well as for iron and 80 percent carbon steel. The reduction was subsequently used by Rosen [26] in connection with heat conduction in solid crystalline molecular hydrogen. It is of interest to note that a particular case of the reduction was retrieved via a Lie-Bäcklund transformation approach by Bluman and Kumei [27].

Under the reciprocal transformation (3.20)

$$\frac{\partial x^*}{\partial x} = \Phi(T),$$

$$\frac{\partial x^*}{\partial t} = \kappa(T)\frac{\partial T}{\partial x} = \int_0^x \frac{\partial}{\partial x}\left[\kappa(T)\frac{\partial T}{\partial x}\right]dx + \kappa(T)\frac{\partial T}{\partial x}\bigg|_{x=0}$$

$$= \int_0^x \frac{\partial}{\partial t}[\Phi(T)]dx + U(t),$$

so that

$$x^*(x,t) = \int_0^x \Phi(T)dx + \Theta(t) - \Theta(0), \tag{3.23}$$

where $\dot{\Theta} = U(t)$ and we have taken $x^*(0,0) = 0$. Accordingly, the fixed boundary condition (3.13) becomes

$$\kappa^* \frac{\partial T^*}{\partial x^*} = -UT^* \qquad \text{on } x^* = \Theta(t^*) - \Theta(0) \tag{3.24}$$

Moreover,

$$\frac{\partial x^*}{\partial t} = \int_{X(t)}^{x} \frac{\partial}{\partial x}\left[\kappa(T)\frac{\partial T}{\partial x}\right]dx + \kappa(T)\frac{\partial T}{\partial x}\bigg|_{x=X(t)}$$

$$= \int_{X(t)}^{x} \frac{\partial}{\partial t}[\Phi(T)]dx + L\rho\dot{X}(t),$$

so that we have an alternative expression to (3.23), namely

$$x^*(x,t)' = \int_{X(t)}^{x} \Phi(T)dx + [\Phi(T_f) + L\rho]X. \tag{3.25}$$

Thus, the moving boundary conditions (3.14)—(3.15) become

$$\left.\begin{aligned} \kappa^*\frac{\partial T^*}{\partial x^*} &= -L\rho T^*\dot{X}, \\ T^* &= \frac{1}{\Phi(T_f)}, \end{aligned}\right\} \text{ on } x^* = X^* \tag{3.26}$$

where

$$X^* = x^*|_{x=X(t)} = [\Phi(T_f) + L\rho]X(t) \tag{3.27}$$

so that the initial condition (3.16) becomes

$$X^*(0) = 0. \tag{3.28}$$

Furthermore, on $x^* = X^*$ it is seen that

$$dX^*/dt^* = (1/T^*)dX/dt - \kappa^* T^*_{x^*}/T^* \tag{3.29}$$

so that the nonlinear boundary condition (3.26), becomes

$$\kappa^*\frac{\partial T^*}{\partial x^*} = -\frac{L\rho dX^*/dt^*}{\Phi(T_f)[\Phi(T_f) + L\rho]} \quad \text{on } x^* = X^*. \tag{3.30}$$

Thus, to summarize, under the reciprocal transformation (3.20), the moving boundary problem given by (3.12)—(3.16) reduces, subject to the condition (3.22) to the reciprocal problem

$$\frac{\partial T^*}{\partial t^*} = \kappa^*\frac{\partial^2 T^*}{\partial x^{*2}},$$

$$\kappa^* \frac{\partial T^*}{\partial x^*} = -UT^* \qquad \text{on } x^* = \Theta(t^*) - \Theta(0),$$

$$\left.\begin{aligned} \kappa^* \frac{\partial T^*}{\partial x^*} &= -\frac{L\rho dX^*/dt^*}{\Phi(T_f)[\Phi(T_f) + L\rho]} \\ T^* &= \frac{1}{\Phi(T_f)} \end{aligned}\right\} \text{on } x^* = X^* \qquad (3.31)$$

$$X^*(0) = 0.$$

If $T^*(x^*,t^*)$ is the solution of the problem (3.31) then it is readily shown that the solution of the original problem (3.12)—(3.16) is given parametrically by the relations

$$\left.\begin{aligned} T &= \Phi^{-1}(1/T^*) \\ x &= \int_{\Theta(t^*)-\Theta(0)}^{x^*} T^* dx_*, \\ t &= t^*, \end{aligned}\right\} \qquad (3.32)$$

while the evolution of the boundary $x = X$ is given in terms of that of $x^* = X^*$ by the simple relation (3.27).

Attention is now specialised to the class of moving boundary problems (3.12)—(3.16) with

$$U(t) = U_0/\sqrt{t}, \qquad X(t) = \sqrt{2\gamma t}. \qquad (3.33), (3.34)$$

The similarity variable

$$\xi^* = x^*/\sqrt{2\gamma t^*} \qquad (3.35)$$

is now introduced into the reciprocal boundary value problem (3.31) and solutions are sought of the type

$$T^* = \phi^*(x^*/\sqrt{2\gamma t^*}) \qquad (3.36)$$

so that

$$\gamma\xi^* \frac{d\phi^*}{d\xi^*} + \kappa^* \frac{d^2\phi^*}{d\xi^{*2}} = 0 \tag{3.37}$$

whence

$$\phi^* = A \operatorname{erf}\left[\sqrt{\frac{\gamma}{2\kappa^*}}\, \xi^*\right] + B. \tag{3.38}$$

The linear boundary conditions require that

$$\kappa^* \frac{d\phi^*}{d\xi^*} = -U_0\sqrt{2\gamma}\phi^* \qquad \text{on } \xi^* = U_0\sqrt{2/\gamma} \tag{3.39}$$

and

$$\phi^* = \frac{1}{\Phi(T_f)} \qquad \text{on } \xi^* = \Phi(T_f) + L\rho \tag{3.40}$$

so that, A and B are given by

$$A\sqrt{\frac{2\gamma\kappa^*}{\pi}}\, e^{-U_0^2/\kappa^*} = -U_0\sqrt{2\gamma}\left[A \operatorname{erf}\left[\frac{U_0}{\sqrt{\kappa^*}}\right] + B\right] \tag{3.41}$$

$$A \operatorname{erf}\left[\sqrt{\frac{\gamma}{2\kappa^*}}\, (\Phi(T_f) + L\rho)\right] + B = \frac{1}{\Phi(T_f)} \tag{3.42}$$

The constant γ which determines the speed $\dot{X}$ of the moving boundary $x = X(t)$ is obtained from the remaining boundary condition which yields

$$\kappa^* \frac{d\phi^*}{d\xi^*} = -\frac{L\rho\gamma}{\Phi(T_f)} \qquad \text{on } \xi^* = \Phi(T_f) + L\rho \tag{3.43}$$

whence

$$A\sqrt{\frac{2\gamma\kappa^*}{\pi}}\, e^{-\frac{\gamma}{2\kappa^*}(\Phi(T_f)+L\rho)^2} = -L\rho\gamma/\Phi(T_f). \tag{3.44}$$

The solution of the original boundary value problem is now given parametrically by the relations

$$T = \Phi^{-1}\left[\frac{1}{A\ \mathrm{erf}\left(\sqrt{\frac{\gamma}{2\kappa^*}}\,\xi^*\right) + B}\right], \tag{3.45}$$

$$\xi = \int_{U_0\sqrt{2/\gamma}}^{\xi^*} \left[A\ \mathrm{erf}\left(\sqrt{\frac{\gamma}{2\kappa^*}}\,\sigma\right) + B\right] d\sigma. \tag{3.46}$$

It is noted that the above class of exact solutions has the property that $T = \text{constant} = T_0$ on the boundary $x = 0$ where

$$\Phi(T_0) = \frac{1}{A\ \mathrm{erf}\left[U_0/\sqrt{\kappa^*}\right] + B}. \tag{3.47}$$

The analysis is valid for any member of the class of materials with $c_p(T)$, $\kappa(T)$ constrained by (3.22).

Two-phase Stefan problems corresponding to melting in simple metals were investigated by Rogers [28]. On the other hand, the single phase analysis presented above has been recently extended to incorporate a degree of inhomogeneity [29].

4. Application to Wave Propagation in Nonlinear Visco-Elasticity

Here, an extension of the reciprocal transformation R_4 of the previous section is presented which allows the reduction of certain important *nonlinear* hyperbolic equations to the *linear* telegraph equation. The result is used in the analysis of finite amplitude waves for a class of nonlinear visco-elastic materials (Varley and Seymour [30]).

A generalised version of the reciprocal transformation (3.5) is now introduced, namely [31, 32]

$$\left.\begin{aligned} dx^* &= [aT + b]dx - [aF + c]dt, \\ t^* &= et + h(u) \\ 0 &< |J(x^*,t^*;x,t)| < \infty \end{aligned}\right\} R_7 \tag{4.1}$$

corresponding to

$$A = \begin{pmatrix} b & -c \\ 0 & e \end{pmatrix}, \quad B = \begin{pmatrix} a & 0 \\ 0 & 1 \end{pmatrix},$$

$$C = \begin{pmatrix} (1-b^2)/a & c(b+e)/a \\ 0 & 1-e^2 \end{pmatrix}, \quad D = \begin{pmatrix} -b & c/a \\ 0 & -e \end{pmatrix} \tag{4.2}$$

together with

$$P = \begin{pmatrix} T(\frac{\partial}{\partial x}, \frac{\partial}{\partial t}; u) & -F(\frac{\partial}{\partial x}, \frac{\partial}{\partial t}; u) \\ h'(u)u_x & h'(u)u_t \end{pmatrix} \tag{4.3}$$

The involutory requirement $R^2 = I$ embodied in (2.14) is readily shown to i.e.,

$$aT^* + b = (e + h'u_t)/\Delta, \qquad aF^* + c = -(aF + c)/\Delta \qquad (4.4), (4.5)$$

$$e^2 = +1, \qquad eh(u) + h(u^{-1}) = 0 \qquad (4.6), (4.7)$$

where

$$\Delta := J(x^*,t^*;x,t) = (aT+b)(e+h'u_t) + (aF+c)h'u_x. \tag{4.8}$$

The relations (4.6), (4.7) show that either

$$e = +1, \qquad h(u) = \Psi(\ln|u|), \qquad \Psi \text{ odd} \tag{4.9}$$

or

$$e = -1, \qquad h(u) = \Theta(\ln|u|), \qquad \Theta \text{ even}. \tag{4.10}$$

A specialisation of the above leads to the following result [31].

<u>Theorem</u>

The *nonlinear* hyperbolic equation

$$\frac{\partial}{\partial t}\left[\gamma \frac{\partial \Phi}{\partial t}(u) + \Phi(u)\right] - \frac{\partial}{\partial x}\left[\kappa(u)\frac{\partial u}{\partial x}\right] = 0, \tag{4.11}$$

is reduced to the *linear* telegraph equation

$$p_0 \frac{\partial^2 \bar{T}^*}{\partial t^{*2}} + q_0 \frac{\partial \bar{T}^*}{\partial t^*} - \frac{\partial^2 \bar{T}^*}{\partial x^{*2}} = 0, \qquad p_0/q_0 = \gamma,\ q_0 \neq 0 \tag{4.12}$$

via the reciprocal transformation

$$\left.\begin{aligned} dx^* &= [a(\gamma\Phi_t + \Phi) + b]dx + a\kappa(u)u_x\, dt, \\ t^* &= t + \gamma \ln|a\Phi + b| + r_0, \\ 0 &< |J(x^*,t^*;x,t)| > \infty \end{aligned}\right\} R_8 \tag{4.13}$$

where

$$\bar{T}^* = \frac{1}{q_0[a\Phi + b]}, \tag{4.14}$$

and

$$\frac{\Phi'}{q_0(a\Phi + b)^2} = \kappa(u). \tag{4.15}$$

The specialisation $\gamma = 0$ produces the reduction due to Storm and employed in the previous section in connection with solidification in simple metals. In the sequel, we present an application of a more general result with $\gamma \neq 0$ to a boundary value problem in nonlinear visco-elasticity [30].

The propagation of plane stress waves in a material whose response is governed by a nonlinear visco-elastic constitutive law is considered. Thus, if (x,t) denote distance and time measures while (v,σ) designate material velocity and stress then we consider the one-dimensional propagation of visco-elastic stress waves for materials with constitutive laws of the type

$$\epsilon_t = d(\sigma)\sigma_t + e(\sigma). \tag{4.16}$$

The momentum equation is

$$\rho v_t = \sigma_x, \tag{4.17}$$

while the strain-rate relation yields

$$\epsilon_t = v_x. \tag{4.18}$$

In what follows, ρ is taken to be unity.

Attention is restricted to the case when

$$d(\sigma) = \frac{d_0}{(1+\kappa e_0\sigma)^2}, \qquad e(\sigma) = \frac{e_0}{(1+\kappa e_0\sigma)}. \tag{4.19), (4.20}$$

where $d_0 > 0$ and κ, e_0 are arbitrary constants. Accordingly, (4.16) yields

$$\epsilon_t = \frac{d_0\sigma_t}{(1+\kappa e_0\sigma)^2} + \frac{e_0\sigma}{(1+\kappa e_0\sigma)}. \tag{4.21}$$

Here, we consider a standing wave problem wherein it is required to solve [30]

$$\left.\begin{aligned} &v_x = \frac{d_0\sigma_t}{(1+\kappa e_0\sigma)^2} + \frac{e_0\sigma}{(1+\kappa e_0\sigma)}, \qquad v_t = \sigma_x,\ 0 < x < L \\ &v = 0 \text{ on } x = 0, \qquad t \geq 0 \\ &v = 0 \text{ on } x = L, \qquad t \geq 0 \\ &v = v_0(x),\ \sigma = 0 \text{ at } t = 0,\ 0 \leq x \leq L \end{aligned}\right. \tag{4.22}$$

for $t > 0$ on the interval $0 < x < L$. It is assumed that $v_0(x)$ is prescribed and is such that $v_0(0) = v_0(L) = 0$.

Introduction of the reciprocal transformation

$$\left.\begin{aligned} dx^* &= \left[1 - k\left\{\frac{d_0\sigma_t}{(1+\kappa e_0\sigma)^2} + \frac{e_0\sigma}{(1+\kappa e_0\sigma)}\right\}\right]dx - \kappa\sigma_x dt, \\ &= [1 - \{\gamma\Phi_t + \Phi\}]dx - \kappa\sigma_x dt, \\ t^* &= t + \gamma \ln|a\Phi + b| = t - \frac{d_0}{e_0}\ln|1 + \kappa e_0\sigma| \end{aligned}\right\} R_9 \tag{4.23}$$

$$0 < |J(x^*,t^*;x,t)| < \infty,$$

together with

$$\overline{T}^* = \frac{1}{a\Phi+b} = 1 + \kappa e_0\sigma, \tag{4.24}$$

so that $a = -1$, $b = 1$, and $\gamma = d_0/e_0$ reduces the nonlinear problem (4.22) to the classical linear visco-elastic problem

$$\begin{gathered} v_{t^*} = \sigma_{x^*}, \; v_{x^*} = d_0\sigma_{t^*} + e_0\sigma, \qquad 0 < x^* < L \\ v = 0 \quad \text{on} \quad x^* = 0, \\ v = 0 \quad \text{on} \quad x^* = L \\ \sigma = 0, \; v = v_0^*(x^*) \text{ at } t^* = 0. \end{gathered} \tag{4.25}$$

In the above $v_0^*(x^*)$ is given by

$$v_0^*(x^*) = v_0(x), \tag{4.26}$$

where, under R_9

$$\frac{\partial x^*}{\partial x} = 1 - (\gamma\Phi_t + \Phi),$$

$$\begin{aligned} \frac{\partial x^*}{\partial t} &= -\kappa\sigma_x = -\int_0^x \frac{\partial}{\partial x}[\kappa\sigma_x]dx - \kappa\sigma_x|_{x=0} \\ &= -\int_0^x \frac{\partial}{\partial t}[\gamma\Phi_t + \Phi]dx - \kappa v_t|_{x=0} \end{aligned}$$

whence, since $v_t|_{x=0} = 0$,

$$x^*(x,t) = x - \int_0^x [\gamma\Phi_t + \Phi]dx = x - \kappa v \tag{4.27}$$

where we have taken the arbitrary constant of integration to be such that $x = 0$ corresponds to $x^* = 0$. Thus, for specified $v_0(x)$, $v_0^*(x^*)$ is given by the functional equation

$$v_0^*(x^*) = v_0(x^* + \kappa v_0^*(x^*)). \tag{4.28}$$

The solution of the boundary value problem (4.25) is [30]

$$v = e^{-w_0 t^*} \sum_{n=1}^{\infty} v_n\left[\cos(w_n t^*) + \frac{w_0}{w_n}\sin(w_n t^*)\right]\sin(\frac{n\pi x^*}{L}) \tag{4.29}$$

where

$$v_n = \frac{2}{L}\int_0^L v_0^*(x^*)\sin(\frac{n\pi x^*}{L})dx \tag{4.30}$$

and

$$w_n = \begin{cases} \dfrac{1}{2\gamma} & n = 0 \\ \dfrac{1}{2d_0}\,[4n^2\pi^2 d_0 - e_0^2]^{1/2} & n = 1,2,\ldots \end{cases} \tag{4.31}$$

The associated stress distribution is given by

$$\sigma = \frac{Lw_0}{\pi}\, e^{-w_0 t^*} \sum_{n=1}^{\infty} \frac{v_n}{n}\left[\frac{w_0}{w_n} + \frac{w_n}{w_0}\right]\sin(w_n t^*)\cos(\frac{n\pi x^*}{L}). \tag{4.32}$$

In particular, if the initial velocity distribution is given by

$$v_0(x) = a\,\sin(\frac{\pi x}{L}) \tag{4.33}$$

then (4.28) becomes

$$v_0^*(x^*) = a\,\sin\left[\frac{\pi}{L}(x^* + \kappa v_0^*(x^*))\right] \tag{4.34}$$

with solution

$$v_0^*(x^*) = \sum_{n=1}^{\infty} v_n^*\,\sin(\frac{n\pi x^*}{L}) \qquad 0 \le x^* \le L \tag{4.35}$$

where

$$v_n^* = \frac{2aJ_n(\epsilon n)}{\epsilon n}, \qquad \epsilon = \frac{\kappa a}{L}. \tag{4.36), (4.37}$$

The variation of σ with t on the boundary $x = 0$ is accordingly determined by

$$\sigma|_{x=0} = \frac{2ac}{r} e^{-w_0 t^*} \sum_{n=1}^{\infty} \frac{J_n(\epsilon n)}{\epsilon n^2} \left[\frac{w_0}{w_n} + \frac{w_n}{w_0}\right] \sin(w_n t^*) \qquad (c = d_0^{-1/2}) \tag{4.38}$$

$$w_0 t = w_0 t^* + \frac{1}{2} \ln|1 + 2\pi \frac{\epsilon}{r} \sigma|_{x=0}| \tag{4.39}$$

where

$$r = 2\pi(Lce_0)^{-1} \tag{4.40}$$

is a measure of the dissipation upon the deformation; ϵ provides a measure of the influence of nonlinearity on the deformation with small values of ϵ corresponding to small nonlinearity.

In conclusion, it is noted that evolution equations of the type (4.11) arise also in nonlinear Cattaneo models of heat conduction [33], electromagnetic theory [34] and in the analysis of van der Pol wave propagation [35].

5. Additional Applications of Reciprocal Transformations

The range of nonlinear initial/boundary value problems that may be solved via reciprocal transformations, either acting alone or conjugated with other types of transformations is broad indeed. Thus, in atmospheric models, the genesis of zones of sharp horizontal temperature change is of considerable importance. Attempts to explain the formation of such fronts have been commonly based on a class of horizontal velocity fields of deformation-type which arise as local features of large-scale horizontal wave motions (Bergeron [36], Stone [37]). Such velocity fields tend to lead to regions with accentuated temperature gradient (Pedlovsky [38]) and were incorporated in the fronto-genesis theory of Hoskins and Bretherton [39]. In particular, if the Ertel potential vorticity is constant then a model may be derived based on a Monge-Ampère equation. It may be shown that the latter can be reduced to Laplace's equation via a novel reciprocal transformation [40]. An important nonlinear boundary value problem which arises in the model may be thereby solved parametrically. There are also interesting applications of single reciprocal transformations in soil mechanics. Thus, the reciprocal transformations R_4 of Section 3 has recently been employed in an analysis of the freezing of soil columns [41-43]. Application to sedimentation columns as separation devices have been subject to investigation [44]. On the other hand, in nonlinear elasticity, the Adkin's duality principle is a reciprocal transformation of hodograph-type which leaves invariant a 4th order plane strain system [45].

Reciprocal transformations may also be combined with other types of transformation to solve nonlinear boundary value problems. In particular, certain nonlinear hierarchies may be reduced via reciprocal-type transformations to intermediate Burgers' hierarchies: under appropriate conditions the latter may be reduced, along with accompanying boundary conditions to linear canonical form on the application of generalised Cole-Hopf transformations [32]. Such methods have been exploited in the important area of oil/water transport through soils [46-47]. Recent extensions allow the analysis of certain such problems on bounded regions [48]. The Carrier-Greenspan linearization of a shallow water system may be revealed to be a combination of hodograph and reciprocal-type transformations [49-50]. In an interesting recent development, it has been shown that a Coleman-Fabrizio-Owen heat conduction model which incorporates relaxation can be linearized by a combination of such reciprocal transformations [51].

Inverse scattering schemes and associated initial value problems on $-\infty < x < \infty$ for privileged nonlinear evolution equations have been the subject of much research. Reciprocal transformations have been shown to provide a key link between the AKNS and IKW scattering schemes as well as between various integrable hierarchies [52-55]. Reciprocal transformations in 2+1-dimensions as recently introduced in [56] constitute a link between the Kadomtsev-Petviashvili equation and the Harry Dym equation in 2+1-dimensions [57]. A link between extensions to 2+1-dimensions of the AKNS and IKW inverse scattering schemes via such reciprocal transformations is suggested but remains to be investigated.

The role of reciprocal transformations in the classification of 1+1-dimensional integrable equations as described in [23] has already been mentioned. The higher-dimensional reciprocal transformations alluded to above present the possibility of extending such classification to 2+1-dimensions.

It is observed that the literature on inverse scattering methods to regions other than $-\infty < x < \infty$ is relatively sparse. However, a recent contribution by Fokas [58] on an initial/boundary value problem for the cubic Schrödinger equation has shown that such problems previously considered intractable may be amenable to exact solution. In this connection, it is noted that application of reciprocal-type transformations to an *initial value problem* on $-\infty < x < \infty$ for the Korteweg-deVries equation lead to a *boundary value problem* for the associated Harry Dym equation [59].

In conclusion, it is noted that the role of reciprocal transformations in the revelation of symmetry structure of integrable hierarchies has recently been described by Carillo and Fuchssteiner [60].

Acknowledgement

The author wishes to acknowledge with gratitude his support under Natural Sciences and Engineering Research Council of Canada Grant No.: A0879.

References

[1] Bateman, H., Proc. Nat. Acad. Sci. USA 24, 246-251 (1938).

[2] Bateman, H., Quart. Appl. Math. 1, 281-295 (1943-44).

[3] Power, G. and Smith, P., J. Math. Mech. 10, 349-361 (1961).

[4] Tsien, H. S., J. Aeronaut. Sci. 6, 399-407 (1939).

[5] Rogers, C. and Shadwick, W. F., Bäcklund Transformations and Their Applications, Mathematics in Science and Engineering, Academic Press, New York (1982).

[6] von Mises, R., Mathematical Theory of Compressible Fluid Flow, Academic Press, New York (1958).

[7] Shapiro, A. H., The Dynamics and Thermodynamics of Compressible Fluid Flow, Ronald Press, New York (1953).

[8] Rogers, C., Arch. Rat. Mech. Anal. 47, 36-46 (1972).

[9] Baker, J. A. and Rogers, C., J. Mécanique, Théorique et Appliquée 1, 563-568 (1982).

[10] Prim, R. C., Arch. Rat. Mech. Anal. 1, 425-497 (1952).

[11] Kingston, J. G. and Rogers, C., Quart. Appl. Math. 51, 423-432 (1984).

[12] Rogers, C., Z. angew. Math. Phys. 19, 58-63 (1968).

[13] Rogers, C., Z. angew. Math. Phys. 20, 370-382 (1969).

[14] Gaffet, B., Physica 11D, 287-308 (1984).

[15] Gaffet, B., J. Math. Phys. 25, 245-255 (1984).

[16] Gaffet,B., Physica 26D, 123-139 (1987).

[17] Mikhailov, A. V., Physica 3D, 73-117 (1981).

[18] Zhiber, A. V. and Shabat, A. B., DAN SSSR No. 5, 247 (1975).

[19] Rogers, C., and Kingston, J. G., J. Mécanique 15, 185-192 (1976).

[20] Rogers, C., Kingston, J. G. and Shadwick, W. F., J. Math. Phys. 21, 395-397 (1980).

[21] Power, G. and Walker, D., Zeit. angew. Math. Phys. 16, 803-817 (1965).

[22] Kingston, J. G. and Rogers, C., Phys. Lett. 92A, 261-264 (1982).

[23] Clarkson, P. A., Fokas, A. S., Ablowitz, M. J., Inst. Nonlinear Studies, Clarkson University Preprint (1987).

[24] Rogers, C., Int. J. Nonlinear. Mech. 21, 249-256 (1986).

[25] Storm, M. L., J. Appl. Phys. 22, 940-951 (1951).

[26] Rosen, G., Phys. Rev. B19, 2398-2399 (1979).

[27] Bluman, G. W. and Kumei, S., J. Math. Phys. 21, 1019-1023 (1980).

[28] Rogers, C., J. Phys. A: Mathematical and General 18, L105-L109 (1985).

[29] Rogers, C., and Broadbridge, P., Zeit. ang. Math. Phys. to be published (1988).

[30] Varley, E. and Seymour, B., Stud., Appl. Math. 72, 241-262 (1985).

[31] Rogers, C. and Ruggeri, T., Lett. Il. Nuovo. Cimento 44, 289-296 (1985).

[32] Rogers, C., J. Math. Phys. 26, 393-395 (1985).

[33] Ruggeri, T., Acta Mechanica 47, 167-183 (1983).

[34] Katayev, I. G., Electromagnetic Shock Waves, Iliffe, London (1966).

[35] Lardner, R. W. and Nicklason, G., SIAM J. Appl. Math. 4, 480-492 (1981).

[36] Bergeron, T., Geofys. Publikasjoner 5, 1-111 (1928).

[37] Stone, P. H., J. Atmos. Sci. 23, 455-465 (1966).

[38] Pedlovsky, J., Geophysical Fluid Dynamics, Springer Verlag, New York (1979).

[39] Hoskins, B. J. and Bretherton, F. P., J. Atmos. Sci. 29, 11-37 (1972).

[40] Hoskins, B. J., in Rotating Fluids in Geophysics, Academic Press, New York, ed. P. H. Roberto and A. M. Sowerd, 170-203 (1978).

[41] Mohamed, F. A. and Guenther, R. B., J. Math. Anal. Appl. 111, 1-13 (1985).

[42] Mohamed, F. A. and Guenther, R. B., J. Math. Anal. Appl. 111, 475-512 (1985).

[43] Mohamed, F. A. and Guenther, R. B., J. Math. Anal. Appl. 111, 513-534 (1985).

[44] Rogers, C., Hill, J. M. and Broadbridge, P., Univ. of Waterloo, Dept. Applied Mathematics, Preprint (1987).

[45] Adkins, J. E., J. Mech. Phys. Solids 6, 267-275 (1958).

[46] Fokas, A. S. and Yortsos, Y. C., SIAM J. Appl. Math. 42, 318-332 (1982).

[47] Rogers, C., Stallybrass, M. P. and Clements, D. L., Nonlinear Analysis, Theory, Methods and Applications 7, 785-799 (1983).

[48] King, M. J., J. Math. Phys. 26, 870-877 (1985).

[49] Carrier, G. F. and Greenspan, H. P., J. Fluid Mech. 4, 97-109 (1958).

[50] Rogers, C., in Nonlinear Evolution Equations and Dynamical Systems Meeting, Proceedings, Balaruc-Les-Bains, France, Ed. J. Léon (1987).

[51] Fusco, D., Universitá di Messina, Dipartimento di Matematica, Preprint (1987).

[52] Rogers, C. and Wong, P., Physica Scripta 30, 10-14 (1984).

[53] Rogers, C. and Nucci, M. C., Physica Scripta 33, 289-292 (1986).

[54] Rogers, C., Nucci, M. C., and Kingston, J. G., Il. Nuovo. Cimento 96, 55-63 (1986).

[55] Rogers, C. and Carillo, S., Physica Scripta 36, 865-869 (1987).

[56] Rogers, C., J. Phys. A. Math. Gen. 49, L491-L496 (1986).

[57] Rogers, C., Phys. Lett. A. 120, 15-18 (1987).

[58] Fokas, A. S., Inst. Nonlinear. Studies, Clarkson University Preprint (1987).

[59] Guo, B. Y. and Rogers, C., to be published Scientia Sinica.

[60] Carillo, S. and Fuchssteiner, B., University of Paderborn, Preprint (1987).

[61] Castell, S. P. and Rogers, C., Quart. Appl. Math. 32, 241-251 (1975).

AN UNUSUAL REALIZATION OF A SYMMETRY GROUP ACTING ON A NONLINEAR DIFFERENTIAL EQUATION†

Yvan SAINT-AUBIN
CRM and Département de mathématiques et de statistique,
Université de Montréal, C.P. 6128, succ. A, Montréal, Canada H3C 3J7

Abstract

The space of solutions of the $\mathbf{CP}^n$ model on two-dimensional Euclidean space has been extensively described by Din and Zakrzewski. One of its peculiar features is that it is the reunion of $(n+1)$ disconnected sets. In these lecture notes, I introduce a group of symmetries whose explicit form depend on the particular set it is acting on. This group can be used to study various topological properties of the solution space.

1. Introduction

The classical theory of infinitesimal transformations acting on the solution space of a given equation is well developed and thoroughly described in Olver's book [1]. It is still a very active subject and its recent applications to nonlinear partial differential equations has led to various interesting results. The usual description of these infinitesimal transformations is done using vector fields expressed in terms of derivatives with respect to independent and dependent variables. The functional form of the vector fields is uniquely determined over the whole solution space and the algebraic structure they span is obtained by calculating their Lie brackets.

In the present course, I would like to report on a group of symmetries dis-

† Supported in part by the Natural Sciences and Engineering Research Council of Canada and by the "Fonds FCAR pour l'aide et le soutien à la recherche"

covered recently for a particular two-dimensional nonlinear equation, the so-called $\mathbf{CP}^n$ σ model.[2–3] Its formulation departs from the usual one by the fact that, even though the group structure is $SL(n+1, \mathbf{C})$ over the whole space, the explicit functional form of the vector fields is different on the various non-connected subsets of the space. This peculiar behavior is possible due to the topological structure of the solution space of these models. The solution space of the Euclidean $\mathbf{CP}^n$ model has been thoroughly described by Din and Zakrzewski [4]. An integer l, $0 \leq l \leq n$, can be naturally attached to each solution. Hence, the solution space can be decomposed into $(n+1)$ (infinite-dimensional) subsets. (When restricted to finite action solutions, each of these subsets can be further decomposed into finite-dimensional subsets characterized by an additional integer $i \in \mathbf{Z}$ associated to the instanton number.) Even though there is no a priori topology on the solution space of a differential equation, it is rather natural to consider the whole solution space as the disconnected union of these $(n+1)$ infinite-dimensional subsets. Various σ models were known to have an infinite-dimensional algebra of symmetries. Restricted to the Euclidean sector of these models, the algebra becomes finite-dimensional. What I shall show in the following is that the functional form of some vector fields does differ from one connected subset to another.

The first section will provide an explicit description of the solution space of the Euclidean $\mathbf{CP}^n$ model. Since this construction is so thorough, one might wonder what is the use of symmetries. Indeed, one of the main interests of symmetries is to generate new solutions from old ones. Here, as the second section will briefly indicate, the symmetries allow instead for a decomposition into strata and orbits of the solution space.

2. The Equation and its Solutions

2.1 The Equation of the $\mathbf{CP}^n$ Model

Contrarily to many nonlinear equations like the Korteweg-de Vries, Kadomtsev-Petviashvili or Sine-Gordon equations, the dependent variable of the $\mathbf{CP}^n$ model is not a single function but a unit length complex vector $z = z(x_1, x_2)$:

$$z \in \mathbf{C}^{n+1} \qquad \text{with} \qquad |z| = 1. \tag{2.1}$$

(The name of the $\mathbf{CP}^n$ model stems from the fact that the hyperbolic space $\mathbf{CP}^n$ is the set of all unit length complex vectors in $\mathbf{C}^{n+1}$.) As we shall restrict the discussion to the Euclidean sector, we introduce the variables $x_\pm \equiv x_1 \pm ix_2$. The equation of the $\mathbf{CP}^n$ model is:

$$D_+D_-z + (z^\dagger\, D_+D_-z)z = 0 \tag{2.2}$$

where the covariant derivative $D_\pm$ is defined by:

$$D_\pm \equiv \partial_\pm \; - (z^\dagger \partial_\pm z)\;. \tag{2.3}$$

(Notice that one has to be careful when dealing with the complex variables x_+ and x_- as $\partial_+(z^\dagger) = (\partial_- z)^\dagger$.)

Exercise 1: Show that, for any function $\alpha(x_+, x_-)$ such that $|\alpha(x_+, x_-)| = 1$, the covariant derivative has the property: $D_\pm(\alpha z) = \alpha D_\pm z$. Hence, if z is a solution of (2.2), αz is also a solution. (This is known as the $U(1)$ gauge invariance of the $\mathbf{CP}^n$ model.)

Exercise 2: Show, by writing out all the derivatives that (2.2) is:

$$\partial_+\partial_-z - (z^\dagger\partial_+\partial_-z)z - (z^\dagger\partial_+z)\partial_-z - (z^\dagger\partial_-z)\partial_+z + 2(z^\dagger\partial_+z)(z^\dagger\partial_-z)z = 0 \quad . \tag{2.4}$$

Exercise 3: Show that: $D_-D_+ \neq D_+D_-$. *Show however that:*

$$D_-D_+z + (z^\dagger\, D_-D_+z)z = 0$$

is equivalent to (2.2).

It is natural to introduce an $(n+1)\times(n+1)$-matrix $P(x_+, x_-)$ related to z by:

$$P \equiv \frac{z\, z^\dagger}{|z|^2}\ . \tag{2.5}$$

The matrix P is the *projector* on the subspace spanned by z:

$$P = P^\dagger = P^2, \qquad Pz = z \qquad \text{and} \qquad \text{rk}\ P = 1\ . \tag{2.6}$$

The correspondence between the P's and the z's is one-to-one up to the $U(1)$ invariance.

Exercise 4: Show that the equation (2.2) on z *is equivalent to the following equation on* P*:*

$$[\partial_+\partial_- P, P] = 0\ . \tag{2.7}$$

A distinguished set of solutions of the $\mathbf{CP}^n$ model is constituted by the solutions having a sufficiently regular behavior on the whole plane (x_1, x_2) (in fact, on $\mathbf{R}^2 \cup \{\infty\}$). The condition to be used is:

$$\int (|D_+z|^2 + |D_-z|^2)\, d^2x \ < \infty \tag{2.8}$$

where the integral is over the real plane (x_1, x_2). Any solution satisfying this condition is said to have a *finite action.*

2.2 The (Anti-)Self-Duality Condition and the Simplest Solutions

Among the space of solutions of the $\mathbf{CP}^n$ model, there is a distinguished subset called the (anti-)self-dual solutions. The self-duality condition (resp. the anti-self-duality) reads:

$$D_{\overset{(-)}{+}} z = 0 \,. \tag{2.9}$$

Observe that a solution satisfying the (anti-)self-duality condition is automatically a solution of the $\mathbf{CP}^n$ model.

Exercise 5: Show that the (anti-)self-duality equation can be rewritten as:

$$D_\mu z = {}_{(}\mp_{)} i\epsilon_{\mu\nu} D_\nu z \tag{2.10}$$

where $\epsilon_{12} = -\epsilon_{21} = 1, \epsilon_{11} = \epsilon_{22} = 0$. *(Notice the similarity with the (anti-)self-duality equations in four-dimensional Yang-Mills theory:* $F_{\mu\nu} = {}_{(}\pm_{)} \epsilon_{\mu\nu\rho\sigma} F_{\rho\sigma}$.*)*

Even though it is still non-linear, the self-duality equation is easy to solve. Suppose $f \equiv f(x_+) \in \mathbf{C}^{n+1}$ is a holomorphic vector of x_+ (i.e. $\partial_- f = 0$). Then:

$$z = \frac{f}{|f|} \tag{2.11}$$

is a solution of $D_- z = 0$. Indeed, one obtains directly:

$$\begin{aligned} \partial_- z &= -\frac{f}{|f|^2} \partial_- |f| \,, \\ z^\dagger \partial_- z &= -\frac{1}{|f|} \partial_- |f| \end{aligned}$$

and hence:

$$\begin{aligned} D_- z &\equiv \partial_- z - (z^\dagger \partial_- z) z \\ &= -\frac{f}{|f|^2} \partial_- |f| + \frac{\partial_- |f|}{|f|} \frac{f}{|f|} \\ &= 0 \,. \end{aligned}$$

This is in fact the general solution as the following exercise shows.

Exercise 6: Let z be a solution of $D_- z = 0$. Show that the ratio of any two components of z is independent of x_-:

$$\partial_- (z_i / z_j) = 0, \qquad \forall i, j.$$

Hence z is of the form (2.11).

Instantons and anti-instantons are finite-action self- and anti-self-dual solutions, respectively. If only single-valued solutions are considered, the finite action condition (2.8) requires that the components of f be rational functions of x_+ without common zero. Moreover, due to the expression of z in terms of f, one can restrict oneself to polynomial components for f without loss of generality.

Hence, the instantons of the $\mathbf{CP}^n$ model are given by (2.11) for f's whose components are polynomials in x_+ (without common zero). (The anti-instantons are obtained similarly from f's holomorphic in x_-.)

Exercise 7: Show that the (anti-)self-duality condition (2.9) on z is equivalent to the following one on P:

$$P\partial_{(\pm)}P = 0\,. \tag{2.12}$$

Exercise 8: Show that the four following equations are equivalent for a projector P:

$$P\partial_+P = 0\,, \qquad \partial_-P\cdot P = 0\,, \qquad \partial_+P\cdot P = \partial_+P \qquad \text{and} \qquad P\partial_-P = \partial_-P\,. \tag{2.13}$$

2.3 The Whole Space of Finite Action Solutions

There are solutions of the $\mathbf{CP}^n$ model which are neither instantons nor anti-instantons. As will be shown in this paragraph, every finite action solution is however related naturally to both an instanton and an anti-instanton.

Let the operators P_+ and P_- acting on a vector $f \in \mathbf{C}^{n+1}$ be defined by:

$$P_\pm f = \partial_\pm f - \frac{(f^\dagger \partial_\pm f)}{|f|^2} f\,. \tag{2.14}$$

Exercise 9: Show that $P_\pm(\alpha f) = \alpha P_\pm f$ for any function $\alpha \equiv \alpha(x_+, x_-)$.

Let now f be a holomorphic vector of $\mathbf{C}^{n+1}$: $f = f(x_+)$. Define the vectors:

$$z_0 = \frac{f}{|f|}, \quad z_1 = \frac{P_+ f}{|P_+ f|}, \quad \ldots, \quad z_k = \frac{P_+^k f}{|P_+^k f|} \tag{2.15}$$

where the integer k is such that $|P_+^k f| \neq 0$ and $|P_+^{k+1} f| = 0$. ($P_+^{l+1} f \equiv P_+(P_+^l f)$.) (As it will be shown later, the integer k always exists.) The set $\{z_0, z_1, \ldots, z_k\}$ will be called the *family* associated to f.

Din and Zakrzewski showed the following results:

$$(i) \quad z_i^\dagger z_j = \delta_{ij} \quad \text{for any} \quad 0 \le i, j \le k\,, \tag{2.16}$$

$$(ii) \quad \partial_-(P_+^i f) = -\frac{|P_+^i f|^2}{|P_+^{i-1} f|^2} P_+^{i-1} f\,, \tag{2.17}$$

$$(iii) \quad \partial_+ \left(\frac{P_+^{i-1} f}{|P_+^{i-1} f|^2} \right) = \frac{1}{|P_+^{i-1} f|^2} P_+^i f\,, \tag{2.18}$$

(iv) $k \le n$,

(v) z_0 and z_k are respectively a self-dual and an anti-self-dual solution,

(vi) the z_i's, $0 \le i \le k$, are solutions of the $\mathbf{CP}^n$ model,

(vii) any solution of the $\mathbf{CP}^n$ model belongs to such a family.

This remarkable result provides a very handy description of the solution space of the $\mathbf{CP}^n$ model. Since any solution z belongs to a family $\{z_0, z_1, \ldots, z_k\}$, the knowledge of the full set of self-dual solutions given in the preceding paragraph leads, by the use of the operator P_+, to all solutions.

The proof of the above properties is rather lengthy. However it relies essentially on linear algebraic methods. The following exercises lead the reader through it. (The first proof was given by Din and Zakrzewski in ref. [4] using results from Borchers and Graber [5]. Later Zakrzewski [6] and Sasaki [7] simplified some of the steps.) Before the following exercise provides an example of a family in $\mathbf{CP}^2$.

Exercise 10: Let f be the following holomorphic vector in $\mathbf{C}^3$:

$$f = \begin{bmatrix} 1 \\ x_+ \\ x_+^2 \end{bmatrix}.$$

(a) By using the definition of P_+, compute the family $\{z_0, z_1, z_2\}$:

$$z_0 = \frac{1}{\sqrt{1+x^2+x^4}} \begin{bmatrix} 1 \\ x_+ \\ x_+^2 \end{bmatrix}$$

$$z_1 = \frac{1}{\sqrt{1+5x^2+6x^4+5x^6+x^8}} \begin{bmatrix} -x_-(1+2x^2) \\ 1-x^4 \\ x_+(2+x^2) \end{bmatrix}$$

$$z_2 = \frac{1}{\sqrt{1+4x^2+x^4}} \begin{bmatrix} x_-^2 \\ -2x_- \\ 1 \end{bmatrix}$$

where $x^2 = x_+x_-$. *(This family has $k = 2$.)*

(b) Show that the family $\{z_0, z_1, z_2\}$ satisfies the property (i): $z_0^\dagger z_1 = z_1^\dagger z_2 = z_0^\dagger z_2 = 0$.

(c) Verify that z_1 solves equation (2.2). (This is a long exercise.)

To prove property (i), it is of course enough to prove that $(P_+^i f)^\dagger (P_+^j f) = 0$ if $i \neq j$. Observe first that, by definition of P_+, $P_+^i f$ is a linear combination of $f, \partial_+ f, \ldots, \partial_+^i f$, this last vector $\partial_+^i f$ occuring with a unit coefficient:

$$P_+^i f = \partial_+^i f + \sum_{j=0}^{i-1} a_j^i \partial_+^j f. \tag{2.19}$$

We shall prove now that the $P_+^i f$'s are precisely the vectors obtained from the frame $\{f, \partial_+ f, \ldots, \partial_+^k f\}$ by the Gram-Schmidt orthogonalization process. It should be pointed out that, for the proof of property (i), the holomorphic character of f is not required. Hence the set $\{z, P_+z/|P_+z|, P_+^2 z/|P_+^2 z|, \ldots, P_+^k z/|P_+^k z|\}$

always contains orthonormal vectors for any starting vector $z = z(x_+, x_-)$. (z does not even have to solve the $\mathbf{CP}^n$ model!) Let ω_i be these vectors:

$$\omega_i = \partial_+^i f - \sum_{l=0}^{i-1} v_l (v_l^\dagger \partial_+^i f) \tag{2.20}$$

where:

$$v_l = \frac{\omega_l}{|\omega_l|}.$$

By construction $\omega_i^\dagger \omega_j = 0$ if $i \neq j$, $0 \leq i, j \leq k$. The statement that $\omega_i = P_+^i f$ is proved recursively. It obviously holds for $i = 0$ and 1 since:

$$\omega_0 = f = P_+^0 f$$

and

$$\omega_1 = \partial_+ f - \frac{f}{|f|}\left(\frac{f^\dagger}{|f|}\partial_+ f\right) \equiv P_+ f\,. \tag{2.21}$$

Suppose now that:

$$\omega_l = P_+^l f \qquad \text{for} \quad 0 \leq l \leq i\,. \tag{2.22}$$

Then:

$$P_+^{i+1} f = P_+(\omega_i).$$

Exercise 11: Using the fact that $z_l = v_l$, $0 \leq l \leq i$, show, by induction, that:

$$P_+^{i+1} f = \omega_{i+1} - (1 - z_i z_i^\dagger) \sum_{l=0}^{i-1} \partial_+(z_l z_l^\dagger) \partial_+^i f\,. \tag{2.23}$$

The proof of property (i) thus becomes equivalent to proving that:

$$(1 - z_i z_i^\dagger) \sum_{l=0}^{i-1} \partial_+(z_l z_l^\dagger) \partial_+^i f \tag{2.24}$$

vanishes. Because $\partial_+ z_l \in \mathrm{Span}\{z_0, z_1, \ldots, z_{l+1}\}$, the above expression has components only along the z_l's, $0 \leq l \leq i-1$.

Exercise 12: Prove that the component along z_l, $0 \leq l \leq i-1$ of (2.24) is:

$$((\partial_- z_l)^\dagger \partial_+^i f) + \sum_{j=0}^{i-1} (z_l^\dagger \partial_+ z_j)(z_j^\dagger \partial_+^i f).$$

To end the proof, these functions have to be zero. This is however an immediate consequence of the following formulae:

$$\partial_+ z_l = (z_{l+1}^\dagger \partial_+ z_l) z_{l+1} + (z_l^\dagger \partial_+ z_l) z_l \qquad \text{for } 0 \leq l \leq i-1 \qquad (2.25)_+$$

and

$$\begin{cases} \partial_- z_l = (z_{l-1}^\dagger \partial_- z_l) z_{l-1} + (z_l^\dagger \partial_- z_l) z_l + v_\perp & \text{for } 1 \leq l \leq i \\ \partial_- z_0 = (z_0^\dagger \partial_- z_0) z_0 + v_\perp \end{cases} \qquad (2.25)_-$$

where $v_\perp$ belongs to the subspace orthogonal to $\{z_0, z_1, \ldots, z_i\}$.

Exercise 13: (a) Use the recursion hypothesis (2.22) to prove that $\partial_+ z_l$, $0 \leq l \leq i-1$, has components along z_l and z_{l+1} only, hence showing $(2.25)_+$.

(b) Use the orthogonality relation $z_j^\dagger z_l = \delta_{jl}$, $0 \leq j, l \leq i$ to prove $(2.25)_-$.

The property (iv) is an immediate consequence of (i). Since the $P_+^i f$'s are proportional to the z_i which are all orthonormal, the maximum number of $P_+^i f$'s is the dimension of $\mathbf{C}^{n+1}$, i.e. $n+1$. Hence the integer k always exists and $k \leq n$.

We now turn to the proof of (ii) and (iii). Observe first that (2.18) for $i = 1$ is an immediate consequence of the analyticity of f ($\partial_- f = 0$):

$$\begin{aligned} \partial_+ \left(\frac{f}{|f|^2} \right) &= \frac{\partial_+ f}{|f|^2} - \frac{f}{|f|^4} \partial_+ |f|^2 \\ &= \frac{P_+ f}{|f|^2} \end{aligned} \qquad (2.18)_1$$

Exercise 14: Prove by a direct calculation that:

$$\partial_-(P_+ f) = -\frac{|\partial_+ f|^2}{|f|^2} f + \frac{(f^\dagger \partial_+ f)((\partial_+ f)^\dagger f)}{|f|^4} f = -\frac{|P_+ f|^2}{|f|^2} f. \qquad (2.17)_1$$

This proves (2.17) for $i = 1$.

The proof will now proceed inductively. The two following formulae will prove to be useful:

$$\partial_+ \left(\frac{P_+^{l-1} f}{|P_+^{l-1} f|^2} \right) = \frac{P_+^l f}{|P_+^{l-1} f|^2} - \frac{(\partial_- P_+^{l-1} f)^\dagger P_+^{l-1} f}{|P_+^{l-1} f|^4} P_+^{l-1} f, \qquad l \geq 1 \qquad (2.26)_l$$

and

$$\partial_-(P_+^l f) = -\frac{P_+^{l-1} f}{|P_+^{l-1} f|^2} |P_+^l f|^2 + \alpha_l P_+^l f, \qquad l \geq 1 \qquad (2.27)_l$$

where α_l is a function to be determined.

Exercise 15: (a) Prove $(2.26)_l$ by direct derivation.

(b) Differentiate the following relation (equivalent to (2.16)):

$$(P_+^l f)^\dagger \left(\frac{P_+^i f}{|P_+^i f|^2} \right) = 0, \qquad i \neq l$$

with respect to x_- to obtain $(2.27)_l$.

Exercise 16: (a) Suppose now that $(2.17)_i$ and $(2.18)_i$ hold for i such that $1 \leq i \leq l$. Then prove that it implies, by eq. $(2.26)_{l+1}$, that $(2.18)_{l+1}$ holds also.

(b) Suppose finally that $(2.17)_i$ holds for i such that $1 \leq i \leq l$ and $(2.18)_i$ for i such that $1 \leq i \leq l+1$. Write $\partial_-(P_+^{l+1} f)$ as:

$$\partial_-(P_+^{l+1} f) = \partial_- \left\{ |P_+^l f|^2 \partial_+ \left(\frac{P_+^l f}{|P_+^l f|^2} \right) \right\}.$$

By direct differentiation, show that $\partial_-(P_+^{l+1} f)$ has components along $P_+^{l-1} f$ and $P_+^l f$. Comparing with $(2.27)_{l+1}$, eq. $(2.17)_{l+1}$ follows.

This ends the recursive proof of properties (ii) and (iii).

We have already seen in the preceding paragraph that $z_0 \equiv f/|f|$ is a self-dual solution. The fact that z_k is an anti-self-dual solution follows from the fact that, on z_k, $D_+ z_k = P_+ z_k = 0$. Hence property (v) is proved.

Property (vi) can be now easily proved using properties (ii) and (iii).

Exercise 17: (a) Writing the projector P_l as:

$$P_l = \frac{P_+^l f}{|P_+^l|^2}(P_+^l f)^\dagger,$$

show that, for $1 \leq l \leq k-1$:

$$\partial_- P_l = \frac{P_+^{l+1} f}{|P_+^l f|^2}(P_+^l f)^\dagger - \frac{P_+^l f (P_+^{l-1} f)^\dagger}{|P_+^{l-1} f|^2} \tag{2.28}$$

and that:

$$\partial_+ \partial_- P_l = \frac{|P_+^{l+1} f|^2}{|P_+^l f|^2}(P_{l+1} - P_l) + \frac{|P_+^l f|^2}{|P_+^{l-1} f|^2}(P_{l-1} - P_l). \tag{2.29}$$

(b) Show that $[\partial_+\partial_- P_l, P_l] = 0$, $1 \leq l \leq k-1$, which, by exercise 4, implies property (vi).

The final property (vii) relies on the observation made earlier that property (i) holds for any starting vector z. Let z be any solution of the $\mathbf{CP}^n$ model. By applying the operator P_+ on z repeatedly, one obtains orthogonal vectors $\{z, P_+ z, \ldots, P_+^k z\}$. Their number cannot exceed $(n+1)$ and hence, there exists k such that $P_+^{k+1} z = 0$ with $P_+^k z \neq 0$. In other words, $(P_+^k z)$ is an anti-self-dual solution. The reader can convince him- or herself that a family $\{g, P_- g, P_-^2 g, \ldots, P_-^{k'} g\}$ can be constructed from the anti-self-dual solution $g = \frac{P_+^k z}{|P_+^k z|}$ by the repeated action of P_- instead of P_+. Properties (i)-(vi) hold of course with the interchange of all "+" and "−" signs. The question arises now to whether the initial z belongs to this family. The answer is yes and is a consequence of the following exercice.

Exercise 18: Show that, for any solution z of the $\mathbf{CP}^n$ model (2.2), one has the following identity:

$$|z|^2 P_- P_+ z = -z|P_+ z|^2 \tag{2.30}.$$

Hence $z \propto P_-^k g$ where $g = P_+^k z$ is an anti-self-dual solution. (The proportionality constant is never zero as one can see from (2.30).)

This ends the long proof of the properties (i)-(vii).

A few integers can be naturally defined to characterize the solutions of $\mathbf{CP}^n$ models. First, there is the integer k, introduced in (2.15). The family of a solution z has $(k+1)$ elements. If the solution z has finite action, one can show that the instanton of the family of z has also finite action. Hence, one can introduce another integer characterizing z as the maximum degree of polynomials in the associated instanton. This integer is known as the instanton number. Another important integer l one may use to characterize a solution z is the number of times one has to apply P_- to z to build an instanton: $P_-^l z/|P_-^l z|$ is an instanton. (Eq. (2.30) insures the existence of this integer.) This integer l is in the range 0 to n. With respect to l, the solution space of the $\mathbf{CP}^n$ model is splitted into $(n+1)$ sets. We shall see in the next section that it plays a major role in the definition of the action of the symmetry group.

Exercise 19: (a) Calculate the integer k for the instanton $z = f/|f|$ where f is the holomorphic vector:

$$f = \begin{pmatrix} 1 \\ x_+^3 - 1 \\ x_+^3 \end{pmatrix}.$$

(b) What is the instanton number of z?

3. Symmetries of the $\mathbf{CP}^n$ Model

In the two first paragraphs of this section, we introduce a $SL(n+1,\mathbf{C})$ action on the whole solution space of the $\mathbf{CP}^n$ model. The last paragraph will outline how this action leads to a better understanding of the topology of the solution space.

3.1 An $SL(n+1,\mathbf{C})$ Action on the Subspace of Self-Dual Solutions

The self-dual condition reads:

$$D_- z = 0 \tag{3.1}$$

i.e.:

$$\partial_- z - (z^\dagger \partial_- z) z = 0. \tag{3.2}$$

An obvious symmetry of this equation is the following:

$$z \rightarrow z' = gz \tag{3.3}$$

where g is a unitary matrix ($\in SU(n+1)$). In fact, this action can be generalized to any $g \in SL(n+1,\mathbf{C})$ by writing:

$$z \rightarrow z' = \frac{gz}{|gz|}. \tag{3.4}$$

For a non-unitary $g \in SL(n+1,\mathbf{C})$, this is far from an obvious symmetry. However it can be shown easily to leave the equation (3.2) satisfied. Indeed, recall that the most general solution of (3.2) is a holomorphic vector, up to a common (point-dependent) factor: $z(x_+,x_-) = f(x_+)/|f|$. If $g \in SL(n+1,\mathbf{C})$ then gf is also holomorphic and the action (3.4) is a symmetry.

Exercise 20: (a) Show that, if $f(x_+)$ has polynomial components without common zeroes, then $gf(x_+)$, for $g \in SL(n+1,\mathbf{C})$, contains also polynomials without common zeroes.

(b) Show that the action (3.4) on instantons preserves the number of instantons defined at the end of section 2.

Before turning to the whole solution space, we recast the symmetry (3.4) in terms of the projector $P = zz^\dagger$. This will turn out to be extremely useful in the sequel. The symmetry (3.4) acting on the projector P associated with z is:

$$P \rightarrow P' = gP[(g^\dagger g)^{-1}(1-P) + P]^{-1} g^{-1}. \tag{3.5}$$

Exercise 21: What does eq.(3.5) become if g is a unitary matrix? Show that the actions defined by (3.4) and (3.5) are then equivalent and that P' satisfies the self-duality condition: $P'\partial_+P' = 0$. (See eq. (2.12).)

For a general $g \in SL(n+1, \mathbf{C})$, the proof of the equivalence between the transformations (3.4) and (3.5) proceeds in three steps. We shall prove that: (i) the matrix $[(g^\dagger g)^{-1}(1-P)+P]$ is always invertible, (ii) P' is a projector: $P' = P'^2 = P'^\dagger$, (iii) P' projects on z'. It is interesting to underline the fact that the proof of (i) and (ii) can be done for any projector P with no restrictions on its rank.

Let Λ be a unitary matrix diagonalizing P: $\Lambda P \Lambda^{-1} = \mathrm{diag}[\mathbf{1}_p, 0]$ where $p =$ rk P. Let us write $(g^\dagger g)^{-1}$ in this basis:

$$\Lambda (g^\dagger g)^{-1} \Lambda^{-1} = \begin{pmatrix} \gamma_1 & \gamma_2 \\ \gamma_3 & \gamma_4 \end{pmatrix}$$

where $\gamma_1, \gamma_2, \gamma_3$ and γ_4 are block-matrices with γ_1 a $p \times p$-matrix. Then:

$$\Lambda[(g^\dagger g)^{-1}(1-P)+P]\Lambda^{-1} = \begin{pmatrix} \mathbf{1} & \gamma_2 \\ 0 & \gamma_4 \end{pmatrix}.$$

The matrix $(g^\dagger g)^{-1}$ is a positive definite matrix. The block γ_4, being a submatrix of $(g^\dagger g)^{-1}$, is positive definite and invertible. Hence $[(g^\dagger g)^{-1}(1-P)+P]$ is invertible.

The second step is to prove that P' is indeed a projector. The hermiticity condition can be proved as follows:

$$
\begin{aligned}
P'^{\dagger} &= g^{\dagger -1}[(1-P)(g^{\dagger}g)^{-1} + P]^{-1}Pg^{\dagger} \\
&= g[(1-P) + P(g^{\dagger}g)]^{-1}Pg^{\dagger} \\
&\qquad \times \{g[(g^{\dagger}g)^{-1}(1-P) + P][(g^{\dagger}g)^{-1}(1-P) + P]^{-1}g^{-1}\} \\
&= g[(1-P) + P(g^{\dagger}g)]^{-1}P[(1-P) + (g^{\dagger}g)P][(g^{\dagger}g)^{-1}(1-P) + P]^{-1}g^{-1} \\
&= g[(1-P) + P(g^{\dagger}g)]^{-1}[(1-P) + P(g^{\dagger}g)]P[(g^{\dagger}g)^{-1}(1-P) + P]^{-1}g^{-1} \\
&= gP[(g^{\dagger}g)^{-1}(1-P) + P]^{-1}g^{-1} \\
&= P' \quad .
\end{aligned}
\tag{3.6}
$$

Now notice that:

$$[(g^{\dagger}g)^{-1}(1-P) + P]P = P$$

and hence:

$$P = [(g^{\dagger}g)^{-1}(1-P) + P]^{-1}P.$$

Then:

$$
\begin{aligned}
P'^2 &= gP\{[(g^{\dagger}g)^{-1}(1-P) + P]^{-1}(g^{-1}g)P\}[(g^{\dagger}g)^{-1}(1-P) + P]^{-1}g^{-1} \\
&= gPP[(g^{\dagger}g)^{-1}(1-P) + P]^{-1}g^{-1} \\
&= P'
\end{aligned}
\tag{3.7}
$$

Exercise 22: (a) Prove that P' projects on gz if P projects on z. (Step (iii).)

(b) Prove that the action (3.5) preserves the rank of the projector: $\mathrm{rk}\ P' = \mathrm{rk}\ P$.

Exercise 23: Prove directly that P' defined in (3.5) satisfy the self-duality equation if P does. (The proof should hold whatever the rank of P is.)

3.2 An $SL(n+1, \mathbf{C})$ Action on the Whole Space of Solutions

The previous paragraph has introduced a rather natural $SL(n+1, \mathbf{C})$ action on the subset of self-dual solutions. How can this action be extended to the whole solution space? If g is unitary, the action $z \to z' = gz$ can be shown to be a symmetry for any solution z. (See eq. (2.4).) However, for a general $g \in SL(n+1, \mathbf{C})$, $g^\dagger g \neq 1$ and there is no reason why $\tilde{z} = gz/|gz|$ should solve (2.4) if z is not an instanton. In fact, it does not!

Let $l \equiv l(z)$ be the integer such that $P_-^l z/|P_-^l z|$ is a self-dual solution. We shall say that z is in the l-th subset. If z is self-dual, $l = 0$. Only anti-self-dual solutions can have $l = n$. The $SL(n+1, \mathbf{C})$ action can be extended to the whole solution space of the $\mathbf{CP}^n$ model by the following prescription: if z is in the l-th subset, then:

$$z \to z' = \frac{P_+^l(g P_-^l z)}{|P_+^l(g P_-^l z)|}, \qquad \text{for } g \in SL(n+1, \mathbf{C}). \tag{3.8}$$

This action depends explicitly on the integer l. For the self-dual solutions (0th subset), it obviously coincides with (3.4).

Exercise 24: Show that, if $g \in SU(n+1)$, then (3.8) becomes simply:

$$z \to z' = gz, \qquad \text{for } g \in SU(n+1) \text{ and } \forall l. \tag{3.9}$$

Exercise 25: Suppose $g = e^T$ with T hermitian: $T = T^\dagger$. Compute the infinitesimal action corresponding to (3.8) for the self-dual solutions and for the first subset. (Do the calculation to first order in T, i.e. $e^T \approx 1 + T + \mathcal{O}(T^2)$.)

Exercise 26: Show that the action (3.8) is a group action, i.e. the solution z'' obtained by acting on z first by g_1, and then by g_2: $z \xrightarrow{g_1} z' \xrightarrow{g_2} z''$ is equal to the solution built from z by action of $(g_2 g_1)$.

In order to proceed further, it is useful to introduce a set of projectors $\{\Sigma_0, \Sigma_1, \ldots, \Sigma_{k+1}\}$ that contains the same information as the family $\{z_0, z_1, \ldots, z_k\}$. We introduced already the projector P associated with a solution z (eq.

(2.5)). Let P_l be the projector associated to the element z_l of the family:

$$P_l \equiv \frac{z_l z_l^\dagger}{|z_l|^2}, \qquad 0 \leq l \leq k. \tag{3.10}$$

The projectors Σ_l, $0 \leq l \leq k+1$, are defined by:

$$\begin{cases} \Sigma_0 \equiv 0, \\ \Sigma_l \equiv \sum_{i=0}^{l-1} P_i, \quad \text{for } 1 \leq l \leq k+1 \end{cases} \tag{3.11}$$

and hence : rk $\Sigma_l = l$. The following properties on the family $\{P_1, P_2, \ldots, P_k\}$:

$$P_l^2 = P_l^\dagger = P_l \tag{3.12a}$$

$$P_i P_j = \delta_{ij} P_i \tag{3.12b}$$

$$[\partial_+ \partial_- P_l, P_l] = 0 \tag{3.12c}$$

$$\text{if } P_l \text{ projects on } z_l, \text{ then } P_{l+1} \text{ projects on } P_+ z_l \tag{3.12d}$$

are equivalent to the following ones on the set $\{\Sigma_0, \Sigma_1, \ldots, \Sigma_{k+1}\}$:

$$\Sigma_l^2 = \Sigma_l^\dagger = \Sigma_l, \qquad \text{for } 0 \leq l \leq k+1 \tag{3.13a}$$

$$\Sigma_l \Sigma_m = \Sigma_m \Sigma_l = \Sigma_l, \qquad \text{for } 0 \leq l \leq m \leq k+1 \tag{3.13b}$$

$$\Sigma_l \partial_+ \Sigma_l = 0, \qquad \text{for } 0 \leq l \leq k+1 \tag{3.13c}$$

$$\Sigma_{l+1} \partial_+ \Sigma_l = \partial_+ \Sigma_l, \qquad \text{for } 0 \leq l \leq k. \tag{3.13d}$$

Notice that the conditions on the P_l's are sufficient to insure that the set of eigenvectors $\{z_0, z_1, \ldots, z_k\}$ of the P_l's (with eigenvalue 1) is a family.

Exercise 27: Using the definition (3.11), verify the equivalence of (3.12a,b) and (3.13a,b).

To prove the equivalence of (3.12c) and (3.13c), observe that $\partial_+ P_l$ can be expressed as:

$$\partial_+ P_l = \frac{(P_+^{l+1} f)(P_+^l f)^\dagger}{|P_+^l f|^2} - \frac{(P_+^l f)(P_+^{l-1} f)^\dagger}{|P_+^{l-1} f|^2}, \qquad 1 \leq l \leq n \tag{3.14$_l$}$$

and

$$\partial_+ P_0 = \frac{(P_+ f) f^\dagger}{|f|^2} \tag{3.14$_0$}$$

where $f \equiv z_0$ is an eigenvector of P_0 with eigenvalue 1. Eq. (3.13) is obtained by direct differentiation of (3.10) and by using eq. (2.30).

Exercise 28: (a) Using (3.14), show that:

$$\partial_+ \Sigma_l = \frac{(P_+^l f)(P_+^{l-1} f)^\dagger}{|P_+^{l-1} f|^2} \tag{3.15}$$

(b) Show then that: $\Sigma_l \partial_+ \Sigma_l = 0$.

Exercise 29: Conversely, show that $\Sigma_l \partial_+ \Sigma_l = 0$ *implies the field equation for* P_l*:* $[\partial_+ \partial_- P_l, P_l] = 0$. *(Hint: The differential consequences of (3.13b) have to be used.)*

Exercise 30: Use the results of Exercise 28 to show that (3.13d) is a consequence of the properties on the P_l*'s.*

To end the proof of equivalence between both sets of conditions on the P_l's and Σ_l's, we still have to prove that the conditions (3.13) imply (3.12d). Observe first that:

$$(1 - P_l)\partial_+ z_l = \partial_+ z_l - \frac{z_l z_l^\dagger}{|z_l|^2} \partial_+ z_l = P_+ z_l$$

and then:

$$\begin{aligned} P_{l+1}(P_+ z_l) &= P_{l+1}(1 - P_l)\partial_+ z_l = P_{l+1}\partial_+ z_l = -(\partial_+ P_{l+1}) z_l \\ &= -(\partial_+ \Sigma_{l+2} - \partial_+ \Sigma_{l+1})(\Sigma_{l+1} z_l) \\ &= (\partial_+ \Sigma_{l+1}) z_l \end{aligned}$$

where (3.13b,d) have been used. Then:

$$\begin{aligned} P_{l+1}(P_+ z_l) &= \partial_+(\Sigma_{l+1} z_l) - \Sigma_{l+1}\partial_+ z_l \\ &= \partial_+ z_l - (P_l + \Sigma_l)\partial_+ z_l \\ &= (1 - P_l)\partial_+ z_l - \partial_+(\Sigma_l z_l) + (\partial_+\Sigma_l)\Sigma_l z_l \\ &= P_+ z_l \end{aligned}$$

which proves that eqs. (3.13) implies that P_{l+1} projects on $P_+ z_l$.

Because of the equivalence between eqs. (3.12) and eqs. (3.13), the set $\{\Sigma_0, \Sigma_1, \ldots, \Sigma_{k+1}\}$ will be called henceforth a family. Two questions can be raised at this point.

First, it is striking that the field equation for the Σ_l's is the self-duality equation (2.12). As it was noticed in paragraph 3.1, the proof that the transformation:

$$P \to P' = gP[(g^\dagger g)^{-1}(1 - P) + P]^{-1} g^{-1}$$

is a symmetry of the self-dual equation is independent of the rank of P. Hence this $SL(n+1, \mathbf{C})$ action transforms any Σ_l into another solution Σ_l' of the self-dual equation. What are the corresponding transformations on the P_l's or on the family of z_0?

The second question goes in the other direction: how can one translate in terms of the Σ_l's the action (3.8) defined on the z_l's?

As the reader might have guessed the transformation (3.8) on the family $\{z_0, z_1, \ldots, z_k\}$ and the transformations (3.5) on the family $\{\Sigma_0, \Sigma_1, \ldots, \Sigma_{k+1}\}$ are equivalent. We shall devote the remaining of this paragraph to the proof of this statement. Observe first that since $\Sigma_1 = P_0$, the action:

$$\Sigma_1 \to \Sigma_1' = g\Sigma_1[(g^\dagger g)^{-1}(1 - \Sigma_1) + \Sigma_1]^{-1} g^{-1} \tag{3.16}$$

corresponds to the transformation (3.4) on z_0:

$$z_0 \to z_0' = \frac{g z_0}{|g z_0|}.$$

The above statement is equivalent to prove that if $\{\Sigma_0, \Sigma_1, \ldots, \Sigma_{k+1}\}$ is **a family,** then the set $\{\Sigma'_0, \Sigma'_1, \ldots, \Sigma'_{k+1}\}$ with:

$$\Sigma'_l = g\Sigma_l[(g^\dagger g)^{-1}(1-\Sigma_l)+\Sigma_l]^{-1}g^{-1} \tag{3.17}$$

is also a family. Properties (3.13a) and (3.13c) of a Σ-family have been shown to hold for the Σ'_l, $0 \le l \le k+1$ if they are satisfied by the original projectors Σ_l, $0 \le l \le k+1$. (See eqs. (3.6), (3.7) and Exercise 23.) Let us now prove property (3.13b) for the primed Σ'_l. If $\Sigma_i\Sigma_j = \Sigma_j\Sigma_i = \Sigma_j$ for $i > j$, then:

$$[(g^\dagger g)^{-1}(1-\Sigma_i)+\Sigma_i]\Sigma_j = \Sigma_j$$

and hence:

$$\Sigma_j = \Sigma_i\Sigma_j = \Sigma_i[(g^\dagger g)^{-1}(1-\Sigma_i)+\Sigma_i]^{-1}\Sigma_j.$$

Then:

$$\begin{aligned}\Sigma'_i\Sigma'_j &= g\,\underbrace{\Sigma_i[(g^\dagger g)^{-1}(1-\Sigma_i)+\Sigma_i]^{-1}g^{-1}g\Sigma_j}_{\Sigma_j}[(g^\dagger g)^{-1}(1-\Sigma_j)+\Sigma_j]^{-1}g^{-1}\\ &= g\Sigma_j[(g^\dagger g)^{-1}(1-\Sigma_j)+\Sigma_j]^{-1}g^{-1}\\ &= \Sigma'_j \qquad\qquad \text{if } i > j.\end{aligned} \tag{3.18}$$

Moreover, since the Σ'_l are hermitian:

$$\Sigma'_j = (\Sigma'_j)^\dagger = (\Sigma'_i\Sigma'_j)^\dagger = \Sigma'_j\Sigma'_i. \tag{3.19}$$

Property (3.13d) is left as an exercise.

Exercise 31: Prove by direct calculation that $\Sigma'_{l+1}\partial\Sigma'_l = \Sigma'_l$ *for* $0 \le l \le k$. *(Hint: The following property:* $\Sigma_{l+1}[(g^\dagger g)^{-1}(1-\Sigma_{l+1})+\Sigma_{l+1}]^{-1}\Sigma_{l+1} = \Sigma_{l+1}$ *might be useful. It is equivalent to* $(\Sigma'_{l+1})^2 = \Sigma'_{l+1}$*.)*

Hence the action (3.8) has been recasted into a relatively simple action on projectors (eq. (3.17)). Simple as it is, the action is however plagued by the same computational drawback as the direct action on the z's: for z in the l-th subset, one has to know explicitly all the z_i's, $0 \leq i \leq l-1$, and all the z_i''s, $0 \leq i \leq l-1$, in order to compute z'. This difficulty seems to be tied to the complexity of the action on solutions that are neither self-dual nor anti-self-dual.

3.3 Orbits and Strata in the Solution Space

Usually, symmetry groups are used, among other things, to construct new solutions from known ones. Here the whole solution space is well understood and the interest of obtaining new solutions is rather limited. Although it might look complicated on a solution of the l-th subset, it is simply a linear transformation acting on the instanton of the associated family. In this last paragraph, I state a few properties of the orbits and the strata under the $SL(n+1, \mathbf{C})$ action presented above. It will become clear that the symmetry group provides here a better understanding of the topological structure of the solution space. (The proofs will be omitted. They have been given in reference [3].)

Among the properties (i)-(vii) introduced by Din and Zakrzewski (see §2.3) there is the existence of an integer k characterizing the number of solutions z_l's in a given family. (The number of solutions is $(k+1)$.) Because the symmetries (3.4) form a group, one can show easily that *the number k is invariant under this $SL(n+1, \mathbf{C})$ action.* For a reason to become clear later on, a family will be called *generic* if $k = n$. In other words, a generic family is an orthonormal frame of $\mathbf{C}^{n+1}$ containing solutions of the $\mathbf{CP}^n$ model. A family which is not generic is said to be degenerate.

Degenerate families are easy to construct. Indeed choose as starting self-dual solution $z_0 = f/|f|$ a solution with $(n-m)$ vanishing components:

$$f = \begin{pmatrix} p_0(x_+) \\ p_1(x_+) \\ \vdots \\ p_m(x_+) \\ 0 \\ \vdots \\ 0 \end{pmatrix} \tag{3.20}$$

where the p_i's, $0 \leq i \leq m$, are non-vanishing holomorphic functions without common zeroes. Due to the definition of the operator P_+, all the elements of the family of z_0 will have $(n-m)$ vanishing components. Hence the family cannot contain more than $m+1$ elements and, if $m \neq n$, this family is not generic. The degenerate families do not have necessarily the above form. However, one can show that for a degenerate family containing k solutions (or a k-family, for short), the projector Σ_{k+1} is constant. This observation is the key element in the proof of the following result: for any degenerate k-family, there exists always a group element $g \in SU(n+1) \subset SL(n+1, \mathbf{C})$ such that all the elements of the family transformed by the action (3.8) have their $(n-m)$ components equal to zero. In other words, a m-family of the $\mathbf{CP}^n$ model lies on the $SU(n+1)$-orbit of a generic family of the $\mathbf{CP}^m$ model trivially imbedded in the $\mathbf{CP}^n$ model.

With the previous results, the determination of the various isotropy groups is easy. (The *isotropy group* of a solution z is the subgroup of $SL(n+1, \mathbf{C})$ which acts trivially on z. The set of solutions sharing the same isotropy group is called a *stratum* under the group action.) The isotropy subgroup of an m-family in the $\mathbf{CP}^n$ model under the $SL(n+1, \mathbf{C})$ action is $SL(n-m, \mathbf{C}) \times (\mathbf{C}\backslash\{0\})$ if $m < n$ or the trivial subgroup $\{\mathbf{1}\}$ if $m = n$. Hence, there are precisely $(n+1)$ strata in the solution space of the $\mathbf{CP}^n$ model, each one being labelled by the integer k introduced through the properties (i)-(vii).

An important property can be added to the description of the solution space if one is interested in finite action solutions. For these solutions, the study can be restricted to the instanton since any family of finite action solutions contains one. An instanton with instanton number $l < n$ always leads to a degenerate family of the $\mathbf{CP}^n$ model. One can define a natural topology on the space of l-instantons as the topology induced by the usual topology on a vector space of polynomials. With this topology, the subset of generic l-instantons can be shown to be dense in the set of all l-instantons if $l \geq n$.

Work is in progress to identify and understand a similar group action on the solution space of the full principal σ model whose fields take their values in $G = SU(n)$.[8] Again, the symmetries appear to be realized in the unusual way described in theses lecture notes. It would be of great interest to find other nonlinear systems with such a symmetry group as it would broaden both the concept and the scope of this useful tool.

ACKNOWLEDGEMENTS

I would like to thank my collaborators Guy Arsenault and Michel Jacques for stimulating discussions.

REFERENCES

[1] Olver, P.J., *Applications of Lie Groups to Differential Equations,* Graduate Texts in Mathematics, vol. 107, Springer-Verlag, New York (1986).

[2] Arsenault, G., Jacques, M., and Saint-Aubin, Y., 'Collapse and Exponentiation of Infinite Symmetry Algebras of Euclidean Projective and Grassmannian σ Models', to appear in J. Math. Phys. **29** (1988).

[3] Arsenault, G., Jacques, M., and Saint-Aubin, Y., *Lett. Math. Phys.* **15, 65** (1988).

[4] Din, A., and Zakrzewski, W., *Nucl. Phys.* **B174,** 397 (1980).

[5] Borchers, M.J., and Garber, W.D., *Commun. Math. Phys.* **71,** 299 (1980); *Commun. Math. Phys.* **72,** 77 (1980).

[6] Zakrzewski, W., *J. Geom. Phys.* **1,** 39 (1984).

[7] Sasaki, R., *Z. Phys. C* **24,** 163 (1984).

[8] Arsenault, G., and Saint-Aubin, Y., in preparation.

MULTIDIMENSIONAL EQUATIONS AND DIFFERENTIAL GEOMETRY

Keti Tenenblat*
UNIVERSIDADE DE BRASILIA

Introduction

The sine-Gordon equation,

$$v_{tt} - v_{xx} = \sin v,$$

was known in the classical literature of differential geometry for being associated to surfaces of constant negative curvature contained in the euclidean space $\Re^3$. In physics it arises in the study of Josephson junctions, particle physics, stability of fluid motions, etc. In 1875, Backlund [3] studied the geometry of the surfaces mentioned above and obtained, what is now called a Backlund transformation which provides solutions for the sine-Gordon equation from a given one. Later Bianchi [8] obtained a permutability theorem, which provides a superposition formula for the sine-Gordon equation.

In 1967 Gardner, Greene, Kruskal and Miura [10] discovered that a Cauchy problem, with suitably decaying initial data on the line, associated with the Korteweg-de Vries equation could be solved by using what is now referred to as the inverse scattering transform. Subsequently a number of nonlinear equations of physical interest, including the sine-Gordon equation, have been solved by this method. See [2,4,5,6] for this method.

Looking for a differential equation with n independent variables, to which the inverse scattering transform might be applied, a generalised

[1]Partially supported by CNPq

sine-Gordon equation (GSGE) was introduced by Tenenblat and Terng in [12,13]. This equation reduces to the clasical sine-Gordon equation,when $n = 2$. Solutions for the GSGE are orthogonal matrix functions which correspond to n-dimensional submanifolds of constant sectional curvature -1 (hyperbolic manifolds) contained in the euclidean space $\Re^{2n-1}$. The generalisations of Backlund's and Bianchi's results were obtained in [12,13], and provided a Backlund transformation and a superposition formula for the GSGE.

These results were extended by the author in [11] to n-dimensional submanifolds of constant sectional curvature contained in the unit sphere $S^{2n-1} \subset \Re^{2n}$ and the hyperbolic space H^{2n-1}. Moreover a generalised wave equation (GWE) was introduced. This equation for $n = 2$ reduces to

$$v_{tt} - v_{xx} = 0,$$

and for $n \geq 3$, it is nonlinear and it is a homogeneous version of the GSGE. Solutions for the GWE are orthogonal matrix functions which correspond to n-dimensional flat submanifolds (zero sectional curvature) of the unit sphere S^{2n-1}.

In order to apply the inverse scattering transform one needs a one-parameter linear problem whose compatibility condition is the GWE or the GSGE. Such linear problems were obtained by Ablowitz, Beals and Tenenblat in [1], where an initial-boundary value problem with small data for both equations was solved by the inverse scattering method. These results were extended to the case of large data by Beals and Tenenblat in [7]. Moreover, the scattering data was related to the Backlund transformation.

In the following sections we summarise the results mentionad above. Further details can be found in the references. We begin in section 1 with the geometric theory for the GWE, GSGE, their Backlund transformations and superposition formulae. In section 2 we relate solutions of both equations to the scattering data and sketch the solution of the Cauchy problem with suitably decaying initial data. In section 3 we interpret the Backlund transformation in terms of scattering data. Moreover, the one-soliton solutions of the GWE and the GSGE, associated to the trivial solution by a Backlund transfomation, are given explicitly. By using the superposition formula one obtains n-soliton solutions.

1 The geometric theory for the GWE and the GSGE

In this section we consider n-dimensional Riemannian manifolds, with constant sectional curvature k, isometrically immersed in a $(2n-1)$-dimensional simply connected, complete, space form $\tilde{M}^{2n-1}$ of curvature K, such that $k < K$. We show that such immersions are in correspondence with solutions of a system of partial differential equations which is the GWE whenever $k = 0$ and the GSGE when $k \neq 0$. Moreover, the geometric theory of these immersions, when interpreted analytically, provide a Backlund transformation and a superposition formula for the GWE and the GSGE.

Let M^n be an n-dimensional Riemannian manifold isometrically immersed in a space form $\tilde{M}^{2n-1}$ of constant sectional curvature K. Let $e_1, \ldots, e_{2n-1}$ be a moving orthonormal frame on an open set of $\tilde{M}$, so that at points of M the vectors $e_1, \ldots, e_n$ are tangent to M. Let ω_A be the dual coframe and consider ω_{AB} defined by

$$de_A = \sum_{B=1}^{2n-1} \omega_{AB} e_B, \qquad \text{if } 1 \leq A, B \leq 2n-1$$

The structure equations of M are

$$d\omega_A = \sum_{B=1}^{2n-1} \omega_B \wedge \omega_{BA}, \qquad \omega_{AB} + \omega_{BA} = 0;$$

$$d\omega_{AB} = \sum_{C=1}^{2n-1} \omega_{AC} \wedge \omega_{CB} - K\omega_A \wedge \omega_B.$$

Restricting these forms to M we have $\omega_\alpha = 0$, for $n+1 \leq \alpha \leq 2n-1$, and the structure equations of the immersion are obtained by considering $\omega_\alpha = 0$ in the equations above. M has constant sectional curvature k, if and only if,

$$\sum_{\alpha=n+1}^{2n-1} \omega_{I\alpha} \wedge \omega_{\alpha J} = (K-k)\omega_I \wedge \omega_J, \text{for } I \leq I, J \leq n. \qquad (1)$$

The first and second fundamental forms of M are denoted by

$$I=\sum_{I=1}^{n}\omega_I^2,\qquad II=\sum_{I=1,\alpha=n+1}^{n,2n-1}\omega_{I,\alpha}\omega_I e_\alpha.$$

Whenever M has constant curvature $k<K$, there exist, locally, coordinates $(x_1,\ldots,x_n)$ on M and a moving orthonormal frame $e_1,\ldots,e_{2n-1}$ on $\bar{M}$ such that

$$\begin{aligned}\omega_I &= a_{1I}dx_I, && \text{for } i\le n,\\ \omega_{I,n+i-1} &= a_{iI}dx_I, && \text{for } 2\le i\le n,\end{aligned}\tag{2}$$

where $a_{1I}\neq 0$ and $\sum_{I=1}^{n}a_{1I}^2=1$. Moreover,the unit vector $\sum_{I=1}^{n}a_{1I}e_I=\sum_{I=1}^{n}\partial/\partial x_I$ is asymptotic,i.e. it annihilates the second fundamental form. Therefore,

$$\sum_{I=1}^{n}a_{iI}a_{1I}=0,\qquad \text{where } 2\le i\le n,\tag{3}$$

and from (1) we have

$$\begin{aligned}&\textstyle\sum_{i=2}^{n}a_{iI}a_{iJ}=(k-K)a_{1I}a_{1J}, && I\neq J, 1\le I,J\le n.\\ &\textstyle\sum_{i=2}^{n}a_{iJ}a_{iJ}=(K-k)(1-a_{1J}^2), && \text{for } 1\le J\le n.\end{aligned}\tag{4}$$

When the constant $K-k=1$, the equations (3) (4) imply that $a=(a_{IJ})$ is an orthogonal matrix. (When the constant $K-k\neq 1$, multiplying a by a constant matrix we get an orthogonal matrix). We consider the two cases $k=-1$, $K=0$ (hyperbolic submanifolds of the euclidean space) and $k=0$, $K=1$ (flat submanifolds of the sphere). The structure equations for M imply that the matrix a satisfies the following equations

$$\begin{aligned}&\frac{\partial}{\partial x_i}\left(\frac{1}{a_{1i}}\frac{\partial a_{1j}}{\partial x_i}\right)+\frac{\partial}{\partial x_j}\left(\frac{1}{a_{1j}}\frac{\partial a_{1i}}{\partial x_j}\right)+\sum_{l\neq i,j}\frac{1}{a_{1l}^2}\frac{\partial a_{1i}}{\partial x_l}\frac{\partial a_{1j}}{\partial x_l}=-ka_{1i}a_{1j}, i\neq j;\\ &\frac{\partial}{\partial x_l}\left(\frac{1}{a_{1j}}\frac{\partial a_{1i}}{\partial x_j}\right)=\frac{1}{a_{1l}a_{1j}}\frac{\partial a_{1i}}{\partial x_l}\frac{\partial a_{1l}}{\partial x_j}, \qquad i,j,l, \text{ distinct};\\ &\frac{\partial a_{jl}}{\partial x_i}=\frac{a_{ji}}{a_{1i}}\frac{\partial a_{1l}}{\partial x_i}, \qquad \text{for } i\neq l.\end{aligned}\tag{5}$$

This system of equations is called a *generalised wave equation* (GWE) when $k=0$ and a *generalised sine-Gordon equation* (GSGE) when $k=-1$.

Conversely, given an orthogonal matrix function a, with $a_{1i} \neq 0, 1 \leq i \leq n$, which satisfies the GWE (resp. GSGE), there exist a manifold M^n of constant curvarature $k = 0$ (resp.$k = -1$) isometrically immersed in the unit sphere $S^{2n-1} \subset \Re^{2n}$ (resp. in the euclidean space $\Re^{2n-1}$) and an orthogonal adapted frame field $e_1, \ldots, e_{2n-1}$ such that the dual forms ω_I, and the connection forms, $\omega_{I,n+j-1}, 1 \leq I \leq n, 2 \leq j \leq n$ are given by (2); see [11] for details.

Observe that for $n = 2$, the above system of equations reduces to

$$v_{x_1x_1} - v_{x_2x_2} = -k \sin v, \tag{6}$$

which is the homogeneous wave equation when $k = 0$ and the sine-Gordon equation when $k \neq 0$. In fact, we may consider

$$a(x_1, x_2) = \begin{pmatrix} \cos\frac{v}{2} & \sin\frac{v}{2} \\ -\sin\frac{v}{2} & \cos\frac{v}{2} \end{pmatrix}.$$

where $v(x_1, x_2)$ is a differentiable function of x_1, x_2. Then (5) reduces to (6).

In order to give the geometrical results whose analytic interpretation provides the Backlund transformation for (5), we recall the classical results on surfaces of constant negative curvature contained in $\Re^3$. In [3] Backlund considerd the following. Let M and M' be surfaces in $\Re^3$ and $l : M \to M'$ be a diffeomorphism such that for any point p in M and corresponding point $p' = l(p)$ one has the following:

a) the line determined by p and p' is tangent to M and M' at p and p' respectively;

b) the distance $d(p, p') = r > 0$ is a constant independent of p.

c) the angle between the normal vectors $N(p)$ and $N'(p')$ to the surfaces is a constant θ independent of p.

Backlund proved that under these conditions the surfaces M and M' have constant Gaussian curvature $k = k' = -\frac{\sin^2\theta}{r^2}$. Moreover, he showed that given any surface $M \subset \Re^3$ with constant negative curvature, there exists a two-parameter family of surfaces M', with the same curvature, related to M by diffeomorphisms which satisfy a)–c). Such a diffeomorphism is called a ***pseudo-spherical line congruence.***

This definition generalises as follows. A ***pseudo-spherical geodesic congruence*** between two n-dimensional submanifolds M and M' of a $2n-1$-dimensional space form $\bar{M}$ is a diffeomorphism $l : M \to M'$ such that for each $p \in M$ and $p' = l(p)$ one has

a') a unique geodesic γ in $\bar{M}$ joining p and p' whose tangent vectors at p and p' are in T_pM and $T_{p'}M'$ respectively;

b') the distance between p and p' on $\bar{M}$, is a constant r independent of p;

c') the $n-1$ angles between the normal spaces ν_p and $\nu'_{p'}$ are all equal to a constant θ, independent of p;

d') the normal bundles ν and ν' are flat;

e') the bundle map $\nu \to \nu'$ given by the orthogonal projection commutes with the normal connections.

Without loss of generality, we consider the space form $\bar{M}$ with constant sectional curvature $K = 0, 1$ or-1,i.e. $\bar{M}$ is respectively the $(2n-1)$-dimensional euclidean space, unit sphere S^{2n-1} or the hyperbolic space H^{2n-1}. Backlund's results generalise as follows (see [11,12]).

Suppose there is a pseudo-spherical geodesic congruence $l : M \to M'$ between two n-dimensional submanifolds of $\bar{M}_K^{2n-1}$, with constants r and $\theta \neq 0$. Then, both M and M' have constant sectional curvature k, where

$$k = \begin{cases} -\sin^2\theta/r^2, & \text{if } K = 0; \\ 1 - \sin^2\theta/\sin^2 r, & \text{if } K = 1; \\ -1 - \sin^2\theta/\sinh^2 r, & \text{if } K = -1; \end{cases}$$

Moreover, given an n- dimensional submanifold M of a space form $\bar{M}_K^{2n-1}$, with constant sectional curvature $k < K$, there exists an n-parameter family of submanifolds M', which are related to M by pseudo-spherical geodesic congruences.

Bianchi's permutability theorem for surfacess also generalises [11,13]. Let M, M', M'' be n-dimensional submanifolds of a space form $\bar{M}_K^{2n-1}$. Suppose there exist pseudo-spherical geodesic congruences $l_1 : M \to M'$

and $l_2 : M \to M''$ with constants r_1, θ_1 and r_2, θ_2 respectively, $\theta_1 \neq \theta_2$. Then there exists $M^* \subset \bar{M}$ and pseudo-spherical geodesic congruences $l_2^* : M' \to M^*, l_1^* : M'' \to M^*$ with constants r_2, θ_2 and r_1, θ_1 respectively such that $l_2^* \circ l_1 = l_1^* \circ l_2$.

These geometric results can be interpreted in terms of solutions for the system of equations (5). If a manifold M corresponds to an orthogonal matrix function a which satisfies (5) then a manifold M', associated to M by a pseudo-spherical geodesic congruence , corresponds to a matrix function X which satisfies the integrable system of equations

$$dX + XA_z^t X = A_z - XC, \qquad \text{BT}(z)$$

where z is a real parameter and $dX = \sum_{j=1}^n \frac{\partial X}{\partial x_j} dx_j$,

$$A_z = \sum_{j=1}^n \beta(z) a_j dx_j, \qquad C = \sum_{j=1}^n \gamma_j dx_j, \tag{7}$$

$$\beta(z) = \begin{cases} zI & \text{for GWE} \\ \frac{1}{2}(zI + z^{-1}u) & \text{for GSGE} \end{cases} \tag{8}$$

$$a_j = ae_j, \quad e_j \text{ is the matrix}, \quad (e_j)_{ik} = \delta_{ij}\delta_{jk}, \tag{9}$$

$$u = e_1 - (e_2 + \cdots + e_n) = \text{diag}(1, -1, \ldots, -1), \tag{10}$$

$$(\gamma_j)_{ik} = (1 - \delta_{ij}) \frac{1}{a_{1i}} \frac{\partial a_{1j}}{\partial x_i} \delta_{kj} - (1 - \delta_{kj}) \frac{1}{a_{1k}} \frac{\partial a_{1j}}{\partial x_k} \delta_{ij}. \tag{11}$$

Thus X is uniquely determined by BT(z) and its value at any given point, say $x = 0$. Moreover, if $X(0)$ is in $O(n)$ then $X^t X = I$. Hence, given a which solves GWE or GSGE we get a new solution X which is *associated to a by* BT(z). We note that the parameter z in BT(z) is related to the constant θ of the pseudo-spherical geodesic congruence by $z = \cot\theta$ for GWE and $z = \tan\frac{\theta}{2}$ for GSGE.

The permutability theorem provides a superposition formula for the GWE and the GSGE. Let a be a solution of (5) and a_i, $i = 1, 2$, solutions associated to a by BT(z_i), $z_1 \neq z_2$. Then a fourth solution a^* can be obtained by solving the algebraic relation

$$a^* a^t = u(\beta(z_1) a_2 a_1^t - \beta(z_2))(\beta(z_2) a_2 a_1^t - \beta(z_1))^{-1}. \tag{12}$$

We observe that BT(z) is an overdetermined system with parameter z whose compatibility conditions are precisely the GWE or GSGE for a, wherever the a_{1j} do not vanish.

The Backlund transformation is a matrix Ricatti equation. Linearisation of such an equation can be performed (see [14]) introducing $X = PQ^{-1}$, where P and Q take values in the space $M_n(\Re)$ of $n \times n$ real matrices. This leads to the linear system

$$\begin{pmatrix} dP \\ dQ \end{pmatrix} = \begin{pmatrix} 0 & A_z \\ A_z^t & C \end{pmatrix} \begin{pmatrix} P \\ Q \end{pmatrix}$$

The compatibility conditions for this system are the same as those for BT(z). Thus, if a satisfies the GWE or the GSGE, there exists a fundamental matrix solution for

$$d\psi(x,z) = \begin{pmatrix} 0 & A_z \\ A_z^t & C \end{pmatrix} \psi, \tag{13}$$

with ψ an invertible $M_{2n}(\Re)$-valued function. Since $\det \psi(x,z)$ is constant, we may assume that ψ has values in $Sl(2n,\Re)$.

The relationship between solutions of (13) and solutions of BT(z) is the following. Let a be an orthogonal matrix function which is a solution of the GWE (resp. GSGE). For real z, let $\psi(\cdot,z)$ be a solution of (13) with values in $Sl(2n,\Re)$. Consider v_1, v_2 elements of $M_n(\Re)$ and set

$$\begin{pmatrix} P \\ Q \end{pmatrix} = \psi(x,z) \begin{pmatrix} v_1 \\ v_2 \end{pmatrix}. \tag{14}$$

If Q is invertible and $X = PQ^{-1}$ has values in $O(n)$, then X is associated to a by BT(z). Conversely, for any X associated to a by BT(z), there are elements v_1, v_2 of $M_n(\Re)$, such that $X = PQ^{-1}$, with P and Q given by (14).

In the following sections the GWE and the GSGE will be treated separately. The GWE as compatibility conditions for the system (11) can be written

$$a^t a = I, \qquad \gamma_j + \gamma_j^t = 0, \tag{15}$$

$$\frac{\partial a_i}{\partial x_j} + a_i \gamma_j = \frac{\partial a_j}{\partial x_i} + a_j \gamma_i, \tag{16}$$

$$\frac{\partial \gamma_i}{\partial x_j} + \gamma_i \gamma_j = \frac{\partial \gamma_j}{\partial x_i} + \gamma_j \gamma_i. \tag{17}$$

It follows from (15) and (16) that γ_j is determined by a. In fact, set $\alpha_i = -a^t \frac{\partial a}{\partial x_i}$, then

$$\gamma_j = -\alpha e_j + e_j \alpha^t, \tag{18}$$

where $\alpha = \sum_{i=1}^{n} e_i \alpha_i$. Thus we regard either a alone or $(a, \gamma_1, \ldots, \gamma_n)$ as the unknown functions in GWE. Moreover, if a solves GWE and if $l \in O(n)$ is fixed, then $\hat{a} = la$ also solves GWE. Generically, $\hat{a}$ will represent a different geometric solution (where a_{1j} and $\hat{a}_{1j}$ do not vanish) because the associated first fundamental forms will differ.

The GSGE can be written

$$a^t a = I, \qquad \gamma_j + \gamma_j^t = 0, \tag{19}$$

$$\frac{\partial a_i}{\partial x_j} + a_i \gamma_j = \frac{\partial a_j}{\partial x_i} + a_j \gamma_i, \tag{20}$$

$$\frac{\partial \gamma_i}{\partial x_j} + \gamma_i \gamma_j + \frac{1}{2} a_i^t u a_j = \frac{\partial \gamma_j}{\partial x_i} + \gamma_j \gamma_i + \frac{1}{2} a_j^t u a_i \tag{21}$$

where, the constant $k = -1$ in (5), and u is defined in (10). As before γ_j is given in terms of a by (18). In this case, if a is a solution of GSGE and $\hat{a} = la$ with $l \in O(n)$, $\hat{a}$ will only be a solution if $lu = ul$, i.e.$l_{11} = \pm 1$. If l satisfies this condition, then a and $\hat{a}$ represent the same geometric solution: the first fundamental form will be identical and l simply represents a different choice of a normal frame for the immersion.

2 Scattering data for the GWE and GSGE

In this section we relate solutions of the GWE and GSGE to the so called "scattering data" and we sketch the solution of an initial-boundary value problem (see[1,7]). We start with the GWE.

The forward scattering problem, for a matrix a satisfying the GWE, consists in obtaining normalised solutions ψ of (13), for $z \in \mathbb{C}$,i.e.

$$\frac{\partial \psi}{\partial x_j}(x,z) = z \begin{pmatrix} 0 & a_j \\ a_j^t & 0 \end{pmatrix} \psi(x,z) + \begin{pmatrix} 0 & 0 \\ 0 & \gamma_j \end{pmatrix} \psi(x,z). \tag{22}$$

The normalisation will take the form of asymptotic conditions on certain directions of the space $\Re^n$ of variables x.

A direction $\varsigma = (\varsigma_1, \ldots, \varsigma_n) \in \Re^n$ is said to be *oblique* if $\pm\varsigma_i, 1 \leq i \leq n$, are all distinct. We say that the matrix function a is *asymptotically constant* in the direction ς, if the matrix

$$\alpha_j = -a^t \frac{\partial a}{\partial x_j}, \; 1 \leq j \leq n,$$

is rapidly decreasing at infinity along each line

$$L(\varsigma, y) = \{y + s\varsigma; s \in \Re\},$$

uniformly with respect to the vectors y orthogonal to ς.

For $z \in C$, we want a solution for (22) of the form

$$\psi(x, z) = U(x)m(x, z)e^{zx \cdot J}, \tag{23}$$

where

$$U = \begin{pmatrix} a & 0 \\ 0 & I \end{pmatrix} U_2, \qquad U_2 = \frac{1}{\sqrt{2}} \begin{pmatrix} I & -I \\ I & I \end{pmatrix}, \tag{24}$$

$$x \cdot J = \sum_{j=1}^{n} x_j J_j, \qquad J_j = \begin{pmatrix} e_j & 0 \\ 0 & -e_j \end{pmatrix}, \tag{25}$$

and I denotes the $n \times n$ identity matrix. Therefore, it follows from (22) that the function m of (23) must satisfy the equations

$$\frac{\partial m}{\partial x_j} = z[J_j, m] + Q_j m, \tag{26}$$

where

$$Q_j = U_2^{-1} \begin{pmatrix} \alpha_j & 0 \\ 0 & \gamma_j \end{pmatrix} U_2.$$

We want m to satisfy

$$\sup_{x} | m(x, z) | < \infty \tag{27}$$

Moreover, since the solution for (26) (27) is not unique, we impose the additional condition

$$\lim_{s \to -\infty} m(y + s\varsigma, z) = 1, \qquad \text{for } y \perp \varsigma. \tag{28}$$

For each $z \in C \setminus i\Re$, there exists at most one solution of (26) — (28). When the matrix function a is asymptotically constant, this solution exists and it is holomorphic with respect to z, for $z \in C \setminus (i\Re \cup D)$, where D is a discrete and bounded subset of $C \setminus i\Re$.

If certain L^1 norms of Q_j over the lines $L(y, \varsigma)$ are small, then $D = \emptyset$, (see[1]), and for each $z \in i\Re$, there exist the limits

$$m_{\pm}(x, z) = \lim_{\epsilon \to 0_+} m(x, z \pm \epsilon), \tag{29}$$

and a matrix $V(z)$ such that

$$m_+(x, z) = m_-(x, z) e^{zx \cdot J} V(z) e^{-zx \cdot J}. \tag{30}$$

Moreover

$$\lim_{z \to \infty} m(x, z) = 1. \tag{31}$$

If the condition on the L^1 norm does not hold, for "generic" matrices a, i.e. for a dense and open subset of matrices a which are asymptotically constant in the direction ς, the limits (29) exist and satisfy (30), moreover D is finite, $m(\cdot, z_0)$ has a simple pole at $z_0 \in D$, characterised by a matrix $V(z_0)$,which satisfies

$$V^2(z_0) = 0, \tag{32}$$

$$m(x, z)(1 - (z - z_0)^{-1} e^{z_0 x \cdot J} V(z_0) e^{-z_0 x \cdot J}) \text{ has a removable singularity at } z_0. \tag{33}$$

The function $V : i\Re \cup D \to M_{2n}(C)$ is called the *matrix of scattering data for a* . V has the following properties

$$V - I, \text{ for } z \in i\Re, \text{ belongs to the Schwartz class } S(i\Re; M_{2n}(C)). \tag{34}$$

$$V(-z) = V(z)^t = (V(z)^{-1})^\sigma, \quad \text{for } z \in i\Re; \tag{35}$$

$$V(-z_0) = V(Z_0)^t = V(z_0)^\sigma \quad \text{for } z_0 \in D, \tag{36}$$

where

$$B^\sigma = \begin{pmatrix} 0 & I \\ I & 0 \end{pmatrix} B \begin{pmatrix} 0 & I \\ I & 0 \end{pmatrix}.$$

The symmetries (35) and (36) are a consequence of the symmetries of m

$$m(x, -z) = (m(x, z)^{-1})^t = m(x, z)^\sigma,$$

which also give $D = -D$. When the matrix a is real there are additional symmetries $D = \bar{D}$,

$$V(\bar{z}) = \overline{V(z)}, \quad z \in i\Re;$$
$$V(\bar{z}_0) = \overline{V(z_0)}, \quad z_0 \in D \tag{37}$$

These follow from the additional symmetry of m,

$$m(x, \bar{z}) = \overline{m(x, z)}.$$

The inverse scattering problem consists in obtaining the solution a of the GWE (by first obtaining $m(x, \cdot)$) from the scattering data V. This is a Riemann-Hilbert factorization problem, with the singularities determined by V. An important result obtained in [7] shows that any function $V : i\Re \cup D \to M_{2n}(C)$, which satisfies (32), (34) - (37) with $D = \bar{D} = -D$, is the scattering data for a real solution a of the GWE. The matrix function a is unique up to left multiplication by a fixed orthogonal matrix.

Now we sketch the solution of an initial-boundary value problem for the GWE. Let ς be an oblique direction in $\Re^n$. Suppose $a(s), \gamma(s)$ are smooth mappings defined on a line $L(\varsigma, y_0)$ such that $a(s) \in O(n), \gamma^t + \gamma = 0$ and $\alpha(s) = -a^t \frac{\partial a}{\partial s}, \gamma(s)$, are Schwartz functions of s. Suppose $a(s)$ is asymptotically constant, we want to obtain a solution $a(x), x \in \Re^n$, of the GWE which extends $a(s)$. Initially, we consider from (26) a differential equation $\frac{\partial m}{\partial s}$, for the directional derivative of m in the direction ς on the line $L(y_0, \varsigma)$. The solution of this equation $m(s, z), s \in \Re$, with the corresponding conditions (27) and (28), provide the scattering data $V : i\Re \cup D \to M_{2n}(C)$. The inverse scattering problem for V gives the solution $a(x)$ which we are looking for.

The scattering data for the GSGE is obtained in analogous way. The forward scattering problem for a matrix a which satisfies the GSGE consists in obtaining normalized solutions ψ of (13),for $z \in C$,i.e.

$$\frac{\partial \psi}{\partial x_j}(x, z) = \frac{1}{2} z \begin{pmatrix} 0 & a_j \\ a_j^t & 0 \end{pmatrix} \psi + \frac{1}{2z} \begin{pmatrix} 0 & ua_j \\ a_j^t u & 0 \end{pmatrix} \psi + \begin{pmatrix} 0 & 0 \\ 0 & \gamma_j \end{pmatrix} \psi. \tag{38}$$

Again the normalization is given as an asymptotic condition on an oblique direction $\varsigma = (\varsigma_1, \ldots, \varsigma_n) \in \Re^n$ and here we also assume that

$$|\varsigma_1| > |\varsigma_2| > \cdots > |\varsigma_n|.$$

We assume that a is assymptotically constant in the direction ς.

For $\varsigma \in C$, we want a solution for (38) of the form

$$\psi(x,z) = U(z)m(x,z)e^{x\cdot J(z)}, \tag{39}$$

where U is given by (28) and

$$x \cdot J(z) = \sum_{j=1}^{n} x_j J_j(z),$$

$$J_j(z) = \frac{1}{2} z J_j + \frac{1}{2z} J_j^{\#},$$

$$J_1^{\#} = J_1, \quad J_j^{\#} = -J_j \text{ for } 2 \leq j \leq n$$

It follows from (38) that the function m of (39) must satisfy the equations

$$\frac{\partial m}{\partial x_j} = [J_j(z), m] + Q_j m \tag{40}$$

where

$$Q_j(x,z) = U_2^{-1} \begin{pmatrix} \alpha_j & 0 \\ 0 & \gamma_j \end{pmatrix} U_2 + \frac{1}{2z}(U^{-1} B_j U - J_j^{\#}),$$

$$\alpha_j = -a^t \frac{\partial a}{\partial x_j}, \qquad B_j = \begin{pmatrix} 0 & u a_j \\ a_j^t u & 0 \end{pmatrix}.$$

Again we want

$$\sup \mid m(x,z) \mid < \infty, \tag{41}$$

$$\lim m(y + s\varsigma, z) = 1, \qquad \text{for } y \perp \varsigma. \tag{42}$$

For "generic" matrices a the solution $m(\cdot, z)$ exists and is holomorphic with respect to z, for $z \in C \setminus (\Sigma \cup D)$ where $D \subset C \setminus \Sigma$ is finite and

$$\Sigma = \begin{cases} i\Re, & \text{if } n = 2, \\ i\Re \cup \{z; \mid z \mid = 1\} & \text{if } n > 2 \end{cases}$$

The singularities of m are characterised by a function $V : \Sigma \cup D \to M_{2n}(C)$ which is the *matrix of scattering data for a.* V has the following properties; Let

$$\Omega_{\pm} = \begin{cases} \{z; \pm\Re z > 0\}, & \text{if } n = 2 \\ \{z; \pm\Re z > 0, \mid z \mid > 1\} \cup \{z; \pm\Re z > 0, \mid z \mid < 1\}, & \text{if } n > 2 \end{cases}$$

and

$$m_{\pm}(x,z) = \lim_{z'\to z, z'\in\Omega_{\pm}} m(x,z'), \qquad z \in \Sigma.$$

Then

$$m_{+}(x,z) = m_{-}(x,z)e^{x\cdot J(z)}V(z)e^{-x\cdot J(z)}, z \in \Sigma; \tag{43}$$

$$m(x,z)[1-(z-z_0)^{-1}e^{x\cdot J(z_0)}V(z)e^{-x\cdot J(z_0)}] \text{ has a removable singularity at } z=z_0. \tag{44}$$

We now change the normalisation (42) to

$$\lim_{z\to\infty} m(x,z) = I. \tag{45}$$

This can be accomplished by multiplying on the right by $m(\cdot,\infty)^{-1}$; we continue to denote this renormalised function by m. V has properties analogous to those of GWE scattering data:

$$\begin{aligned} &V(z)-I, \text{ and} V(z^{-1})-I \text{ belong to the Schwartz space } S(i\Re; M_{2n}(C)) \quad \text{if } n=2;\\ &V(\pm is)-I, \text{ and} V(\pm is^{-1})-I \text{ extend so as to belong to } S([1,\infty); M_{2n}(C)) \quad \text{if, } n>2; \end{aligned} \tag{46}$$

$$V(z_0)^2 = 0, \qquad \text{for } z_0 \in D; \tag{47}$$

$$\sup \mid e^{x\cdot J(z)}V(z)e^{-x\cdot J(z)} \mid < \infty, \quad z \in \Sigma \tag{48}$$

condition (48) is superfluous on $i\Re$, where the exponentials are unitary, while on $\{\mid z \mid = 1, z \neq \pm 1\}$ it is equivalent to

$$V(z)_{jk} = 0 \text{ if } j=k \text{ and } j \text{ or } k=1 \text{ or } n, z \in \Sigma \setminus i\Re. \tag{49}$$

For $n > 2$,there is an additional condition at the juncture of the imaginary axis and the unit circle. Let V_j denote the restriction of V to Σ_j, where

$$\Sigma_1 = \{\mid z \mid = 1, \Re z > 0\}, \quad \Sigma_2 = i\Re \cap \{\mid z \mid > 1\},$$

$$\Sigma_3 = \{\mid z \mid = 1, \Re z < 0\}, \quad \Sigma_4 = i\Re \cap \{\mid z \mid < 1\}.$$

Let $\tilde{V}_j(\pm i)$ denote the Taylor expansion of V_j at $\pm i$, considered as a formal power series. Then, the condition is

$$\tilde{V}_1(i)\tilde{V}_2(i)\tilde{V}_3(i)\tilde{V}_4(i) = 1 = \tilde{V}_4(-i)\tilde{V}_3(-i)\tilde{V}_2(-i)\tilde{V}_1(-i). \qquad (50)$$

Moreover, since $V(0) = I, m$ is continuous at $z = 0$.

The normalised solution m has the symmetries

$$m(x,-z) = (m(x,z)^{-1})^t = m(x,z)^*;$$

$$m(x,z) = \overline{m(\bar{x},\bar{z})}, \text{if and only if } a, \gamma_1, \ldots, \gamma_n \text{ are real;}$$

Moreover,

$$[m(x,0)^{-1}m(x,z^{-1})]^{\#} = m(x,z),$$

where

$$B^{\#} = \frac{1}{4}\begin{pmatrix} I+u & I-u \\ I-u & I+u \end{pmatrix} B \begin{pmatrix} I+u & I-u \\ I-u & I+u \end{pmatrix}.$$

These symmetries of m are equivalent to

$$\begin{aligned} &V(-z) = V(z)^t = (V(z)^{-1})^*, \quad V(z_{-1}) = (V(z)^{-1})^{\#}, \\ &V(-z_0) = V(z_0)^t = -V(z_0)^*, \quad V(z_0^{-1}) = -z_0^{-2}V(z_0)^{\#} \end{aligned} \qquad (51)$$

where $z \in \sum, z_0 \in D$. Reality of $a, \gamma_1, \ldots, \gamma_n$ is equivalent to

$$V(\bar{z}) = \overline{V(z)}, \qquad z \in \textstyle\sum \cup D. \qquad (52)$$

As for GWE, there is an inversion theorem for scattering data in the real case. Suppose $V : \sum \cup D \to M_{2n}(C)$ satisfies (46) – (52), where D is a finite subset of $C \setminus \sum$ which is closed under the involutions $z \to -z, z \to \bar{z}, z \to z^{-1}$. Then, V is the scattering data for a real solution a of the GSGE. The matrix function a is unique up to left multiplication by a constant orthogonal matrix which commutes with u.

The solution of an initial-boundary-value problem for the GSGE is obtained following the same steps as in the case of the GWE.

3 The Backlund transformation in terms of scattering data

In this section we summarise the effect of the Backlund transformation on the scattering data (see [7]). Going from a given solution of the GWE (or GSGE) into another, via the Backlund transformation, corresponds to inserting or deleting a discrete singularity in the scattering data, while conjugating the continuous data by a simple matrix valued function. In order to state the result explicitly we introduce the following concept. A *permutation matrix* $\Pi \in M_{2n}(\Re)$ *is regular* if

$$\Pi = \begin{pmatrix} \Pi_1 & \Pi_2 \\ \Pi_2 & \Pi_1 \end{pmatrix}, \tag{53}$$

where Π_1 and Π_2 are matrices which correspond to orthogonal projections on complementary coordinate subspaces of $\Re^n$. In particular Π_1 and Π_2 are diagonal and

$$\Pi = \Pi^* = \Pi^t = \Pi^{-1}.$$

Let a be a solution of the GWE with associated eigenfunctions m and ψ (i.e. m is a solution of (26) - (28) and ψ is given by (23) and let V be the matrix of scattering data for a defined on $i\Re \cup D$. Suppose $z_0 \notin (i\Re \cup D)$ and let $\hat{a}$ be a solution of the GWE associated to a by $\mathrm{BT}(z_0)$. Then there exists a regular permutation matrix $\Pi \in M_{2n}(\Re)$ and $v \in M_n(C)$ with $v + v^t = 0$ such that the scattering data $\hat{V}$ associated to $\hat{a}$ are given by

$$\begin{aligned} \hat{V}(z) &= J(z + z_0\Pi J\Pi)V(z)(z + z_0\Pi J\Pi)^{-1}J, \; z \in (i\Re \cup D), \\ \hat{V}(z_0) &= J\Pi \begin{pmatrix} 0 & -2z_0 v \\ 0 & 0 \end{pmatrix} \Pi J = \hat{V}(-z_0)^t, \end{aligned} \tag{54}$$

where

$$J = \begin{pmatrix} I & 0 \\ 0 & -I \end{pmatrix}.$$

Conversely, given matrices Π and v as above define

$$\begin{pmatrix} P \\ Q \end{pmatrix} = \psi(x, z_0)\Pi \begin{pmatrix} v \\ I \end{pmatrix}. \tag{55}$$

Then Q is invertible for any $x \in \Re^n$ and

$$\hat{a} = PQ^{-1} \tag{56}$$

is a solution of the GWE associated to a by BT(z_0).

The eigenfunction $\hat{m}$ associated to $\hat{a}$ is given by

$$\hat{m}(x,z) = J(z - z_0T(x))m(x,z)(z + z_0\Pi J\Pi)^{-1}J,$$

where

$$T(x) = U_2^{-1}\begin{pmatrix} 0 & Y \\ Y^t & 0 \end{pmatrix}U_2,$$

and

$$Y = a^tPQ^{-1}.$$

The solutions $\hat{a}$ of the GWE associated to $a = I, \gamma_1 = \cdots = \gamma_n = 0$ by BT(z_0) are one-soliton solutions. As a consequence of the above result, these are described by

$$\hat{a}(x, z_0) = (\Pi_2 - \Pi_1)(L_1v - L_1^{-1})(L_1v + L_1^{-1})^{-1}, \tag{57}$$

where $z_0 \in C \setminus 0, v \in M_n(C), v + v^t = 0$, Π is a regular permutation matrix as in '(53) and

$$L_1 = \Pi_1L + \Pi_2L^{-1},$$
$$L(x, z_0) = \exp(z_0 \textstyle\sum_{j=1}^n x_je_j). \tag{58}$$

We observe that applying the Backlund transformation BT(z_0) to a real solution a of GWE will provide a new real solution $\hat{a}$, given by (55) (56) if z_0 and the matrix v characterising the singularities are real. More generally, starting from a real solution, we can obtain a new real solution in two steps, inserting singularities at $\pm z_0 \notin \Re$ and then at $\pm\bar{z}_0$.

Similarly, applying the Backlund transformation to solutions of the GSGE correspond to introducing singularities at $\pm z_0$ and $\pm z_0^{-1}$. Let a be a solution of the GSGE with associated eigenfunctions m and ψ (i.e. m is a solution of (40) (41) (45) and ψ is given by (39). Let V be the matrix function of scattering data for a defined on $\sum \cup D$. Suppose $z_0 \notin \sum \cup D$. Let $\hat{a}$ be a solution of the GSGE associated to a by BT(z_0). Then, there exists a regular permutation matrix $\Pi \in M_{2n}(\Re)$ and $v \in M_n(C)$ with $v + v^t = 0$, such that the scattering data $\hat{V}$ associated to $\hat{a}$ are given by

$$\begin{aligned}\hat{V}(z) &= J\Delta V(z)\Delta^{-1}J, \text{ for } z \in (\sum \cup D)\\ \hat{V}(z_0) &= J\Pi \begin{pmatrix} 0 & -2z_0 v \\ 0 & 0 \end{pmatrix} \Pi J = \hat{V}(-z_0)^t = \\ &= -z_0^2 \hat{V}(z_0^{-1}) = -z_0^2 \hat{V}(-z_0^{-1})^t.\end{aligned} \tag{59}$$

where

$$\Delta = (z + z_0^{-1}\Pi J \Pi)(z + z_0 \Pi J \Pi)$$

Conversely, given matrices Π and v as above, define

$$\begin{pmatrix} P \\ Q \end{pmatrix} = \psi(x, z_0)\Pi \begin{pmatrix} v \\ I \end{pmatrix}. \tag{60}$$

Then Q is invertible for each $x \in \Re^n$ and

$$\hat{a} = PQ^{-1} \tag{61}$$

is a solution of the GSGE associated to a by $\mathrm{BT}(z_0)$.

The eigenfunction $\hat{m}$ associated to $\hat{a}$ is given by

$$\hat{m}(x, z) = J(z - z_0^{-1}T_2)(z - z_0 T_1)m(x, z)(z + z_0 \Pi J \Pi)^{-1}(z + z_0^{-1}\Pi J \Pi)^{-1},$$

where

$$\begin{aligned}T_1(x) &= U_2^{-1} \begin{pmatrix} 0 & Y_1 \\ Y_1^t & 0 \end{pmatrix} U_2,\\ Y_1 &= a^t P Q^{-1}.\end{aligned}$$

$$T_2^{\#}(x) = m(x, 0)^{-1} T_1(x) m(x, 0).$$

The solutions $\hat{a}$ of the GSGE associated to $a = I, \gamma_1 = \cdots = \gamma_n = 0$ by $\mathrm{BT}(z_0)$ are one-soliton solutions. As a consequence of the above result, these are described by

$$\hat{a}(x, z_0) = (\Pi_2 - \Pi_1)(E_1 v - E_1^{-1})(E_1 v + E_1^{-1})^{-1}, \tag{62}$$

where $z_0 \in C \setminus 0, v \in M_n(C), v + v^t = 0$, Π is a regular permutation matrix as in (53) and

$$\begin{aligned}E_1 &= \Pi_1 E + \Pi_2 E^{-1},\\ E(x, z_0) &= \exp\left(\textstyle\sum_{j=1}^n \frac{1}{2} x_j (z_0 I + z_0^{-1} u) e_j\right).\end{aligned} \tag{63}$$

As in the case of the GWE, we may preserve the reality of solutions of the GSGE associated by BT(z_0) under the appropriate assumptions: if a is real and if z_0 and the matrix v characterising the new singularity are real, then $\hat{a}$ given by (60) (61) will be real. More generally, one can go from real a to real $\hat{a}$ in two steps, inserting singularities at $\pm z_0 \notin \Re \setminus 0, \pm z_0^{-1}$ and then at $\pm \bar{z}_0, \pm \bar{z}_0^{-1}$.

It is clear that the Backlund transformation may delete singularities (instead of inserting them). In fact, we only need to interchange the scattering data V and $\hat{V}$ in the above procedure in the case of the GWE or the GSGE.

Finally, we observe that the permutability theorems can also be interpreted in terms of scattering data (see [7]). Moreover, n-soliton solutions for GWE and GSGE are explicitly obtained by successive use of the superposition formula (12) starting from (57) (58) and (62) (63) respectively. In particular the following matrix function a^* is a two-soliton solution:

$$a^* = u(\beta(z_1)a_2a_1^t - \beta(z_2))(\beta(z_2)a_2a_1^t - \beta(z_1))^{-1},$$

where for $i = 1, 2$, $z_i \in C \setminus 0$, $z_1 \neq z_2$, $\beta(z_i)$ is given by (8)

$$a_i = (\Pi_2^i - \Pi_1^i)(W_iv_i - W_i^{-1})(W_iv_i + W_i^{-1})^{-1},$$

Π^i is a regular pertmutation matrix as in (53), $v_i \in M_n(C), v_i + v_i^t = 0$, and

$$W_i = \Pi_1^i\tilde{W}_i + \Pi_2^i\tilde{W}_i^{-1},$$

$$\tilde{W}_i = \begin{cases} \exp(z_i \sum_{j=1}^n x_je_j), & \text{for GWE,} \\ \exp(\sum_{j=1}^n \frac{1}{2}x_j(z_iI + z_i^{-1}u)e_j), & \text{for GSGE} \end{cases}$$

References

[1] Ablowitz,M.J.; Beals,R.; Tenenblat,K., *On the solution of the generalized wave and generalized sine-Gordon equations*, Stud. Appl. Math. 74,(1986),177–203.

[2] Ablowitz,M.J.; Kaup,D.J.; Newell,A.C.; Segur,H., *The inverse scattering transform — Fourier analysis for nonlinear problems*, Stud. Appl.Math.53,(1974),249–315.

[3] Backlund,A.V.*Concerning surfaces with constant negative curvature* translated by E.M.Coddington, New Era Printing Co., Lancaster pa,1905.

[4] Beals,R.; Coifman,R.R.,*Scattering and inverse scattering for first order systems*, Comm.Pure Appl. Math. 37,(1984),39–90.

[5] Beals,R.; Coifman,R.R.,*Inverse scattering and evolution equations* , Comm. Pure Appl.Math. 38,(1985),29–42.

[6] Beals,R.; Coifman,R.R.,*Scattering and inverse scattering for first order systems,II,Inverse problems* to appear.

[7] Beals,R.; Tenenblat,K.,*Inverse scattering and the Backlund transformation for the generalized wave and generalized sine-Gordon equations*,to appear.

[8] Bianchi,L.*Lezioni di geometria differenziale* Ed. Nicola Zanchelli, Bologna 1927.

[9] Chern,S.S.; Terng,C.L.,*An analogue of Backlund's theorem in affine geometry* Rocky Mountain Math. 10,(1980),105–124.

[10] Gardner,C.S.; Greene,J.M.; Kruskal,M.D.; Miura,R.M.,*Method for solving the Korteweg-de Vries equation*, Phys. Rev. Lett. 19 (1967), 1095–1097.

[11] Tenenblat,K.,*Backlund's theorem for submanifolds of space forms and a generalized wave equation*, Bol.Soc. Bras. Mat. 16, (1985), 69–94.

[12] Tenenblat,K.; Terng,C.L.,*Backlund theorem for n-dimensional submanifolds of* $\Re^{2n-1}$, Ann.of Math.,111,(1980),477–490.

[13] Terng,C.L., *A higher dimensional generalization of the sine-Gordon equation and its soliton theory*, Ann.of Math. 111, (1980), 491–510.

[14] Winternitz,P., Lecture Notes in Phisics, 189, Proceedings of the CIFMO School and Workshop held at Oaxtepec, Mexico, (K. B. Wolf Ed.), Springer Verlag, 1982.

SUPERSYMMETRIES IN QUANTUM MECHANICS - AN OVERVIEW

Luc Vinet

Laboratoire de Physique Nucléaire
Université de Montréal
C.P. 6128, Succ. "A"
Montréal, Québec,
H3C 3J7 Canada

Supersymmetry is one of the most original ideas that emerged in physics recently. Traditionally this concept is discussed in field theory where it has become more and more essential. The novel feature of these symmetry transformations is the fact that they mix bosonic and fermionic degrees of freedom. Also, they involve Grassmann parameters and have led to the development of the superspace formalism. The simplest context in which these salient aspects of supersymmetry can be exposed is that of classical mechanics which can be viewed as field theory in 0+1 dimensions with the particle's coordinates considered as scalar fields of the one dimensional time variable.

Suppose we have a theory involving bosonic fields ϕ and fermionic fields ψ. It is well known that canonical quantization is achieved by imposing equal-time anticommutation relation of the form

$$\{\psi_i(t,\vec{x}),\, \psi_j^\dagger(t,\vec{y})\} = \hbar\, \delta_{ij}\, \delta(\vec{x}-\vec{y})$$

on the Fermi fields. (Analogous relations involving commutators are imposed on the Bose fields.) In the classical limit $\hbar \to 0$, the fermion fields turn into anticommuting variables. This description of "classical" fermions with the help of Grassmann variables also proves necessary to obtain the correct path integral quantization. We see that it requires the introduction of an underlying Grassmann algebra with the bosonic fields and the fermionic fields respectively taking their values in its even and odd parts. It is therefore clear that linear symmetry operations transforming bosons into fermions (or vice-versa) must involve anticommuting parameters if the gradings are to be respected.

Let us discuss an example in 0+1 dimensions. Consider the Lagrangian

$$L = \frac{1}{2}\dot{\phi}_i(t)\,\dot{\phi}_i(t) + \frac{i}{2}\psi_i(t)\,\dot{\psi}_i(t) \qquad i = 1,2,3$$

with ϕ_i and ψ_i respectively commuting and anticommuting (real) variables. We shall see shortly that it describes a non-relativistic free spin 1/2 particle localized at the point $\vec{\phi}$ in 3-space. The corresponding Hamiltonian is given by $H = p_i p_i/2$ with $p_i = \partial L/\partial\dot{\phi}_i = \dot{\phi}_i$. Examine for the moment the following infinitesimal supersymmetry transformations:

$$\delta_\alpha \phi_i = i\,\alpha\,\psi_i \qquad\qquad \delta_\alpha \psi_i = -\,\alpha\,\dot{\phi}_i$$

with α an anticommuting constant parameter. Note that they mix the positions with the fermionic variables. Under these variations, the Lagrangian changes by a total time derivative: $\delta L = d/dt(i\alpha\dot{\phi}_i\psi_i/2)$ and we therefore have a supersymmetric theory. The

conserved Noether charge Q corresponding to the symmetry is proportional to $\dot{\phi}_i\psi_i$. The angular momentum

$$J_i = \varepsilon_{ijk}\phi_j\dot{\phi}_k + S_i \qquad \text{with} \qquad S_i = -\frac{i}{2}\,\varepsilon_{ijk}\psi_j\psi_k$$

is also conserved owing to the invariance of L under the rotations

$$\delta_\omega\phi_i = \varepsilon_{ijk}\omega_j\phi_k \qquad\qquad \delta_\omega\psi_i = \varepsilon_{ijk}\omega_j\psi_k\ .$$

Quantization is achieved by turning the dynamical variables into operators and by demanding that these operators satisfy

$$[\phi_i\,,\,p_j] = i\hbar\delta_{ij} \qquad\qquad \{\psi_i\,,\,\psi_j\} = \hbar\delta_{ij}\ .$$

These canonical relations are realized by setting

$$p_i = -\,i\hbar\frac{\partial}{\partial\phi_i} \qquad , \qquad \psi_i = \sqrt{\frac{\hbar}{2}}\,\sigma_i$$

with σ_i the Pauli matrices. The quantum Hamiltonian and angular momentum operators are then given by

$$H = -\frac{\hbar^2}{2}\vec{\nabla}^2_\phi \qquad , \qquad J = -i\hbar\vec{\phi}\times\vec{\nabla}_\phi + \hbar\frac{\vec{\sigma}}{2}$$

and are readily seen to correspond to those of a free non-relativistic spin-1/2 particle. In retrospect, we now see why L with its anticommuting variables, provides a "classical" description of such a particle. In the quantum theory, the supersymmetry generator becomes

$$Q = i\sqrt{\frac{\hbar}{2}}\,\sigma_i\,\nabla_i\ .$$

We then observe that $H = Q^2$. The fact that the Hamiltonian can be written as a quadratic expression in the supersymmetry generators is characteristic of supersymmetric quantum-mechanical systems. It should not come as a surprise in view of the fact that a time translation results if the supersymmetry transformations are performed twice: $\delta_\alpha(\delta_\beta\phi_i) = -i\alpha\beta\dot{\phi}_i$, $\delta_\alpha(\delta_\beta\psi_i) = -i\alpha\beta\dot{\psi}_i$.

Our simple example can also be used to illustrate the main aspects of the superspace formalism. The idea is to enlarge space-time by the addition of (a finite number of) odd "coordinates". In the present case we simply adjoin to t the odd variable θ (with $\theta^2 = 0$). Superfields are then functions of superspace defined through their (necessarily finite) power expansions in the odd variables. For our supersymmetric point **particle, the** appropriate supervariables are

$$Z_i(t, \theta) = \phi_i(t) + i\theta\psi_i(t)$$

It is practical to introduce the so-called super-derivative

$$\mathcal{D} = \frac{\partial}{\partial\theta} - i\theta\frac{\partial}{\partial t} \qquad \text{with} \qquad \mathcal{D}^2 = -i\frac{\partial}{\partial t} \; .$$

Recalling the rules of Berezin integration: $\int d\Theta = 0, \int d\Theta\, \Theta = 1$, one can show that the action $I = \int dt\, L$ can be rewritten as

$$I = \int dt\, d\theta\, (-\frac{1}{2}\, \mathcal{D}^2 Z_i\, \mathcal{D} Z_i) \; .$$

Further, with $Q = \partial/\partial\Theta + i\Theta\partial/\partial t$ one observes that $\delta_\alpha Z = \alpha QZ$. The reader will certainly appreciate the notational economy of this formalism and its potential usefulness.

We have illustrated so far the main aspects of supersymmetry with the help of an extremely simple example namely the free spin-1/2 particle. Interactions can obviously be included and many more systems of greater complexity exhibit supersymmetry. The first lesson to be learned is that quantum mechanics represents a very nice framework for the pedagogical illustration of supersymmetry. It should be stressed however that the uselfulness of supersymmetric quantum mechanics does not stop there. We would like to point out that superalgebras may play an important rôle in the study of multicomponent (Schrödinger) equations.

Consider for instance the Schrödinger-Pauli equation for a free particle in one space dimension. We shall now denote by x the space coordinate. Through a procedure which is by now standard, one finds the symmetries. Let us put a grading on the vector space that these span and let us therefore group them as follows:

Grading 0 (bosonic)

$$H = i\, \partial/\partial t$$
$$D = tH + (i/2)(x\, \partial/\partial x + 1/2)$$
$$K = -t^2 H + 2tD + x^2/2$$
$$P = -i\, \partial/\partial x$$
$$G = -tP + x$$
$$Y = (1/4)\, \sigma_3$$
$$I = 1$$

Grading 1 (fermionic)

$$Q_a = (1/\sqrt{2})\,\sigma_a P$$
$$S_a = (1/\sqrt{2})\,\sigma_a G$$
$$T_a = (1/\sqrt{2})\,\sigma_a \qquad a = 1, 2.$$

It is easy to see that these generators close under the graded Lie-bracket. (The reader is undoubtely aware of the fact that superalgebras involve both commutators and anticommutators.) The structure relations are those of the so-called Schrödinger superalgebra:

$$[D, H] = -iH \qquad [K, H] = -2iD \qquad [D, K] = iK$$

$$[P, H] = 0 \qquad [G, H] = iP$$
$$[D, G] = G/2 \qquad [D, P] = -(i/2)P$$
$$[G, K] = 0 \qquad [K, P] = iK$$

$$[G, P] = i\,I$$

$$[Y, H] = [Y, D] = [Y, K] = [Y, G] = [Y, P] = 0$$

$$\{Q_a, Q_b\} = 2\delta_{ab} H \qquad \{T_a, T_b\} = \delta_{ab} I$$
$$\{S_a, S_b\} = 2\delta_{ab} K \qquad \{T_a, Q_b\} = \delta_{ab} P$$
$$\{Q_a, S_b\} = -2\delta_{ab} D + 2\varepsilon_{ab} Y \qquad \{T_a, S_b\} = \delta_{ab} G$$

$$[T_a, H] = 0 \qquad [Q_a, H] = 0 \qquad [S_a, H] = iQ_a$$
$$[T_a, D] = 0 \qquad [Q_a, D] = (i/2)Q_a \qquad [S_a, D] = -(i/2)S_a$$
$$[T_a, K] = 0 \qquad [Q_a, K] = -iS_a \qquad [S_a, K] = 0$$
$$[T_a, G] = 0 \qquad [Q_a, G] = -iT_a \qquad [S_a, G] = 0$$
$$[T_a, P] = 0 \qquad [Q_a, P] = 0 \qquad [S_a, P] = iT_a$$
$$[T_a, Y] = -(i/2)\varepsilon_{ab}T_b \qquad [Q_a, Y] = -(i/2)\varepsilon_{ab}Q_b \qquad [S_a, Y] = -(i/2)\varepsilon_{ab}T_b$$

$$\varepsilon_{12} = -\varepsilon_{21} = 1$$

This algebra possesses a number of important subalgebras. One can check that the set $\{H, D, K, Y, Q_1, Q_2, S_1, S_2, \}$ closes upon itself under the graded Lie product. This subalgebra is identified as OSp(2,1). The notation comes from the fact that the bosonic subalgebra of OSp(2,1) is O(2) ⊕ Sp(1). (We denote by Sp(n) the symplectic algebra of rank n.) Note also that {H, D, K, Q_a, S_a with a = 1 or 2} generate OSp(1,1) subalgebras.

The invariance algebra of the free Schrödinger-Pauli equation is thus a superalgebra. We might therefore suspect that this will still be the case in other more complicated instances where spin degrees of freedom come into play. This possibility has been thoroughly investigated in recent years. We have found that there are many systems which exhibit dynamical supersymmetry. When this is so, we know that the space of solutions supports representations of the corresponding superalgebra. By constructing these representations, it proves in general possible to obtain an algebraic determination of the spectrum and wave functions.

Since a review of our results is soon to be available(1), we shall here simply list our findings and refer the reader to the survey article quoted above or the original literature for the details.

1. We have shown that a non-relativistic spin-1/2 particle in the field of a Dirac magnetic monopole possesses an OSp(1,1) dynamical superalgebra(2) and provided an analysis in the superspace formalism(3).

2. We further observed that the above system can be generalized to accomodate a $1/r^2$ - potential and pointed out the presence of an OSp(2,1) symmetry in such instances(4).

3. We demonstrated that a supermultiplet of two spin 0 and one spin-1/2 particles in a Coulomb potential admit an OSp(2,1) dynamical algebra(5,6).

4. We have obtained the constants of motion responsible for the spectrum degeneracy of a non-relativistic spin-1/2 particle in the field of a dyon(7,8). We have shown that two vector operators are conserved in addition to the angular momentum: one generalizes the Runge-Lenz vector while the other has no counter part for spinless systems. These operators are part of the U(2/2)⊕SU(2)⊕SU(2) invariance superalgebra of tho spin-1/2 in dyon fields. (The U(2/2) superalgebra can be defined as the dynamical algebra of the 2-dimensional supersymmetric harmonic oscillator

$$H = \sum_{i=1}^{2} (a_i^\dagger a_i + b_i^\dagger b_i)$$

with $[a_i, a^\dagger_j] = \{b_i, b^\dagger_j\} = \delta_{ij}$.

5. **We have** provided an algebraic treatment of the supersymmetric quantum mechanics of particles in one-dimensional Morse and Pöschl-Teller potentials[9].

Acknowledgements

I wish to thank D. Levi, E. Posada, G. Violini and P. Winternitz for giving me the opportunity to present this material and the Centro Internacional de Fisica, Universidad Nacional de Columbia for their marvellous hospitality. I am grateful to M. Mayrand for his help in the preparation of the manuscript. This work has been supported in part through funds provided by the Natural Science and Engineering Research Council (NSERC) of Canada and the Fonds FCAR of the Quebec Ministry of Education.

References

1. E. D'Hoker, V.A. Kostelecky and L. Vinet in "Dynamical Groups and Spectrum Generating Algebras" edited by A. Barut, A. Bohm and Y. Ne'eman, World Scientific (to appear).
2. E. D'Hoker and L. Vinet, Phys. Lett. 137B, 72 (1984).
3. E. D'Hoker and L. Vinet, Lett. Math. Phys. 8, 439 (1984).
4. E. D'Hoker and L. Vinet, Comm. Math. Phys. 97, 391 (1985).
5. E. D'Hoker and L. Vinet, Nucl. Phys. B260, 79 (1985)
6. L. Benoit and L. Vinet, LPNUM preprint (1988).
7. E. D'Hoker and L. Vinet, Phys. Rev. Lett. 55, 1043 (1985).
8. E. D'Hoker and L. Vinet, Lett. Math. Phys. 12, 71 (1986).
9. B. Piette and L. Vinet, Phys. Lett. A125, 380 (1987).

SYMMETRIES AND BERRY CONNECTIONS

Luc Vinet

Laboratoire de Physique Nucléaire
Université de Montréal
C.P. 6128, Succ. "A"
Montréal, Québec,
H3C 3J7 Canada

CONTENTS

INTRODUCTION

The adiabatic and Born-Oppenheimer approximations are both methods for obtaining approximate solutions of the Schrödinger equation. Involved in these approximation is the rate of change in time of the Hamiltonian. The adiabatic approximation applies when the Hamiltonian changes very slowly with time. In such instances, solutions to the time dependent equation can be approximated by means of stationary energy eigenfunctions of the instantaneous Hamiltonian. One then relies on the adiabatic theorem which says that a system initially in a state described by a particular eigenfunction will be found at a later time in a state described by the corresponding eigenfunction. The Born-Oppenheimer approximation which is closely related to the adiabatic approximation arises in problems where there are two different time scales. The typical example is the molecule. Systems that lend themselves to this approximation scheme therefore possess fast and slow variables. For the molecule these are respectively the electronic and nuclear variables. In the Born-Oppenheimer approximation, instead of treating the slow variables as adiabatically evolving in a prescribed way, one quantizes these after the dynamics of the fast variables has been resolved.

These methods are very standard and usually covered in undergraduate Quantum Mechanics courses [(1)]. It was therefore a surprise when a few years ago, Berry [(2)] made new observations which amended the textbook treatment of these topics. In particular, it is now appreciated that these adiabatic descriptions often require the presenee of external gauge potentials which we shall refer to as Berry connections. These connections will be the subject of this course.

We shall first review Berry's analysis in the context of both the adiabatic and the Born-Oppenheimer approximation. We shall then discuss a few standard quantum-mechanical examples and also briefly indicate how Berry connections manifest themselves in Quantum Field Theory through the anomaly phenomenon. Subsequently, we will focus on the symmetry properties of the Berry connections and explain how the theory of invariant Yang-Mills potentials can advantageously be used to compute these connections. We will conclude by discussing in this light another example,namely the so-called generalized harmonic oscillator.

1. Review of Berry's analysis [3].

A. The adiabatic approximation

Let $H(\vec{\mathcal{R}}(t))$ be a Hamiltonian depending on parameter $\vec{\mathcal{R}} = (X, Y, Z, ...)$ which vary slowly. The state $|t>$ evolves according to

$$i\hbar \frac{\partial}{\partial t} |t> = H(\vec{\mathcal{R}}(t))\, |t> . \tag{1.1}$$

At any instant, for $\vec{\mathcal{R}} = \vec{\mathcal{R}}(t)$, the instantaneous eigenstates satisfy

$$H(\vec{\mathcal{R}})\, |n, \vec{\mathcal{R}}> = E_n(\vec{\mathcal{R}})\, |n, \vec{\mathcal{R}}> \tag{1.2}$$

$$<n, \vec{\mathcal{R}}\, |\, n, \vec{\mathcal{R}}> = 1$$

We will consider first the case where $|n, \vec{\mathcal{R}}>$ is non-degenerate. The adiabatic theorem states that a system prepared at $t = 0$ in one of these states, e.g. $|n, \vec{\mathcal{R}}(0)>$, will be in the state $|n, \vec{\mathcal{R}}(t)>$ at t. It follows that an approximate solution to the time dependent Schrödinger equation (1.1) will be given by

$$|t> = e^{-\frac{i}{\hbar}\int_0^t dt'\, E_n(\vec{\mathcal{R}}(t'))}\; e^{i\gamma_n(t)}\; |n, \vec{\mathcal{R}}(t)> \tag{1.3}$$

Since physical states correspond to rays in a Hilbert space, we have introduced an extra phase in addition to the standard dynamical factor. Substituting in (1.1), we find for γ the equation:

$$\dot{\gamma}_n(t) = i <n, \vec{\mathcal{R}}\, |\, (\frac{\partial}{\partial t} |n, \vec{\mathcal{R}}>)$$

$$= i\, \vec{A}(\vec{\mathcal{R}}) \cdot \dot{\vec{\mathcal{R}}} \tag{1.4}$$

where

$$\vec{A}(\vec{\mathcal{R}}) = <n, \vec{\mathcal{R}}\, |\, \vec{\nabla}\, |\, n, \vec{\mathcal{R}}> \tag{1.5}$$

and the dot stands for differentiation with respect to time. Note that γ_n is real since $|n, \vec{\mathcal{R}}>$ is normalized. We see that $\vec{A}$ is some kind of vector potential on parameter space. The notation brings out the fact that there is a "gauge" freedom. Indeed, if the phases of of the instantaneous states are changed to

$$|\, n, \vec{\mathcal{R}} > \;\rightarrow\; e^{i\,\theta(\vec{\mathcal{R}})}\, |\, n, \vec{\mathcal{R}} > \tag{1.6a}$$

$\vec{A}(\vec{\mathcal{R}})$ is transformed into

$$\vec{A}(\vec{\mathcal{R}}) \;\rightarrow\; \vec{A}(\vec{\mathcal{R}}) + i\,\vec{\nabla}\theta(\vec{\mathcal{R}}) \tag{1.6b}$$

In textbook treatments (see for instance Ref.1) the phase $e^{i\gamma_n}$ is usually omitted on the basis that it can be absorbed into the instantaneous eigenstates $|\, n, \vec{\mathcal{R}} >$. In other words, it is argued that a smooth choise of "gauge" for the $|\, n, \vec{\mathcal{R}} >$ can be made so that $\vec{A}(\vec{\mathcal{R}}) = 0$. Berry's main contribution was to observe that this cannot always be done globally and that there are systems for which we cannot dispense with the vector potential $\vec{A}(\vec{\mathcal{R}})$. In this respect, consider a situation where the time evolution is periodic so that the parameters return to their original values $\vec{\mathcal{R}}(0)$ after some large time T, i.e. $\vec{\mathcal{R}}(T) = \vec{\mathcal{R}}(0)$. According to (1.4), the accumulated phase over the cycle referred to as Berry's phase, is given by

$$\gamma_n = \int_0^T dt\, \dot{\gamma}_n(t) = \int_0^T dt\, \dot{\vec{\mathcal{R}}}(t) \cdot \vec{A}(\vec{\mathcal{R}}(t))$$

$$= i \oint_{C=\partial S} d\vec{\mathcal{R}} \cdot \vec{A}(\vec{\mathcal{R}}) \tag{1.7}$$

Converting the above line integral into a surface integral through Stokes theorem, one has

$$\gamma_n = i \int_S d\vec{S} \cdot \vec{F} \tag{1.8}$$

with

$$\vec{F}(\vec{\mathcal{R}}) = \vec{\nabla} \times \vec{A}(\vec{\mathcal{R}})\,. \tag{1.9}$$

This is an invariant object which is unaffected by choices of phases of the wavefunctions. As we shall see through some examples, there are instances where Berry's phase does not vanish and where non-trivial holonomy occurs. We shall call $\vec{A}$ the Berry connection and F the associated curvature.

B. The Born-Oppenheimer approximation

In the Born-Oppenheimer approximation, rather then treating the parameters $\vec{\mathcal{R}}$ adiabatically, one quantizes them. Suppose for illustration that the kinetic energy for the slow variables is of the form $P^2/2M$, the full Hamiltonian $\mathcal{H}$ is then given by

$$\mathcal{H} = \frac{\vec{P}^2}{2M} + H(\vec{\mathcal{R}}(t)) \tag{1.10}$$

One now seeks eigenfunctions of $\mathcal{H}$ in the form

$$\Psi(\vec{\mathcal{R}}) = \psi(\vec{\mathcal{R}})\, |\, n, \vec{\mathcal{R}} > \ . \tag{1.11}$$

By projecting the complete eigenvalue equation

$$\mathcal{H}\Psi = E\Psi \tag{1.12}$$

on $|\, n, \vec{\mathcal{R}}>$ and neglecting the off-diagonal elements, one obtains for $\psi(\vec{\mathcal{R}})$ the following equation where $|n, \vec{\mathcal{R}}>$ is non-degenerate:

$$\left[\frac{1}{2M}(\vec{P} + i\,\vec{A})^2 + V(\vec{\mathcal{R}})\right]\psi(\vec{\mathcal{R}}) = E\,\psi(\vec{\mathcal{R}}) \tag{1.13}$$

with

$$V(\vec{\mathcal{R}}) = E_n(\vec{\mathcal{R}}) + \frac{1}{2M}\sum_{n' \neq n} < \nabla n\,,\, \mathcal{R}\,|\, n', \mathcal{R} > < n', \mathcal{R}\,|\, \nabla n\,,\, \mathcal{R} > \tag{1.14}$$

The effective Born-Oppenheimer potential consists of the fast energy eigenvalue while the kinetic term is analogous to that of a particle in a magnetic field. This equation is again derived in many textbooks but typically one assumes that the phases of the states $|\, n, \vec{\mathcal{R}}>$ can be chosen so that these wavefunctions are real in which case $\vec{A} = 0$. From Berry's work we know that it might be impossible to remove the phases in this fashion and that we should stick with equation (1.13).

C. Degenerate cases - Yang-Mills potentials

So far, we have only discussed situations where $|\, n, \vec{\mathcal{R}} >$ is non-degenerate. Suppose now that there are degeneracies and let the index a label the degenerate instantaneous states of a certain eigensubspace of $H(\vec{\mathcal{R}})$: $| n, \vec{\mathcal{R}}, a >$, a = 1, 2 .... . The results in this case are similar to the ones presented above except that the connection is now a matrix [4]:

$$A_{ab} = < n, \vec{\mathcal{R}}, a | \vec{\nabla} | n, \vec{\mathcal{R}}, b >. \tag{1.15}$$

The phase freedom is replaced by unitary transformations

$$| n, \vec{\mathcal{R}}, a > \rightarrow U_{a'a} \; | n, \vec{\mathcal{R}}, a' > \tag{1.16}$$

under which $\vec{A}$ transforms as

$$\vec{A}_{ab} \rightarrow U^{-1}_{aa'} \vec{A}_{a'b'} U_{b'b} + U^{-1}_{aa'} \vec{\nabla} U_{a'b} \tag{1.17}$$

(Summation over repeated indices is understood.) Finally, the curvature acquires an additional term beyond the curl:

$$F_{ab} = \vec{\nabla} \times \vec{A}_{ab} + (\vec{A} \times \vec{A})_{ab} \tag{1.18}$$

Let us now examine a few examples where this gauge structure will appear.

2. Examples

We want to evaluate the Berry phase

$$\gamma_n = i \oint_{C = \partial S} d\vec{\mathcal{R}} \cdot < n | \vec{\nabla} n > \tag{2.1}$$

in a simple case where it is non-trivial. Before we do so, let us obtain following Berry [2], an expression for γ_n that is more tractable than (2.1). Indeed, a direct evaluation of the above line integral requires locally single-valued $| n >$ and can be awkward. It is easier to use the surface integral

$$\gamma_n = \int_S d\vec{S} \cdot i \vec{F} \tag{2.2}$$

where

$$i \vec{F} = - \mathrm{Im} \; \vec{\nabla} \times < n | \vec{\nabla} n > \tag{2.3}$$

Inserting a complete set of states, one has

$$i\vec{F} = -\operatorname{Im} < \vec{\nabla} n \,|\times|\, \vec{\nabla} n >$$

$$= -\operatorname{Im} \sum_{m \neq n} < \vec{\nabla} n \,|\, m > \times < m \,|\, \vec{\nabla} n > \qquad (2.4)$$

The exclusion in the summation is justified by the fact that $< n \,|\, \vec{\nabla} n >$ is imaginary. From $H(\vec{\mathcal{R}})\,|\, n > = E_n(\vec{\mathcal{R}})|\, n >$ one obtains for the off-diagonal matrix elements:

$$< m \,|\, \vec{\nabla} n > = \frac{1}{E_n - E_m} < m \,|\, \vec{\nabla} H \,|\, n > \qquad (2.5)$$

Substituting in (2.4), we can finally express iF as

$$i\vec{F} = -\operatorname{Im} \sum_{m \neq n} \frac{< n \,|\, \vec{\nabla} H \,|\, m > \times < m \,|\, \vec{\nabla} H \,|\, n >}{(E_n - E_m)^2} \qquad (2.6)$$

A. Berry's paradigm example

Let us here turn to specific problems. The paradigm example has been given by Berry (2). Take

$$H = \frac{1}{2} \vec{\sigma} \cdot \vec{\mathcal{R}} \qquad (2.7)$$

with $\vec{\mathcal{R}} = (X, Y, Z)$ and $\vec{\sigma} = (\sigma_x, \sigma_y, \sigma_z)$ the standard Pauli matrices. Since $H^2 = 1/4|\vec{\mathcal{R}}|^2$, the eigenvalues of H are simply $E_\pm(\vec{\mathcal{R}}) = \pm\, \mathcal{R}/2$ $(\mathcal{R} = |\vec{\mathcal{R}}|)$. In this case, the quantum numbers n only take two values which we shall denote by $+$ and $-$. Note that $\vec{\nabla} H = \vec{\sigma}/2$. It is convenient to first rotate the axes so that the Z-axis points along $\vec{\mathcal{R}}$; we get $H = \mathcal{R}\sigma_z/2$. The eigenvectors $|\pm>$ satisfy

$$\sigma_z |\pm> = \pm |\pm>, \sigma_x |\pm> = |\mp>, \sigma_y |\pm> = \pm i |\mp> \qquad (2.8)$$

We shall consider the adiabatic evolution of the state $|+>$. With respect to the rotated axes, we get from (2.6):

$$i F_x = -\operatorname{Im} \frac{<+|\sigma_y|-><-|\sigma_z|+>}{2\mathcal{R}^2} = 0 \qquad (2.9a)$$

$$i F_y = -\,\mathrm{Im}\,\frac{\langle + | \sigma_z | - \rangle \langle - | \sigma_x | + \rangle}{2\mathcal{R}^2} = 0 \qquad (2.9b)$$

$$i F_z = -\,\mathrm{Im}\,\frac{\langle + | \sigma_x | - \rangle \langle - | \sigma_y | + \rangle}{2\mathcal{R}^2} = -\frac{1}{2\mathcal{R}^2} \qquad (2.9c)$$

Returning to unrotated axes, we obtain

$$i\vec{F} = -\frac{\vec{\mathcal{R}}}{2\mathcal{R}^3} \qquad (2.10)$$

The Berry phase for this problem is thus given by

$$\gamma_+ = -\frac{1}{2}\int_S d\vec{S}\cdot\frac{\vec{\mathcal{R}}}{\mathcal{R}^3} \qquad (2.11)$$

It is in general non-zero and equal to the flux through $C = \partial S$ of the magnetic field of a monopole with strength -1/2 ! We shall see later that the computations that led to this result can sometimes be short-circuited by invoking the theory of invariant gauge potentials. It is instructive however to see first Berry phases are determined without recourse to these more powerful techniques.

The case of a particle with spin S interacting with a varying magnetic field $\vec{B}$ is completely similar (2) to the example just treated. Here the Hamiltonian is (up to multiplicative constants):

$$H = \vec{S}\cdot\vec{B} \qquad (2.12)$$

and the energy eigenvalues are

$$E_n(\vec{B}) = n\,|\vec{B}| \qquad -S \leq n \leq S \qquad (2.13)$$

The Berry curvature is found to be

$$i\vec{F}(\vec{B}) = -n\frac{\vec{B}}{B^3} \qquad (2.14)$$

and γ_n is thus equal to the flux of the magnetic field of a monopole of charge - n at the origin of magnetic field space.

B. Diatomic molecules

In a diatomic molecule, the electric field of the nuclei has axial symmetry about the axis connecting the nuclei. Hence the projection Λ on this axis of the total angular momentum of the electrons is conserved. A reflection in a plane passing through the axis of the molecule does not alter the energy but, under such a reflection the sign of the electronic orbital angular momentum about the symmetry axis changes. We already remarked that non-abelian gauge potentials might be needed in the adiabatic treatment of situations of that kind. Moody, Shapere and Wilczek [5] have shown that it is indeed the case and that the effective Born-Oppenheimer Hamiltonian for a diatom in $\Lambda = 1/2$ states contains the U(2) Yang-Mills potential [6]

$$\vec{A}(\vec{R}) = \frac{1+\kappa}{2} \frac{\vec{R}\times\vec{\sigma}}{R^2} \tag{2.15}$$

with κ a parameter.

C. Chiral Schwinger model and anomalies

In order to illustrate the universality of Berry connections and their wide applicability let us also mention an example in the context of quantum field theory where the occurence of Berry phases is related to the anomaly phenomena [7]. The electrodynamics of massless Weyl fermions in one dimension is governed by the following Hamiltonian

$$H = \int dx \left\{ \frac{1}{2} E^2 + \psi^\dagger h \psi \right\} \tag{2.16}$$

with $h = id/dx + A(x)$. This is the so-called chiral Schwinger model. Classically this theory is gauge invariant: indeed, a gauge transformation of the electromagnetic potential

$$A(x) \rightarrow A(x) + \frac{d}{dx}\theta(x) \tag{2.17}$$

can be compensated by the following transformation of the fermion field:

$$\psi(x) \rightarrow e^{i\theta(x)}\psi(x) \tag{2.18}$$

Quantizing the theory, if one treats the fermion fields as fast variables and the gauge field as slow variables, one obtains the following Berry connection with respect to the fermionic vacuum [8]:

$$\mathcal{A}(x) = \langle 0 | \frac{\delta}{\delta A(x)} | 0 \rangle = -\frac{i}{8\pi}\int dy \, \mathrm{sign}(x-y)\, A(y) \qquad (2.19)$$

Naively, we would expect the quantum theory to possess the same symmetries as its classical counterpart and to be gauge invariant. In particular, this would mean for the fermionic vacuum to be stable under the transformation (2.17). We see however that parallel transport of this state over a cycle in the space of gauge potentials consisting of adiabatic gauge transforms of the initial potential will lead because of (2.19) to non-trivial holonomy with the result that the vacuum will acquire a Berry phase and will not be invariant. The conclusion is that gauge invariance is lost at the quantum level. This is an example of anomalous symmetry, that is a classical symmetry which is destroyed by quantum effects [9].

3. Symmetry properties

In all the examples that we have considered in Section 2, the fast Hamiltonians are invariant under transformations of the dynamical (fast) variables if these are accompanied by appropriate transformations of the parameters, the slow backgrounds.

In Berry's paradigm example, the Hamiltonian $H(\vec{\mathcal{R}}) = \vec{\sigma}\cdot\vec{\mathcal{R}}/2$ clearly does not change under simultaneous rotations of $\vec{\sigma}$ and of $\vec{\mathcal{R}}$. The total Hamiltonian for a diatomic molecule is invariant under rotations of both the nuclear and the electronic variables; this remains true if the kinetic energy of the nuclei is omitted. Finally, the invariance of the Hamiltonian (2.16) under gauge transformations has already been pointed out. We therefore have, associated to these symmetries, group actions on parameter space. It is thereafter natural to investigate the transformation properties of the Berry connections under these actions. It is found that the transformed Berry connections coincide with the original one apart from a gauge transformation.

For Berry's example, the adiabatic connection is identical to the vector potential of a Dirac magnetic monopole. It is well known that the effect of a rotation on this potential can be compensated by a gauge transformation. That the same is true for the connection arising in the treatment of the diatom has been shown by Jackiw [6]. In our field theory example, if we effect the transformation $A(x) \rightarrow A(x) + d/dx\theta(x)$, the connection $\mathcal{A}(x)$ given in

(2.19) is transformed into $\mathfrak{a}(x) + \delta/\delta A(x)\,(2\int dy\, \theta(y)A(y))$ an expression which again differs from $\mathfrak{a}(x)$ by a pure gauge term.

Let us here introduce the following definition. Suppose that the connection A is an nxn matrix. Let g be a transformation of the parameters: $\vec{\mathcal{R}} \to \vec{\mathcal{R}}^{g}$ and let $g^* A(\vec{\mathcal{R}})$ be the transform of $A(\vec{\mathcal{R}})$ under g. If

$$g^* \vec{A}(\vec{\mathcal{R}}) = h^{-1}(g, \vec{\mathcal{R}})\, \vec{A}\, h(g, \vec{\mathcal{R}}) + h^{-1}(g, \vec{\mathcal{R}})\, \vec{\nabla}\, h(g, \vec{\mathcal{R}}) \qquad h \in U(n) \qquad (3.1)$$

we shall say that A is invariant under g. That this is a proper definition of invariance should be clear from the fact that two connections that only differ by a gauge transformation describe the same physical situation. In all our examples, according to this definition, the Berry connection is invariant under the action of the symmetry group. A question that immediately comes to mind at this point is the following: Is this a general fact ? Are Berry connections always invariant whenever there is a symmetry group ? We shall answer in the affirmative [10,11].

Although this result can be established generally [12], we shall here provide a proof in the following context. Denote by $\mathfrak{g}$ the Lie algebra of the transformation group G, by $\{\mathcal{X}_i\}$ an antihermitian representation of a basis of $\mathfrak{g}$ and by $\{ c_\ell, \ell = 1, \ldots \text{rand}\, \mathfrak{g} \}$ an antihermitian representation of a basis of a Cartan subalgebra of $\mathfrak{g}$. With ε_ℓ imaginary constants, form

$$H_D = \sum_\ell \varepsilon_\ell c_\ell \qquad (3.2)$$

and denote by $\mathfrak{C}(H_D)$ the centralizer of H_D in G that is, the symmetry group of H_D. On the coset space, $G/\mathfrak{C}(H_D)$ use local coordinates $\{\alpha\}$ and represent by

$$\sigma : G/\mathfrak{C}(H_D) \to G \qquad (3.3)$$

sections of $G \to G/\mathfrak{C}(H_D)$. Consider the Hamiltonian

$$H[\alpha(t)] = \sigma(\alpha(t))\, H_D\, \sigma^{-1}(\alpha(t)) \,. \qquad (3.4)$$

As no interesting effect would arise from varying the ε we shall henceforth keep these fixed and only vary the α. The parameter space is thus identified with $G/\mathfrak{C}(H_D)$.

Let $| n, a >_D$, $a = 1, \ldots d_n$ stand for the orthonormalized eigenstates of H_D. (We are allowing for a d_n-fold degeneracy of the nth level.) These states form a representation space for

$$\lambda ;\ \mathfrak{C}(H_D) \to U(d_n) \qquad (3.5)$$

A convenient choice for the instantaneous eigenstates of $H(\alpha)$ is then

$$| n, a, \alpha(t) > \; = \; \sigma(\alpha(t)) \, | n, a >_D \tag{3.6}$$

Observe now that there is a natural G-action, $\alpha \to \alpha^g$ on parameter space defined by

$$g^{-1} H(\alpha^g) \, g \; = \; H(\alpha) \tag{3.7}$$

It is identified with left-multiplication on $G/\mathfrak{C}(H_D)$. To sum up, we have an Hamiltonian that belong to the Lie algebra of a group G. It depends on a set of parameters $\{\alpha\}$ that we shall vary adiabatically. In general we might thus expect the occurence of a non-trivial Berry connection given by

$$A_{ab} \; = \; < n, \alpha, a \,|\, d \,\, | n, \alpha, b > \tag{3.8}$$

The system moreover is invariant under G with the action on the parameters given through the adjoint representation of G. The stage is now set for the analysis of the transformation properties of A_{ab} under $\alpha \to \alpha^g$. Note that in (3.8), we have written the Berry connection as a 1-form by replacing the gradient $\vec{\nabla}$ with the exterior derivative d. We shall use this coordinate-independent notation from now on.

The pull-back of Berry's connection under $\alpha \to \alpha^g$ is simply obtained by substituting $| n, \alpha >_g \equiv g \, | n, \alpha >$ for $| n, \alpha >$ in (3.8). Hence,

$$(g^*A)_{ab} \; = \; {}^g< n, \alpha, a \,|\, d \,\, | n, \alpha, b >^g \tag{3.9}$$

where

$$| n, \alpha, a >^g \; \equiv \; g \, | n, a, \alpha > \; = \; g \, \sigma(\alpha) \, | n, a >_D \; . \tag{3.10}$$

Let $h: G \times G/\mathfrak{C}(H_D) \to \mathfrak{C}(H_D)$ denote the transition function which relates σ and its image under G; that is

$$g \, \sigma(\alpha) \; = \; \sigma(\alpha^g) \, h(g, \alpha) \tag{3.11}$$

For consistency under group composition, h must satisfy the two-cocycle condition

$$h(g_1, \alpha^{g_2}) \, h(g_2, \alpha) \; = \; h(g_1 g_2, \alpha) \qquad g_1, g_2 \in G \tag{3.12}$$

From (3.9) we then have

$$(g^*A)_{ab} \; = \; {}_D< n, a \,|\, h^{-1}(g, \alpha) \, \sigma^{-1}(\alpha^g) \; d \; [\, \sigma(\alpha^g) \, h(g, \alpha)] \, | n \, , b >_D$$

$$= \lambda^{-1}(h)_{ac} \; {}_D< n, c \,|\, \sigma^{-1}(\alpha^g) \; d \; [\sigma(\alpha^g)] \, | n, d > \lambda(h)_{db} + \lambda^{-1}(h)_{ac} d \, \lambda(h)_{cb}$$

$$= \lambda^{-1}(h)_{ac}\, A_{cd}\, \lambda(h)_{db} + \lambda^{-1}(h)_{ac}\, d\, \lambda(h)_{cb} \tag{3.13}$$

This completes the proof of our assertion: whenever a symmetry group acts on parameter space, the Berry connections are invariant up to gauge transformations under this action.

4. Application of the theory of invariant gauge potentials to the evaluation of Berry connections

Whenever symmetries are identified in a given problem we know that the Berry connections will satisfy a condition like (3.13). We can then turn the reasoning around and determine these Berry connections from the requirement of invariance. This procedure has the major advantage of bypassing the necessity to compute the instantaneous eigenstates.

A few years ago, John Harnad, Stephen Shnider and myself have shown (13) how one can characterize the most general Yang-Mills potentials that are invariant (up to gauge transformations) under a given group action. I shall briefly summarize how the procedure applies to the case at hand (10).

In order to solve condition (3.13), one needs first to obtain all possible accompanying gauge transformations, i.e. all admissible $\lambda[h(g, \alpha)]$. This it turns out is tantamount to classifying the $U(d_n)$ - bundle over $G/\mathfrak{C}(H_D)$ that admit lifts of the G-left multiplication on the cosets. The local 1-forms A define connection 1-forms ω on these bundles and condition (3.13) simply entails the (strict) invariance of these ω under the lifted G-action. We then appeal to Wang's theorem (14) to obtain the invariant ω and we recover the invariant local forms A by pulling-back the invariant connection under some choice of section of the principal bundle. In order to express the result, we decompose $\mathfrak{g}$ according to

$$\mathfrak{g} = \mathfrak{c}(H_D) \oplus \mathfrak{m} \tag{4.1}$$

and let $\omega^\sigma = \sigma^{-1} d\sigma$ stand for the pullback under $\sigma(\alpha)$ of the left-invariant Maurer Cartan forms on G. Relative to the reductive decomposition (4.1) of $\mathfrak{g}$ we write

$$\omega^\sigma = \omega^\sigma_{\mathfrak{c}} + \omega^\sigma_{\mathfrak{m}} \tag{4.2}$$

The invariant (local) connections are then given by

$$A = \lambda_* \omega^{\sigma}_{\mathfrak{c}} + \Phi \cdot \omega^{\sigma}_{\mathfrak{m}} \tag{4.3}$$

where $\lambda_* : \mathfrak{c} \rightarrow \mathfrak{u}(d_n)$ is the differential of λ and $\Phi : \mathfrak{m} \rightarrow \mathfrak{u}(d_n)$ a linear map satisfying the constraints

$$\Phi[\mathcal{X}_{\mathfrak{c}}, \mathcal{X}] = [\lambda_*(\mathcal{X}_{\mathfrak{c}}), \Phi(\mathcal{X})] \qquad \forall\ \mathcal{X}_{\mathfrak{c}} \in \mathfrak{c} \tag{4.4}$$

In the specific case of Berry connections, we further have $\Phi(x) = \Psi(\pi\ x\ \pi)$ with π the projection on the λ-representation space and Ψ an application into $\mathfrak{u}(d_n)$ of the corresponding restriction of the representation of $\mathfrak{g}$ on the eigenstates of H_D.

5. Another example: the generalized harmonic oscillator

In order to illustrate the method of Section 4 for computing Berry connection we consider the one-dimensional generalized harmonic oscillator (15) with time varying parameters. The Hamiltonian is

$$H = \frac{1}{2}\left[X(t)\, q^2 + Y(t)\,(qp + pq) + Z(t)\, p^2\right] \tag{5.1}$$

We shall limit ourselves to the bound-state situation by demanding that the parameters satisfy

$$XZ - Y^2 > 0 \tag{5.2}$$

The Berry connection associated to this model is non-trivial and we shall show that it can straight forwardly be obtained from symmetry considerations.

It is well known (16) that the operators q^2, $qp + pq$ and p^2 realize the $\mathfrak{so}(2,1)$ (or $\mathfrak{sp}(1)$) algebra. Indeed, let

$$U_1 = -\frac{i}{4}(p^2 - q^2) \quad U_2 = \frac{i}{4}(qp + pq) \quad U_3 = \frac{i}{4}(p^2 + q^2) \tag{5.3}$$

it is not hard to check that the following commutation relations are satisfied :

$$[U_1, U_2] = -U_3 \quad , \quad [U_2, U_3] = U_1 \quad , \quad [U_3, U_1] = U_2 \tag{5.4}$$

Note that U_3 is the compact rotation generator. In the U-basis, the Hamiltonian reads as

$$H(z) = z_i(t)\, U_i \tag{5.5}$$

with $z_1 = -i\,(X - Z)$, $z_2 = -2\,i\,Y$, $z_3 = -i\,(X + Z)$. It obviously belongs to the class of operator that we have examined in detail in Section 3. The quantity

$$z_1^2 + z_2^2 + z_3^2 = 4\,(XZ - Y^2) \equiv -\Delta^2 > 0 \tag{5.6}$$

is invariant under the adjoint action of SO(2,1) and excluding the (positive) energy scale, the set of parameters is in correspondence with the points on one sheet of the two-sheeted hyperboloid SO(2,1)/SO(2). A convenient parametrization is given by

$$z_1 = \Delta \sin\theta \,\mathrm{sh}\,\beta\ , \quad z_2 = \Delta\cos\theta\,\mathrm{sh}\,\beta\ , \quad z_3 = \Delta\,\mathrm{ch}\,\beta \tag{5.7}$$

$$0 \leq \theta < 2\pi\ , \quad -\infty \leq \beta \leq \infty$$

The (singular) section

$$\sigma(\theta, \beta) = e^{-\theta U_3}\, e^{-\beta U_1} \tag{5.8}$$

of SO(2,1) $\rightarrow$ SO(2,1)/SO(2) maps the reference point $(0, 0, \Delta)$ into (z_1, z_2, z_3). We further observe that

$$H(\Delta, \theta, \beta) = z_i\, U_i = \sigma(\theta, \beta)\, \Delta\, U_3\, \sigma^{-1}(\theta, \beta) \tag{5.9}$$

The eigenvalues of U_3 are essentially those of an harmonic oscillator with unit frequency : $U_3 = (i/4)(p^2+q^2) = (i/2)(n+1/2)$ and from (5.9), the energy spectrum is therefore given by

$$E_n = \sqrt{XY - Y^2}\,(n + \tfrac{1}{2}) \qquad n \in \mathbf{N}\,. \tag{5.10}$$

(Positivity of the energy implies that $\Delta = -2i\sqrt{XZ - Y^2}$.)

The centralizer of $H_D = \Delta U_3$ in SO(2,1) is the 2-dimensional rotation group generated by U_3. It is represented on the eigenstates $|n\rangle_D$ of H_D by phase multiplication :

$$\lambda(e^{\phi U_3})\, |\, n \rangle_D = e^{\frac{i}{2}\phi(n + \frac{1}{2})}\, |\, n\rangle_D \tag{5.11}$$

Since $\mathfrak{so}(2,1)$ is simple, the only map $\Phi : \{U_1, U_2\} \rightarrow \mathfrak{u}(1)$ that satisfy the constraints (4.4) is $\Phi = 0$. According to formula (4.3), the Berry connection will thus be proportional to the pull-back under σ of the canonical connection on the U(1) - fibration of $SO(2,1)/SO(2)$:

$$A = \frac{i}{2}(n+\frac{1}{2})\,\sigma^{-1} d\sigma \big|_{U_3} . \tag{5.12}$$

The explicit expression for the U_3 component of the form $\sigma^{-1}d\sigma$ is easily obtained and one finds (16)

$$A = -\frac{i}{2}(n+\frac{1}{2})\,\mathrm{ch}\beta\, d\theta . \tag{5.13}$$

6. Conclusions

Let us summarize to conclude, the main points of our discussion. We have seen that gauge potentials in the form of Berry connections arise in the adiabatic treatment of quantum mechanical problems. We have stressed the universal character of the approach and its wide applicability. Whenever symmetries are present, we have shown that the Berry connections are invariant (up to gauge transformations) under the corresponding transformations. We have finally indicated how this observation allows one to use to great advantage the theory of invariant Yang-Mills potentials to compute these Berry connections.

Acknowledgements

I wish to thank D. Levi, E. Posada, G. Violini and P. Winternitz for giving me the opportunity to present this material and the Centro Internacional de Fisica, Universidad Nacional de Columbia for their marvellous hospitality. I am grateful to M. Mayrand for his help in the preparation of the manuscript. This work has been supported in part through funds provided by the Natural Science and Engineering Research Council (NSERC) of Canada and the Fonds FCAR of the Quebec Ministry of Education.

References

1. L. Schiff, "Quantum Mechanics", McGraw-Hill, New York, 1955.

2. M.V. Berry, Proc. R. Soc. A392, 45 (1984). See also B. Simon, Phys. Rev. Lett. 51, 2167 (1983).

3. Most of the material in this section is borrowed from R. Jackiw, Comments At. Mol. Phys. 21, 71 (1988).

4. F. Wilczek and A. Zee, Phys. Rev. Lett. 52, 2111 (1984).

5. J. Moody, A. Shapere and F. Wilczek, Phys. Rev. Lett. 56, 893 (1986).

6. R. Jackiw, Phys. Rev. Lett. 56, 2779 (1986).

7. P. Nelson and L. Alvarez-Gaumé, Comm. Math. Phys. 99, 103 (1985); H. Sonoda, Phys. Lett. 156B, 220 (1985), Nucl. Phys. B266, 410 (1986); A. Niemi and G. Semenoff, Phys. Rev. Lett. 55, 927 (1985); 56, 1019 (1986); A. Niemi, G. Semenoff and Y.-S. Wu, Nucl. Phys. B276, 173 (1986).

8. G. Semenoff in "Super Field" edited by H.C. Lee, V. Elias, G. Kunstatter, R.B. Mann and K.S. Vishwanathan, Plenum, New York (1987).

9. For a review see "Current Algebra and Anomalies", S. Treiman, R. Jackiw, B. Zumino and E. Witten eds., Princeton University Press / World Scientific, Princeton, N.J./ Singapore, (1985).

10. L. Vinet, Phys. Rev. D37, xxx (1988).

11. R. Jackiw, Lecture at the workshop on "Non-Integrable Phase in Dynamical Systems", University of Minnesota (1987); P. Gerbert, M.I.T. preprint (1987).

12. J. Harnad and L. Vinet in preparation.

13. J. Harnad, S. Shnider and L. Vinet , J. Math. Phys. 21, 2719 (1980); see also P. Forgács and N. Manton, Comm.Math. Phys. 72,15 (1980) and R. Jackiw, Acta Phys. Austriaca Suppl. XXII, 383 (1980).

14. H. Wang, Nagaya Math. J. 13, 1 (1958).

15. J.H. Hannay, J. Phys. A18, 221 (1985); H.V. Berry, ibid. A18, 15 (1985).

16. V. DeAlfaro, S. Fubini and G. Furlan, Nuovo Cimento A34, 569 (1976); R. Jackiw, Ann. Phys. 129, 183 (1980); M. Moshinsky and P. Winternitz, J. Math. Phys. 21, 1667 (1980).

17. SO(2,1)-invariant gauge potentials have also been discussed in E. D'Hoker et L. Vinet, Ann. Phys. 162, 413 (1985) and R. Floreanini and L. Vinet, Phys. Rev. D36, 1731 (1987).

KAC–MOODY–VIRASORO SYMMETRIES OF INTEGRABLE NONLINEAR PARTIAL DIFFERENTIAL EQUATIONS

Pavel Winternitz
Centre de recherches mathématiques,
Université de Montréal,
C.P. 6128, A, Montréal, Québec, H3C 3J7
Canada

ABSTRACT

The local point symmetry groups of integrable equations in 2+1 dimensions are all infinite–dimensional and their Lie algebras have the structure of infinite dimensional subalgebras of Kac–Moody–Virasoro algebras. The case of the Kadomtsev–Petviashvili equation is treated in detail. It is shown how such symmetry groups can be used to perform symmetry reduction, to recognize equivalent equations and to generate families of equations having the same symmetry group.

1. INTRODUCTION

The purpose of this contribution is to discuss the Lie algebras of the Lie groups of local point transformations leaving invariant soliton equations in 2+1 dimensions. The points that we wish to bring out are the following ones. First of all, it has been established that the symmetry algebras of the Kadomtsev–Petviashvili question[1)] (KP), the modified KP equation[2)], the potential KP equation[3)], the cylindrical KP equation[4)], the Davey–Stewartson equation[5)], the integrable three–wave equations[6,7)] and an integrable dispersive long wave equation[8,9)] all have infinite dimensional symmetry algebras. Moreover, all these algebras have a very specific Kac–Moody–Virasoro loop algebra structure. Secondly, these symmetry algebras and the corresponding symmetry groups can be put to good use. Thus, they can be used to perform symmetry reduction and to generate large families of solutions that are invariant under one– or two–dimensional subgroups of the symmetry group. A further application of the symmetry algebras is to identify mutually equivalent equations, i.e.equations that can be transformed into each other by Lie point transformations. An example of such an equivalence is given below, namely a transformation between the KP and the cylindrical KP equations[4]. Finally, once the symmetry algebra of a given differential equation is established, it is possible to obtain an entire class of equations, having the same symmetry algebras. This will be illustrated below using the example of the symmetry group of the KP equation.

2. KAC–MOODY–VIRASORO ALGEBRAS AND THEIR SUBALGEBRAS

A Kac–Moody–Virasoro algebra is an infinite dimensional Lie algebra with a basis

$$\{L_m, T_m^a, C, K\}, \quad 1 \le a \le N, \quad m \in \mathbb{Z} \tag{1}$$

(N is a positive integer), satisfying the commutation relations

$$[L_m, L_n] = (m-n)L_{m+n} + \frac{1}{2}m(m^2-1)\delta_{m,-n}C \tag{2a}$$

$$[T_m^a, T_n^b] = f^{abc}T_{m+n}^c + m\delta^{ab}\delta_{m,-n}K \tag{2b}$$

$$[L_m, T_n^a] = -nT_{m+n}^a \tag{2c}$$

$$[C, L_m] = [C, T_m^a] = [K, L_m] = [K, T_m^a] = [K, C] = 0. \tag{2d}$$

The real or complex constants f^{abc} are the structure constants of some finite dimensional simple Lie algebra L. The elements $\{T_m^a, K\}$ form the basis of a Kac–Moody algebra, $\{L_m, C\}$ form the Virasoro algebra; K and C are central elements, i.e. they commute with all other elements, and with each other. We see from formula (2c) that the Kac–Moody algebra is an ideal in the entire structure (1).

Recent reviews of the theory of Kac–Moody–Virasoro algebras exist[10,11,12], giving references to the original literature. Here we just recall that the Virasoro algebra $\{L_m, C\}$ was introduced into physics in the context of dual resonance models and string theory in particle physics. Kac–Moody algebras $\{T_m^a, K\}$ also figure in particle physics in current algebra and hence in quark models. The Kac–Moody–Virasoro algebras (2) now play an important role in many applications, ranging from critical phenomena in statistical mechanics, to superstring theories and the theory of completely integrable systems[10,11,12].

A simple realization of a Kac–Moody–Virasoro algebra is obtained by putting

$$L_m = -\lambda^{m+1}\frac{d}{d\lambda}, \quad T^a_m = X^a\lambda^m, \quad C=0, \quad K=0 \tag{3}$$

where $\{X^a\}$ is a basis for a finite dimensional simple Lie algebra and λ is some scalar parameter. Since the two central elements C and K are represented trivially in (3), we actually have a representation of an affine loop algebra.

We shall see below how infinite dimensional subalgebras of Kac–Moody–Virasoro algebras occur as symmetry algebras of soliton equations in 2+1 dimensions.

Methods exist for classifying subalgebras of finite dimensional Lie algebras into conjugacy classes under the action of the group of inner automorphisms (see e.g. Ref. 13,14). These methods can to a large degree also be applied to infinite dimensional Lie algebras. Since the Kac–Moody–Virasoro algebras (2) have a semidirect product structure we can distinguish two types of subalgebras, "splitting" and "nonsplitting" ones. For splitting subalgebras there exist bases such that every basis element lies either entirely in the ideal (i.e. the Kac–Moody algebra), or entirely in the factor–algebra (i.e. the Virasoro algebra). For nonsplitting subalgebras there exist, in any basis, elements with nonzero projections onto both the ideal and the factor–algebra, such that neither part can be transformed away by inner automorphisms.

The symmetry algebras that arise in the study of soliton equations typically have bases that in the realization (3) can be written as

$$\begin{aligned} V_n &= L_n + \sum_{j=0}\sum_a \tau_{pa} T^a_{n-j} \\ Q^P_n &= \sum_{j=0}^{P}\sum_a \rho_a T^a_{n-j} \end{aligned} \tag{4}$$

where τ_{pa} and ρ_{pa} are real numbers, such that (4) is indeed a subalgebra and satisfies the commutation relations

$$\begin{aligned} [V_m, V_n] &= (m-n)V_{m+n} \\ [Q,Q] \subset Q, &\quad [V,Q] \subseteq Q. \end{aligned} \tag{5}$$

We have denoted $V \equiv \{V_n\}$, $Q = \{Q^P_n\}$. If $\tau_{pa} = 0$ for all p and a (or if all τ_{pa} can be transformed into zero by inner automorphisms of the algebra), then we have a splitting subalgebra; otherwise it is nonsplitting. The elements V_n thus generate an algebra isomorphic to a centerless Virasoro algebra. The elements Q_n^P form a subalgebra of a centerless Kac–Moody algebra. As we shall see below this subalgebra can be interpreted as a loop algebra, based on a solvable or a nilpotent Lie algebra, rather than a simple one.

More specifically, let L_0 be a solvable finite–dimensional Lie alegbra. It can, according to Ado's theorem [15], always be realized as a subalgebra of $s\ell(n, \mathbb{C})$ (or $s\ell(n, \mathbb{R})$) for some sufficiently large n. Moreover, since L_0 is solvable, it can (at least over the field of complex numbers) be represented by upper triangular matrices. We can now introduce a grading on L_0 and choose a basis respecting this grading, namely one in which each basis element has a degree, corresponding to its distance from the main diagonal of the matrix. We denote this basis of L_0 $\{Y^{pq}\}$ with $0 \le p \le n-1$ (for $L_0 \subset s\ell(n, \mathbb{R})$ denoting the degree in the grading and q enumerating different elements having the same degree. We attribute the degree n to the monomial t^n and construct a basis for the considered subalgebra, in which each basis element has a well defined degree in the grading, namely

$$\begin{aligned} V_n^{(\tau)} &= -\lambda^{n+1}\partial_\lambda + \sum_{p=0}^{n_0} \sum_q \tau_{pq} Y^{pq} \lambda^{n-p}, \qquad 0 \le n_0 \le n \\ Q_n^{(\rho)} &= \sum_{p=0}^{n_0} \sum_q \rho_{pq} Y^{pq} \lambda^{n-p}. \end{aligned} \tag{6}$$

The set $\{V_n(\tau), Q_n(\rho), n \in \mathbb{Z}\}$ will, under certain conditions imposed on the constants τ_{pq} and ρ_{pq}, form an infinite dimensional subalgebra of a centerless Kac–Moody–Virasoro algebra.

The symmetry algebras of all known integrable nonlinear evolution equations in 2+1 dimensions have precisely this structure. We shall demonstrate this below, using the example of the KP equation.

3. LIE POINT SYMMETRIES OF DIFFERENTIAL EQUATIONS

We now turn to the question of finding the symmetry algebra of a given system of differential equations. More precisely, we are interested in finding the Lie algebra of the Lie group of local point transformations leaving the equation, or system of equations, invariant.

Thus, consider a system of N differential equations for m unknown functions $u = (u_1,...,u_m)$ of n independent variables $x = (x_1,...,x_n)$

$$\Sigma^k(x,u,u^{(1)},\ldots,u^{(M)}) = 0, \quad 1 \le k \le N \tag{7}$$

where $u^{(j)}$ denotes all partial derivatives of order j. A group of local point transformations

$$\tilde{x} = \Omega_g(x,u), \quad \tilde{u} = \Lambda_g(x,u) \tag{8}$$

will be called the symmetry group of equation (7) if

$$\tilde{u}(\tilde{x}) = g \circ u(\tilde{x}) \tag{9}$$

is a solution of (7) for $g \in G$ such that $g \circ u$ is defined and for all $u(x)$ that are solutions of (7).

The symmetry group G will transform solutions of the considered equations amongst each other, without necessarily leaving any individual solutions invariant. The fact that the functions Ω_g and Λ_g in (8) depend only on x and u and not on derivatives, such as $u^{(1)}, u^{(2)},\ldots,$ is an important restriction. If we allow first derivatives to be present, we obtain contact transformations, rather than only point transformations. The presence of higher derivatives would lead to so called generalized symmetries[16], also called Lie–Bäcklund symmetries[17]. Here we restrict ourselves to Lie point symmetries, as in (8), (9).

Algorithms exist for finding the symmetry group of a given system of equations. They go back to the origins of Lie group theory and are described in various contemporary books[16,18]. They provide the Lie algebra L of vector fields, corresponding to the Lie group G. The vector fields have the form

$$V = \sum_{i=1}^{n} \eta^i(x,u)\frac{\partial}{\partial x_i} + \sum_{a=1}^{m} \phi^a(x,u)\frac{\partial}{\partial u_a}, \tag{10}$$

where the fact that we are considering point transformations is reflected in the fact that η^i and ϕ^a do not depend on derivatives of u. The coefficients $\eta^i(x,u)$ and $\phi^a(x,u)$ in (10) are determined from the condition

$$\mathrm{pr}^{(M)}V \bullet \Sigma^k(x,u,u^{(1)},\ldots,u^{(M)})\Big|_{\Sigma^j=0} = 0 \qquad \begin{matrix} k=1,\ldots,N \\ j=1,\ldots,N \end{matrix} \tag{11}$$

where $\mathrm{pr}^{(M)}V$ is the M–th prolongation of V. Condition (11) states that the prolonged vector fields must annihilate the equation on the solution set of the equation.

The procedure is entirely algorithmic and has been implemented as REDUCE[19,20] and MACSYMA[2] packages. The packages calculate the M–th prolongation (where M is the order of the equation), impose conditions (11) and obtain a system of first order linear partial differential equations for the coefficients in (1). These equations, called the determining equations, are solved, to a varying degree, by the packages, or are printed out and then solved by hand. Three different situations can occur. Thus, the determining equations may be incompatible, i.e. their only solution is $\eta^i = 0$, $\phi^a = 0$; the equations then have no nontrivial continuous symmetry group. The second possibility is that the solutions of the determining equations depend on a finite number K of significant integration constants. We obtain a K–dimensional symmetry algebra and symmetry group. Finally, the solutions η^i and ϕ^a may depend on some arbitrary functions of the independent and dependent variables and we obtain an infinite dimensional Lie algebra and Lie group.

As stated above, this last situation occurs for all integrable nonlinear partial differential equations. Moreover, the obtained symmetry algebras all have the Kac–Moody–Virasoro structure, described in Section 2.

We now turn to the example of the Kadomtsev–Petviashvili equation.

4. SYMMETRIES OF THE KADOMTSEV–PETVIASHVILI EQUATION

We write the KP equation in the form

$$(u_t + \frac{3}{2} u u_x + \frac{1}{4} u_{xxx})_x + \frac{3}{4}\sigma u_{yy} = 0, \quad \sigma = \pm 1 \tag{12}$$

and calculate the Lie algebra of its symmetry group, using the algorithm described above (and the program[2]). The result is that the algebra is realized by vector fields[1,20]

$$V = T(f) + Y(g) + X(h) \tag{13}$$

$$\begin{aligned} T(f) = {} & f\partial_t + [\tfrac{1}{3}x\dot{f} - \tfrac{2}{9}\sigma y^2 \ddot{f}]\partial_x + \tfrac{2}{3} y\dot{f}\partial_y \\ & - [(4\sigma/27)y^2 \dddot{f} - \tfrac{2}{9}x\ddot{f} + \tfrac{2}{3}u\dot{f}]\partial_u \end{aligned} \tag{14a}$$

$$Y(g) = g\partial_y - \frac{2}{3}\sigma y\dot{g}\partial_x - (4\sigma/9)y\ddot{g}\,\partial_u \tag{14b}$$

$$X(h) = h\partial_x + \frac{2}{3}\dot{h}\partial_u \tag{14c}$$

where f, g, and h are arbitrary $C^\infty(I)$ functions of time t (where I is some open subset of $\mathbb{R}$).

Let us now expand the functions f(t), g(t), and h(t) into formal Laurent series and write the vector fields (14) in the form (6). We first introduce the finite–dimensional solvable algebra L_0, which in this case is 8–dimensional. As a basis of L_0 we use

$$\begin{aligned} Y^{0,1} &= x\partial_x + 2y\partial_y - 2u\partial_u, \\ Y^{1,1} &= -\sigma y^2\partial_x + x\partial_u, \quad Y^{1,2} = \partial_y \\ Y^{2,1} &= y^2\partial_u, \quad Y^{2,2} = y\partial_x \\ Y^{3,1} &= y\partial_u, \quad Y^{3,2} = \partial_x, \quad Y^{4,1} = \partial_u \end{aligned} \tag{15}$$

and we have a grading

$$[Y^{p,q}, Y^{p',q'}] \subseteq Y^{p+p',q''} \quad (\text{mod } 4). \tag{16}$$

The subalgebra obtained by dropping $Y^{0,1}$ is the nilradical (maximal nilpotent ideal) of L_0. The elements $\{Y^{2,k}, Y^{3,k}$ and $Y^{4,1}\}$ form a maximal abelian subalgebra (and ideal) of L_0. We obtain a basis for the symmetry algebra of the KP equation in the form

$$\begin{aligned} T(t^n) &= t^n\partial_t + \frac{1}{3}nY^{0,1}t^{n-1} + \frac{2}{9}n(n-1)Y^{1,1}t^{n-2} \\ &\quad - (4\sigma/27)\,n(n-1)(n-2)\,Y^{2,1}t^{n-3} \\ Y(t^n) &= Y^{1,2}t^n - \frac{2}{3}\sigma nY^{2,2}t^{n-1} - (4\sigma/9)\,n(n-1)\,Y^{3,1}t^{n-2} \\ X(t^n) &= Y^{3,2}t^n + (2n/3)\,Y^{4,1}t^{n-1}. \end{aligned} \tag{17}$$

The smallest $s\ell(n, \mathbb{R})$ algebra into which we can imbed the solvable algebra L_0 of (15) is $s\ell(5, \mathbb{R})$. On the other hand, the nilpotent subalgebra $\{Y^{1,2}, Y^{2,2}, Y^{3,1}, Y^{3,2}, Y^{4,1}\}$ figuring in $Q = \{Y(t^n), X(t^n)\}$, i.e. in the Kac–Moody part of the symmetry algebra, can be imbedded into $s\ell(4, \mathbb{R})$. Thus the Lie algebra given by (17), i.e. the symmetry algebra of the KP equation, can be interpreted in two complementary manners. On one hand it is a nonsplitting subalgebra of the algebra

$$V \ni \widehat{s\ell}(5, \mathbb{R}) \tag{18}$$

where V denotes the standard realization $\{L_m\}$ of the Virasoro algebra, as in eq. (3), and $\hat{s\ell}(n, \mathbb{R})$ denotes the loop algebra, based on $s\ell(n, \mathbb{R})$. On the other hand, the algebra (17) can be viewed as a splitting subalgebra of

$$\tilde{V} \ni \hat{s\ell}(4, \mathbb{R}) \tag{19}$$

where $\tilde{V}$ denotes a realization of the Virasoro algebra by the operators $T(t^n)$ of (17).

The "physically obvious" symmetries of the KP equation are obtained by restricting the arbitrary functions f, g and h in (14) to be first order polynomials. Thus, T(1), Y(1) and X(1) are translations in t, y and x, respectively, and T(t), and X(t) are dilations and Galilei transformations in the x direction, respectively, whereas Y(t) could be called a "pseudorotation" – a mixture of a rotation and a Galilei boost in the y–direction. Indeed, integrating Y(t) we obtain the finite transformation

$$t' = t, \quad y' = y + \lambda t, \quad x' = x - \frac{2}{3}\sigma\lambda y - \frac{1}{3}\sigma\lambda^2 t, \quad u' = u \tag{20}$$

The group transformations corresponding to the vector fields (13), (14) for general f(t), g(t) and h(t) have also been obtained[1]. We shall not reproduce the general result here, but just give some simple special cases.

Thus, for $f(t) = 0$, $g(t) = 0$, $h(t) \neq 0$ we have

$$\begin{aligned} x' &= x + \lambda h(t), \quad y' = y, \quad t' = t \\ u'(t', x', y') &= u(t', x' - \lambda h(t'), y') + \frac{2}{3}\lambda \dot{h}(t') \end{aligned} \tag{21}$$

and we see that (21) is, for a general $h(t)$ with $\dot{h}(t) \neq 0$ a non–Galileian boost to a frame moving along the x axis with an arbitrary (variable) velocity $\lambda \dot{h}(t)$.

For $f(t) = 0$, $g(t) \neq 0$ the transformation is

$$\begin{aligned} x' &= x - \lambda[\frac{2}{3}\sigma\dot{g}(t)y - h(t)] - \frac{1}{3}\sigma g(t)\,\dot{g}(t)\lambda^2 \\ y' &= y + \lambda g(t), \quad t' = t \\ u'(t', x', y') &= u(t', x' + \frac{2}{3}\lambda\sigma\dot{g}(t')y' - \lambda h(t'), y' - \lambda g(t')) \\ &\quad + [\frac{2\dot{h}(t')}{3} - \frac{4\sigma}{9}\ddot{g}(t')(y' - \lambda g(t'))]\lambda - \frac{2\sigma}{9} g(t')\ddot{g}(t')\lambda^2 . \end{aligned} \tag{22}$$

The group transformations for $f(t) \neq 0$ are more complicated[1] and we refer for them to the original article. Instead, we turn to applications of the symmetry group of the KP equation.

5. SYMMETRY REDUCTION

The most standard application of the symmetry group of a partial differential equation is to perform symmetry reduction, i.e. to reduce the equation to one with fewer independent variables. This is done by looking for particular solutions that are invariant under some subgroup G_0 of the symmetry group G. Such solutions will only be compatible with specific boundary conditions: it is necessary for the boundaries to be invariant under the subgroup G_0. The symmetry group acts on the product space $X \otimes U$, where X and U are spaces of the independent and dependent variables, respectively. In general if we wish to reduce a PDE in n variables to one in k variables $(1 \leq k \leq n)$, we must impose invariance under a subgroup G_0 with generic orbits of codimension k in X–space. Usually, low–dimensional subgroups of G are of the greatest interest in this context. A systematic application of symmetry reduction requires a classification of subgroups of the symmetry group into conjugacy classes under the action of the group of inner automorphisms. Different conjugacy classes in general provide reductions to different equations.

Consider a PDE for m unknown functions $(u_1, \ldots, u_m) = u$ of n independent variables $(x_1, \ldots, x_n) = x$. Let G be its symmetry group of local point transformations. A subgroup $G_0 \subset G$ with generic orbits of codimension K in the space $X \otimes U$ will have K invariants in this space, say $I_1(x, u), \ldots, I_K(x, u)$. The invariants are usually not difficult to find explicitly. Indeed, they are obtained by solving a system of first order PDE's

$$L_i Q(x, u) = 0, \quad 1 \leq i \leq \dim L_0 \tag{23}$$

where $\{L_i\}$ is a basis for the Lie algebra L_0 of the group G_0, realized by vector fields of the form (10).

The group $G_0 \subset G$ leads to invariant solutions and to a symmetry reduction of the considered PDE, if the structure of the orbits is such that the K invariants $I_1, \ldots, I_k$ satisfy the following conditions:

1. It is possible to choose $k = K - m < n$–invariants $\xi_1, \ldots, \xi_k$, that depend only on the independent variables $x_1, \ldots, x_n$.

2. The remaining invariants $\Omega_1(x, u), \ldots, \Omega_n(x, u)$ define a locally invertible mapping $u \to \Omega$, i.e. the Jacobian

$$J = \left\{ \frac{\partial \Omega_i}{\partial u_k} \right\}, \quad \det J \neq 0 \tag{24}$$

is of rank m.

In this situation we can replace the variables $(x_1, \ldots, x_n)$ by

$$\xi_1(x_1), \ldots, \xi_k(x), \quad \eta_{k+1}(x), \ldots, \eta_n(x) \tag{25}$$

where $\eta_a(x)$ are some variables that are chosen so as to complement $\xi_1, \ldots, \xi_k$ to a (local) coordinate system for X. Moreover, we can now express the dependent variables as

$$u_i = F_i(\xi, \eta), \quad \xi = (\xi_1, \ldots, \xi_k), \quad \eta = \eta_{k+1}, \ldots, \eta_n, \quad i = 1, \ldots, m \tag{26}$$

where F_i are some functions. Substituting the expression (26) back into the original equations, we obtain a set of PDE's for the functions F_i, involving the variables ξ_i only. Indeed, invariance of the equations guarantees that the noninvariant quantities η_a drop out.

The general reduction procedure described above is equally applicable to infinite dimensional symmetry groups, as it is to finite ones. Moreover, it turns out to be particularly simple for the Kac–Moody–Virasoro type symmetries that occur for integrable PDE's.

Indeed, let us return to the KP equation and its symmetry algebra (13,14) and let us consider symmetry reduction to a PDE in two variables. This will be achieved by requiring that the solution $u(x, y, t)$ be invariant under a one–dimensional subgroup of the symmetry group. Relation (23) for one operator L_i of the form (13) then yields relations (26) in the form

$$\xi_a = \xi_a(t, x, y), \quad u(t, x, y) = \alpha(t, x, y) q(\xi_1, \xi_2) + \beta(t, x, y), \quad a = 1, 2 \tag{27}$$

where ξ_1, ξ_2, α and β are known and a PDE in the two invariant variables ξ_1 and ξ_2 is obtained for the function $q(\xi_1, \xi_2)$, i.e. the the third invariant, by substituting (27) into the KP equation. The specific form of ξ_1, ξ_2, α and β depends on the choice of the subgroup of the KP group.

All one–, two– and three–dimensional subalgebras of the KP symmetry algebra have been classified into conjugacy classes, under the adjoint action of the symmetry group of the equation[1]. The final result for one–dimensional subalgebras of the KP algebra is that there exist precisely three conjugacy classes, represented by

$$L_{1,1} = \{T(1)\}, \quad L_{1,2} = \{Y(1)\}, \quad L_{1,3} = \{X(1)\}. \tag{28}$$

A non-zero element V of the form (13) is conjugate to $T(1)$ if $f(t) \neq 0$ (in some interval of t), to $Y(1)$ if $f(t) = 0$, $g(t) \neq 0$, and to $X(1)$ if $f(t) = g(t) = 0$, $h(t) \neq 0$.

Symmetry reduction by a one-dimensional subgroup with its Lie algebra in class $L_{1,1}$ leads to the solutions

$$
\begin{aligned}
u(t,x,y) &= f^{-2/3} q(\xi,\eta) + 2\dot{f}x(9f)^{-1} \\
&\quad + 4\sigma(2g\dot{f} - 3f\dot{g})y(27f^2)^{-1} + 4\sigma(2\dot{f}^2 - 3f\ddot{f})y^2(9f)^{-2} \\
&\quad + 2\sigma g^2(3f)^{-2} + 2h(3f)^{-1} \qquad (29) \\
\xi &= [x + 2\sigma g y(3f)^{-1} + 2\sigma\dot{f}y^2(9f)^{-1}]f^{-1/3} \\
&\quad - \int_0^t [2\sigma g^2(s)3^{-1} f^{-7/3}(s) + h(s)f(s)^{-4/3}]ds \\
\eta &= y f^{-2/3} - \int_0^t g(s) f^{-5/3}(s)ds
\end{aligned}
$$

where $f(t) \neq 0$, $g(t)$ and $h(t)$ are arbitrary sufficiently smooth functions and $q(\xi,\eta)$ satisfies the Boussinesq equation

$$\sigma q_{\eta\eta} + (q^2)_{\xi\xi} + \frac{1}{3} q_{\xi\xi\xi\xi} = 0 \qquad (30)$$

(we replace ξ_1, ξ_2 of (27) by ξ, η).

Algebras of the class $L_{1,2}$ lead to the solutions

$$
\begin{aligned}
u(t,x,y) &= g^{-1/2} q(\xi,\eta) + \dot{g}x(3g)^{-1} + (2\dot{h}g - \dot{g}h)y(3g^2)^{-1} \\
&\quad + \sigma(\dot{g}^2 - 2g\ddot{g})y^2(3g)^{-2} - \sigma h^2(2g^2)^{-1} \\
\xi &= g^{-1/2}[x - hyg^{-1} + \sigma\dot{g}y^2(3g)^{-1}] \qquad (31) \\
\eta &= \int_0^t g^{-3/2}(s)ds,
\end{aligned}
$$

where q satisfies the once-differentiated Korteweg–de Vries equation

$$[q_\eta + \frac{3}{2}qq_\xi + q_{\xi\xi\xi}]_\xi = 0. \qquad (32)$$

Finally, algebras of the class $L_{1,3}$ lead to a linear equation that can be solved directly and hence to the solutions

$$u(t,x,y) = 2\dot{h}x(3h)^{-1} - 4\sigma\ddot{h}y^2(9h)^{-1} + K(t)y + L(t). \qquad (33)$$

The function $h(t)$ figures in the Lie algebra, the functions $K(t)$ and $L(t)$ come from the integration of the reduced (linear) equation.

For reductions to ordinary differential equations, using two–dimensional subgroups of the symmetry group, see Ref. 1).

6. RECOGNITION OF EQUIVALENT EQUATIONS: THE GENERALIZED CYLINDRICAL KP–EQUATION

An important application of group theory in the study of differential equations is the problem of recognizing mutually equivalent equations. For a recent systematic approach to this problem, we refer to papers by Kamran, Shadwick and collaborators[21,22].

Let us consider two different partial differential equations, both with the same number of dependent and independent variables. A necessary condition for a Lie point transformation to exist, transforming one of the equations into the other, is that their Lie point symmetry groups be locally isomorphic. This is a very useful criterion, even if it is not a sufficient one, and has been used, Kumei and Bluman[23] to determine, when a nonlinear equation is equivalent, under a Lie point transformation, to a linear one.

Here we wish to show how Kac–Moody–Virasoro symmetry algebras can be used in the same context[4]. Let us consider a recently proposed class of integrable nonlinear partial differential equations which we shall call the generalized cyclindrical Kadomtsev–Petviashvili equation (GCKP)[24]

$$[u_t + 6uu_x + u_{xxx} + \frac{1}{2t}u + (H+yG)u_x + Fu_y]_x + \frac{\alpha}{4t^2}u_{yy} = 0. \qquad (34)$$

In (34), $H(t)$, $G(t)$ and $F(t)$ are arbitrary functions of time t and α is a constant ($\alpha = -1$ or $\alpha = +1$ for waves propagating in a media with positive or negative dispersion, respectively). For

$H = G = F = 0$ we obtain the usual cylindrical KP equation[25,26,27]. Using the MACSYMA program[2] we find that the symmetry algebra of the GCKP is realized by vector fields of the form $V = \tilde{T}(f) + \tilde{Y}(g) + \tilde{X}(h)$, with

$$\begin{aligned}
\tilde{T}(f) &= f[\partial_t + \alpha[2t(F + t\dot{F}) - y]y\partial_x - yt^{-1}\partial_y \\
&\quad + 6^{-1}[(t^{-1}G - \dot{G} + 6\alpha\dot{F}t + 2\alpha t^2\ddot{F})y - \dot{H} + 2\alpha Ft(F + t\dot{F})]\partial_u\} \\
&\quad + 9^{-1}\dot{f}\{3(x + 2\alpha t^2 yF + 3\alpha ty^2)\partial_x + 6y\partial_y \\
&\quad + [-6u + (-2G + 8\alpha tF + 4\alpha t^2\dot{F})y - H + \tfrac{1}{2}t^2F]\partial_u \\
&\quad + (18)^{-1}\ddot{f}\{-12\alpha t^2y^2\partial_x + (x - \alpha ty^2 - 2\alpha t^2Fy)\partial_u\} - 9^{-1}\dddot{f}\,\alpha t^2y^2\partial_u \\
\tilde{Y}(g) &= g(\partial_y - 6^{-1}G\partial_u) - \dot{g}\alpha t3^{-1}[6ty\partial_x + (2y + tF)\partial_u] - 3^{-1}\alpha\ddot{g}t^2y\partial_u \\
\tilde{X}(h) &= h\partial_x + 6^{-1}\dot{h}\partial_u
\end{aligned} \tag{35}$$

where $f(t)$, $g(t)$ and $h(t)$ are arbitrary $C^{\infty}(I)$ functions of time t.

Expanding f, g and h into power series in t, we easily convince ourselves that (35) has the structure of a Kac–Moody–Virasoro algebra. Moreover, performing an appropriate change of basis

$$T(f) = \tilde{T}(f), \quad Y(g) = \left(\frac{\alpha}{2}\right)^{1/2}[\tilde{Y}(g/t) + 2\alpha\tilde{X}(gFt)], \quad X(h) = \tilde{X}(h) \tag{36}$$

we find that the basis elements $T(f)$, $Y(g)$ and $X(h)$ have the same commutation relations as the basis elements (14) for the KP symmetry algebra. The fact that the KP and GCKP symmetry algebras are isomorphic provides motivation to look for a point transformation, that transforms the two sets of vector fields into each other and possibly also transforms the GCKP into the KP equation. Such a transformation does indeed exist and is actually quite simple, namely

$$\begin{aligned}
u(x, y, t) &= v(\xi, \eta, \tau) - \frac{1}{6}[(G - 2t\alpha F)y + H] \\
\xi = x - \alpha ty^2, \quad \eta &= 2(\sigma\alpha)^{1/2}[ty - \int_0^t sF(s)ds], \quad \tau = t.
\end{aligned} \tag{37}$$

Thus, if v satisfies the KP equation written as

$$(v_\tau + 6vv_\xi + v_{\xi\xi\xi})_\xi + \sigma v_{\eta\eta} = 0, \quad \sigma = \pm 1 \tag{38}$$

then u satisfies the GCKP equation and vice–versa.

For the cylindrical KP equation itself (G = F = H = 0) the transformation (37) preserves the asymptotic behavior of the solutions and in particular takes solitons into solitons.

7. EQUATIONS INVARIANT UNDER A GIVEN SYMMETRY GROUP

Given a system of differential equations, we can determine its symmetry group in a unique manner, e.g. by following the algorithm described above in Section 3. The inverse is certainly not true. Given a symmetry group of local point transformations and a specific realization of its Lie algebra by vector fields, we can in general construct many systems of differential equations of the same order, invariant under this group.

The techniques for doing this are, at least conceptually, quite simple. Let us be given a Lie algebra of vector fields of the form (10) and let us construct the most general system of differential equations of some order M, invariant under the corresponding Lie group of local point transformations. All we need to do is to construct the prolongation of order M of the vector fields and to find their invariants $I_\mu(x, u, u^{(1)}, \ldots, u^{(M)})$, i.e. find a basis for the set of all invariants. The invariant equations then have the form

$$\Sigma^k(I_1, \ldots, I_k) = 0, \quad 1 \le k \le N \tag{39}$$

where Σ^k is a set of arbitrary sufficiently smooth functions.

Let us now as an example again consider the symmetry group of the KP equation and its Lie algebra (13), (14). We wish to find a general fourth order evolution equation, invariant under the KP symmetry group. We make the restriction that the equation should have the form[28]

$$\Phi(x, y, t, u, u_i, u_{ik}, u_{abc}, u_{abcd}) = 0 \tag{40}$$

where the subscripts denote partial derivatives, with $\{i, k\} = \{x, y, t\}$, $\{a, b, c, d\} = \{x, y\}$. We now impose the condition $pr^{(4)}V \bullet \Phi = 0$, for general functions $f(t)$, $g(t)$ and $h(t)$. This means that we assume that all the derivatives of these functions that figure in $pr^{(4)}V$ are independent. Since $pr^{(4)}V$ contains $f, f', \ldots, f^{(v)}$, $g, g', \ldots, g^{(iv)}$, h, h', h'', and h''', we obtain 15 first order PDE's.

Solving these equations we obtain ten functionally independent invariants which we choose to be

$$
\begin{aligned}
I_1 &= [u_{xt} + \tfrac{3}{2}(uu_x)_x + \tfrac{3}{4}\sigma u_{yy}]u_{xx}^{-3/2} \\
I_2 &= u_{xxxx}u_{xx}^{-3/2} \\
I_3 &= u_{xxx}u_{xx}^{-5/4} \\
I_4 &= (u_{xx}u_{xxy} - u_{xy}u_{xxx})u_{xx}^{-5/2} \\
I_5 &= (u_{xxx}u_{xyy} - u_{xxy}^2 - 2\sigma u_x u_{xx}u_{xxx})u_{xx}^{-3} \qquad (41) \\
I_6 &= [u_{xxx}^2 u_{yyy} + 2u_{xxy}^3 - 3u_{xxx}u_{xyy}u_{xxy} \\
&\quad + 6\sigma u_x u_{xxx}(u_{xx}u_{xxy} - u_{xy}u_{xxx})]u_{xx}^{-9/2} \\
I_7 &= (u_{xx}u_{xxxy} - u_{xy}u_{xxxx})u_{xx}^{-11/4} \\
I_8 &= (u_{xxxx}u_{xxyy} - u_{xxxy}^2 - 2\sigma u_x u_{xxx}u_{xxxx})u_{xx}^{-7/2} \\
I_9 &= [(u_{xxxx}^2 u_{xyyy} + 2u_{xxxy}^3 - 3u_{xxxx}u_{xxyy}u_{xxxy} \\
&\quad + 6\sigma u_x u_{xxxx}(u_{xxx}u_{xxxy} - u_{xxy}u_{xxxx}))]u_{xx}^{-21/4} \\
I_{10} &= [u_{xxxx}u_{yyyy} + 3u_{xxyy}^2 - 4u_{xyyy}u_{xxxy} \\
&\quad + 12u_x^2(u_{xx}u_{xxxx} + u_{xxx}^2) \\
&\quad - 12\sigma u_x(u_{xyy}u_{xxxx} - 2u_{xxy}u_{xxxy} + u_{xxx}u_{xxyy})]u_{xx}^{-4}.
\end{aligned}
$$

If we require that we have an evolution equation, i.e. that a time derivative be present, we obtain an invariant PDE that we can write as

$$I_1 = F(I_2, \ldots, I_{10}), \qquad (42)$$

where F is arbitrary.

If we add some further requirements, namely that the equation should be of first degree in u_{xt} and polynomial in u and all of its derivatives, we obtain a much more restricted equation, namely

$$u_{xt} + \frac{3}{2}(uu_x)_x + \frac{3}{4}\sigma u_{yy} + \kappa u_{xxxx} = a \qquad (43)$$

where κ and a are constant. For $a = 0$, $\kappa \neq 0$ we obtain the KP equation, for $\kappa = 0$, $a = 0$ we obtain the Zabolotskaya–Khoklov equation[29)], that can be linearized by a generalized hodograph transformation[30)].

An interesting question is whether one can find other equations of the form (42) that are integrable, in the same sense as the KP equations. For example, it would be useful to find integrable equations that have the form of a KP equation with additional terms, involving only lower derivatives than the KP equation itself. If such an equation is invariant under the KP group, then it would have the form

$$u_{xt} + \frac{3}{2}(uu_x)_x + \frac{3}{4}\sigma u_{yy} + u_{xxxx} = u_{xx}^{3/2} H(I_3, I_4, I_5, I_6) \qquad (44)$$

where H is some function. If we require that $u_{xx}^{3/2} H$ be rational in u and its derivatives, we obtain e.g.

$$H = \alpha I_3^2 + \beta I_4 + \gamma I_6 + \delta \frac{I_5}{I_3^2}, \qquad (45)$$

where α, β,γ, and δ are constants, not all equal to zero. Equation (44) does not have the Painlevé property for any nonzero values of the constants in (45).

Thus, on one hand, all presently known integrable equations in 2+1 dimensions do have Kac–Moody–Virasoro algebras as Lie algebras of local point symmetry groups. On the other hand it is possible to construct large families of equations that have the same symmetry groups, but are not integrable. The question of providing a group theoretical characterization of integrability remains open.

8. CONCLUSIONS

Using the Kadomtsev–Petviashvili equation as a vehicle, we have shown how subalgebras of Kac–Moody–Virasoro algebras occur as symmetries of soliton equations in 2+1 dimensions. These infinite dimensional symmetry algebras and the corresponding infinite dimensional symmetry groups can be used in the same way as finite dimensional symmetry groups are used to generate particular solutions, recognize equivalent equations, study symmetry breaking, etc.

Many fundamental questions in this context remain open. Among them the most important is the exact relationship between symmetries and integrability. An important mathematical problem is that of classifying, in a systematic manner, infinite dimensional subalgebras of Kac–Moody–Virasoro algebras. In the context of nonlinear equations an equally important problem is that of classifying realizations of these subalgebras in terms of vector fields. Such a classification would also provide a classification of invariant differential equations.

ACKNOWLEDGEMENTS

This article is an extended version of a lecture presented at the School on Symmetries and Nonlinear Phenomena, sponsored by the Centro Internacional de Fisica in Bogota, Colombia.

I thank my collaborators, the joint work with whom is reported here. Numerous discussions in particular with D. Levi are much appreciated. Discussions with B. Grammaticos and A. Ramani greatly helped in establishing that equations of the form (44), (45) do not have the Painlevé property.

Research support from NSERC of Canada and FCAR du Gouvernement du Québec is greatfully acknoweldged.

REFERENCES

1) David, D., Kaman, N., Levi, D. and Winternitz, P., Phys. Rev. Lett. 55, 2111 (1985), J. Math. Phys. 27, 1225 (1986).

2) Champagne, B. and Winternitz, P., preprint CRM-1278, Montreal (1985).

3) David, D., Levi, D. and Winternitz, P., Phys. Lett. A118, 390 (1986).

4) Levi, D. and Winternitz, P., Phys. Lett. A (1988) to appear.

5) Champagne, B. and Winternitz, P., J. Math. Phys. 29, 1 (1988).

6) Leo, R.A., Martina, L. and Soliani, G., J. Math. Phys. 27, 2623 (1986).

7) Martina, L. and Winternitz, P., to be published.

8) Boiti, M., Leon, J.J.P. and Pempinelli, F., Inverse Problems 3, 371 (1987).

9) Paquin, G. and Winternitz, P., to be published.

10) Kac, V.G., "Infinite Dimensional Lie Algebras" , Birkhauser, Boston (1983).

11) Goddard, P. and Olive, D., Int. J. Mod. Phys. A1, 303 (1986).

12) Jimbo, M. and Miwa, T., Publ. RIMS, Kyoto Univ. 19, 943 (1983).

13) Patera, J., Winternitz, P. and Zassenhaus, H., J. Math. Phys. 16, 1597 (1975); 16, 1615 (1975) 17; 717 (1976).

14) Patera, J., Sharp, R.T., Winternitz, P. and Zassenhaus, H., J. Math. Phys. 18, 2259 (1977).

15) Ado, I.D., Usp. Mat. Nauk. 2, 159 (1947).

16) Olver, P.J., "Applications of Lie Groups to Differential Equations", Springer, New York (1986).

17) Anderson, R.L. and Ibragimov, N.J., "Lie–Bäcklund Transformations in Applications", SIAM, Philadelphia, (1979).

18) Bluman, G.W. and Cole, J.D., "Similarity Methods for Differential Equations", Springer, New York (1974).

19) Schwarz, F., Computing 34, 91 (1985).

20) Schwarz, F., J. Phys. Soc. Jpn. 51, 2387 (1982).

21) Kamran, N. and Shadwick, W.F., in "Field Theory, Quantum Gravity and Strings", Lecture Notes in Physics 246, Springer, New York (1986).

22) Kamran, N., these Proceedings.

23) Kumei, S. and Bluman, G.W., SIAM J. Appl. Math. 42, 1157 (1982).

24) Levi, D., Ragnisco, O. and Santini, P.M., to be publisehd.

25) Johnson, R.S., J. Fluid Mech. 97, 701 (1980).

26) Dryuma, V.S., Sov. Math. Dokl. 27, 6 (1983).

27) Lipovski, V.D., Sov. Phys. Dokl. 31, 31 (1986).

28) David, D., Levi, D. and Winternitz, P., Phys. Lett. A (1988) to appear.

29) Zabolotskaya, E.A. and Khokhlov, R.V., Sov. Phys. Acoustics 15, 35 (1969).

30) Gibbons, J. and Kodama, Y., preprint DPNU–87–44, Nagoya (1987).

Nonlinearity in Aberration Optics

KURT BERNARDO WOLF

Instituto de Investigaciones en Matemáticas Aplicadas y en Sistemas — Cuernavaca

Universidad Nacional Autónoma de México
Apartado Postal 20–726, 01000 México DF, México

ABSTRACT: We see nonlinearity in geometric optics. In the very foundation of Snell's law, is phase space. The Hamilton equations of optical evolution in a medium with a refraction index gradient are obtained in a succint manner out of simple considerations. Yet, transformations due to a refracting surface do not follow from them, but must be complemented by a discontinuous *root* transformation embodying transparent symmetry. Dynamics stems from the Euclidean group; through deformation we may reach also Lorentz transformations leading to an apparently new coma effect. The nonlinearity of the phase space mapping we describe in terms of an aberration *group*, whose structure and visible action we determine. It has subgroup chains of finite dimension with the labels of rank, symplectic spin, and Seidel axis, in complete analogy with the structure of the three-dimensional quantum harmonic oscillator. Spot diagrams show the beautiful faces of Aberration.

THE purpose of this contribution to the School proceedings is to assemble the foundations, results, and perspectives in Hamilton–Lie optics, to make them accessible to graduate students with training in engineering, physics, and mathematics, and easy reading for researchers of nonlinear phenomena that use symmetry methods. The magic of our TEX processor allows us to include several results and speculations, some published, some not, some philosophical, the majority not, that should gradually develop into a more comprehensive text where the reader in mind is the graduate student of the scientific disciplines.

Certainly, the bold outline of the green ecuatorial Andes under the cold, somber skies of Colombia inspired this project. I thank Dr. Galileo Violini, Director of the Centro Internacional de Física, and the Organizers of this School/Workshop, for the opportunity to start it.

CONTENTS:

1. Descartes and Snell, Huygens and Fermat

"I was introduced to the Court of Aranjuez, in the month of March of 1799. The King was kind to receive me in good disposition. I presented the motives that incited me to start a journey to the New Continent... The interest that he [caballero de Urquijo, *minister to Charles IV*] *constantly demonstrated for the fulfillment of my plans had no other reason but his love for Science. It is both a duty and a satisfaction for me to acknowledge in this work the good memory of the services received."* [1]

Descartes reflected upon the Nature of Light and found Snell's law. The paternity of the Law may be reopened to dispute, but the loser is apparent from its very name. Snell found the sine law empirically somewhat earlier than Descartes. Nevertheless, the cartesian Mind abstracted it from a Symmetry of Nature, stated as a conservation law —albeit with the wrong physical intepretation, but very much in the spirit of Sophus Lie and his followers.

[1] *Relation historique du Voyage aux régions équinoxiales du Nouveau Continent, fait en 1799, 1800, 1801, 1803 et 1804 par A. de Humboldt et A. Bopland.* 3 Vols. (Paris, 1814 – 1819); Introductory Remarks.

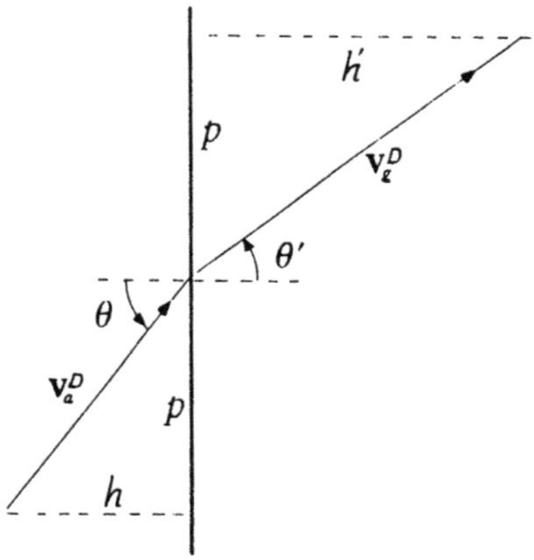

FIGURE 1.1 The Descartes rendering of the refraction process: A change in direction takes place at the vertical surface separating air from glass. The process is such that it conserves the vertical component p (optical momentum). The two direction vectors $\mathbf{v}_a^D$ and $\mathbf{v}_g^D$ are of lengths are n_a and n_g, the refractive indices of the two media. The horizontal components, h and h' are the values of the Hamiltonian before and after refraction. Descartes perceived this construction as *local* (rightly), and pertaining (wrongly) light velocities. This is not a *position* diagram at all, but a *momentum* diagram. Light travels *slower* than c in glass, not faster as a the figure would imply for moving particles.

1.1. The demons of Descartes

It is easy to visualize the sun's light as it enters through blinds a darkened, smoky room. The Greeks loved geometry and turned them into straight lines, as ideal as they could possibly be. Let us call them light *rays*. In Alexandria, Hero arrived at the law of *reflection* in plane mirrors, observing that the angle of incidence of the light ray equals the angle of reflection, both measured, say, from the normal to the surface. Moreover, the three lines lie in a plane. It is quite remarkable that this geometry also applies to the bouncing of point billiard balls off their table edges. The analogy is a powerful one, in spite of the very different nature of the objects involved.

As best told by Stavroudis [1], the conceptual gap between mechanical mass-points and light rays, was bridged by Descartes through the expedient of assigning to the latter the nature of a *tendency* to motion, subject to the same laws as motion itself. By that time, the phenomenon of refraction had been typified, and the French philosopher gave a reasoning that predicted it quantitatively. Strictly, a corpuscularian must say something about the *velocity* of light particles; we need only say that homogeneous media should keep it *constant*. Suppose a light billiard in air has velocity v_a^D and in glass v_g^D —the index D is for Descartes, who realized that the argument only depends on their *ratio*: the values themselves could be both infinite. Let us now state the cartesian law of refraction through Figure 1.1, where light enters glass. Again, the two rays and the surface normal lie in a plane, so we draw the figure in that plane with no loss of information.

Since the origin of Light is in the heavy elements such that kindle Fire, It must surely display a tendency to rush back to them, as if little *dæmons* of the Maxwell

School [1] were receiving them with tennis rackets, and accelerating them into the glass. In Descartes' time, "force" and "momentum" were imprecise concepts so it was thought that glass simply "attracts" photons. In any case, the local demons, since they only distinguished up *vs.* down directions, had no good reason to affect the component of the motion *parallel* to the refracting surface. So, in Figure 1.1, they would respect the component marked by p. The conservation of that component of the velocity vector yields, by elementary trigonometry

$$v_a^D \sin\theta_a = v_g^D \sin\theta_g, \tag{1.1}$$

where θ_a and θ_g are the angles between the ray in air and in glass with the surface normal. The normal component, $h = n\cos\theta$, would change to $h' = n\cos\theta'$, while respecting $p^2 + h^2 = n^2$ and $p'^2 + h'^2 = n'^2$.

Figure 1.1 embodies the high-school method of construction of the refraction process. Indeed, Snell himself had found (1.1) with little attention to the meaning of the refractive index. Taken at face value, nevertheless, the result of Descartes is quite impressive since it is *consistent* with the previously known result on reflection, and it is *exact*. Where Descartes failed was merely in the *interpretation* of the constants deferentially called v^D. These cannot be velocities, but they come close.

1.2. Huygens' ranks and files

Snell's law is extremely resistant to modelling: it may *also* be derived seeing light as a wave phenomenon. This point of view is due to Huygens by (imperfect) analogy with the surface waves in a two-dimensional pond with changing depth. Consider a better analogy shown in the familiar figure that appears in textbooks of optics, and that more than one helicoptered general must have noticed, the ranks of his soldiers *refract* when the army passes from tarmac to swamp at an angle. When he imagines the fronts to extend to infinity and the soldiers to have no individuality, the direction of progress of each soldier is irrelevant* and the direction of motion of the *army* will be perpendicular to its broken ranks. This may be drawn on top of Figure 1.1, for all that matters, with the arrow lengths given by the time between passages of succesive wavefronts, *i.e.*, *inversely* to army velocity. It will lead again to Snell's law. If v_a^H and v_g^H are the army velocities in air and glass (tarmac and mud), proportional to the distance between the ranks, then the common segment of two right triangles leads to Huygens' (H) rendering of Snell's law:

$$\frac{1}{v_a^H}\sin\theta = \frac{1}{v_g^H}\sin\theta'. \tag{1.2}$$

Newton based much of his *Opticks* on the point-mass view of light, but did so reluctantly since it was known to him that mere gravitation at the surface would not

* There will always be a Good Soldier Schweick... (P.W., private communication).

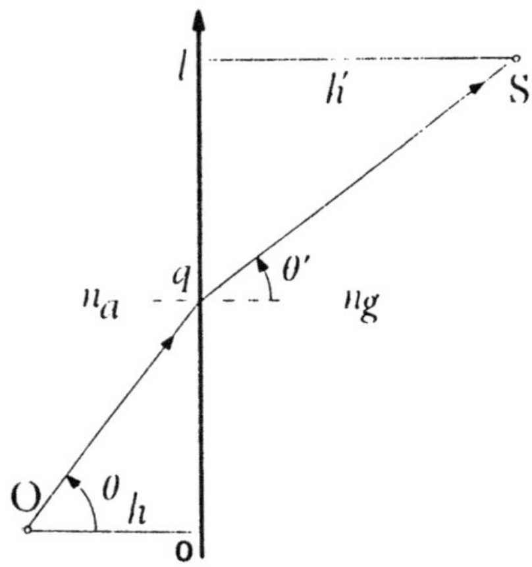

FIGURE 1.2 The Fermat rendering of refraction. This is a *position* diagram where a light ray from O arrives at S by choosing minimal transit time or number of intervening wavefronts. The arrows lengths mean nothing in particular, but we have chosen them to have ratio $n_a : n_g$ for the purpose of comparison with the previous figure.

account for the Descartes demons' precise ball play. The issue of whether (1.1) or (1.2) is the correct interpretation of the subject was open to experiment by the mid-nineteenth century, when light velocities could be compared in different media. Huygens' picture, improbable as it may seem to geometrical opticists, won: light moved *slower* in glass than in air, not faster, as Descartes would have thought.

1.3. Minimal action

A third approach to the matter of light, that is in truth bound neither to the corpuscular nor wave pictures, is the principle of minimal action. It borders with philosophy: *"Je reconnais premièrement ... la verité de ce principe, que la Nature agit toujours par les voies les plus courtes"*. Fermat proposed that a light ray between two points, O and S in Figure 1.2, follows a path such that something, such as the *time* employed in a billiard flight, or the *number* of soldier rows between the points, be minimal. That *something* is proportional to the geometrical distance along the path of the ray multiplied by some local factor inherent to the propagating medium. It is called *action*. The local (dimensionless) factor is called *refractive index*, and by convention is $n = 1$ for vacuum.

Let n_a and n_g denote the refractive indices for air and glass, and let us eschew variational in favor of simpler differential calculus. Figure 1.2 expresses the action A

as a function of the intersection point q in the diagram as

$$\begin{aligned} A(q) &= n_a\sqrt{h^2+q^2} + n_g\sqrt{h'^2+(\ell-q)^2} \\ &= n_a h \sec\theta + n_g h' \sec\theta'. \end{aligned} \tag{1.3}$$

The minimum of $A(q)$ is found at the vanishing of the first derivative,

$$\begin{aligned} \frac{dA(q)}{dq} &= n_a \frac{q}{\sqrt{h^2+q^2}} + n_g \frac{q-\ell}{\sqrt{h'^2+(\ell-q)^2}} \\ &= n_a \sin\theta - n_g \sin\theta'. \end{aligned} \tag{1.4}$$

The last expression is zero when the incidence and refraction angles are bound by Snell's law. Fermat thus also, again, found The Law.

Figures 1.1 and 1.2 are similar to the point of confusion. Yet they were drawn in two very different spaces: *(a)* for Descartes it was a velocity diagram —more correctly, it is a *momentum* diagram, while *(b)*, for Fermat it is a *position* diagram. Their duality harbors geometrical optics phase space, as we shall argue in the next section. Huygens' figure will deserve corresponding comments when we regard wave optics —not here, because purpose and space forbid, but elsewhere.

1.4. A conservation statement and Descartes' first sphere

Let us repeat Snell's law in local form one last time, writing it as a *vector* equality, referred to the refracting surface *normal* $\vec{Z}$ at the point of incidence. We denote by $\vec{n}$ and $\vec{n}'$ the vectors tangent to the directions of the incident and refracted rays (*at the same point in position space*) such that their lengths are n and n' respectively, the indices of their two media. Then [1],

$$\vec{Z} \times \vec{n} = \vec{Z} \times \vec{n}'. \tag{1.5}$$

The change in ray direction, $\vec{n} - \vec{n}'$, is thus always normal to the surface. We may draw Figure 1.3 for the case $n < n'$ and "right-moving" rays. The *magnitude* of (1.5) is the familiar formula $n \sin\theta = n' \sin\theta'$ where θ is the angle between $\vec{Z}$ and $\vec{n}$, etc.

The left-hand side of equation (1.5) depends only on things in the first medium, and the right-hand only on things in the second. The equation embodies a statement of *conservation* of the two components of $\vec{n}$ normal to $\vec{Z}$. It is independent of the "size" of the index discontinuity $n - n'$ and the orientation of $\vec{n}$ on the sphere. These remarks are not quite trivial: the first one will lead in the next section to the Hamilton equations for optics, and the second permits the use of the Descartes sphere of (in general changing) radius n as a homogeneous space for the action of the Euclidean group of rotations and translations, embodying the dynamics of optical systems; this we shall illustrate in Section 3 for the simplest case of homogeneous media.

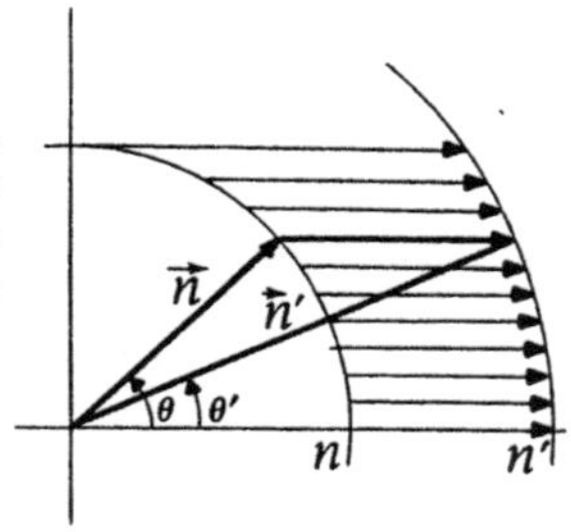

FIGURE 1.3 Refraction seen as a mapping between the Descartes spheres of rays in two media, of radii $n < n'$, the respective refractive indices in the 'initial' and 'final' medium. All rays can transit into the latter. The difference between the two vectors, $\vec{n} - \vec{n}'$, is perpendicular to the surface at the point of incidence.

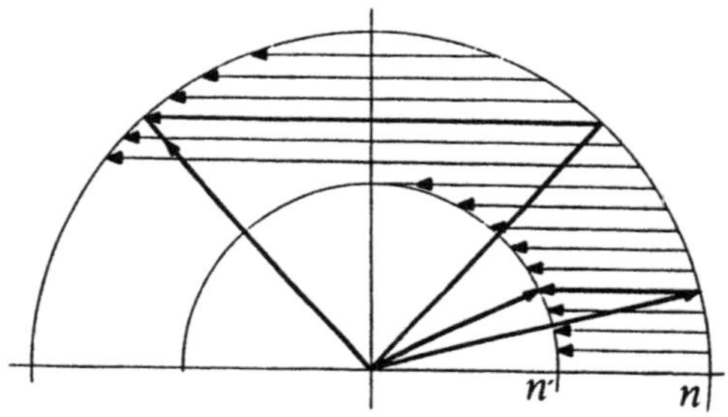

FIGURE 1.4 When $n > n'$, as in glass-to-air transit, a cone of the 'initial' rays $\vec{n}$ map on the smaller sphere of radius n', emerging into air as $\vec{n}'$; others suffer total internal reflection, remaining in the first medium.

Descartes' diagram, Figure 1.3, is *global* over the full direction sphere. This is important, but shows mathematical difficulties when we try to describe things analitically. In Figure 1.4 we show the case when $n > n'$, as in *glass-to-air* interfaces: the right half of the larger sphere of "initial, right-moving" rays projects both on the smaller sphere of refracted, right-moving rays and, beyond a *critical* angle, on part of its *left* half: rays that remain in the denser medium after total internal reflection, moving now to the left. Finally —and not shown in the figure— rays in glass in the "shadow cap" of the larger sphere can only come from initial rays in the left half of the smaller sphere, *i.e.*, *backward* rays from air to glass. *Both* Descartes spheres are needed, hence, to describe the full refraction process. Such is indispensable when we try to *wavize* the system, but that subject will not be attempted in the present notes, dedicated only to nonlinearity in geometric optics.

In the next section we shall express these concepts in coordinates and find the Hamilton equations of motion. It will be in section 3 that we shall again refer to cartesian view of optics as determined by the geometry of phase space under Euclidean and other transformations. Section 4 holds a second Descartes sphere, and we are interested in its harmonics as a basis to study and classify aberrations. Those topics are left for sections 5 and 6.

2. Optical phase space, Hamilton's equations, and refraction

> *"Mutis put at out disposition [in Santa Fé de Bogotá] a house near his, and treated us with exceptional friendship. He is an aged, venerable eclessiastic of almost seventy-two years and a wealthy man.* The King has given ten thousand piastres per year for the Botanical Expedition. Since fifteen years, thirty artists work in Mutis' workshop, who has from two to three thousand plates in major folio that are miniatures. After the Banks collection, I have never seen another botanical library as large as that of Mutis."*[II]

We restate the Descartes scheme of *pre*-phase space, fittingly, in cartesian coordinates. A light ray is described through *(a)*, one vector to any point *in* the ray —a *position* vector $\vec{P}(z) = (q_x(z), q_y(z), z)$ (disregarding rays in $z =$ constant planes), and *(b)*, a second vector, $\vec{n} = (p_x, p_y, h)$ that indicates the *direction* of the ray on a sphere of radius $|\vec{n}| = n(\vec{P})$, the refractive index of the medium at $\vec{P}$. In doing so, we are distinguishing an *optical center* and an *optical axis*, origin, respectively of positions and directions. We may *translate* the former without affecting the latter, but when we *rotate* the latter, the former also rotates. (We thus have a semidirect product structure in the group of space motions.)

Next, we restrict attention to the section of the world of rays at a *screen* $z =$ constant, adhering to the conventions of indicating *three*-dimensional vectors with arrows *i.e.* $\vec{v} = (v_x, v_y, v_z)$, *two*-dimensional vectors *in* the plane of the screen by boldface *i.e.* $\mathbf{v} = (v_x, v_y)$, and $v^2 = |\mathbf{v}|^2$. The range of the ray position vector $\mathbf{q}$ is thus the full $\Re^2$ plane of the screen.

2.1. The Hamilton equations of optics

We shall call $\mathbf{p} = (p_x, p_y)$ the *momentum* vector of the ray (we shall show below that it is canonically conjugate to position $\mathbf{q}$). The third component of the direction vector is

$$h = \sigma\sqrt{n(\mathbf{q}, z)^2 - p^2}, \tag{2.1}$$

and the role of σ is to distinguish *forward* ($\sigma = 1$) from *backward* ($\sigma = -1$) rays. The range of $\mathbf{p}$ is the disk δ_n^+ of radius n for *forward* rays $\sigma = +1$, a similar disk δ_n^- for *backward* rays $\sigma = -1$, and sown at the circumference δ_n^0 of $h = 0$ rays parallel to

* José Celestino Mutis (1732–1808) was Professor of Anatomy in Madrid, and Surgeon to Don Pedro Messía de la Cerda, named viceroy to Nueva Granada in 1760. He lectured on Mathematics at the Colegio del Rosario between 1762 and 1766 holding a celebrated controversy over the Copernican system. On November 1, 1783, he is designated leader of the Royal Botanical Expedition ordered by Charles III with the intervention of the viceroy Caballero y Góngora. He introduced the smallpox vaccine to Spanish America. Over ten years after his death, as Santa Fé de Bogotá was recovered by the Spaniards, the Mutis collection was shipped to Spain: *"...16 crates with 5 900 botanic and animal plates, 1 crate of manuscripts, 48 crates with plants, 15 of minerals, 9 of seeds, 8 of woods, 1 of cinnamon, 45 of herbs, 6 of 'various curiosities', and several more, lacking labels with fruits and resins."* The collection was kept at the Jardín Botánico of Madrid, and the Encyclopedist of Nueva Granada, having met the *Aufklärer* of Europe, remained unknown to his contemporaries. The 100 plates he gave as present to Humboldt were shipped to the French Institut National, but never arrived at their destiny. Belatedly, the *savant* received recognition in *Archivo epistolar del sabio naturalista José Celestino Mutis*, by Guillermo Hernández de Alba, 2 Vols. (Bogotá, 1947–1949). Since 1954, the Institutos de Cultura Hispánica initiated a monumental 51-volume edition of his work, with 2 666 plates.

[II] Letter of Alexander to his brother Wilhelm, Baron von Humboldt.

the screen. This can be described economically as the 2-sphere S_2. We shall generally avoid speaking of or writing explicitly σ among the coordinates because, besides being uncomfortable, the separation of rays by σ is truly needed only in near-to-90° beams and in cases of reflection. Reorienting the optical axis is the best cure for the first case, and a mental note is for the second.

The four-dimensional manifold $w = (\mathbf{p}, \mathbf{q})$ is optical *phase space* $\wp = S_2 \times \Re^2$. (Notice that it is quite distinct, globally, from the better-known phase space $\Re^4$ of point-particle dynamics.) We may describe the path of a light ray in terms of a line in $\wp$ with coordinates $w(z) = (\mathbf{p}(z), \mathbf{q}(z))$, seeing the screen move along the z-axis or, equivalently, as a z-parametrized transformation of phase space from the *reference screen* $z = 0$.

Consider an infinitesimal displacement dz as shown in Figure 2.1. The vector $d\vec{P} = (d\mathbf{q}, dz)$ is tangent to $\vec{n}(\vec{P}) = (\mathbf{p}, h)$ and hence

$$\frac{d\mathbf{q}(z)}{dz} = \frac{\mathbf{p}(z)}{h(\mathbf{p}, \mathbf{q}; z)} = -\frac{\partial h}{\partial \mathbf{p}}. \tag{2.2}$$

This expression holds true even when $n(\mathbf{q}, z)$ varies (smoothly) in the interval dz of the figure, because for any change in the direction of $\vec{n}$ in the interim by a quantity proportional to dz, the correction in the outcome $\mathbf{q}(z + dz)$ will be of order $(dz)^2$ and hence vanish from consideration.

The change in the direction three-vector, $d\vec{n} = (d\mathbf{p}, dh)$ after dz, will be along the gradient of the refractive index. See Figure 2.2, where the lettering is incremental. (*This* is Snell's law à la Descartes.) In screen-and-z coordinates we have $\vec{\nabla} n(\mathbf{q}, z) = (\partial n/\partial \mathbf{q}, \partial n/\partial z)$. We write $d\vec{n}/dz = \alpha \vec{\nabla} n$ and determine α through asking that $\vec{n}$ remain on the sphere $p^2 + h^2 = n^2$. Thus,

$$\frac{d}{dz}(p^2 + h^2) = 2\mathbf{p} \cdot \frac{d\mathbf{p}}{dz} + 2h\frac{dh}{dz} = 2\alpha\left(\mathbf{p} \cdot \frac{\partial n}{\partial \mathbf{q}} + h\frac{\partial n}{\partial z}\right), \tag{2.3a}$$

while

$$\frac{d}{dz}n^2 = 2n\frac{dn}{dz} = 2n\left(\frac{\partial n}{\partial \mathbf{q}} \cdot \frac{d\mathbf{q}}{dz} + \frac{\partial n}{\partial z}\right). \tag{2.3b}$$

Replacing $d\mathbf{q}/dz$ from (2.2) into the last expression, we find that $\alpha = n/h$. It follows that $d\vec{n}/dz = n/h\, \vec{\nabla} n$ or, in components,

$$\frac{d\mathbf{p}}{dz} = \frac{n}{h}\frac{\partial n}{\partial \mathbf{q}} = \frac{\partial h}{\partial \mathbf{q}}, \tag{2.4a}$$

$$\frac{dh}{dz} = \frac{n}{h}\frac{\partial n}{\partial z} = \frac{\partial h}{\partial z}. \tag{2.4b}$$

¡Listo! Equations (2.2)–(2.4) are the *Hamilton equations of optics*, as the reader must have recognized, for the canonically conjugate coordinates $\mathbf{p}$ and $\mathbf{q}$, under the

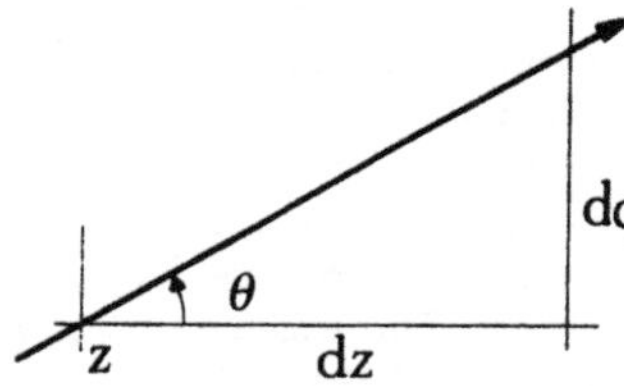

FIGURES 2.1 *(a)* A displacement of the reference screen by dz. The ray position at the screen will increase by $d\mathbf{q}$, with $d\mathbf{q}/dz$ a finite vector of size $\tan\theta$. *(b)* The ray is tangent to its direction vector, and $\mathbf{p}/h$ parallel to $d\mathbf{q}/dz$ and also of size $\tan\theta$, hence Hamilton's first equation holds.

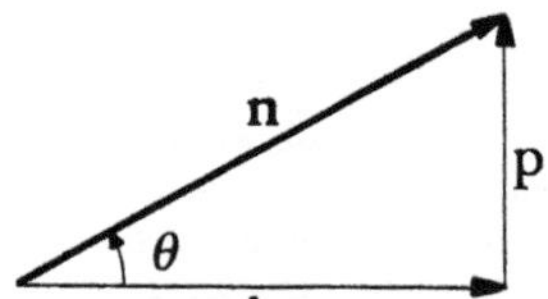

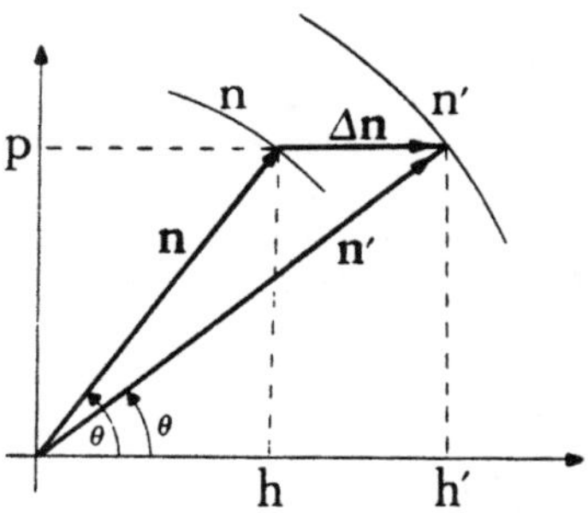

FIGURE 2.2 An incremental refraction from a medium n to a medium n' through a surface with normal $\vec{\nabla} n(\mathbf{q}, z)$. This vector will be parallel to $\Delta\vec{n} = \vec{n}' - \vec{n}$, shown in the figure. The condition that $\vec{n}$ and $\vec{n}'$ remain on their Descartes spheres implies Hamilton's second and third equations.

(minus) Hamiltonian function given in (2.1). The reason for the traditional Hamiltonian function $H(\mathbf{p}, \mathbf{q}) = -h$ to be *minus* the z-component of the ray direction vector is that, upon expanding in powers of p^2, the optical $-h$ takes the form

$$H = -\sqrt{n^2 - p^2} = -n + \frac{p^2}{2n} + \frac{(p^2)^2}{8n^3} + \frac{(p^2)^3}{16n^5} + \frac{5(p^2)^4}{128n^7} + \cdots. \tag{2.5}$$

The $p^2/2n$ term resembles mechanical kinetic energy and $-n(\mathbf{q}, z)$ *looks* like a potential. The mechanical and optical Hamiltonians are by no means identical, though. The former is the *paraxial* approximation of the latter (for $p^2 \ll n^2$), and a basis for its aberration expansions, as we shall show below.

It is not necessary to advocate to this audience the usefulness of the Hamilton equations of motion and of the accompanying symplectic structure of phase space. The equations integrate to a Lie group of phase space transformations that are canonical. Nevertheless, it is very important to note that the Hamiltonian formulation of optics sketched above applies to media where the refractive index $n(\mathbf{q}, z)$ *has* a gradient. We are not entitled to claim description of light optics, unless we can include the very ubiquitous *lens* systems, *i.e.* surfaces $z = \varsigma(\mathbf{q})$ where $n(\vec{P})$ is *dis*continuous.

In fact, the Lie-theoretic description of refracting interfaces was a rather hard computational problem. The result for the third-order approximation case was reported by Alex J. Dragt in his classical Lie optics article [2]. It was in trying to obtain the reported results in a pedestrian way that (after a wrong statement of the solution [3]) we came upon the factorization theorem presented below [4,5]. Miguel Navarro Saad was able then to solve, for the first time, the implicit equations up to ninth aberration order [6]. This was done using symbolic REDUCE computation [7] on a Foonly F2, then working at IIMAS. The result seems worth repeating here since it is presented in a new, simplified, form; its consequences are still relatively unexplored, as we shall come to note.

2.2. The factorization of refraction

Suppose we have our usual reference screen, and a refracting surface $z = \varsigma(\mathbf{q})$ between two homogeneous media with refractive indices n and n'. As usual, we think of the prime to stand for the *second, denser* medium, so as to follow the figures with our intuition of air-to-glass refraction. To simplify matters, let us also assume $\varsigma(\mathbf{q})$ is a one-valued function of $\mathbf{q}$ with all the necessary derivatives. Now look at Figures 2.3. The ray, referred to the screen as $w = (\mathbf{p}, \mathbf{q})$ flies straight and hits the surface ς at $\bar{\mathbf{q}}$; there it refracts and becomes the ray $w' = (\mathbf{p}', \mathbf{q}')$, as referred back to the *same* screen. This forth-and-back argument serves the calculation: the two triangles in Figure 2.3*a* are isolated in Fig. 2.3*b* and noting that $\mathbf{p}/h$ vectorizes $\tan\theta$, lead to

$$\mathbf{q} + \varsigma(\bar{\mathbf{q}})\mathbf{p}/h = \bar{\mathbf{q}} = \mathbf{q}' + \varsigma(\bar{\mathbf{q}})\mathbf{p}'/h'. \tag{2.6}$$

This is the *first* refraction equation. It allows the determination of $\bar{\mathbf{q}}$ through an *implicit* vector equation that may be solved algebraically for $\bar{\mathbf{q}}$ when ς is a conic surface, and algorithmically otherwise.

Finally, consider the ray directions abstracted in Figure 2.3*c*. At the point of incidence $(\bar{\mathbf{q}}, \varsigma(\bar{\mathbf{q}}))$, the surface will be normal to $\vec{Z}(\bar{\mathbf{q}}) = (\nabla\varsigma(\mathbf{q})|_{\mathbf{q}=\bar{\mathbf{q}}}, -1)$, where $\nabla = (\partial_{q_x}, \partial_{q_y})$. Snell's law in vector form (1.5) has then the following components:

$$\begin{pmatrix} \partial_{q_x}\varsigma|_{\bar{\mathbf{q}}} \\ \partial_{q_y}\varsigma|_{\bar{\mathbf{q}}} \\ -1 \end{pmatrix} \times \begin{pmatrix} p_x \\ p_y \\ h \end{pmatrix} = \begin{pmatrix} h\partial_{q_y}\varsigma|_{\bar{\mathbf{q}}} + p_y \\ -p_x - h\partial_{q_x}\varsigma|_{\bar{\mathbf{q}}} \\ p_y\partial_{q_x}\varsigma|_{\bar{\mathbf{q}}} - p_x\partial_{q_y}\varsigma|_{\bar{\mathbf{q}}} \end{pmatrix} = \begin{pmatrix} \partial_{q_x}\varsigma|_{\bar{\mathbf{q}}} \\ \partial_{q_y}\varsigma|_{\bar{\mathbf{q}}} \\ -1 \end{pmatrix} \times \begin{pmatrix} p'_x \\ p'_y \\ h' \end{pmatrix}. \tag{2.7}$$

In components, hence, the refraction process may be written as the conservation of the following barred quantities, the *second* refraction equations:

$$\mathbf{p} + h\nabla\varsigma(\bar{\mathbf{q}}) = \bar{\mathbf{p}} = \mathbf{p}' + h'\nabla\varsigma(\bar{\mathbf{q}}), \tag{2.8}$$

$$-\mathbf{p} \times \nabla\varsigma(\bar{\mathbf{q}}) = \bar{h} = -\mathbf{p}' \times \nabla\varsigma(\bar{\mathbf{q}}), \tag{2.9}$$

where we abbreviate $\nabla\varsigma(\bar{\mathbf{q}}) = \nabla\varsigma(\mathbf{q})|_{\mathbf{q}=\bar{\mathbf{q}}}$. Having solved (2.6) for $\bar{\mathbf{q}}$, thus, we plug the result in (2.8) to obtain $\bar{\mathbf{p}}$; (2.9) only assures us that the change in the screen momenta $\mathbf{p} - \mathbf{p}'$ will be in the direction of the screen projection of the normal $\nabla\varsigma(\bar{\mathbf{q}})$.

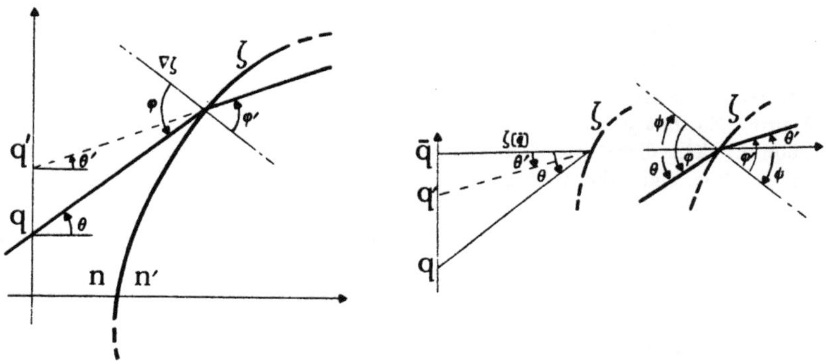

FIGURES 2.3 *(a)* The refraction process. In the plane of the figure, ray (q, θ) hits the surface ς and refracts to the ray (q', θ'). *(b)* The position coordinates showing the factorisation into root transformations through the first refraction equation. *(c)* The angles leading to the second refraction equation.

Equations (2.6) and (2.8) express the $\wp$ transformation $S_{n,n';\varsigma}$ between media n and n' separated by the surface ς, as a two-step, *factorized* process:

$$\begin{pmatrix}\mathbf{p}\\ \mathbf{q}\end{pmatrix} \overset{R_{n;\varsigma}}{\longrightarrow} \begin{pmatrix}\bar{\mathbf{p}}\\ \bar{\mathbf{q}}\end{pmatrix} \overset{R_{n';\varsigma}}{\longleftarrow} \begin{pmatrix}\mathbf{p}'\\ \mathbf{q}'\end{pmatrix}, \tag{2.10}$$

where the *root transformation* $R_{n;\varsigma}$ due to ς in medium n is given by

$$R_{n;\varsigma}\,\mathbf{q} = \bar{\mathbf{q}} = \mathbf{q} + \varsigma(\bar{\mathbf{q}})\,\mathbf{p}/\sqrt{n^2 - \mathbf{p}^2}, \tag{2.11a}$$

$$R_{n;\varsigma}\,\mathbf{p} = \bar{\mathbf{p}} = \mathbf{p} + (\nabla\varsigma)(\bar{\mathbf{q}})\sqrt{n^2 - \mathbf{p}^2}. \tag{2.11b}$$

The action of a refracting surface is thus expressed as a root transformation (related to things in the first medium) times the *inverse* root transformation (related to things in the second medium),

$$S_{n,n';\varsigma} = R_{n;\varsigma}\,R^{-1}_{n';\varsigma}, \tag{2.12}$$

with operators that act always on the *arguments* of functions. [8]

Similar factorization and canonicity properties were proven in Ref. [9] for surfaces of discontinuity within general inhomogeneous media; in Ref. [10] this was applied to a quadratic-shaped "crack" in a quartic-index fiber-like medium, and computed explicitly to third aberration order.

I still find it quite surprising that $\bar{\mathbf{q}}$ and $\bar{\mathbf{p}}$ are actually *canonically conjugate* coordinates. The former is the intersection point of the ray with the surface ς shown in Figure 2.3*b*, but $\bar{\mathbf{p}}$ is *not* a true optical momentum, since the magnitude of $\bar{\mathbf{p}}$ is not bound. The direct proof of this result may be found in the appendix of reference [4]. But, since the product of two canonical transformations is canonical, the net consequence is that the finite refracting surface transformation $S_{n,n';\varsigma}$ is shown to be canonical, a fact not unexpected, but not implied by any Hamilton differential equations.

2.3. The roots of reflection

Another realm in which root transformations build optical elements is *reflection*, a necessarily finite transformation in $\wp$. Indeed, since under reflection only the *sign* of the component normal to the local mirror surface changes, we may cast the onus of this involution onto the formal sign of n.

The transformation $M_{n;\varsigma}$ in a medium n due to a reflecting surface ς may be written also in factorized form:

$$M_{n;\varsigma} = R_{n;\varsigma}\, R^{-1}_{-n;\varsigma}, \tag{2.13}$$

where

$$R_{-n;\varsigma}\, \mathbf{q}' = \bar{\mathbf{q}} = \mathbf{q}' - \varsigma(\bar{\mathbf{q}})\, \mathbf{p}'/\sqrt{n^2 - \mathbf{p}'^2}, \tag{2.14a}$$

$$R_{-n;\varsigma}\, \mathbf{p}' = \bar{\mathbf{p}} = \mathbf{p}' - \nabla\varsigma(\bar{\mathbf{q}})\sqrt{n^2 - \mathbf{p}'^2}, \tag{2.14b}$$

i.e., changing only the sign of the square root factors containing n^2.

It is clear that the root, refracting surface, and/or mirror transformations do **not** form a group. Their differential properties neatly lead to selection rules in the aberration coefficients of revolution [9,11] and asymmetric [12] surfaces, and some global properties for spherical surfaces —such as the existence and location of *aplanatic* points— have been reported [11].

Let me end this section stressing that we still know very little of the coordinate-free characterization of these transformations within the group of all optical symplectomorphisms, of their global properties, the singularities for rays that are tangent to the surface, or have multiple intersections, and the caustics they generate in z or project on a screen.

3. Euclidean dynamics and other transformation groups

"It is admirable that this young American could instruct himself to the point of being able to make the most delicate astronomical observations, with instrtuments built with his own hands... This Señor Caldás is a prodigy in Astronomy. Born in the obscurity of Popayán, and not having travelled farther than Santa Fé, he built his Barometers; a Sector and a gauged Segment made of wood. He is able to draw Meridians and to measure the latitude by Gnomones of 12–15 feet. What could this boy not achieve in a country with means and where the need be not that one learn everything by oneself?!"* ///

In the last section we showed that the position and momentum coordinates of a geometric light ray at a screen are canonically conjugate. It is time now to bring into gear the Poisson bracket formalism [13] that allows us to exploit this property to its full extent.

3.1. Lie operators, Lie transformations

We recall the definition the usual Poisson brackets between any two (differentiable) functions f, g of $\mathbf{p}$ and $\mathbf{q}$:

$$\{f,g\} = \frac{\partial f}{\partial \mathbf{q}} \cdot \frac{\partial g}{\partial \mathbf{p}} - \frac{\partial f}{\partial \mathbf{p}} \cdot \frac{\partial g}{\partial \mathbf{q}}. \tag{3.1a}$$

In this way we build the *Lie operators*† associated to any such function,

$$\hat{f} = \frac{\partial f}{\partial \mathbf{q}} \cdot \frac{\partial}{\partial \mathbf{p}} = \frac{\partial f}{\partial \mathbf{p}} \cdot \frac{\partial}{\partial \mathbf{q}}, \tag{3.1b}$$

Various important and well-known properties follow that are well documented in Ref. [8]. Lie operators are *linear* $(af + bg)\hat{} = a\hat{f} + b\hat{g}$, are a *derivation* $(fg)\hat{} = f\hat{g} + g\hat{f}$, they act on *arguments* of functions $\hat{f}\,g(\mathbf{p},\mathbf{q}) = g(\hat{f}\mathbf{p}, \hat{f}\mathbf{q})$, and intertwine Poisson brackets with *commutators* $\{f,g\}\hat{} = [\hat{f},\hat{g}]$. It follows that we can build Lie algebras out of the set of functions with the Lie bracket given by the Poisson bracket, their enveloping algebras through function product, and apply the corresponding Lie operators on the optical phase space $\wp$ that thereby becomes a homogeneous space for the algebra.

We may lastly define the *Lie transformation* generated by the function f through exponentiating $\hat{f}$ with the formal series,

$$\exp \alpha \hat{f} = \sum_{m=0}^{\infty} \frac{\alpha^m}{m!} \hat{f}^m, \tag{3.2a}$$

*Humboldt met Francisco José de Caldás on the last day of the year of 1801. Born in 1779, Caldás finished degrees in Law in 1793 and 1796, but his work included methods for geographical measurements, studies on the wood of quinine, and service to mines. The *ilustrado* was accused of adhesion to the independist cause and shot in Bogotá, on October 29, 1816.

/// Quoted by Charles Minguet, in *Alexandre de Humboldt historien et géographe de l'Amerique espagnole 1799–1804* (François Maspero, Paris, 1968).

† With the usual apology to all authors that use different notations, such as $f_{\rm op}$ [14], $:f:$ [2], or $\{f,\circ\}$ [8].

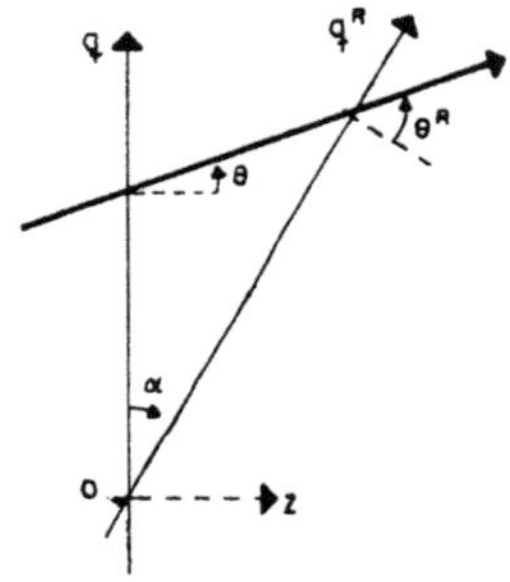

FIGURE 3.1 Rotation of the reference screen O—q by an angle α. The world of light rays (q,θ) becomes (q^R,θ^R). This is a simple Euclidean transformation in spite of the apparent singularity of the transformation for $\alpha+\theta=\frac{1}{2}\pi$.

and act on functions of $\wp$ through

$$\exp \alpha\hat{f}\, g(\mathbf{p},\mathbf{q}) = g(\mathbf{p},\mathbf{q};\alpha) = \sum_{m=0}^{\infty} \frac{\alpha^m}{m!}\{f,\{f,\cdots\{f,g\}\cdots\}\}(\mathbf{p},\mathbf{q}), \tag{3.2b}$$

wherever the series converges appropriately. Optical phase space $\wp$ thus also becomes a homogeneous space for the integrated *group* action.

3.2. Translations and rotations

The prime example of Lie transformations are those generated by the three components of the direction vector $\vec{n}=(\mathbf{p},h)$. Optical momentum $\mathbf{p}$ generates *translations in the screen,*

$$\exp \mathbf{a}\cdot\hat{\mathbf{p}}\, g(\mathbf{p},\mathbf{q}) = g(\mathbf{p},\mathbf{q}-\mathbf{a}), \qquad \mathbf{a}\in\Re^2, \tag{3.3a}$$

and the (minus) Hamiltonian h generates *evolution along the optical z axis,*

$$\exp z\hat{h}\, g(\mathbf{p},\mathbf{q}) = g(\mathbf{p},\mathbf{q}+z\mathbf{p}/h), \qquad z\in\Re. \tag{3.3b}$$

This is the *translation* group of $\wp$; it does not affect the direction of the ray, only its position; it maps $\wp$ properly onto itself. In contrast, the Lie operator $\hat{\mathbf{q}}$ when exponentiated, would translate $\mathbf{p}$ out of its proper optical range ($p^2\le n^2$) and hence does *not* lead to a valid Lie transformation, in contradistinction to the $\Re^4$ phase space of point-particle mechanics, where it clearly does as a Galielan boost.

The role of "momentum translation" in $\wp$ is taken by *rotations* of the ray direction sphere S_2. In Fig. 3.1 we show the two-dimensional transformation [11],

$$p\mapsto p' = p\cos\gamma + h\sin\gamma, \tag{3.4a}$$

$$h\mapsto h' = -p\sin\gamma + h\cos\gamma, \tag{3.4b}$$

$$q\mapsto q' = q/(\cos\gamma - p/h\,\sin\gamma), \qquad \gamma\in S_1. \tag{3.4c}$$

The function that generates this Lie transformation is $r = qh$. In three dimensions, there are *three* such functions generating rotations of the direction sphere with the concomitant screen coordinate transformations. They are (*cf.* [11])

$$r_x = q_y h, \qquad r_y = -q_x h, \qquad r_z = q_x p_y - q_y p_x = \mathbf{q} \times \mathbf{p}, \tag{3.5}$$

and close into an $so(3)$ algebra,

$$\{r_x, r_y\} = r_z, \qquad \{r_y, r_z\} = r_x, \qquad \{r_z, r_x\} = r_y. \tag{3.6a}$$

They are *proper* transformations of $\wp$. This does *not* mean that the series (3.2b) is bounded everywhere: take $g(\mathbf{p}, \mathbf{q}) = \mathbf{q}$ and consider those rays that under a given rotation will become parallel to the screen. Their image $\mathbf{q}$ then escapes to infinity. This is clearly an artifice of our "screen 2-world" and does not constitute an impropriety to the 3-world of optical rays.

Let us denote h interchangeably by p_z. The six functions $(p_x, p_y, p_z; r_x, r_y, r_z)$ then close under Poisson brackets; in addition to (3.6a) they satisfy

$$\{r_x, p_y\} = p_z, \qquad \{r_y, p_z\} = p_x, \qquad \{r_z, p_x\} = p_y, \tag{3.6b}$$

$$\{p_x, p_y\} = 0, \qquad \{p_y, p_z\} = 0, \qquad \{p_z, p_x\} = 0. \tag{3.6c}$$

This is the Lie algebra $iso(3)$ of the Euclidean group of motions $ISO(3)$. Since the Hamiltonian of the system is among the generators, we identify it as the *dynamical* group of optics. The realization corresponding to *geometrical* optics is the one obtained from $f \mapsto \hat{f}$, *i.e.*, the Lie operator association defined in (3.2).

3.3. Relativistic transformations and coma

The ray direction sphere S_2, $p^2 + h^2 = n^2$ has a dual role: its coordinates (p_x, p_y, p_z) are the generators of the (abelian) ideal of translations in $ISO(3)$, and it suffers the (rigid) action of the rotation subgroup $SO(3)$. These are the ingredients to build the *deformation* $ISO(3) \Longrightarrow SO(3,1)$ of the Euclidean to the *Lorentz* group of relativity, one particular case of the deformation process between inhomogeneous and homogeneous classical algebras and groups whose beginning dates back to Bargmann's work [15] in 1947. (For the more general cases see Refs. [16].) The generators of the deformed group are *(a)*, the generators of the semisimple subgoup, here $\vec{r} = (r_x, r_y, r_z)$ of $SO(3)$, and *(b)* an ideal $\vec{b} = (b_x, b_y, b_z)$, built as bilinear functions of the original generators with the right transformation properties under the former. For the present case, studied in Ref. [17], we need only consider the three functions $\vec{b} = \vec{r} \times \vec{p}$ or, in components,

$$\mathbf{b} = n\mathbf{q} - \mathbf{p} \cdot \mathbf{q}\mathbf{p}/n, \tag{3.7a}$$

$$b_z = \quad - \mathbf{p} \cdot \mathbf{q} h/n. \tag{3.7b}$$

The six functions $(\vec{r}, \vec{b})$ satisfy the Lie bracket relations of the generators of the relativistic $so(3,1)$ algebra, Eqs. (3.6a) and

$$\{r_x, b_y\} = b_z, \qquad \{r_y, b_z\} = b_x, \qquad \{r_z, b_x\} = b_y, \tag{3.8a}$$

$$\{b_x, b_y\} = -r_z, \qquad \{b_y, b_z\} = -r_x, \qquad \{b_z, b_x\} = -r_y. \tag{3.8b}$$

The two Lorentz invariants are $\vec{b}^2 - \vec{r}^2 = 0$ and $\vec{r} \cdot \vec{b} = 0$.

The Lie transformation generated by the *boost* normal to the screen, b_z, on the coordinates of $\wp$ can be shown to be:

$$\exp \alpha \hat{b}_z \, \mathbf{p} = \frac{\mathbf{p}}{\cosh \alpha + h/n \sinh \alpha}, \tag{3.9a}$$

$$\exp \alpha \hat{b}_z \, \mathbf{q} = (\cosh \alpha + h/n \sinh \alpha) \left(\mathbf{q} - \frac{\mathbf{p} \cdot \mathbf{q} \mathbf{p}/n}{n \sinh \alpha + h \cosh \alpha} \right). \tag{3.9b}$$

As the Euclidean group, the Lorentz group transforms $\wp$ *properly* onto itself, *canonically*, and with the *nested* structure $\mathbf{p}'(\mathbf{p}; \alpha)$, $\mathbf{q}'(\mathbf{p}, \mathbf{q}; \alpha)$. We underline that (3.9) *is* the action of the Lorentz group of special relativity, of which only (3.9*a*) was hereforeto known under the name of *stellar aberration*: due to the earth's motion in orbit, the directions of arrival of the light rays distort on the celestial sphere (identical with Descartes' sphere!). Equation (3.9*b*) complements as its *canonical conjugate*.

To show this, we identify the direction 3-vector $\vec{n} = (\mathbf{p}, h)$ of a geometric ray with the direction $\vec{l} = (\mathbf{l}, l_z)$ of a lightlike 4-vector $(\vec{l}, l_0)$ through $\vec{n} = n\vec{l}/l_0$. Keeping our conventions $p = |\mathbf{p}|$ and $l = |\mathbf{l}|$, and recalling $n = |\vec{n}|$ and $l_0 = |\vec{l}|$, the angle θ between the 3-vectors and the the boost direction (the z-axis) may be written as $p/n = \sin \theta = l/l_0$, $h/n = cos\theta = l_z/l_0$, or

$$\frac{p}{n+h} = \tan \tfrac{1}{2}\theta = \frac{l}{l_0 + l_z}. \tag{3.10}$$

Now, the well-known form of the Lorentz z-boost to velocity $v = c \tanh \alpha$, that is, $l_z \mapsto l'_z = l_z \cosh \alpha + l_0 \sinh \alpha$, $l_0 \mapsto l'_0 = l_z \sinh \alpha + l_0 \cosh \alpha$, leads to the *distortion* of the sphere of ray directions characterized in (3.10) by

$$\tan \tfrac{1}{2}\theta \mapsto \tan \tfrac{1}{2}\theta' = e^{-\alpha} \tan \tfrac{1}{2}\theta, \tag{3.11}$$

or $\sin \theta' = \sin \theta / (\cosh \alpha + \cos \theta \sinh \alpha)$, that is, the magnitude of (3.9*a*). Angles *in* the plane perpendicular to the boost are not affected, of course.

The transformation of optical phase space $\wp$ given by (3.9), and the corresponding closed formulas for $\exp \alpha \hat{b}_z$, have been called the *relativistic coma* aberration in Ref. [17], for the reason that the position space transformation is *purely comatic* in a sense that will be clarified in the following sections. Let us say here that coma is the aberration canonically conjugate to momentum *distortion*, $\mathbf{p}'(\mathbf{p})$ independent of $\mathbf{q}$, and that first-order coma is pure magnification, as will be also borne out below. This seems to be a type of aberration common to *any* phase space where the momentum subspace is a homogeneous space for a group of transformations.

Stellar aberration was discovered by Herschel in 1725, and the formula is (3.10). Telescopes are Fourier transformers in the sense of imaging directions into screen positions, but they receive essentially parallel rays. If we had an image-forming device that collected a bundle of rays with a nonvanishing angular spread from a stationary object

—the proverbial optical candle in front of a wide aperture lens—, then our formalism predicts that the snapshot image of the same object *moving* with respect to the screen would suffer from relativistic coma aberration. Whether or not this phenomenon is beyond mere mathematics and actually observable is a question that may be of interest to elementary particle experimentalists that collect equivalent images of radiating relativistic particles in their beams, for example.

Between the extremes argued in the previous two paragraphs (mathematical *vs.* experimental), a geometrical opticist may baffle and delight in the fact that *motion* is introduced in the ideal world of light rays that has no time variable at all. Lest this appear to be a paradox due to idealization, let me point out to Ref. [18], where the same authors have presented Helmholtz *wave* optics under the same deformation.

3.4. The Snell algebra for refracting surfaces

The previous generators have been built on empty, homogeneous space, or else they must be given a local sense. In correspondence with the previous sections, let us give now some results for *refracting surfaces*, where the conserved quantities $\bar{\mathbf{p}}$ and $\bar{\mathbf{q}}$ and functions thereof are a rich source of interesting algebras. Snell's law referred to the local normal to a surface, $n \sin\theta_{\rm loc} = n' \sin\theta'_{\rm loc}$, we recall, is a conservation law for the tangential component of the ray direction vector. In terms of the notation around Eq. (2.8), with the normal appropriately normalized to unity, in Ref. [11] we defined *Snell's* invariant two-vector

$$\mathbf{k} = \bar{\mathbf{p}}/\sqrt{1+\varsigma'^2}, \qquad \text{where } \varsigma'^2 = [\nabla\varsigma(\bar{\mathbf{q}})]^2, \tag{3.12}$$

whose value before and after refraction is the same. Using the canonical conjugacy of $\bar{\mathbf{p}}$ and $\bar{\mathbf{q}}$, we may show that the Poisson bracket between the two components is

$$\{k_x, k_y\} = \frac{\bar{\mathbf{p}} \times \bar{\nabla}\varsigma'^2}{2(1+\varsigma'^2)^2}, \qquad \text{(general surface)} \tag{3.13a}$$

$$= \left[\frac{d}{d(\bar{\mathbf{q}}^2)}\left(\frac{-1}{1+\varsigma'^2}\right)\right] \bar{\mathbf{q}} \times \bar{\mathbf{p}}, \qquad \begin{pmatrix}\text{axis-symmetric}\\ \text{surface } \varsigma(|\bar{\mathbf{q}}|)\end{pmatrix} \tag{3.13b}$$

$$= R^{-2}\bar{\mathbf{q}} \times \bar{\mathbf{p}} = r_z/R^2, \qquad \begin{pmatrix}\text{spherical surface}\\ \text{of radius } R\end{pmatrix} \tag{3.13c}$$

where $\bar{\nabla}$ is the gradient with respect to $\bar{\mathbf{q}}$.

When the surface is axially symmetric, then the ray *skewness* or *Petzval*,

$$r_z = \mathbf{q} \times \mathbf{p} = \bar{\mathbf{q}} \times \bar{\mathbf{p}} = \mathbf{q}' \times \mathbf{p}', \tag{3.14}$$

is also an invariant and generates rotations around the symmetry axis, intertwining the two Snell invariants:

$$\{r_z, k_x\} = k_y, \qquad \{r_z, k_y\} = -k_x. \tag{3.15}$$

When the refracting surface is a *sphere* tangent to the screen at the optical center, $\varsigma(\mathbf{q}) = R - \sqrt{R^2 - q^2}$, the three functions (Rk_x, Rk_y, r_z) close —predictably enough— into the *so(3)* algebra of rotations around the center of the sphere.

We shall not detail the subject of refraction symmetries, but only note that a number of features clarify considerably using Lie's methods. Global arguments presented in [11] lead to an economical location of the well-known *aplanatic points* of the sphere (a pencil of rays passing through one of a pair, upon refraction, pass through the other). Such arguments may be drawn on other conic surfaces immediately, but the author has not done this yet. Infinitesimal arguments for the sphere and a recursive computation reported in the same reference lead to the determination of its aberration coefficients to ninth order. Interesting problems await research on the global sphere model, such as multiple internal refraction/reflection leading to a group-theoretical understanding of rainbow and glory [19], to name one where wavization should also play a part.

4. The Nth rank aberration group

On February 10, 1840, Humboldt protests in writing against H. Berghaus who had published in Stuttgart, three years earlier, Allgemeine Länder und Volkerkunde..., proposing that the current off the coast of Perú be called after him.

"The current has been known for three hundred years by all fishermen from Chile to Payta; my only merit was being the first to measure the water temperatures of that current: 12°5 Réaumur in it and 21° R. outside."[IV]

Most students are familiar with geometrical *linear* optics from any first course on the subject. This is the approximation to geometrical optics that considers rays with small angles $|\mathbf{p}| \ll n$, and objects and images well within the apertures of the pupils of the system $|\mathbf{q}| \ll 1$. It is called the *paraxial* approximation of optics, and allows the calculations to proceed through matrix algebra and vector sum of (correspondingly small) displacements and tilts on the ray coordinates $w = (\mathbf{p}, \mathbf{q})$. For the group theorist, this approximation means studying the inhomogeneous four-dimensional symplectic group *ISp(4,R)*, that acts, not on the optical phase space $\wp$ described in the previous sections, but on the point-particle mechanical phase space $\Re^4$, with abbreviation of concepts. (Similarly for wave optics we study the Weyl-symplectic group *WSp(4,R)* in its metaplectic representation [20]. The homomorphism allows the usual Schrödinger wavization of paraxial optics.)

A second simplification is to consider optical systems that are *aligned*, *i.e.*, built out of elements such as free spaces, almost-flat refracting surfaces, optical fibers,

---

[IV] Quoted by Charles Minguet, *op. cit.*

and generally media that are invariant under rotations around a *common optical axis*. Except for *magnetic* optics, these elements are also invariant under reflections across any plane containing that axis. Group-theoretically this reduces *ISP(4,R)* to *ISp(4,R)*, the little group of null translations, and thereafter through the chain $Sp(4,R) \supset O(2) \times Sp(2,R)$ with the trivial representation of *O(2)*. Thus, *Sp(2,R)* is our basic group in the realization of geometrical optics.

4.1. The Gaussian (paraxial) group

Within this framework, let us define the following quadratic functions of phase space that are invariant under rotations around the z-axis:

$$\xi_+ = \tfrac{1}{\sqrt{2}}p^2, \qquad \xi_0 = \mathbf{p}\cdot\mathbf{q}, \qquad \xi_- = \tfrac{1}{\sqrt{2}}q^2. \tag{4.1}$$

Their Lie operators generate the Lie algebra *sp(2,R)*:

$$\{\xi_0, \xi_\pm\} = \pm\xi_\pm, \qquad \{\xi_+, \xi_-\} = 2\xi_0. \tag{4.2}$$

The Lie transformation generated by the linear combination of ξ's on $\Re^4$ phase space may be given in matrix form as

$$\begin{aligned} &\exp(A_1\xi_+ + A_0\xi_0 + A_{-1}\xi_-)^\wedge \begin{pmatrix}\mathbf{p}\\ \mathbf{q}\end{pmatrix} \\ &\qquad = \begin{pmatrix} \cosh\omega - A_0\operatorname{sinch}\omega & -\sqrt{2}A_{-1}\operatorname{sinch}\omega \\ \sqrt{2}A_1\operatorname{sinch}\omega & \cosh\omega + A_0\operatorname{sinch}\omega \end{pmatrix}\begin{pmatrix}\mathbf{p}\\ \mathbf{q}\end{pmatrix}, \\ &\omega = \pm\sqrt{A_0^2 - 2A_1A_{-1}}, \qquad \operatorname{sinch}\omega = \omega^{-1}\sinh\omega. \end{aligned} \tag{4.3}$$

We notice that although vectors $\mathbf{p}$ and $\mathbf{q}$ appear, their x- and y-components are not separately transformed; they may be regarded as irreducible variables for momentum and position.

The above formula is all we require to describe paraxial aligned optics through action on $(\mathbf{p}, \mathbf{q})$. The optical elements are represented by matrices that are known to anyone with a first course in optics, except, perhaps for the *root* transformation, that here is all too obvious:

Free propagation by ℓ in a medium of refractive index n:

$$\exp(-\ell p^2/2n)^\wedge \quad \longleftrightarrow \quad \begin{pmatrix} 1 & 0 \\ -\ell/n & 1 \end{pmatrix}. \tag{4.4a}$$

Root transformation by the surface $\varsigma(q) = \varsigma_2 q^2 + \cdots$ in medium n:

$$\exp(-n\varsigma_2 q^2)^\wedge \quad \longleftrightarrow \quad \begin{pmatrix} 1 & -2n\varsigma_2 \\ 0 & 1 \end{pmatrix}. \tag{4.4b}$$

Refracting surface $\varsigma(q) = \varsigma_2 q^2 + \cdots$ from medium n_1 to medium n_2:

$$\exp([n_2 - n_1]\varsigma_2 q^2)^\wedge \quad \longleftrightarrow \quad \begin{pmatrix} 1 & 2(n_2 - n_1)\varsigma_2 \\ 0 & 1 \end{pmatrix}. \tag{4.4c}$$

Propagation in fiber with $n(q) = n_0 - n_2 q^2 + \cdots$ through length ℓ:

$$\exp -\ell(p^2/2n_0 + n_2 q^2)^\wedge \quad \longleftrightarrow \quad \begin{pmatrix} \cos\nu\ell & n_0\nu\sin\nu\ell \\ -\frac{\nu}{2n_2}\sin\nu\ell & \cos\nu\ell \end{pmatrix}, \quad \nu = \sqrt{\frac{2n_2}{n_0}}. \tag{4.4d}$$

Pure magnifier (an ideal optical element) with image reduction factor $e^{-\mu}$:

$$\exp(\mu\mathbf{p}\cdot\mathbf{q})^\wedge \quad \longleftrightarrow \quad \begin{pmatrix} e^{\mu} & 0 \\ 0 & e^{-\mu} \end{pmatrix}. \tag{4.4e}$$

An optical system composed of more than one optical element is obtained by multiplication of the corresponding group elements or of their matrix representatives *in the same order as light rays rays traverse them.* The universal optical convention is along the positive z-axis, extending to the *right.*

4.2. The second Descartes sphere

Implied in (4.1) we have the spherical coordinates of a three-vector that may be written $\xi = (\xi_1, \xi_2, \xi_3)$ in an abstract $\Re^3$ space:

$$\begin{aligned} \xi_+ &= -\tfrac{1}{\sqrt{2}}(\xi_1 + i\xi_2), & \xi_1 &= \tfrac{1}{2}(p^2 - q^2), \\ \xi_0 &= \xi_3, & \xi_2 &= \tfrac{i}{2}(p^2 + q^2), \\ \xi_- &= \tfrac{1}{\sqrt{2}}(\xi_1 - i\xi_2), & \xi_3 &= \mathbf{p}\cdot\mathbf{q}. \end{aligned} \tag{4.5}$$

The norm of this three-vector is an *invariant* under *Sp(2,R)*:

$$\xi^2 = \xi_1^2 + \xi_2^2 + \xi_3^2 = \xi_0^2 - 2\xi_+\xi_- = (\mathbf{p}\cdot\mathbf{q})^2 - p^2q^2 = -(\mathbf{q}\times\mathbf{p})^2, \tag{4.6}$$

since $\{\xi_i, \xi^2\} = 0$. This sphere (4.6) we may call the *second* Descartes sphere; its radius is recognized as (minus) the *Petzval* of classical optics. Also, $\mathbf{q}\times\mathbf{p} = r_z$ was given in (3.5), as the function that generates the *SO(2)* group. This is complementary to *Sp(2,R)* within *Sp(4,R)*. (The corresponding Casimir operators bear out this complementarity, but will not be otherwise needed in what follows.)

The radius of this second Descartes sphere happens to be pure imaginary, but we shall use it below to introduce spherical harmonic functions on phase space, and thus provide the group-theoretical basis for the classification of aberrations.

4.3. The nesting of aberration algebras

Our interest lies here in aberration, since it is a *nonlinear phenomenon*. In the last section we presented two explicit examples of optical phase space transformations that were given through closed formulas: *(a)* the Euclidean transformations, that appear to be nonlinear on $\wp$ only due to the artifact of projecting the 3-world of rays on a flat screen but that are perfectly *rigid* otherwise, and *(b)* the Lorentz transformation of

$\wp$ that are nonlinear by most standards although the same group has, of course, linear representation spaces.

Algebraic nonlinearities are of a mild kind, though, and the phenomenon of optic aberration has been solved since Ludwig Seidel by perturbation expansions.♠ The subject of aberration in optics is daunting in extent and in depth. The pair of books by Buchdahl [22] are the classic references to this subject and have been applied to lens design by a large number of users. It is perhaps fortunate that I happened to approach the subject without prior knowledge of its extensive literature. Instead, from research originally done on linear canonical transformations in classical and quantum mechanics with Prof. Marcos Moshinsky [23], it seems that the subject of aberration may be presented in the following terms, based on *any* connected Lie group Γ with its Lie algebra γ. This is done here concretely for *Sp(2,R)* in the geometric optics realization. We shall refer to Γ and γ as the *Gaussian* group and algebra, respectively, and understand that this is the *first*-order, linear transformation structure of interest.

We build a *nested family* of *enveloping algebras* γ_N out of γ, with $\gamma_1 = \gamma$ and $\gamma_N \subset \gamma_{N+1}$, referring to them as N^{th} rank *aberration* algebras. To this end, an operation of *multiplication* must be introduced between the elements of γ that is distributive with respect to addition and satisfies the Leibnitz identity under the Lie bracket in γ. Ordinary function product under Poisson brackets fulfills these conditions. Thus if a linear basis of generator functions of γ_1 is ξ_i ($i = +,0,-$ or $1,2,3$ for *sp(2,R)*), $\xi_i\xi_j$ and ξ_i modulo degree 3, will be a basis for γ_2, $\xi_i\xi_j\xi_k$, $\xi_i\xi_j$ and ξ_i modulo degree 4 for γ_3, etc.

The enveloping algebras γ_N are built here with commutative multiplication and the Lie bracket that induced by γ *modulo* degree N in the generators of γ. If F_k is the linear space of polynomials $a_k(\xi)$ of homogeneous degree k in $\xi_i \in \gamma$, then $F_{k_1} \cdot F_{k_2} \subset F_{k_1+k_2}$, and $F_n \equiv 0$ for $n > N$. The Leibnitz identity leads from $\{F_1, F_1 \cdot F_1\} \subset F_2$ to $\{F_1, F_k\} \subset F_k$, $k = 1, \ldots, N$, and thence to*

$$\{F_{k_1}, F_{k_2}\} \subset \begin{cases} F_{k_1+k_2-1}, & k_1 + k_2 - 1 \leq N, \\ 0, & \text{otherwise.} \end{cases} \tag{4.7}$$

Taking the operation of multiplication modulo degre $N+1$, we assure that we obtain finite-dimensional algebras γ_N whose generators are in $\uplus_{k=1}^{N} F_k$, (the additive union of functions) and that the algebras γ_k are nested by *rank* as $\gamma = \gamma_1 \subset \gamma_2 \subset \cdots \subset \gamma_N$. Since $\{F_1, F_k\} = F_k$, the original γ is *normal* in any γ_k, and $\gamma_N - \gamma_k$ is an *ideal* under the former, as will the *pure aberration* subalgebra $\gamma_N^{\mathfrak{a}} = \uplus_{k=2}^{N} \gamma_k$. The structure of the algebra will be thus that of a *semidirect sum* $\gamma_N = \gamma_N^{\mathfrak{a}} \niplus \gamma_1$. In the case of geometrical

---

♠ *"Hier zeigt sich nun, dass durch Einführung der Grössen σ und h in den Bruch die Kettengestalt völlig aufgelöst wird, sodass sich bei der Berechnung sämmtlicher Zähler und Nenner seiner Näherungsbrüche Alles auf gewöhnlicher Summen reducirt, welche selbst für den allgemeinsten Fall ohne Weiteres hingeschrieben werden können."* For Seidel [21], who considered only *meridional* rays, σ was proportional to the angle between the ray and the optical axis $|\mathbf{p}|$, and h the distance to the optical center $|\mathbf{q}|$.

* The usual appearance of this relation [9] has been $\{F_n, F_m\} \subset F_{n+m-2}$ due to the fact that the F's were considered homogeneous functions of $\mathbf{p}$ and $\mathbf{q}$, rather than ξ. The former variables correspond to the Gaussian algebra γ being the Heisenberg-Weyl algebra.

optics, since the generating functions ξ_i of γ are quadratic in the phase space variables (p_x, p_y, q_x, q_y), rank N relates to *aberration order* A through $A = 2N - 1$. The latter refers to the highest degree of nonlinearity of the phase-space mapping through Γ_N, as will be given explicitly in the next sections. Ranks $N = 1, 2, 3$, and 4, correspond to aberration orders $A = 1, 3, 5$, and 7; first order is paraxial optics.

4.4. Aberration groups in factored product form

We build the Nth rank aberration *group* Γ_N as the Lie exponential of the rank-N aberration algebra γ_N. Let $a_k \in F_k$; we have found very useful the following generic parametrization of $G \in \Gamma_N$ as a *factored product* [24]:

$$G\{a_N, \ldots, a_3, a_2, a_1\} = \exp \hat{a}_N \cdots \exp \hat{a}_3 \exp \hat{a}_2 \exp \hat{a}_1. \tag{4.8}$$

The rightmost factor $G\{0, \ldots, 0, a_1\} = \exp \hat{a}_1$ is an element of the original Gaussian group Γ. The structure of Γ_N is inherited from that of γ_N: it is a *semidirect product* $\Gamma_N = \Gamma_N^{\mathbf{a}} \rtimes \Gamma_1$, where $\Gamma_N^{\mathbf{a}}$ is the *pure aberration* subgroup $G\{a_N, \ldots, a_2, 0\}$

It is most convenient to handle the $\Gamma_1 = Sp(2,R)$ parameters is using the 2×2 unimodular matrix representation given in (4.3). We may thus write (4.8) further abbreviated to $G\{\mathbf{a}, \mathbf{M}\}$, with $\mathbf{a} = \{a_N, \ldots, a_3, a_2\}$, the aberration *vector*. The composition of two generic elements $G\{\mathbf{a}, \mathbf{M}\}$ and $G\{\mathbf{b}, \mathbf{N}\}$ in Γ_N will thus reflect the semidirect product structure as:

$$\begin{aligned} G\{\mathbf{a}, \mathbf{M}\}G\{\mathbf{b}, \mathbf{N}\} &= G\{\mathbf{a}, \mathbf{1}\}G\{\mathbf{0}, \mathbf{M}\}G\{\mathbf{b}, \mathbf{1}\}G\{\mathbf{0}, \mathbf{N}\} \\ &= G\{\mathbf{a}, \mathbf{1}\}G\{\mathbf{0}, \mathbf{M}\}G\{\mathbf{b}, \mathbf{1}\}G\{\mathbf{0}, \mathbf{M}\}^{-1}G\{\mathbf{0}, \mathbf{M}\}G\{\mathbf{0}, \mathbf{N}\} \\ &= G\{\mathbf{a}, \mathbf{1}\}G\{\mathbf{D}(\mathbf{M}, \mathbf{b}), \mathbf{1}\}G\{\mathbf{0}, \mathbf{MN}\} \\ &= G\{\mathbf{a} \sharp \mathbf{D}(\mathbf{M}, \mathbf{b}), \mathbf{MN}\}. \end{aligned} \tag{4.9}$$

Here $\mathbf{D}(\mathbf{M}, \mathbf{b}) = \{\mathbf{D}^{(N)}(\mathbf{M}, b_N), \ldots, \mathbf{D}^{(3)}(\mathbf{M}, b_3), \mathbf{D}^{(2)}(\mathbf{M}, b_2)\}$ is the result of the adjoint action of the Gaussian group element $\mathbf{M} \in \Gamma$ on the ideal $b_k \in F_k$ already reduced completely by rank. The operation "$\sharp$" of *pure aberration composition* will be seen below. Let us first summarize the results of the action $\mathbf{D}^{(k)}(\mathbf{M}, b_k)$ stating the next step of the construction.

To write explicit expressions, and for concrete calculations, we will refer to a *basis* of the algebras γ_k. The most obvious basis, as has been implied above, is that of simple products of the generator functions of γ_1, *i.e.*, the *monomials*

$$M_{k_+, k_0, k_-} = (p^2)^{k_+} (\mathbf{p} \cdot \mathbf{q})^{k_0} (q^2)^{k_-} = 2^{(k_+ + k_-)/2} \xi_+^{k_+} \xi_0^{k_0} \xi_-^{k_-}. \tag{4.10}$$

Every $a_k(\mathbf{x}) \in F_k$ may be thus written as the linear combination

$$a_k(\mathbf{x}) = \sum_{k_+ + k_0 + k_- = k} A^{\text{mon}}_{k_+, k_0, k_-} M_{k_+, k_0, k_-}. \tag{4.11}$$

This *monomial* basis of aberration coefficients, $\{A^{\text{mon}}_{k_+,k_0,k_-}\}$ is very useful for many calculations and also to find and state theoretical results, particularly those pertaining refracting surfaces. To those familiar with harmonic oscillator physics, it should recall the classification of states according to the number of energy quanta along three orthogonal axes; see Figure 4.1*a*. In the same analogue system, however, we must recall that the classification of states by *total angular momentum* and magnetic quantum numbers has its equivalent share of advantages, particularly in the study of physical phenomena such as spin-orbit coupling in nuclei. In optics the only *prima facie* reason to use one classification basis over another is to achieve concordance with the commonly known aberrations (spherical aberration, coma, astigmatism, curvature of field, distortion, etc.) We suggest an equally powerful reason to propose a different scheme, the analogue of total angular momentum classification, and that is computational efficiency in the concatenation of optical elements and subsystems. This may not be altogether obvious to the veteran opticist, to whom we should present honest apologies; less persuation is needed when presenting this material to researchers with group-theoretical inclinations. In fact, my first reaction upon reading the article of Alex Dragt in JOSA [2], was to try to understand the matter precisely in the way I did before, as a student (with Prof. Moshinsky) on nuclear interactions in the *2s-1d* shell. Presenting one group-theoretically classified aberration basis will make it easier for the reader to shift to other ones better adapted to the particular system under study —such as geometrical aberration optics in fiber-like or other non-imaging media.

4.5. Symplectic classification of aberrations

We shall classify the aberration algebra generators $a_k(\xi)$ into *irreducible representations* under the Gaussian group Γ. As shown in Figure 4.1*b*, for *Sp(2,R)*, the classification of kth degree polynomials in the variables ξ_i is completely analogous to that of the states of a three-dimensional quantum oscillator in multiplets of angular momentum: The 6 generators in F_2 for the third-order aberration algebra γ_2 are divided into a $j = 2$ quintuplet and a $j = 0$ singlet; the 10 generators of F_3 of fifth-order aberrations are a $j = 3$ septuplet and a $j = 1$ triplet; the 15 generators of γ_4 are a nonuplet, a quintuplet, and a singlet of seventh order aberrations,* etc. In this way, the polynomials $a_k(\xi)$ are further classified as $a_{k,j}(\xi)$ by the *symplectic spin* label j, with range $j = k, k-2, \ldots 1$ or 0. These are the finite-dimensional non-unitary irreducible representations of *Sp(2,R)*, in the guise of *SO(3)* spherical harmonics of integral spin j [5,9,11]. *Half*-integral spin objects are the phase space variables $(\mathbf{p}, \mathbf{q})$. In these notes, however, and for reasons purely of length and state of development of the theory, we shall stay away from the spin-$\frac{1}{2}$ philosophy, even though parts of it have appeared in the author's published literature. (For *asymmetrical* optical systems, for example, we

---

* The classic literature on the subject [22], p. 51, counts 5, 9, and 14 aberrations of orders 3, 5, and 7, respectively. This comes from not including the monomials $(q^2)^k$, *i.e.*, ξ_-^k among the generators, since they leave the image (q) space unaffected. They *do* aberrate the conjugate p-space of directions, though.

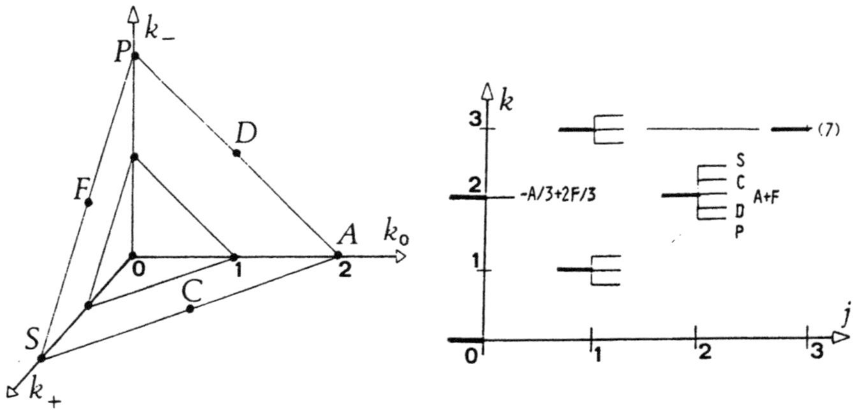

FIGURES 4.1 *(a)* **Classification of *Sp(2,R)* generators and third-order aberrations in the monomial basis. There appear spherical aberration (*S*), coma (*S*), astigmatism (*A*), curvature of field (*F*), distortion (*D*), and pocus (*P*).** *(b)* **Classification of the same objects into symplectic 'angular momentum' multiplets: the three *Sp(2,R)* generators at $k = 1$, and the six third-order aberrations at $k = 2$ decomposing into a singlet and a quintuplet. This separates *astigmature* $-\frac{1}{3}A + \frac{2}{3}F$ from *curvatism* $A + F$.**

must deal with *Sp(4,R)* and, for the same reason, we have *SO(5)* spinor harmonics with *SO(3)* breaking [12]).

The choice of lens systems is to aim at perfect *imaging* in the sense that $\mathbf{q} \mapsto \rho\mathbf{q}$, $\mathbf{p} \mapsto \rho^{-1}\mathbf{p} + \tau(\mathbf{q})$ and, more stringently, $\tau = 0$. In doing so, we are distinguishing the generator $\xi_0 = \mathbf{p}\cdot\mathbf{q}$ by diagonal matrices, as can be seen in (4.4*e*). The "magnetic quantum number" m of oscillator physics follows the spectrum of $\hat{\xi}_0$ in each (k, j) multiplet. We may call this the ***Seidel index*** [25] since it classifies the commonly named Seidel aberrations, although it does not distinguish by itself the generic astigmatism/curvature of field and similar degeneracies discussed in the last section. The traditional names[♡] are given in Table 4.1.

We should add that when the choice of Gaussian system is different from that of pure magnifiers, the appropriate "magnetic axis" should be different. For optical fiber-like systems (including Fourier transformers, *i.e.*, paraxial harmonic oscillators), $-\frac{1}{2}(p^2 + q^2) = -i\xi_2$ is a better choice; it is related to ξ_0 by a Bargmann transformation [26,27], *i.e.* by a rotation of $i\pi/2$ around the ξ_1 axis [25]. Finally, it should be noted

---

♡ Except for the last one, that is new, but has already passed the referees of several respectable journals!

SEIDEL INDEX	DEGENERACY	CLASSICAL NAME
$m = k$	$j = k$	spherical aberration,
$m = k - 1$	$j = k$	circular coma,
$m = k - 2$	$j = k,\ k - 2$	oblique spherical aberration,
$m = k - 3$	$j = k,\ k - 2$	(nameless)
...	...	...
$m = 0$	$j = k,\ k - 2,\ k - 4 \ldots, 1$ or 0	(nameless)
...	...	...
$m = 3 - k$	$j = k,\ k - 2$	elliptical coma,
$m = 2 - k$	$j = k,\ k - 2$	curvature of field / astigmatism,
$m = 1 - k$	$j = k$	distortion,
$m = -k$	$j = k$	pocus.

TABLE 4.1 The Seidel index classification of aberrations.

that for *asymmetric* magnifying systems we may use ξ_0 and the Petzval invariant $\mathbf{p} \times \mathbf{q}$, together with the (broken) j-spin to classify aberrations [12].

4.6. Symplectic and spherical harmonics

Here we use the decomposition and classification of the aberration generator polynomials in the symplectic basis,

$$a_k(\xi) = \sum_{j=k(-2)}^{1 \text{ or } 0} \sum_{m=-j}^{j} A_{kjm}\, {}^k\mathcal{X}^j_m(\xi) = \mathbf{A}_k \cdot {}^k\mathcal{X}(\xi), \tag{4.12}$$

with the *symplectic harmonics* ${}^k\mathcal{X}^j_m(\xi)$. These are given in terms of the solid spherical harmonics $\mathcal{Y}^j_m(\xi)$ [9,11] by:

$${}^k\mathcal{X}^j_m(\xi) = (\xi^2)^{(k-j)/2}\sqrt{\frac{4\pi(2j+1)\,(j+m)!\,(j-m)!}{(2j-1)!!}}\,\mathcal{Y}^j_m(\xi) \tag{4.13a}$$

$$= (\xi^2)^{(k-j)/2}\frac{(j+m)!\,(j-m)!}{2^{m/2}(2j-1)!!}\sum_n \frac{1}{2^n}\frac{\xi_+^{m+n}}{(m+n)!}\frac{\xi_0^{j-m-2n}}{(j-m-2n)!}\frac{\xi_-^n}{n!}. \tag{4.13b}$$

The square root factor in the definition above is used so as to have ${}^k\mathcal{X}^k_k = (p^2)^k$, with unit coefficient, and all other ${}^k\mathcal{X}^k_m$'s with *rational* coefficients of monomials in p^2, $\mathbf{p}\cdot\mathbf{q}$, and q^2, that have the property of summing to unity. In Ref. [9] these were tabulated as functions of p^2, $\mathbf{p}\cdot\mathbf{q}$, and q^2. Below, we write their expressions as functions of the ξ_i's; we need only give them for $m \geq 0$, since

$${}^k\mathcal{X}^j_{-m}(\xi_+, \xi_0, \xi_-) = {}^k\mathcal{X}^j_m(\xi_-, \xi_0, \xi_+). \tag{4.14}$$

The first two ($k = 0, 1$) are a scalar and the *sp(2,R)* generator functions:

$$ {}^0\chi^0_0 = 1; \qquad {}^1\chi^1_{\pm 1} = \sqrt{2}\,\xi_\pm, \quad {}^1\chi^1_0 = \xi_0. \tag{4.15$_1$}$$

The basis functions $k = 2$ are the generators of **third**-order aberrations:

$$\begin{aligned} {}^2\chi^2_2 &= 2\xi_+^2, \\ {}^2\chi^2_1 &= \sqrt{2}\,\xi_+\xi_0, \\ {}^2\chi^2_0 &= \tfrac{2}{3}(\xi_+\xi_- + \xi_0^2), \qquad {}^2\chi^0_0 = 2\xi_+\xi_- - \xi_0^2. \end{aligned} \tag{4.15$_2$}$$

For $k = 3$ we have the generators of **fifth**-order aberrations:

$$\begin{aligned} {}^3\chi^3_3 &= 2\sqrt{2}\,\xi_+^3, \\ {}^3\chi^3_2 &= 2\xi_+^2\xi_0, \\ {}^3\chi^3_1 &= \tfrac{2}{5}\sqrt{2}\,(\xi_+^2\xi_- + 2\xi_+\xi_0^2), \qquad {}^3\chi^1_1 = {}^2\chi^0_0\,{}^1\chi^1_1, \\ {}^3\chi^3_0 &= \tfrac{2}{5}(3\xi_+\xi_0\xi_- + \xi_0^3), \qquad {}^3\chi^1_0 = {}^2\chi^0_0\,{}^1\chi^1_0. \end{aligned} \tag{4.15$_3$}$$

For $k = 4$ we have those of *seventh*-order aberrations:

$$\begin{aligned} {}^4\chi^4_4 &= 4\xi_+^4, \\ {}^4\chi^4_3 &= 2\sqrt{2}\,\xi_+^3\xi_0, \\ {}^4\chi^4_2 &= \tfrac{4}{7}(\xi_+^3\xi_- + 3\xi_+^2\xi_0^2), \qquad {}^4\chi^2_2 = {}^2\chi^0_0\,{}^2\chi^2_2, \\ {}^4\chi^4_1 &= \tfrac{2}{7}\sqrt{2}\,(3\xi_+^2\xi_0\xi_- + 2\xi_+\xi_0^3), \qquad {}^4\chi^2_1 = {}^2\chi^0_0\,{}^2\chi^2_1, \\ {}^4\chi^4_0 &= \tfrac{4}{35}(3\xi_+^2\xi_-^2 + 12\xi_+\xi_0^2\xi_- + 2\xi_0^4), \quad {}^4\chi^2_0 = {}^2\chi^0_0\,{}^2\chi^2_0, \quad {}^4\chi^0_0 = ({}^2\chi^0_0)^2. \end{aligned} \tag{4.15$_4$}$$

4.7. Irreducible Gaussian action

Let us now return to the composition rule in the aberration group, given by equation (4.9). When $a_k(\xi)$ is decomposed by symplectic spin as in (4.12), the action of the Gaussian subgroup $\mathbf{M}$ on the aberration polynomial is given by $\mathbf{D}(\mathbf{M}, a_k)(\xi) = G\{\mathbf{0},\mathbf{M}\}a_k(\xi) = a_k(G\{\mathbf{0},\mathbf{M}\}\xi)$, and will be *independent of rank k*. The irreducible matrices $\mathbf{D}^{(j)}(\mathbf{M}^{-1})$ of spin j will act on the column vector ${}^k\chi^j$ in the last term of that equation thus:

$$\begin{aligned} G\{\mathbf{0},\mathbf{M}\}\,{}^k\chi^j(\xi) &= {}^k\chi^j(G\{\mathbf{0},\mathbf{M}\}\xi) \\ &= {}^k\chi^j(\mathbf{D}^{(1)}(\mathbf{M}^{-1})\xi) = \mathbf{D}^{(j)}(\mathbf{M}^{-1})\,{}^k\chi^j(\xi). \end{aligned} \tag{4.16}$$

We check that $G\{\mathbf{0},\mathbf{M}\}G\{\mathbf{0},\mathbf{N}\}\,{}^k\chi^j(\xi) = \mathbf{D}^{(j)}([\mathbf{MN}]^{-1})\,{}^k\chi^j(\xi)$, so products are well preserved by the representation.

Alternatively, we may characterize the aberration polynomials $a_k(\xi)$ through their *aberration coefficients* A_{kjm} and write the latter as numerical or symbolic group parameters in a column vector. (*This* is what we have done to implement the formalism through symbolic computation algorithms.) The Gaussian group action on that vector is given through the *transpose* matrix $\mathbf{D}^{(j)}(\mathbf{M}^{-1})^\top$. The transformation is now

$$\begin{aligned} a'_{kj}(\xi) &= G\{\mathbf{0},\mathbf{M}\}a_{kj}(\xi) = \mathbf{A}_{kj}\cdot G\{\mathbf{0},\mathbf{M}\}\,{}^k\chi^j(\xi) \\ &= \mathbf{A}_{kj}\cdot \mathbf{D}^{(j)}(\mathbf{M}^{-1})\,{}^k\chi^j(\xi) = [\mathbf{D}^{(j)}(\mathbf{M}^{-1})^\top\mathbf{A}_{kj}]\cdot {}^k\chi^j(\xi) \\ &= \mathbf{A}'_{kj}\cdot {}^k\chi^j(\xi). \end{aligned} \tag{4.17a}$$

Hence

$$\mathbf{A}'_{kj} = \mathbf{D}^{(j)}(\mathbf{M}^{-1})^\top\mathbf{A}_{kj}. \tag{4.17b}$$

Again, we check that under two succesive transformations, $a''_k(\xi) = G\{\mathbf{0},\mathbf{M}\}a'_{kj}(\xi) = G\{\mathbf{0},\mathbf{M}\}G\{\mathbf{0},\mathbf{N}\}a_{kj}(\xi)$ the aberration coefficients follow $A''_{kj} = \mathbf{D}^{(j)}(\mathbf{M}^{-1})^\top\mathbf{A}'_{kj} = \mathbf{D}^{(j)}(\mathbf{M}^{-1})^\top\mathbf{D}^{(j)}(\mathbf{N}^{-1})^\top\mathbf{A}_{kj} = \mathbf{D}^{(j)}([\mathbf{MN}]^{-1})^\top\mathbf{A}_{kj}$, as they should.

The generic matrix elements of $\mathbf{D}^{(j)}(\mathbf{M})$ are, explicitly [5],

$$D^{(j)}_{m,m'}\begin{pmatrix} a & b \\ c & d\end{pmatrix} = \sum_n \begin{pmatrix} j-m \\ j+m'+n\end{pmatrix}\begin{pmatrix} j+m \\ n\end{pmatrix} a^n b^{j+m-n} c^{j+m'-n} d^{n-m-m'}. \tag{4.18}$$

where the matrix $\mathbf{M}$ is unimodular: $ad - bc = 1$.

To close this section, we should underline the first advantage of the symplectic classification of aberrations in the economy of computing the adjoint action of the Gaussian group Γ, *i.e.*, *Sp(2,R)* on the aberration coefficients when concatenating two optical transformations as in (4.9). The action is fully block-diagonal. The *active* matrix elements $[\mathbf{D}^{(j)}(\mathbf{M}^{-1})^\top]_{n,n'}$ we count in $D(M,a)$, for strict aberration orders 3, 5, and 7 are: $5^2+1^2=26$, $7^2+3^2=58$, and $9^2+5^2+1^2=107$. Had we used the *monomial* basis, the matrices of each rank would be *full* and, for the same orders, the number of elements involved for generic $\mathbf{a}$ would be $6^2=36$, $10^2=100$, and $15^2=225$. The computation complexity score for this operation in seventh order thus stands at 191 *vs.* 361, *i.e.*, roughly 1:2 in favor of group theory.

4.8. Composition of aberrations

We may write the composition of two *pure* aberration group elements $\mathbf{a}$ and $\mathbf{b}$ (standing for the column vectors of all their aberration coefficients $\{A_{kjm}\}$ and $\{B_{kjm}\}$), to third pure aberration $\mathbf{c}$,

$$G\{\mathbf{a},\mathbf{1}\}G\{\mathbf{b},\mathbf{1}\} = G\{\mathbf{a}\sharp\mathbf{b},\mathbf{1}\} = G\{\mathbf{c},\mathbf{1}\}. \tag{4.19}$$

In Γ_4 for seventh-order aberrations, in the factored-product form,

$$\exp\hat{c}_4\ \exp\hat{c}_3\ \exp\hat{c}_2 = \exp\hat{a}_4\ \exp\hat{a}_3\ \exp\hat{a}_2\ \cdot\ \exp\hat{b}_4\ \exp\hat{b}_3\ \exp\hat{b}_2, \tag{4.20}$$

the composition is given explicitly[1] by [8]

$$c_2 = a_2 + b_2, \tag{4.21a}$$

$$c_3 = a_3 + b_3 + \tfrac{1}{2}\{a_2, b_2\}, \tag{4.21b}$$

$$\begin{aligned} c_4 = a_4 + b_4 &+ \{a_2, b_3\} \\ &+ \tfrac{1}{3}\{a_2, \{a_2, b_2\}\} - \tfrac{1}{6}\{\{a_2, b_2\}, b_2\}. \end{aligned} \tag{4.21c}$$

For the $\Gamma = Sp(2,R)$ group, the number of parameters in the *pure* aberration subgroup $\Gamma_2^{\rm a}$ is 6, composition is abelian: *cf.* (4.21*a*). Fifth-order pure aberrations $\Gamma_3^{\rm a}$ bring 10 new parameters, now 16; it is no longer abelian; third-order coefficients *compound* to fifth-order ones, bilinearly. Finally $\Gamma_4^{\rm a}$ adds 15 seventh-order coefficients to a grand total of 31, they involve bilinear compounding between a_2 and b_3 (and not a_3 with b_2), and trilinear composition of the third-order coefficients. If nothing is assumed about a polynomial basis, the generic case $\{a_2, b_2\}$ involves $6 \times 7/2 = 21$ Poisson brackets between the basis elements, while $\{a_2, b_3\}$ involves $6 \times 10 = 60$ summands.

Let us take the monomial basis (4.10)–(4.11), $M_{k_+,k_0,k_-}(p^2, \mathbf{p}\cdot\mathbf{q}, q^2)$. For them we have the quite simple-looking Poisson bracket

$$\begin{aligned} \{M_{a,b,c}, M_{a',b',c'}\} = {} & 4(ca' - ac')M_{a+a'-1,b+b'+1,c+c'-1} \\ & + 2(ba' - ab' + cb' - bc')M_{a+a',b+b'-1,c+c'}. \end{aligned} \tag{4.22}$$

Excepting those near the fringes of the k-multiplet, the number of monomials thus rougly duplicate (a graphical algorithm may be given). It seems counterindicated to compute Poisson brackets for the general case through differentiation every time with respect to $\mathbf{p}$, $\mathbf{q}$, or ξ_i. This has to be done "once and for all" to find the *algebraic* relations between the aberration *coefficients*, the set of $\{A^{\rm mon}_{k_+,k_0,k_-}\}$'s. We did that and found a long formula. Through counting (with LENGTH function in muSIMP) we found the visible 6 sums in (4.21*a*), 44 sums in (4.21*b*), and 407 sums in (4.21*c*). The number would decrease if we had some extra *selection rule* for the Poisson brackets.

Enter the symplectic basis. The Poisson bracket between the generic symplectic polynomials can be written as

$$\{{}^k\chi^j_m, {}^{k'}\chi^{j'}_{m'}\} = \sum_{j''} S^{j,j',j''}_{m,m'}\, {}^{k+k'-1}\chi^{j''}_{m+m'}, \tag{4.23}$$

where the Seidel indices simply add, and the sum is over all j'' compatible with $m+m'$, and such that $k+k'+j''$ is *odd*. The coefficients $S^{j,j',j''}_{m,m'}$ have been written [9] in terms of a sum of three *SO(3)* Wigner coefficients, times a reduced one, times common factors. They involve again 6 sums for (4.21*a*), now 42 sums (*vs.* 44) for (4.21*b*) (27 for the

[1] In a preprint by the author [28], Eq. (4.21*c*) here appeared there as (3.2*c*) with a *mis-write* that omitted the bilinear term. Since the product reproduced here was claimed to derive from them, we should add that the tables are correct (we checked they compose correctly), and only the preprint erratum escaped.

septuplet and 15 for the triplet), and the hefty sum of 303 of them (*vs.* 407) for (4.21c) (169 for the nonuplet, 133 for the quintuplet, and 1 for the singlet).

Once the expressions were calculated (by *other* means), they became a muSIMP algebraic function `COMPOSE (A, B)` that takes $A = \{A_{kjm}\}$ and $B = \{B_{kjm}\}$ for arguments (a third, optional argument specifies the aberration order, default is 7), and prints $C = \{C_{kjm}\}$. Adapted to TEX, they are:

Aberration order 3:

$$C_{2,j,m} = A_{2,j,m} + B_{2,j,m}, \qquad j = 2, 0, \quad m = j, j-1, \ldots, -j. \tag{4.24a}$$

Aberration order 5:

$$\begin{aligned}
C_{3,3,3} &= 2A_{2,2,1}B_{2,2,2} - 2A_{2,2,2}B_{2,2,1} + A_{3,3,3} + B_{3,3,3}, \\
C_{3,3,2} &= 4A_{2,2,0}B_{2,2,2} - 4A_{2,2,2}B_{2,2,0} + A_{3,3,2} + B_{3,3,2}, \\
C_{3,3,1} &= 6A_{2,2,-1}B_{2,2,2} + 2A_{2,2,0}B_{2,2,1} - 2A_{2,2,1}B_{2,2,0} - 6A_{2,2,2}B_{2,2,-1} + A_{3,3,1} + B_{3,3,1}, \\
C_{3,3,0} &= 8A_{2,2,-2}B_{2,2,2} + 4A_{2,2,-1}B_{2,2,1} - 4A_{2,2,1}B_{2,2,-1} - 8A_{2,2,2}B_{2,2,-2} + A_{3,3,0} + B_{3,3,0}, \\
C_{3,3,-1} &= 6A_{2,2,-2}B_{2,2,1} + 2A_{2,2,-1}B_{2,2,0} - 2A_{2,2,0}B_{2,2,-1} - 6A_{2,2,1}B_{2,2,-2} + A_{3,3,-1} + B_{3,3,-1}, \\
C_{3,3,-2} &= 4A_{2,2,-2}B_{2,2,0} - 4A_{2,2,0}B_{2,2,-2} + A_{3,3,-2} + B_{3,3,-2}, \\
C_{3,3,-3} &= 2A_{2,2,-2}B_{2,2,-1} - 2A_{2,2,-1}B_{2,2,-2} + A_{3,3,-3} + B_{3,3,-3};
\end{aligned} \tag{4.24b$_1$}$$

$$\begin{aligned}
C_{3,1,1} &= 4/5A_{2,2,-1}B_{2,2,2} - 2/5A_{2,2,0}B_{2,2,1} + 2/5A_{2,2,1}B_{2,2,0} - 4/5A_{2,2,2}B_{2,2,-1} + A_{3,1,1} + B_{3,1,1}, \\
C_{3,1,0} &= 16/5A_{2,2,-2}B_{2,2,2} - 2/5A_{2,2,-1}B_{2,2,1} + 2/5A_{2,2,1}B_{2,2,-1} - 16/5A_{2,2,2}B_{2,2,-2} + A_{3,1,0} + B_{3,1,0}, \\
C_{3,1,-1} &= 4/5A_{2,2,-2}B_{2,2,1} - 2/5A_{2,2,-1}B_{2,2,0} + 2/5A_{2,2,0}B_{2,2,-1} - 4/5A_{2,2,1}B_{2,2,-2} + A_{3,1,-1} + B_{3,1,-1}.
\end{aligned} \tag{4.24b$_2$}$$

Aberration order 7:

$$\begin{aligned}
C_{4,4,4} = {} & 8A_{2,2,1}^2B_{2,2,2} + 32/3A_{2,2,2}^2B_{2,2,0} - 16/3A_{2,2,0}B_{2,2,2}^2 - 32/3A_{2,2,0}A_{2,2,2}B_{2,2,2} - 8A_{2,2,1}A_{2,2,2}B_{2,2,1} + \\
& 4A_{2,2,1}B_{2,2,1}B_{2,2,2} + 6A_{2,2,1}B_{3,3,3} - 4A_{2,2,2}B_{2,2,1}^2 + 16/3A_{2,2,2}B_{2,2,0}B_{2,2,2} - 4A_{2,2,2}B_{3,3,2} + A_{4,4,4} + B_{4,4,4},
\end{aligned}$$

$$\begin{aligned}
C_{4,4,3} = {} & 32A_{2,2,2}^2B_{2,2,-1} - 16A_{2,2,-1}B_{2,2,2}^2 - 32A_{2,2,-1}A_{2,2,2}B_{2,2,2} + 64/3A_{2,2,0}A_{2,2,1}B_{2,2,2} - \\
& 80/3A_{2,2,0}A_{2,2,2}B_{2,2,1} - 8/3A_{2,2,0}B_{2,2,1}B_{2,2,2} + 12A_{2,2,0}B_{3,3,3} + 16/3A_{2,2,1}A_{2,2,2}B_{2,2,0} + \\
& 40/3A_{2,2,1}B_{2,2,0}B_{2,2,2} + 2A_{2,2,1}B_{3,3,2} + 16A_{2,2,2}B_{2,2,-1}B_{2,2,2} - 32/3A_{2,2,2}B_{2,2,0}B_{2,2,1} - 8A_{2,2,2}B_{3,3,1} + \\
& A_{4,4,3} + B_{4,4,3},
\end{aligned}$$

$$\begin{aligned}
C_{4,4,2} = {} & 64/3A_{2,2,0}^2B_{2,2,2} + 8/3A_{2,2,1}^2B_{2,2,0} + 64A_{2,2,2}^2B_{2,2,-2} - 32A_{2,2,-2}B_{2,2,2}^2 - 64A_{2,2,-2}A_{2,2,2}B_{2,2,2} + \\
& 16A_{2,2,-1}A_{2,2,1}B_{2,2,2} - 56A_{2,2,-1}A_{2,2,2}B_{2,2,1} - 20A_{2,2,-1}B_{2,2,1}B_{2,2,2} + 18A_{2,2,-1}B_{3,3,3} - \\
& 4/3A_{2,2,0}B_{2,2,1}^2 - 8/3A_{2,2,0}A_{2,2,1}B_{2,2,1} - 64/3A_{2,2,0}A_{2,2,2}B_{2,2,0} + 32/3A_{2,2,0}B_{2,2,0}B_{2,2,2} + \\
& 8A_{2,2,0}B_{3,3,2} + 40A_{2,2,1}A_{2,2,2}B_{2,2,-1} + 28A_{2,2,1}B_{2,2,-1}B_{2,2,2} + 4/3A_{2,2,1}B_{2,2,0}B_{2,2,1} - 2A_{2,2,1}B_{3,3,1} - \\
& 32/3A_{2,2,2}B_{2,2,0}^2 + 32A_{2,2,2}B_{2,2,-2}B_{2,2,2} - 8A_{2,2,2}B_{2,2,-1}B_{2,2,1} - 12A_{2,2,2}B_{3,3,0} + A_{4,4,2} + B_{4,4,2},
\end{aligned}$$

$$\begin{aligned}
C_{4,4,1} = {} & 16/3A_{2,2,0}^2B_{2,2,1} + 16A_{2,2,1}^2B_{2,2,-1} - 96A_{2,2,-2}A_{2,2,2}B_{2,2,1} - 48A_{2,2,-2}B_{2,2,1}B_{2,2,2} + \\
& 24A_{2,2,-2}B_{3,3,3} - 8A_{2,2,-1}B_{2,2,1}^2 + 160/3A_{2,2,-1}A_{2,2,0}B_{2,2,2} - 16A_{2,2,-1}A_{2,2,1}B_{2,2,1} - \\
& 176/3A_{2,2,-1}A_{2,2,2}B_{2,2,0} - 8/3A_{2,2,-1}B_{2,2,0}B_{2,2,2} + 14A_{2,2,-1}B_{3,3,2} - 16/3A_{2,2,0}A_{2,2,1}B_{2,2,0} + \\
& 16/3A_{2,2,0}A_{2,2,2}B_{2,2,-1} + 88/3A_{2,2,0}B_{2,2,-1}B_{2,2,2} + 8/3A_{2,2,0}B_{2,2,0}B_{2,2,1} + 4A_{2,2,0}B_{3,3,1} -
\end{aligned}$$

$$8/3A_{2,2,1}B_{2,2,0}^2+96A_{2,2,1}A_{2,2,2}B_{2,2,-2}+48A_{2,2,1}B_{2,2,-2}B_{2,2,2}+8A_{2,2,1}B_{2,2,-1}B_{2,2,1}-6A_{2,2,1}B_{3,3,0}-$$
$$80/3A_{2,2,2}B_{2,2,-1}B_{2,2,0}-16A_{2,2,2}B_{3,3,-1}+A_{4,4,1}+B_{4,4,1},$$

$$C_{4,4,0}=40A_{2,2,-1}^2B_{2,2,2}+40A_{2,2,1}^2B_{2,2,-2}-20A_{2,2,-2}B_{2,2,1}^2+160/3A_{2,2,-2}A_{2,2,0}B_{2,2,2}-$$
$$40A_{2,2,-2}A_{2,2,1}B_{2,2,1}-320/3A_{2,2,-2}A_{2,2,2}B_{2,2,0}-80/3A_{2,2,-2}B_{2,2,0}B_{2,2,2}+20A_{2,2,-2}B_{3,3,2}+$$
$$40/3A_{2,2,-1}A_{2,2,0}B_{2,2,1}-80/3A_{2,2,-1}A_{2,2,1}B_{2,2,0}-40A_{2,2,-1}A_{2,2,2}B_{2,2,-1}+20A_{2,2,-1}B_{2,2,-1}B_{2,2,2}-$$
$$20/3A_{2,2,-1}B_{2,2,0}B_{2,2,1}+10A_{2,2,-1}B_{3,3,1}+40/3A_{2,2,0}A_{2,2,1}B_{2,2,-1}+160/3A_{2,2,0}A_{2,2,2}B_{2,2,-2}+$$
$$160/3A_{2,2,0}B_{2,2,-2}B_{2,2,2}+40/3A_{2,2,0}B_{2,2,-1}B_{2,2,1}+20A_{2,2,1}B_{2,2,-2}B_{2,2,1}-20/3A_{2,2,1}B_{2,2,-1}B_{2,2,0}-$$
$$10A_{2,2,1}B_{3,3,-1}-20A_{2,2,2}B_{2,2,-1}^2-80/3A_{2,2,2}B_{2,2,-2}B_{2,2,0}-20A_{2,2,2}B_{3,3,-2}+A_{4,4,0}+B_{4,4,0},$$

$$C_{4,4,-1}=16A_{2,2,-1}^2B_{2,2,1}+16/3A_{2,2,0}^2B_{2,2,-1}+96A_{2,2,-2}A_{2,2,-1}B_{2,2,2}+16/3A_{2,2,-2}A_{2,2,0}B_{2,2,1}-$$
$$176/3A_{2,2,-2}A_{2,2,1}B_{2,2,0}-96A_{2,2,-2}A_{2,2,2}B_{2,2,-1}-80/3A_{2,2,-2}B_{2,2,0}B_{2,2,1}+16A_{2,2,-2}B_{3,3,1}-$$
$$8/3A_{2,2,-1}B_{2,2,0}^2-16/3A_{2,2,-1}A_{2,2,0}B_{2,2,0}-16A_{2,2,-1}A_{2,2,1}B_{2,2,-1}+48A_{2,2,-1}B_{2,2,-2}B_{2,2,2}+$$
$$8A_{2,2,-1}B_{2,2,-1}B_{2,2,1}+6A_{2,2,-1}B_{3,3,0}+160/3A_{2,2,0}A_{2,2,1}B_{2,2,-2}+88/3A_{2,2,0}B_{2,2,-2}B_{2,2,1}+$$
$$8/3A_{2,2,0}B_{2,2,-1}B_{2,2,0}-4A_{2,2,0}B_{3,3,-1}-8A_{2,2,1}B_{2,2,-1}^2-8/3A_{2,2,1}B_{2,2,-2}B_{2,2,0}-14A_{2,2,1}B_{3,3,-2}-$$
$$48A_{2,2,2}B_{2,2,-2}B_{2,2,-1}-24A_{2,2,2}B_{3,3,-3}+A_{4,4,-1}+B_{4,4,-1},$$

$$C_{4,4,-2}=64A_{2,2,-2}^2B_{2,2,2}+8/3A_{2,2,-1}^2B_{2,2,0}+64/3A_{2,2,0}^2B_{2,2,-2}-32/3A_{2,2,-2}B_{2,2,0}^2+$$
$$40A_{2,2,-2}A_{2,2,-1}B_{2,2,1}-64/3A_{2,2,-2}A_{2,2,0}B_{2,2,0}-56A_{2,2,-2}A_{2,2,1}B_{2,2,-1}-64A_{2,2,-2}A_{2,2,2}B_{2,2,-2}+$$
$$32A_{2,2,-2}B_{2,2,-2}B_{2,2,2}-8A_{2,2,-2}B_{2,2,-1}B_{2,2,1}+12A_{2,2,-2}B_{3,3,0}-8/3A_{2,2,-1}A_{2,2,0}B_{2,2,-1}+$$
$$16A_{2,2,-1}A_{2,2,1}B_{2,2,-2}+28A_{2,2,-1}B_{2,2,-2}B_{2,2,1}+4/3A_{2,2,-1}B_{2,2,-1}B_{2,2,0}+2A_{2,2,-1}B_{3,3,-1}-$$
$$4/3A_{2,2,0}B_{2,2,-1}^2+32/3A_{2,2,0}B_{2,2,-2}B_{2,2,0}-8A_{2,2,0}B_{3,3,-2}-20A_{2,2,1}B_{2,2,-2}B_{2,2,-1}-$$
$$18A_{2,2,1}B_{3,3,-3}-32A_{2,2,2}B_{2,2,-2}^2+A_{4,4,-2}+B_{4,4,-2},$$

$$C_{4,4,-3}=32A_{2,2,-2}^2B_{2,2,1}+16/3A_{2,2,-2}A_{2,2,-1}B_{2,2,0}-80/3A_{2,2,-2}A_{2,2,0}B_{2,2,-1}-$$
$$32A_{2,2,-2}A_{2,2,1}B_{2,2,-2}+16A_{2,2,-2}B_{2,2,-2}B_{2,2,1}-32/3A_{2,2,-2}B_{2,2,-1}B_{2,2,0}+8A_{2,2,-2}B_{3,3,-1}+$$
$$64/3A_{2,2,-1}A_{2,2,0}B_{2,2,-2}+40/3A_{2,2,-1}B_{2,2,-2}B_{2,2,0}-2A_{2,2,-1}B_{3,3,-2}-8/3A_{2,2,0}B_{2,2,-2}B_{2,2,-1}-$$
$$12A_{2,2,0}B_{3,3,-3}-16A_{2,2,1}B_{2,2,-2}^2+A_{4,4,-3}+B_{4,4,-3},$$

$$C_{4,4,-4}=32/3A_{2,2,-2}^2B_{2,2,0}+8A_{2,2,-1}^2B_{2,2,-2}-4A_{2,2,-2}B_{2,2,-1}^2-8A_{2,2,-2}A_{2,2,-1}B_{2,2,-1}-$$
$$32/3A_{2,2,-2}A_{2,2,0}B_{2,2,-2}+16/3A_{2,2,-2}B_{2,2,-2}B_{2,2,0}+4A_{2,2,-2}B_{3,3,-2}+4A_{2,2,-1}B_{2,2,-2}B_{2,2,-1}-$$
$$6A_{2,2,-1}B_{3,3,-3}-16/3A_{2,2,0}B_{2,2,-2}^2+A_{4,4,-4}+B_{4,4,-4};$$

$$(4.24c_1)$$

$$C_{4,2,2}=-64/21A_{2,2,0}^2B_{2,2,2}-8/21A_{2,2,1}^2B_{2,2,0}+256/21A_{2,2,2}^2B_{2,2,-2}-128/21A_{2,2,-2}B_{2,2,2}^2-$$
$$256/21A_{2,2,-2}A_{2,2,2}B_{2,2,2}+176/21A_{2,2,-1}A_{2,2,1}B_{2,2,2}-16/3A_{2,2,-1}A_{2,2,2}B_{2,2,1}+$$
$$32/21A_{2,2,-1}B_{2,2,1}B_{2,2,2}+24/7A_{2,2,-1}B_{3,3,3}+4/21A_{2,2,0}B_{2,2,1}^2+8/21A_{2,2,0}A_{2,2,1}B_{2,2,1}+$$
$$64/21A_{2,2,0}A_{2,2,2}B_{2,2,0}-32/21A_{2,2,0}B_{2,2,0}B_{2,2,2}-8/7A_{2,2,0}B_{3,3,2}-64/21A_{2,2,1}A_{2,2,2}B_{2,2,-1}+$$
$$8/3A_{2,2,1}B_{2,2,-1}B_{2,2,2}-4/21A_{2,2,1}B_{2,2,0}B_{2,2,1}+2A_{2,2,1}B_{3,1,1}+24/35A_{2,2,1}B_{3,3,1}+32/21A_{2,2,2}B_{2,2,0}^2+$$
$$128/21A_{2,2,2}B_{2,2,-2}B_{2,2,2}-88/21A_{2,2,2}B_{2,2,-1}B_{2,2,1}-4A_{2,2,2}B_{3,1,0}-24/35A_{2,2,2}B_{3,3,0}+A_{4,2,2}+B_{4,2,2},$$

$$C_{4,2,1}=-16/7A_{2,2,0}^2B_{2,2,1}-88/21A_{2,2,1}^2B_{2,2,-1}+64/3A_{2,2,-2}A_{2,2,1}B_{2,2,2}-704/21A_{2,2,-2}A_{2,2,2}B_{2,2,1}-$$
$$128/21A_{2,2,-2}B_{2,2,1}B_{2,2,2}+96/7A_{2,2,-2}B_{3,3,3}+44/21A_{2,2,-1}B_{2,2,1}^2-32/21A_{2,2,-1}A_{2,2,0}B_{2,2,2}+$$
$$88/21A_{2,2,-1}A_{2,2,1}B_{2,2,1}-32/21A_{2,2,-1}A_{2,2,2}B_{2,2,0}-32/21A_{2,2,-1}B_{2,2,0}B_{2,2,2}+$$
$$16/7A_{2,2,0}A_{2,2,1}B_{2,2,0}+64/21A_{2,2,0}A_{2,2,2}B_{2,2,-1}+16/21A_{2,2,0}B_{2,2,-1}B_{2,2,2}-8/7A_{2,2,0}B_{2,2,0}B_{2,2,1}+$$
$$4A_{2,2,0}B_{3,1,1}-32/35A_{2,2,0}B_{3,3,1}+8/7A_{2,2,1}B_{2,2,0}^2+256/21A_{2,2,1}A_{2,2,2}B_{2,2,-2}+$$
$$352/21A_{2,2,1}B_{2,2,-2}B_{2,2,2}-44/21A_{2,2,1}B_{2,2,-1}B_{2,2,1}-2A_{2,2,1}B_{3,1,0}+48/35A_{2,2,1}B_{3,3,0}-$$
$$32/3A_{2,2,2}B_{2,2,-2}B_{2,2,1}+16/21A_{2,2,2}B_{2,2,-1}B_{2,2,0}-8A_{2,2,2}B_{3,1,-1}-96/35A_{2,2,2}B_{3,3,-1}+A_{4,2,1}+B_{4,2,1},$$

$$C_{4,2,0}=-16/7A_{2,2,-1}^2B_{2,2,2}-16/7A_{2,2,1}^2B_{2,2,-2}+8/7A_{2,2,-2}B_{2,2,1}^2+128/7A_{2,2,-2}A_{2,2,0}B_{2,2,2}+$$
$$16/7A_{2,2,-2}A_{2,2,1}B_{2,2,1}-256/7A_{2,2,-2}A_{2,2,2}B_{2,2,0}-64/7A_{2,2,-2}B_{2,2,0}B_{2,2,2}+48/7A_{2,2,-2}B_{3,3,2}-$$
$$24/7A_{2,2,-1}A_{2,2,0}B_{2,2,1}+48/7A_{2,2,-1}A_{2,2,1}B_{2,2,0}+16/7A_{2,2,-1}A_{2,2,2}B_{2,2,-1}-$$

$$
\begin{aligned}
&8/7A_{2,2,-1}B_{2,2,-1}B_{2,2,2} + 12/7A_{2,2,-1}B_{2,2,0}B_{2,2,1} + 6A_{2,2,-1}B_{3,1,1} - 48/35A_{2,2,-1}B_{3,3,1} - \\
&24/7A_{2,2,0}A_{2,2,1}B_{2,2,-1} + 128/7A_{2,2,0}A_{2,2,2}B_{2,2,-2} + 128/7A_{2,2,0}B_{2,2,-2}B_{2,2,2} - \\
&24/7A_{2,2,0}B_{2,2,-1}B_{2,2,1} - 8/7A_{2,2,1}B_{2,2,-2}B_{2,2,1} + 12/7A_{2,2,1}B_{2,2,-1}B_{2,2,0} - 6A_{2,2,1}B_{3,1,-1} + \\
&48/35A_{2,2,1}B_{3,3,-1} + 8/7A_{2,2,2}B^2_{2,2,-1} - 64/7A_{2,2,2}B_{2,2,-2}B_{2,2,0} - 48/7A_{2,2,2}B_{3,3,-2} + A_{4,2,0} + B_{4,2,0},
\end{aligned}
$$

$$
\begin{aligned}
C_{4,2,-1} = {} & -88/21A^2_{2,2,-1}B_{2,2,1} - 16/7A^2_{2,2,0}B_{2,2,-1} + 256/21A_{2,2,-2}A_{2,2,-1}B_{2,2,2} + \\
&64/21A_{2,2,-2}A_{2,2,0}B_{2,2,1} - 32/21A_{2,2,-2}A_{2,2,1}B_{2,2,0} - 704/21A_{2,2,-2}A_{2,2,2}B_{2,2,-1} - \\
&32/3A_{2,2,-2}B_{2,2,-1}B_{2,2,2} + 16/21A_{2,2,-2}B_{2,2,0}B_{2,2,1} + 8A_{2,2,-2}B_{3,1,1} + 96/35A_{2,2,-2}B_{3,3,1} + \\
&8/7A_{2,2,-1}B^2_{2,2,0} + 16/7A_{2,2,-1}A_{2,2,0}B_{2,2,0} + 88/21A_{2,2,-1}A_{2,2,1}B_{2,2,-1} + 64/3A_{2,2,-1}A_{2,2,2}B_{2,2,-2} + \\
&352/21A_{2,2,-1}B_{2,2,-2}B_{2,2,2} - 44/21A_{2,2,-1}B_{2,2,-1}B_{2,2,1} + 2A_{2,2,-1}B_{3,1,0} - 48/35A_{2,2,-1}B_{3,3,0} - \\
&32/21A_{2,2,0}A_{2,2,1}B_{2,2,-2} + 16/21A_{2,2,0}B_{2,2,-2}B_{2,2,1} - 8/7A_{2,2,0}B_{2,2,-1}B_{2,2,0} - 4A_{2,2,0}B_{3,1,-1} + \\
&32/35A_{2,2,0}B_{3,3,-1} + 44/21A_{2,2,1}B^2_{2,2,-1} - 32/21A_{2,2,1}B_{2,2,-2}B_{2,2,0} - 128/21A_{2,2,2}B_{2,2,-2}B_{2,2,-1} - \\
&96/7A_{2,2,2}B_{3,3,-3} + A_{4,2,-1} + B_{4,2,-1},
\end{aligned}
$$

$$
\begin{aligned}
C_{4,2,-2} = {} & 256/21A^2_{2,2,-2}B_{2,2,2} - 8/21A^2_{2,2,-1}B_{2,2,0} - 64/21A^2_{2,2,0}B_{2,2,-2} + 32/21A_{2,2,-2}B^2_{2,2,0} - \\
&64/21A_{2,2,-2}A_{2,2,-1}B_{2,2,1} + 64/21A_{2,2,-2}A_{2,2,0}B_{2,2,0} - 16/3A_{2,2,-2}A_{2,2,1}B_{2,2,-1} - \\
&256/21A_{2,2,-2}A_{2,2,2}B_{2,2,-2} + 128/21A_{2,2,-2}B_{2,2,-2}B_{2,2,2} - 88/21A_{2,2,-2}B_{2,2,-1}B_{2,2,1} + \\
&4A_{2,2,-2}B_{3,1,0} + 24/35A_{2,2,-2}B_{3,3,0} + 8/21A_{2,2,-1}A_{2,2,0}B_{2,2,-1} + 176/21A_{2,2,-1}A_{2,2,1}B_{2,2,-2} + \\
&8/3A_{2,2,-1}B_{2,2,-2}B_{2,2,1} - 4/21A_{2,2,-1}B_{2,2,-1}B_{2,2,0} - 2A_{2,2,-1}B_{3,1,-1} - 24/35A_{2,2,-1}B_{3,3,-1} + \\
&4/21A_{2,2,0}B^2_{2,2,-1} - 32/21A_{2,2,0}B_{2,2,-2}B_{2,2,0} + 8/7A_{2,2,0}B_{3,3,-2} + 32/21A_{2,2,1}B_{2,2,-2}B_{2,2,-1} - \\
&24/7A_{2,2,1}B_{3,3,-3} - 128/21A_{2,2,2}B^2_{2,2,-2} + A_{4,2,-2} + B_{4,2,-2};
\end{aligned}
\tag{4.24c$_2$}
$$

$$
C_{4,0,0} = A_{4,0,0} + B_{4,0,0}. \tag{4.24c$_3$}
$$

It should not be necessary to apologize the display of the full results for seventh-order aberrations in these notes, for the following reasons:

(a) when the formulas are part of a fast numerical manipulation systems, they may become useful for actual optical design, and copying them from here is still within the bounds of possibility for a determined team;

(b) Long as they are, they still yield to visual examination for specific properties. Both arguments, admittedly, are stretched to their limit. The author is in the best disposition to provide the source files on diskette, as soon as the entire symbolic computation project is reasonably finished.

These product functions have been tested to preserve associativity, a♯(b♯c) = (a♯b)♯c for the generic case of fifth-order aberrations and, due to memory restrictions, for random numerical cases for seventh aberration order. Numerical languages such as FORTRAN are of course faster than muSIMP for these operations, once the formulas have been appropriately translated.

5. The aberration group action on phase space

Alexander von Humboldt left Guayaquil on February 5th, 1802, and reached Acapulco on March 22nd. Indefatigable, on April 11th he determined the height of Cuernavaca to be 1656 meters over sea level, and that of Mexico City, 2 200 m. There, he met and was informed of the scientific works of Joaquín Velásques Cárdenas y León, José Antonio Alzate y Ramíres, and Antonio de León y Gama, at the Colegio Real de Minas. Their acquaintance prompted the Hofdemokrat to write an essay with the following stated purpose, which may read parrochial today but was so foreign to the early nineteenth-century European mind that it was soon forgotten:

"... only with the end of demostrating through their example that the ignorance that European pride pleases to attribute to those born in America is a consequence neither of climate nor of want of spiritual energy; on the contrary, there where it may still be seen, such ignorance is nothing but the fruit of isolation and the defects peculiar to the social institutions in the colonies." [V]

In the last section we constructed the Nth rank aberration group Γ_N out of the Gaussian group $\Gamma = \Gamma_1$. For the latter we have a linear representation space $\rho(\xi)$ built from its Lie algebra γ considered as a linear vector space, with basis ξ_i. The action of $\exp \hat{a}_1 \xi = \xi'(\xi)$ on this space is thus *linear*. Now, the generators of aberrations are in F_k and are of degree k in the ξ_i's and enjoys Lie brackets with the generators of Γ_N, hence the Lie transformation $\exp \hat{a}_k \xi = \xi'(\xi)$ will be *non*-linear, and will yield a series for $\xi'(\xi)$ with terms of degrees 1, k, $2k-1$, $3k-2$, ..., $n(k-1)+1$, ... in the components of ξ. (There is no assurance that the series will converge outside a neighborhood of the origin, though.) Since Lie exponential operators conserve their own Lie bracket, this transformation is formally *canonical* for $g \in \Gamma$, that is, it will conserve Lie brackets: $g\{\xi_1, \xi_2\} = \{g\xi_1, g\xi_2\}$.

5.1. The cutting of homogeneous space by degree

The geometrical optics model imposes a variant of this construction in that the homogeneous space of interest, ρ, is built not out of the algebra, but out of the *phase space* variables of optical momentum and position on a screen, namely $\mathbf{p}$ and $\mathbf{q}$. These are *spinor* variables for the Gaussian group $\Gamma = Sp(2,R)$. The action of the Gaussian group Γ_1 on $\mathbf{p}$, $\mathbf{q}$ is defined and given in (4.3) in terms of the matrix called $\mathbf{M}$ for short; the action of Γ_N on the same space will be then of the form

$$G\{\mathbf{a},\mathbf{M}\}\begin{pmatrix}\mathbf{p}\\ \mathbf{q}\end{pmatrix} = G\{\mathbf{a},\mathbf{1}\}\mathbf{M}^{-1}\begin{pmatrix}\mathbf{p}\\ \mathbf{q}\end{pmatrix} = \mathbf{M}^{-1}\begin{pmatrix}G\{\mathbf{a},\mathbf{1}\}\mathbf{p}\\ G\{\mathbf{a},\mathbf{1}\}\mathbf{q}\end{pmatrix} = \mathbf{M}^{-1}\begin{pmatrix}\alpha(\mathbf{a},\mathbf{p})\\ \alpha(\mathbf{a},\mathbf{q})\end{pmatrix}, \quad (5.1)$$

[V] *Essai politique sur le royaume de la Nouvelle-Espagne. Dédié a S.M. Charles IV.* (Schoell, Paris, 1811), quoted by Charles Minguet, *op. cit.*

where the action of the pure aberration subgroup on $\mathbf{p}$ and $\mathbf{q}$ is

$$\begin{aligned}\alpha(\mathbf{a},\mathbf{w}) &= \exp\hat{a}_N\ \ldots\exp\hat{a}_3\ \exp\hat{a}_2\,\mathbf{w}\\ &= \mathbf{w}+\{a_2,\mathbf{w}\}\\ &\quad+\{a_3,\mathbf{w}\}+\tfrac{1}{2!}\{a_2,\{a_2,\mathbf{w}\}\}\\ &\quad+\{a_4,\mathbf{w}\}+\{a_3,\{a_2,\mathbf{w}\}\}+\tfrac{1}{3!}\{a_2,\{a_2,\{a_2,\mathbf{w}\}\}\}\\ &\quad+\cdots.\end{aligned} \tag{5.2}$$

The transformation will be canonical now in the usual sense of Poisson brackets since α, as all Lie transformations, acts on arguments of functions, $\{\alpha(\mathbf{a},\mathbf{q}),\alpha(\mathbf{a},\mathbf{p})\} = \alpha(\mathbf{a},\{\mathbf{q},\mathbf{p}\}) = \{\mathbf{q},\mathbf{p}\}$, since for any constant c, $\alpha(\mathbf{a},c)=c$.

It is very important to point out here the *difference* between aberrations à la Lie and à la Seidel: [the followers of] Lie consider phase space transformations defined by Lie series (5.2) that are therefore canonical on phase space. Seidel [and *his* followers] rather *cut* after the first significant term in the $\mathbf{q}$-series, $\mathbf{q}+\{a_k,\mathbf{q}\}$, and do not worry about what happens to the conjugate momentum. Of course one may always obtain a transformation of ray direction to $\mathbf{p}_c(\mathbf{p},\mathbf{q})$ that will mantain canonicity; in fact a *whole family* of transformations $\{\mathbf{p}_c+S(\mathbf{q})\}$, related by the addition of an arbitrary function $S(\mathbf{q})$ will do so. This disregard for the specification of ray directions makes the followers of Seidel also generally discard* the aberration generated by ${}^k\chi^k_{-k}$.

To deal with series, nevertheless, it is useful to *cut* the series consistently to some degree D in $\mathbf{p}$ and $\mathbf{q}$. Canonicity will then hold generally up to terms of degree D. In doing so, however, if $D<2N-1$, the group action will not be *effective*, *i.e.*, aberrations of rank $k>D+1$ will leave the chosen space invariant. Here we shall cut $\alpha(\mathbf{a},\mathbf{w})$ to degree $A=2N-1$ in $\mathbf{p}$ and $\mathbf{q}$, *i.e.*, to the *aberration order*. For aberration order 7 we discard the ellipses in Eq. (5.2). In applications [29], however, when the system itself may be known or represented only to some low aberration order, the canonicity of the computed map is very important and needs D's larger than $2N-1$.

5.2. Explicit action on phase space

For the generic pure aberration group element of order 7, when the aberration polynomials a_2, a_3, and a_4 are specified by their symplectic aberration coefficients $\{A_{kjm}\}$ as in equation (4.12) we find the following *nonlinear* action on the position coordinate [28]:

$$\mathbf{q}'(\mathbf{p},\mathbf{q}) = G\{\mathbf{a},\mathbf{1}\}\mathbf{q} = \mathbf{q}+\eta(p^2,\mathbf{p}\cdot\mathbf{q},q^2)\,\mathbf{p}+\varsigma(p^2,\mathbf{p}\cdot\mathbf{q},q^2)\,\mathbf{q}, \tag{5.3a}$$

where

* *cf.* Ref. [22], Sect. 125, where the coefficient appears as "p_6", with some incidental remarks on duality, *i.e.*, Fourier conjugation. See ahead Eqs. (5.4)–(5.7). This coefficient is the Fourier conjugate of spherical aberration, we playfully called *pocus* in [5], since it p-unfocuses a position-perfect image, as will be detailed in Section 6.

$$
\begin{aligned}
\eta = {} & [2A_{2,0,0} + 2/3A_{2,2,0}]p^2 \\
& +[2A_{2,2,-1}]\,\mathbf{p}\cdot\mathbf{q} \\
& +[4A_{2,2,-2}]q^2 \\
& +[2A_{2,0,0}A_{2,2,1} - 4A_{2,2,-1}A_{2,2,2} + 2/3A_{2,2,0}A_{2,2,1} + 2A_{3,1,1} + 2/5A_{3,3,1}](p^2)^2 \\
& +[4A_{2,0,0}A_{2,2,0} - 16A_{2,2,-2}A_{2,2,2} - 2A_{2,2,-1}A_{2,2,1} + 4/3A^2_{2,2,0} + 2A_{3,1,0} + 6/5A_{3,3,0}]p^2\,\mathbf{p}\cdot\mathbf{q} \\
& +[2A_{2,0,0}A_{2,2,-1} - 4A_{2,2,-2}A_{2,2,1} + 2/3A_{2,2,-1}A_{2,2,0} + 4A_{3,1,-1} + 4/5A_{3,3,-1}]p^2q^2 \\
& +[4A_{2,0,0}A_{2,2,-1} - 8A_{2,2,-2}A_{2,2,1} + 4/3A_{2,2,-1}A_{2,2,0} - 2A_{3,1,-1} + 8/5A_{3,3,-1}](\mathbf{p}\cdot\mathbf{q})^2 \\
& +[8A_{2,0,0}A_{2,2,-2} - 16/3A_{2,2,-2}A_{2,2,0} + 2A^2_{2,2,-1} + 4A_{3,3,-2}]\,\mathbf{p}\cdot\mathbf{q}\,q^2 \\
& +[6A_{3,3,-3}](q^2)^2 \\
& +[-8/3A^2_{2,2,0}A_{2,2,2} + 3A_{2,0,0}A^2_{2,2,1} - 8A_{2,0,0}A_{2,2,0}A_{2,2,2} + 2A_{2,0,0}A_{3,3,2} + 64/3A_{2,2,-2}A^2_{2,2,2} - \\
& 4/3A_{2,2,-1}A_{2,2,1}A_{2,2,2} - 12A_{2,2,-1}A_{3,3,3} + A_{2,2,0}A^2_{2,2,1} + 2/3A_{2,2,0}A_{3,3,2} + 2A_{2,2,1}A_{3,1,1} + \\
& 2/5A_{2,2,1}A_{3,3,1} + 2A_{4,2,2} + 2/7A_{4,4,2}](p^2)^3 \\
& +[4/3A^2_{2,2,0}A_{2,2,1} - 24A_{2,0,0}A_{2,2,-1}A_{2,2,2} + 4A_{2,0,0}A_{2,2,0}A_{2,2,1} + 4A_{2,0,0}A_{3,3,1} + \\
& 112/3A_{2,2,-2}A_{2,2,1}A_{2,2,2} - 48A_{2,2,-2}A_{3,3,3} + 5/3A_{2,2,-1}A^2_{2,2,1} - 56/3A_{2,2,-1}A_{2,2,0}A_{2,2,2} - \\
& 10A_{2,2,-1}A_{3,3,2} + 4A_{2,2,0}A_{3,1,1} + 32/15A_{2,2,0}A_{3,3,1} + 2A_{2,2,1}A_{3,1,0} + 6/5A_{2,2,1}A_{3,3,0} + 2A_{4,2,1} + \\
& 6/7A_{4,4,1}](p^2)^2\,\mathbf{p}\cdot\mathbf{q} \\
& +[-4/3A^2_{2,0,0}A_{2,2,0} - 20/3A^2_{2,2,-1}A_{2,2,2} - 4/9A_{2,0,0}A^2_{2,2,0} - 16A_{2,0,0}A_{2,2,-2}A_{2,2,2} + \\
& 2A_{2,0,0}A_{2,2,-1}A_{2,2,1} + 2A_{2,0,0}A_{3,1,0} + 6/5A_{2,0,0}A_{3,3,0} + 10/3A_{2,2,-2}A^2_{2,2,1} - 12A_{2,2,-2}A_{3,3,2} + \\
& 2/3A_{2,2,-1}A_{2,2,0}A_{2,2,1} - 2A_{2,2,-1}A_{3,1,1} - 2/5A_{2,2,-1}A_{3,3,1} + 2/3A_{2,2,0}A_{3,1,0} + 2/5A_{2,2,0}A_{3,3,0} + \\
& 4A_{2,2,1}A_{3,1,-1} + 4/5A_{2,2,1}A_{3,3,-1} - 4/3A^3_{2,0,0} - 4/81A^3_{2,2,0} + 4A_{4,0,0} + 4/3A_{4,2,0} + 12/35A_{4,4,0}](p^2)^2q^2 \\
& +[4/3A^2_{2,0,0}A_{2,2,0} - 64/3A^2_{2,2,-1}A_{2,2,2} + 40/9A_{2,0,0}A^2_{2,2,0} - 32A_{2,0,0}A_{2,2,-2}A_{2,2,2} - 8A_{2,0,0}A_{2,2,-1}A_{2,2,1} - \\
& 2A_{2,0,0}A_{3,1,0} + 24/5A_{2,0,0}A_{3,3,0} + 56/3A_{2,2,-2}A^2_{2,2,1} - 32A_{2,2,-2}A_{3,3,2} - 8/3A_{2,2,-1}A_{2,2,0}A_{2,2,1} + \\
& 8A_{2,2,-1}A_{3,1,1} - 32/5A_{2,2,-1}A_{3,3,1} + 10/3A_{2,2,0}A_{3,1,0} + 4A_{2,2,0}A_{3,3,0} - 2A_{2,2,1}A_{3,1,-1} + \\
& 8/5A_{2,2,1}A_{3,3,-1} + 4/3A^3_{2,0,0} + 112/81A^3_{2,2,0} - 4A_{4,0,0} + 2/3A_{4,2,0} + 48/35A_{4,4,0}]p^2(\mathbf{p}\cdot\mathbf{q})^2 \\
& +[-4A^2_{2,0,0}A_{2,2,-1} - 14/3A^2_{2,2,-1}A_{2,2,1} - 8A_{2,0,0}A_{2,2,-2}A_{2,2,1} + 4/3A_{2,0,0}A_{2,2,-1}A_{2,2,0} + \\
& 4A_{2,0,0}A_{3,1,-1} + 24/5A_{2,0,0}A_{3,3,-1} - 160/3A_{2,2,-2}A_{2,2,-1}A_{2,2,2} + 16A_{2,2,-2}A_{2,2,0}A_{2,2,1} - \\
& 8A_{2,2,-2}A_{3,1,1} - 128/5A_{2,2,-2}A_{3,3,1} + 8/9A_{2,2,-1}A^2_{2,2,0} + 28/3A_{2,2,0}A_{3,1,-1} + 16/5A_{2,2,0}A_{3,3,-1} + \\
& 4A_{2,2,1}A_{3,3,-2} + 4A_{4,2,-1} + 12/7A_{4,4,-1}]p^2\,\mathbf{p}\cdot\mathbf{q}\,q^2 \\
& +[-8A^2_{2,0,0}A_{2,2,-2} - 32A^2_{2,2,-2}A_{2,2,2} - A^2_{2,2,-1}A_{2,2,0}/3 - A_{2,0,0}A^2_{2,2,-1} + 8/3A_{2,0,0}A_{2,2,-2}A_{2,2,0} + \\
& 2A_{2,0,0}A_{3,3,-2} + 16/9A_{2,2,-2}A^2_{2,2,0} - 12A_{2,2,-2}A_{3,1,0} - 36/5A_{2,2,-2}A_{3,3,0} + 8A_{2,2,-1}A_{3,1,-1} + \\
& 8/5A_{2,2,-1}A_{3,3,-1} + 2/3A_{2,2,0}A_{3,3,-2} + 6A_{2,2,1}A_{3,3,-3} + 6A_{4,2,-2} + 6/7A_{4,4,-2}]p^2(q^2)^2 \\
& +[4A^2_{2,0,0}A_{2,2,-1} - 20/3A^2_{2,2,-1}A_{2,2,1} - 16A_{2,0,0}A_{2,2,-2}A_{2,2,1} + 8/3A_{2,0,0}A_{2,2,-1}A_{2,2,0} - \\
& 4A_{2,0,0}A_{3,1,-1} + 16/5A_{2,0,0}A_{3,3,-1} - 64/3A_{2,2,-2}A_{2,2,-1}A_{2,2,2} + 32/3A_{2,2,-2}A_{2,2,0}A_{2,2,1} + \\
& 16A_{2,2,-2}A_{3,1,1} - 64/5A_{2,2,-2}A_{3,3,1} + 16/9A_{2,2,-1}A^2_{2,2,0} + 6A_{2,2,-1}A_{3,1,0} - 12/5A_{2,2,-1}A_{3,3,0} - \\
& 16/3A_{2,2,0}A_{3,1,-1} + 64/15A_{2,2,0}A_{3,3,-1} - 2A_{4,2,-1} + 8/7A_{4,4,-1}](\mathbf{p}\cdot\mathbf{q})^3 \\
& +[8A^2_{2,0,0}A_{2,2,-2} - 128/3A^2_{2,2,-2}A_{2,2,2} - 4/3A^2_{2,2,-1}A_{2,2,0} + 4A_{2,0,0}A^2_{2,2,-1} - 32/3A_{2,0,0}A_{2,2,-2}A_{2,2,0} + \\
& 8A_{2,0,0}A_{3,3,-2} + 128/9A_{2,2,-2}A^2_{2,2,0} - 64/3A_{2,2,-2}A_{2,2,-1}A_{2,2,1} + 20A_{2,2,-2}A_{3,1,0} - 24A_{2,2,-2}A_{3,3,0} - \\
& 2A_{2,2,-1}A_{3,1,-1} + 8/5A_{2,2,-1}A_{3,3,-1} + 32/3A_{2,2,0}A_{3,3,-2} - 4A_{4,2,-2} + 24/7A_{4,4,-2}](\mathbf{p}\cdot\mathbf{q})^2q^2 \\
& +[-32A^2_{2,2,-2}A_{2,2,1} + 12A_{2,0,0}A_{3,3,-3} + 8A_{2,2,-2}A_{2,2,-1}A_{2,2,0} + 8A_{2,2,-2}A_{3,1,-1} - 112/5A_{2,2,-2}A_{3,3,-1} + \\
& 10A_{2,2,-1}A_{3,3,-2} + 16A_{2,2,0}A_{3,3,-3} - A^3_{2,2,-1} + 6A_{4,4,-3}]\,\mathbf{p}\cdot\mathbf{q}\,(q^2)^2 \\
& +[-16/3A^2_{2,2,-2}A_{2,2,0} + 2A_{2,2,-2}A^2_{2,2,-1} - 12A_{2,2,-2}A_{3,3,-2} + 18A_{2,2,-1}A_{3,3,-3} + 8A_{4,4,-4}](q^2)^3,
\end{aligned}
\tag{5.3b}
$$

$$
\begin{aligned}
\zeta = {} & [A_{2,2,1}]p^2 \\
& +[-2A_{2,0,0} + 4/3A_{2,2,0}]\,\mathbf{p}\cdot\mathbf{q} \\
& +[A_{2,2,-1}]q^2 \\
& +[-4A_{2,2,0}A_{2,2,2} + 3/2A^2_{2,2,1} + A_{3,3,2}]p^2q^2
\end{aligned}
$$

$$
\begin{aligned}
&+[-2A_{2,0,0}A_{2,2,1} - 8A_{2,2,-1}A_{2,2,2} + 4/3A_{2,2,0}A_{2,2,1} - 2A_{3,1,1} + 8/5A_{3,3,1}]p^2\,\mathbf{p}\cdot\mathbf{q}\\
&+[-4/3A_{2,0,0}A_{2,2,0} - 8A_{2,2,-2}A_{2,2,2} + A_{2,2,-1}A_{2,2,1} - 2A^2_{2,0,0} - 2/9A^2_{2,2,0} + A_{3,1,0} + 3/5A_{3,3,0}]p^2q^2\\
&+[-8/3A_{2,0,0}A_{2,2,0} - 2A_{2,2,-1}A_{2,2,1} + 2A^2_{2,0,0} + 8/9A^2_{2,2,0} - 3A_{3,1,0} + 6/5A_{3,3,0}](\mathbf{p}\cdot\mathbf{q})^2\\
&+[-6A_{2,0,0}A_{2,2,-1} - 2A_{3,1,-1} + 8/5A_{3,3,-1}]\,\mathbf{p}\cdot\mathbf{q}\,q^2\\
&+[-8A_{2,0,0}A_{2,2,-2} + 4/3A_{2,2,-2}A_{2,2,0} - A^2_{2,2,-1}/2 + A_{3,3,-2}](q^2)^2\\
&+[16A_{2,2,-1}A^2_{2,2,2} - 28/3A_{2,2,0}A_{2,2,1}A_{2,2,2} - 12A_{2,2,0}A_{3,3,3} + 3A_{2,2,1}A_{3,3,2} + 5/2A^3_{2,2,1} + A_{4,4,3}](p^2)^3\\
&+[-16A^2_{2,2,0}A_{2,2,2} - 3A_{2,0,0}A^2_{2,2,1} + 8A_{2,0,0}A_{2,2,0}A_{2,2,2} - 2A_{2,0,0}A_{3,3,2} + 128/3A_{2,2,-2}A^2_{2,2,2} +\\
&16/3A_{2,2,-1}A_{2,2,1}A_{2,2,2} - 24A_{2,2,-1}A_{3,3,3} + 14/3A_{2,2,0}A^2_{2,2,1} - 20/3A_{2,2,0}A_{3,3,2} - 2A_{2,2,1}A_{3,1,1} +\\
&28/5A_{2,2,1}A_{3,3,1} - 2A_{4,2,2} + 12/7A_{4,4,2}](p^2)^2\,\mathbf{p}\cdot\mathbf{q}\\
&+[-2A^2_{2,0,0}A_{2,2,1} - 2/9A^2_{2,2,0}A_{2,2,1} + 8A_{2,0,0}A_{2,2,-1}A_{2,2,2} - 4/3A_{2,0,0}A_{2,2,0}A_{2,2,1} -\\
&4A_{2,0,0}A_{3,1,1} - 4/5A_{2,0,0}A_{3,3,1} + 8/3A_{2,2,-2}A_{2,2,1}A_{2,2,2} - 24A_{2,2,-2}A_{3,3,3} + 17/6A_{2,2,-1}A^2_{2,2,1} -\\
&20/3A_{2,2,-1}A_{2,2,0}A_{2,2,2} - A_{2,2,-1}A_{3,3,2} - 16/3A_{2,2,0}A_{3,1,1} - 16/15A_{2,2,0}A_{3,3,1} + 3A_{2,2,1}A_{3,1,0} +\\
&9/5A_{2,2,1}A_{3,3,0} + A_{4,2,1} + 3/7A_{4,4,1}](p^2)^2q^2\\
&+[2A^2_{2,0,0}A_{2,2,1} + 8/9A^2_{2,2,0}A_{2,2,1} + 16A_{2,0,0}A_{2,2,-1}A_{2,2,2} - 8/3A_{2,0,0}A_{2,2,0}A_{2,2,1} + 4A_{2,0,0}A_{3,1,1} -\\
&16/5A_{2,0,0}A_{3,3,1} + 32A_{2,2,-2}A_{2,2,1}A_{2,2,2} + 6A_{2,2,-1}A^2_{2,2,1} - 80/3A_{2,2,-1}A_{2,2,0}A_{2,2,2} - 16A_{2,2,-1}A_{3,3,2} +\\
&4/3A_{2,2,0}A_{3,1,1} - 16/15A_{2,2,0}A_{3,3,1} - 5A_{2,2,1}A_{3,1,0} + 6A_{2,2,1}A_{3,3,0} - 3A_{4,2,1} + 12/7A_{4,4,1}]p^2(\mathbf{p}\cdot\mathbf{q})^2\\
&+[-8/3A^2_{2,0,0}A_{2,2,0} - 40/3A^2_{2,2,-1}A_{2,2,2} - 20/9A_{2,0,0}A^2_{2,2,0} + 48A_{2,0,0}A_{2,2,-2}A_{2,2,2} +\\
&2A_{2,0,0}A_{2,2,-1}A_{2,2,1} - 6A_{2,0,0}A_{3,1,0} - 18/5A_{2,0,0}A_{3,3,0} + 32/3A_{2,2,-2}A^2_{2,2,1} - 64/3A_{2,2,-2}A_{2,2,0}A_{2,2,2} -\\
&16A_{2,2,-2}A_{3,3,2} + 8/3A_{2,2,-1}A_{2,2,0}A_{2,2,1} - 14A_{2,2,-1}A_{3,1,1} - 24/5A_{2,2,-1}A_{3,3,1} + 2A_{2,2,1}A_{3,1,-1} +\\
&32/5A_{2,2,1}A_{3,3,-1} + 4/3A^3_{2,0,0} - 32/81A^3_{2,2,0} - 4A_{4,0,0} + 2/3A_{4,2,0} + 48/35A_{4,4,0}]p^2\,\mathbf{p}\cdot\mathbf{q}\,q^2\\
&+[-2A^2_{2,0,0}A_{2,2,-1} + 5/6A^2_{2,2,-1}A_{2,2,1} + 8A_{2,0,0}A_{2,2,-2}A_{2,2,1} - 4/3A_{2,0,0}A_{2,2,-1}A_{2,2,0} - 8A_{2,0,0}A_{3,1,-1} -\\
&8/5A_{2,0,0}A_{3,3,-1} - 40/3A_{2,2,-2}A_{2,2,-1}A_{2,2,2} + 4/3A_{2,2,-2}A_{2,2,0}A_{2,2,1} - 16A_{2,2,-2}A_{3,1,1} -\\
&16/5A_{2,2,-2}A_{3,3,1} - 2/9A_{2,2,-1}A^2_{2,2,0} - A_{2,2,-1}A_{3,1,0} - 3/5A_{2,2,-1}A_{3,3,0} + 4/3A_{2,2,0}A_{3,1,-1} +\\
&4/15A_{2,2,0}A_{3,3,-1} + 3A_{2,2,1}A_{3,3,-2} + A_{4,2,-1} + 3/7A_{4,4,-1}]p^2(q^2)^2\\
&+[8/3A^2_{2,0,0}A_{2,2,0} - 32/3A^2_{2,2,-1}A_{2,2,2} - 16/9A_{2,0,0}A^2_{2,2,0} + 4A_{2,0,0}A_{2,2,-1}A_{2,2,1} + 6A_{2,0,0}A_{3,1,0} -\\
&12/5A_{2,0,0}A_{3,3,0} + 16/3A_{2,2,-2}A^2_{2,2,1} + 8A_{2,2,-1}A_{3,1,1} - 32/5A_{2,2,-1}A_{3,3,1} - 4A_{2,2,0}A_{3,1,0} +\\
&8/5A_{2,2,0}A_{3,3,0} - 4A_{2,2,1}A_{3,1,-1} + 16/5A_{2,2,1}A_{3,3,-1} - 4/3A^3_{2,0,0} + 32/81A^3_{2,2,0} + 4A_{4,0,0} - 8/3A_{4,2,0} +\\
&32/35A_{4,4,0}](\mathbf{p}\cdot\mathbf{q})^3\\
&+[2A^2_{2,0,0}A_{2,2,-1} - 2A^2_{2,2,-1}A_{2,2,1} + 16A_{2,0,0}A_{2,2,-2}A_{2,2,1} - 8/3A_{2,0,0}A_{2,2,-1}A_{2,2,0} + 8A_{2,0,0}A_{3,1,-1} -\\
&32/5A_{2,0,0}A_{3,3,-1} - 32A_{2,2,-2}A_{2,2,-1}A_{2,2,2} + 16/3A_{2,2,-2}A_{2,2,0}A_{2,2,1} + 8A_{2,2,-2}A_{3,1,1} -\\
&32/5A_{2,2,-2}A_{3,3,1} + 8/9A_{2,2,-1}A^2_{2,2,0} - 5A_{2,2,-1}A_{3,1,0} - 6A_{2,2,-1}A_{3,3,0} - 16/3A_{2,2,0}A_{3,1,-1} +\\
&64/15A_{2,2,0}A_{3,3,-1} + 8A_{2,2,1}A_{3,3,-2} - 3A_{4,2,-1} + 12/7A_{4,4,-1}](\mathbf{p}\cdot\mathbf{q})^2q^2\\
&+[-64/3A^2_{2,2,-2}A_{2,2,2} + 2/3A^2_{2,2,-1}A_{2,2,0} - 3A_{2,0,0}A^2_{2,2,-1} + 8A_{2,0,0}A_{2,2,-2}A_{2,2,0} - 10A_{2,0,0}A_{3,3,-2} -\\
&8/3A_{2,2,-2}A_{2,2,-1}A_{2,2,1} - 8A_{2,2,-2}A_{3,1,0} - 24/5A_{2,2,-2}A_{3,3,0} - 6A_{2,2,-1}A_{3,1,-1} - 16/5A_{2,2,-1}A_{3,3,-1} +\\
&20/3A_{2,2,0}A_{3,3,-2} + 12A_{2,2,1}A_{3,3,-3} - 2A_{4,2,-2} + 12/7A_{4,4,-2}]\,\mathbf{p}\cdot\mathbf{q}\,(q^2)^2\\
&+[-12A_{2,0,0}A_{3,3,-3} - 4/3A_{2,2,-2}A_{2,2,-1}A_{2,2,0} - 8A_{2,2,-2}A_{3,1,-1} - 8/5A_{2,2,-2}A_{3,3,-1} - A_{2,2,-1}A_{3,3,-2} +\\
&8A_{2,2,0}A_{3,3,-3} + A^3_{2,2,-1}/2 + A_{4,4,-3}](q^2)^3.
\end{aligned}
\tag{5.3c}
$$

5.3. The Fourier transformation

It is not necessary to give above the results corresponding for $\mathbf{p}'(\mathbf{p},\mathbf{q})$, since they can be obtained from the above results through *Fourier conjugation*,

$$F = G\{\mathbf{0},\mathbf{F}\}, \qquad \mathbf{F} = \begin{pmatrix} 0 & 1 \\ -1 & 0 \end{pmatrix}. \tag{5.4}$$

The basic observables of phase space and monomials undergo the transformations

$$(\mathbf{p},\mathbf{q}) \mapsto (-\mathbf{q},\mathbf{p}), \tag{5.5a}$$

$$M(p^2,\mathbf{p}\cdot\mathbf{q},q^2) \mapsto M(q^2,-\mathbf{p}\cdot\mathbf{q},p^2). \tag{5.5b}$$

The aberration *coefficients* appearing in (5.3) may be found from the representation $\mathbf{D}^{(j)}(\mathbf{F})$, given by the antidiagonal matrix

$$D^{(j)}_{m,m'}(\mathbf{F}) = (-1)^{j-m}\delta_{m',-m}. \tag{5.6}$$

On the symplectic function basis, thus, Fourier conjugation effects

$$G\{\mathbf{0},\mathbf{F}\}\,{}^k\chi^j_m = {}^k\tilde{\chi}^j_m = (-1)^{j-m}\,{}^k\chi^j_{-m} \tag{5.7}$$

and, on the corresponding aberration *coefficients* in (5.3),

$$G\{\mathbf{0},\mathbf{F}\}\ :\ A_{k,j,m} \mapsto (-1)^{j-m}A_{k,j,-m}. \tag{5.8}$$

Note, finally, that in order to obtain the action of the *general* aberration group element $G\{\mathbf{a},\mathbf{M}\}\in\Gamma_4$ on phase space, it is necessary to multiply the two-vector $(\mathbf{p}',\mathbf{q}')$ by the matrix $\mathbf{M}^{-1}$, as demanded by equation (5.1).

The formulas given above were obtained from manipulation of the mathematical objects described in Sections 3 and 4 through symbolic computation algorithms developed by the author and Miguel Navarro Saad, first in REDUCE-2 and, after the Foonly F-2 at IIMAS stopped operations in July 1986, all over again in muSIMP-83 4.12 available for home microcomputers. The muSIMP output files were formatted with some help from MINCE replace string command, and turned into TEX files that have been printed interspersed with the normal text. We can be reasonably sure that they contain none of the ubiquitous mistakes that plague human formula typing.

5.4. Common group/optical elements

The problem *inverse* to that of finding the nonlinear action of the aberration group on phase space, is the following: *given* a nonlinear mapping of phase space $(\mathbf{p}'(\mathbf{p},\mathbf{q}),\mathbf{q}'(\mathbf{p},\mathbf{q}))$, *solve* equations (5.1)–(5.3) for the aberration coefficients $\{A_{k,j,m}\}$ (*provided*, of course, that the mapping is canonical). This was done by Navarro Saad [6] for the *root* transformation seen in Section 2, and again by the author in the general case that extracts the aberration coefficients out of a given transformation on $\mathbf{p}$ and $\mathbf{q}$. It allows the display of the of the results below for the common elements of imaging optical systems.

We repeat the Gaussian part for the elements given in Eqs. (4.4) (except for the graded-profile fiber still in process), and organize the aberration coefficients of a_k in rows-within-rows as

$$\begin{aligned} {}^k\mathbf{a}\ :\ \{&\{A_{k,k,k},A_{k,k,k-1},\ldots,A_{k,k,1-k},A_{k,k,-k}\},\\ &\{A_{k,k-2,k-2},\ldots,A_{k,k-2,2-k}\},\\ &\begin{cases}\{A_{k,1,1},A_{k,1,0},A_{k,1,-1}\}\}, & k \text{ odd},\\ \{A_{k,0,0}\}\}, & k \text{ even}.\end{cases}\end{aligned} \tag{5.9}$$

We write:

Free propagation $F^\ell = G\{\mathbf{f}^\ell, \mathbf{F}^\ell\}$ **by length** ℓ **in medium** n.

The gaussian part is $\mathbf{F}^\ell = \begin{pmatrix} 1 & 0 \\ -\ell/n & 1 \end{pmatrix}$.

All but the *spherical aberration* coefficients vanish. They are:

$$f^\ell_{2,2,2} = -\ell/(8n^3), \quad f^\ell_{3,3,3} = -\ell/(16n^5), \quad f^\ell_{4,4,4} = -5\ell/(128n^7). \tag{5.10a}$$

The root transformation $R_{n;\varsigma} = G\{\mathbf{r}, \mathbf{R}\}$ **in medium** n **by the surface** $\varsigma(q^2) = \varsigma_2 q^2 + \varsigma_4 (q^2)^2 + \varsigma_6 (q^2)^3 + \varsigma_8 (q^2)^4$.

The gaussian part is $\mathbf{R} = \begin{pmatrix} 1 & -2n\varsigma_2 \\ 0 & 1 \end{pmatrix}$.

The aberration multiplets are:

$$\begin{aligned}
{}^2\mathrm{r} &= \{\{0,0,-\varsigma_2/(2n),0,n\varsigma_4\},\{-\varsigma_2/(3n)\}\},\\
{}^3\mathrm{r} &= \{\{0,0,-\varsigma_2/(8n^3),-\varsigma_2^2/(2n^2),-\varsigma_4/(2n),2\varsigma_2\varsigma_4,n\varsigma_6\},\{-\varsigma_2/(10n^3),-\varsigma_2^2/(5n^2),-2/5\varsigma_4/n\}\},\\
{}^4\mathrm{r} &= \{\{0,0,-\varsigma_2/(16n^5),-\varsigma_2^2/(4n^4),-\varsigma_4/(8n^3)-5/6\varsigma_2^3/n^3,-2\varsigma_2\varsigma_4/n^2,\\
&\quad -\varsigma_6/(2n)+16/3\varsigma_2^2\varsigma_4/n,6\varsigma_2\varsigma_6,8/3n\varsigma_2\varsigma_4^2+n\varsigma_8\},\{-3/56\varsigma_2/n^5,-\varsigma_2^2/(7n^4),\\
&\quad -\varsigma_4/(7n^3)-2/7\varsigma_2^3/n^3,-8/7\varsigma_2\varsigma_4/n^2,-3/7\varsigma_6/n-16/21\varsigma_2^2\varsigma_4/n\},\{-\varsigma_4/(15n^3)\}\}.
\end{aligned} \tag{5.10b}$$

The refracting surface transformation $S_{n,m;\varsigma} = G\{\mathbf{s}, \mathbf{S}\}$ **from medium** n **to medium** m, **by the surface** $\varsigma(q^2) = \varsigma_2 q^2 + \varsigma_4 (q^2)^2 + \varsigma_6 (q^2)^3 + \varsigma_8 (q^2)^4$.

The gaussian part is $\mathbf{S} = \begin{pmatrix} 1 & 2m\varsigma_2 - 2n\varsigma_2 \\ 0 & 1 \end{pmatrix}$.

The aberration multiplets are:

$$\begin{aligned}
{}^2\mathrm{s} &= \{\{0,0,\varsigma_2/(2m)-\varsigma_2/(2n),2n\varsigma_2^2/m-2\varsigma_2^2,-m\varsigma_4+2m\varsigma_2^3+n\varsigma_4-4n\varsigma_2^3+2n^2\varsigma_2^3/m\},\{\varsigma_2/(3m)-\varsigma_2/(3n)\}\},\\
{}^3\mathrm{s} &= \{\{0,0,\\
&\quad \varsigma_2/(8m^3)-\varsigma_2/(8n^3),\\
&\quad n\varsigma_2^2/m^3-\varsigma_2^2/(2m^2)-\varsigma_2^2/(2n^2),\\
&\quad 3n^2\varsigma_2^3/m^3-3n\varsigma_2^3/m^2+\varsigma_4/(2m)+2\varsigma_2^3/m-\varsigma_4/(2n)-2\varsigma_2^3/n,\\
&\quad -2m\varsigma_2\varsigma_4/n+4m\varsigma_2^4/n-2\varsigma_2\varsigma_4+4n^3\varsigma_2^4/m^3-6n^2\varsigma_2^4/m^2+4n\varsigma_2\varsigma_4/m+4n\varsigma_2^4/m-6\varsigma_2^4,\\
&\quad -m\varsigma_6+6m\varsigma_2^2\varsigma_4-2m\varsigma_2^5+n\varsigma_6-12n\varsigma_2^2\varsigma_4+4n\varsigma_2^5+2n^4\varsigma_2^5/m^3-4n^3\varsigma_2^5/m^2+6n^2\varsigma_2^2\varsigma_4/m\},\\
&\quad \{\varsigma_2/(10m^3)-\varsigma_2/(10n^3),\\
&\quad 2/5n\varsigma_2^2/m^3-\varsigma_2^2/(5m^2)-\varsigma_2^2/(5n^2),\\
&\quad 2/5n^2\varsigma_2^3/m^3-2/5n\varsigma_2^3/m^2+2/5\varsigma_4/m-2/5\varsigma_2^3/m-2/5\varsigma_4/n+2/5\varsigma_2^3/n\}\},
\end{aligned}$$

$$\begin{aligned}
{}^4\mathrm{s} &= \{\{0,0,\\
&\quad \varsigma_2/(16m^5)-\varsigma_2/(16n^5),\\
&\quad 3/4n\varsigma_2^2/m^5-\varsigma_2^2/(4m^4)-\varsigma_2^2/(4m^3n)-\varsigma_2^2/(4n^4),\\
&\quad 15/4n^2\varsigma_2^3/m^5-5/2n\varsigma_2^3/m^4+\varsigma_4/(8m^3)-5/12\varsigma_2^3/m^3-\varsigma_4/(8n^3)-5/6\varsigma_2^3/n^3,\\
&\quad 10n^3\varsigma_2^4/m^5-10n^2\varsigma_2^4/m^4+3n\varsigma_2\varsigma_4/m^3+8/3n\varsigma_2^4/m^3-2\varsigma_2\varsigma_4/m^2-4\varsigma_2^4/m^2+\varsigma_2\varsigma_4/(mn)+4\varsigma_2^4/(mn)-\\
&\quad 2\varsigma_2\varsigma_4/n^2-8/3\varsigma_2^4/n^2,
\end{aligned}$$

$$
\begin{aligned}
&-16/3m\varsigma_2^2\varsigma_4/n^2+32/3m\varsigma_2^5/n^2+15n^4\varsigma_2^5/m^5-20n^3\varsigma_2^5/m^4+15n^2\varsigma_2^2\varsigma_4/m^3+6n^2\varsigma_2^5/m^3-62/3n\varsigma_2^2\varsigma_4/m^2-\\
&\quad 20/3n\varsigma_2^5/m^2+\varsigma_6/(2m)+19\varsigma_2^2\varsigma_4/m+11\varsigma_2^5/m-\varsigma_6/(2n)-8\varsigma_2^2\varsigma_4/n-16\varsigma_2^5/n,\\
&-6m\varsigma_2\varsigma_6/n+116/3m\varsigma_2^3\varsigma_4/n-52/3m\varsigma_2^6/n+4\varsigma_2\varsigma_6+12n^5\varsigma_2^6/m^5-20n^4\varsigma_2^6/m^4+28n^3\varsigma_2^3\varsigma_4/m^3-4/3n^3\varsigma_2^6/m^3-\\
&\quad 176/3n^2\varsigma_2^3\varsigma_4/m^2+64/3n^2\varsigma_2^6/m^2+2n\varsigma_2\varsigma_6/m+72n\varsigma_2^3\varsigma_4/m-36n\varsigma_2^6/m+4n\varsigma_4^2/m-80\varsigma_2^3\varsigma_4+124/3\varsigma_2^6-4\varsigma_4^2,\\
&8m\varsigma_2\varsigma_4^2-m\varsigma_8+2m\varsigma_2^2\varsigma_6+86/3m\varsigma_2^4\varsigma_4-124/3m\varsigma_2^7-24n\varsigma_2\varsigma_4^2+n\varsigma_8-4n\varsigma_2^2\varsigma_6-40n\varsigma_2^4\varsigma_4+200/3n\varsigma_2^7+4n^6\varsigma_2^7/m^5-\\
&\quad 8n^5\varsigma_2^7/m^4+18n^4\varsigma_2^4\varsigma_4/m^3-20/3n^4\varsigma_2^7/m^3-152/3n^3\varsigma_2^4\varsigma_4/m^2+112/3n^3\varsigma_2^7/m^2+40/3n^2\varsigma_2\varsigma_4^2/m+\\
&\quad 2n^2\varsigma_2^2\varsigma_6/m+164/3n^2\varsigma_2^4\varsigma_4/m-188/3n^2\varsigma_2^7/m+8/3m^2\varsigma_2\varsigma_4^2/n-32/3m^2\varsigma_2^4\varsigma_4/n+32/3m^2\varsigma_2^7/n\},\\
&\{3/56\varsigma_2/m^5-3/56\varsigma_2/n^5,\\
&3/7n\varsigma_2^2/m^5-\varsigma_2^2/(7m^4)-\varsigma_2^2/(7m^3n)-\varsigma_2^2/(7n^4),\\
&9/7n^2\varsigma_2^3/m^5-6/7n\varsigma_2^3/m^4+\varsigma_4/(7m^3)-\varsigma_2^3/(7m^3)-\varsigma_4/(7n^3)-2/7\varsigma_2^3/n^3,\\
&12/7n^3\varsigma_2^4/m^5-12/7n^2\varsigma_2^4/m^4+12/7n\varsigma_2\varsigma_4/m^3-8/7n\varsigma_2^4/m^3-8/7\varsigma_2\varsigma_4/m^2+12/7\varsigma_2^4/m^2+4/7\varsigma_2\varsigma_4/(mn)-\\
&\quad 12/7\varsigma_2^4/(mn)-8/7\varsigma_2\varsigma_4/n^2+8/7\varsigma_2^4/n^2,\\
&16/21m\varsigma_2^2\varsigma_4/n^2-32/21m\varsigma_2^5/n^2+6/7n^4\varsigma_2^5/m^5-8/7n^3\varsigma_2^5/m^4+20/7n^2\varsigma_2^2\varsigma_4/m^3-20/7n^2\varsigma_2^5/m^3-\\
&\quad 64/21n\varsigma_2^2\varsigma_4/m^2+104/21n\varsigma_2^5/m^2+3/7\varsigma_6/m-12/7\varsigma_2^2\varsigma_4/m-18/7\varsigma_2^5/m-3/7\varsigma_6/n+8/7\varsigma_2^2\varsigma_4/n+16/7\varsigma_2^5/n\},\\
&\{\varsigma_4/(15m^3)-\varsigma_4/(15n^3)\}\}.
\end{aligned}
\tag{5.10c}
$$

The curved mirror transformation $M_{n;\varsigma} = G\{\mathbf{m},\mathbf{M}\}$ in medium n, by the surface $\varsigma(q^2) = \varsigma_2 q^2 + \varsigma_4(q^2)^2 + \varsigma_6(q^2)^3 + \varsigma_8(q^2)^4$.

The gaussian part is $\mathbf{M} = \begin{pmatrix} 1 & -4n\varsigma_2 \\ 0 & 1 \end{pmatrix}$.

The aberration multiplets are:

$$
\begin{aligned}
{}^2\mathfrak{m} &= \{\{0,0,-\varsigma_2/n,-4\varsigma_2^2,2n(\varsigma_4-4\varsigma_2^3)\},\{-2/3\varsigma_2/n\}\},\\
{}^3\mathfrak{m} &= \{\{0,0,-\varsigma_2/(4n^3),-2\varsigma_2^2/n^2,-(\varsigma_4+10\varsigma_2^3)/n,-4\varsigma_2(\varsigma_4+6\varsigma_2^3),2n(\varsigma_6-12\varsigma_2^2\varsigma_4)\},\\
&\quad \{-\varsigma_2/(5n^3),-4/5\varsigma_2^2/n^2,-4/5\varsigma_4/n\}\},\\
{}^4\mathfrak{m} &= \{\{0,0,-\varsigma_2/(8n^5),-\varsigma_2^2/n^4,-(3\varsigma_4+80\varsigma_2^3)/(12n^3),-4/3\varsigma_2(6\varsigma_4+25\varsigma_2^3)/n^2,\\
&\quad -[3\varsigma_6+4\varsigma_2^2(43\varsigma_4+64\varsigma_2^3)]/(3n),8/3(\varsigma_2[3\varsigma_6+8\varsigma_2^2(-13\varsigma_4+4\varsigma_2^3)]-3\varsigma_4^2),\\
&\quad 2/3n(3\varsigma_8+4\varsigma_2(\varsigma_2[-3\varsigma_6+4\varsigma_2^2(-19\varsigma_4+20\varsigma_2^3)]-16\varsigma_4^2))\},\\
&\quad \{-3/28\varsigma_2/n^5,-4/7\varsigma_2^2/n^4,-2/7(\varsigma_4+8\varsigma_2^3)/n^3,16/7\varsigma_2(-2\varsigma_4+\varsigma_2^3)/n^2,\\
&\quad 2/21[-9\varsigma_6+8\varsigma_2^2(-5\varsigma_4+16\varsigma_2^3)]/n\},\{-2/15\varsigma_4/n^3\}\}.
\end{aligned}
\tag{5.10d}
$$

These are muSIMP functions called

```
FREEFLIGHT (L, N),

ROOT (N, Z2, Z4, Z6, Z8),

SURF (N, NP, Z2, Z4, Z6, Z8),
```

etc., that return the above Lie transformations. An extra optional parameter specifies aberration order, with default value 7.

6. The face of aberration: spot diagrams

"By a fortunate chaining of causes and effects, often and without man's foresight, the True, the Beautiful, and the Good are often bound to the Useful"[VI]

When an optical system composed of lenses and empty spaces is designed according to Gaussian (first-order) calculations, it will contain image aberrations, as compositions of (5.10) show. Each geometrical object ray will still map into a single image ray, but a *pencil* of such rays spreading from an object point will not in general map into an image pencil converging at a single image point. The result will be a smudge with very interesting properties.

6.1. Spot diagrams and their symmetry

Imagine one geometric ray passing through a point at the object plane, drawing out polar coordinates, circles and radii of the direction plane, with the origin showing the direction parallel to the optical axis. After passing through the optical system, the image at the *screen* will draw out the ***spot diagram*** of the system at that point, as in the figures of this Section. Here we have drawn them for the individual pure aberrations in the symplectic basis. In actual optical systems, a linear combination of them will be present. In a perfect *imaging* system, the spots may be reduced to *points* on the screen, but their location may still change due to magnification or distortion of the system. Magnification is a first-order effect, generated by $\xi_0 = {}^1\chi_0^1$, in which we are not really interested here. We would like to see pure Aberration show us her* naked face.

In the figures that follow [28] we plot $\mathbf{q}'(\mathbf{p},\mathbf{q}) = \exp {}^k\chi_m^j\, \mathbf{q}$ as a function of $\mathbf{p}$, with $\mathbf{q}$ fixed, for each of the pure symplectic-classified aberrations in turn, to seventh aberration order. Since we deal with the aberrations of systems that are assumed to be invariant under rotations around the optical axis, and symmetric under reflections across planes containing that axis, we may rotate the *object* point $\mathbf{q}$ (together with $\mathbf{p}$) to lie on the x-axis of the screen (so we can stack the figures into vertical columns), and may reflect $y \leftrightarrow -y$ (so we show only the upper half). We let $\mathbf{p}$ run over a polar coordinate grid of five circles and eight diameter lines.

One basic consideration about inversion symmetry is the following: opposite ray directions, $\mathbf{p}$ and $-\mathbf{p}$, correspond to opposite signs of $\xi_0 = \mathbf{p}\cdot\mathbf{q}$; from (4.13*b*) we see that $\xi_0 \mapsto -\xi_0$ implies ${}^k\chi_m^j \mapsto (-1)^{j-m}\,{}^k\chi_m^j$; hence $\{{}^j\chi_m^k, \mathbf{w}\} \mapsto (-1)^{j-m+1}\{{}^j\chi_m^k, \mathbf{w}\}$ (for $\mathbf{q}$ since the Poisson bracket involves a p-derivative, and for $\mathbf{p}$ since $\mathbf{p}$ changes sign). When $j-m$ is *odd*, therefore, the terms in the series $\mathbf{q}' - \mathbf{q} = \{{}^k\chi_m^j, \mathbf{q}\} + \frac{1}{2!}\{{}^k\chi_m^j, \{{}^k\chi_m^j, \mathbf{q}\}\} + \frac{1}{3!}\{{}^k\chi_m^j, \{{}^k\chi_m^j, \{{}^k\chi_m^j, \mathbf{q}\}\}\} + \cdots$ will all be *invariant* under $\mathbf{p} \mapsto -\mathbf{p}$, and the spot figures will be *two-to-one* mappings of the polar grid of the $\mathbf{p}$-plane on the $\mathbf{q}'$ plane; these give rise to the *comet*-like spots, thereafter called *comatic*. (All

---

[VI] A. v. Humboldt, *Kosmos, Entwurf einer physischen Weltbeschreibung*, 5 Vols. (Cotta, Stutgart, 1845–1862); quoted by Charles Minguet, *op. cit.*

* In all Indo-european languages having gender, it seems that *Aberration* is femenine.

RANK:	$k=2$		$k=3$		$k=4$		
SPIN:	$j=2,$	0	$j=3,$	1	$j=4,$	$2,$	0
$m=4$					⊙		
3			⊙		○▷		
2	⊙		○▷		⋈	8	
1	○▷		⋈	8	⬭▷	▷	
0	⬭	|	⬭▷	▷	⋈	8′	|
−1	↼		⬭	|	⬭▷	▷	
−2	•		↼		⬭	|	
−3			•		↼		
−4					•		
ORDER:	THIRD		FIFTH		SEVENTH		

TABLE 6.1 The aberration multiplets to seventh order.

the comas extend to the left.) For $j-m$ even, on the other hand, the nth term in the series $\mathbf{q}'-\mathbf{q}$ will multiply by $(-1)^n$ so the map will distinguish opposite ray directions. Note that for ${}^k\chi^k_k$ (termed rank-k *spherical aberration*) the series will consist of a *single* term, $\mathbf{q}'-\mathbf{q}=\{{}^k\chi^k_k,\mathbf{q}\}=-2k(p^2)^{k-1}\mathbf{p}$; that family of spots is therefore symmetric under reflections across the vertical axis. Since we are working to seventh degree, all aberrations of orders 5 and 7 have their exponential series in (5.2) cut to a single term $\mathbf{q}'=\mathbf{q}+\{{}^k\chi^j_m,\mathbf{q}\}$; hence *only* the third-order aberration ${}^2\chi^2_0$ exhibits a spot diagram that is not symmetric.

6.2. Generic shapes and classification

To organize the aberrations visibly, we recall the basic structure of the aberration multiplets of orders 3, 5, and 7, and asign some *ad hoc* symbols for them, shown above in Table 6.1.

With friendly reference to the classical aberration names given in Section 4, we may indicate by $\odot_2$, $\odot_3$, and $\odot_4$ the third, fifth, and seventh-order spherical aberrations on the upper fringe of the multiplets $m=k$; circular comas are the next diagonal, ○▷$_2$, ○▷$_3$, and ○▷$_4$, are also nondegenerate. They will be plotted together in the same figure for easy comparison. The distinctions among the *degenerate* aberrations (such as ⬭▷$_3$ and ▷$_3$ that cohabitate the $k=3$, $m=0$ niche) will be analyzed below. Two aberrations in each k-multiplet will not be shown in the figures since their spot diagrams are *points*: k^{th} rank *distortion*, ↼$_k$, generated by ${}^k\chi^k_{1-k}=\mathbf{p}\cdot\mathbf{q}(q^2)^{k-1}$, that effects only image displacement $\mathbf{q}\mapsto\mathbf{q}'(\mathbf{q})=\mathbf{q}-(q^2)^{k-1}\mathbf{q}+\cdots$, and k^{th} rank *pocus*, $\bullet_k$ generated by ${}^k\chi^k_{-k}=(q^2)^k$ affecting $\mathbf{q}$ not at all.

A word of caution when comparing with the classical optics spot diagrams is the following: there, the consideration holds that they are made out of the images of those rays only, that pass through a pupil stop in the apparatus. The size of the aberration depends on the pupil diameter. *Here* we have a mathematical construct that maps the full $\mathbf{p}$ *plane* with no such limitation; the expansion in aberration rank is Taylor's philosophy, and the validity of this approximation is violated certainly beyond $|\mathbf{p}| = n$. In contradistinction to the relativistic coma effect seen in section 3 that was exact, cutting series to polynomials may hide the divergences. Our claim to properly represent the aberration phenomena is valid only in a finite neighborhood of the phase space origin. We retain all and only the terms written in (5.2).

Finally, the question of *scales* is important for the figures. Take for instance $\odot_k$ (spherical aberration of rank k, *i.e.*, of order $2k-1$): as we remarked above, its spot diagram plots $\exp {}^k\chi^k_k \mathbf{q} - \mathbf{q} = -2k(p^2)^{k-1}\mathbf{p}$, with an intrinsic scale factor of $2k$ that must be visible; on the other hand, the smallness of the factor $(p^2)^{k-1}$ for 'reasonable' rays (in vacuum, a ray at 30° has $p_{\max} = 0.5$) would shrink high-k spot diagrams out of sight. We have chosen $\mathbf{q}$ to lie at $1/\sqrt{10} \approx 0.316228$ and have let $\mathbf{p}$ trace out circles up to $p_{\max} = 1/\sqrt{10}$, or $18°\,26'$. In this way the aberrations of orders 3, 5, 7, carry extrinsic factors of 10^{-3}, 10^{-4}, 10^{-5}, that are corrected in the figures by scales of 0.1, 0.01, and 0.001; the spots are thus shown on a 'uniform' scale.

The same figures will yield further information when subject to Fourier conjugation (5.5). They show aberration in direction $\mathbf{p}$-space through $\mathbf{p}'(\mathbf{p}, \mathbf{q})$ where we choose to fix a certain ray *direction* $\mathbf{p}$, and transport the ray parallel to itself around the object screen, drawing a polar-coordinate grid in $\mathbf{q}$. The image we obrtain is given by its Fourier conjugate aberration, *i.e.*, that obtained by inverting the sign of its Seidel number, m in ${}^k\chi^j_m$. When $\mathbf{q}$ undergoes aberration by ${}^k\chi^j_m$, $\mathbf{p}$ concurrently aberrates by ${}^k\chi^j_{-m}$.

6.3. Nondegenerate aberration families

Spherical aberration: $\odot_k$ This is generated by ${}^k\chi^k_k$, has been mentioned before, and is shown in Figures 6.1 identified by the $(k\ j\ m)$ labels. Here the Lie exponential series has the same action as the Seidel approximation, namely $\mathbf{q} \mapsto \mathbf{q}' = \mathbf{q} - 2kp^2\mathbf{p}$, $\mathbf{p} \mapsto \mathbf{p}' = \mathbf{p}$. The smudging of image is *homogeneous* over the screen, *i.e.*, the spot $\odot_k$ looks the same when we translate the object in $\mathbf{q}$ (since $\mathbf{q}' - \mathbf{q}$ is independent of $\mathbf{q}$) and it has a $(2k-1)$-power growth in the circles' radii. There is no change in momentum space, *i.e.*, in ray direction.

Pocus: $\bullet_k$ This is the Fourier conjugate of spherical aberration, and the statements above apply, *mutatis mutandis* under $\mathbf{q} \mapsto \mathbf{p}$, $\mathbf{p} \mapsto -\mathbf{q}$. Rays converging to image points far from the optical axis do so at larger angles. Small movements of the screen will result in rapid loss of focus at the edges, with a $(2k-1)$-power decreasing depth of field. (My cheap telephoto has this kind of aberration: it has decreasing depth of field at increasing distance from the center. It does very good portraits for this reason, though.)

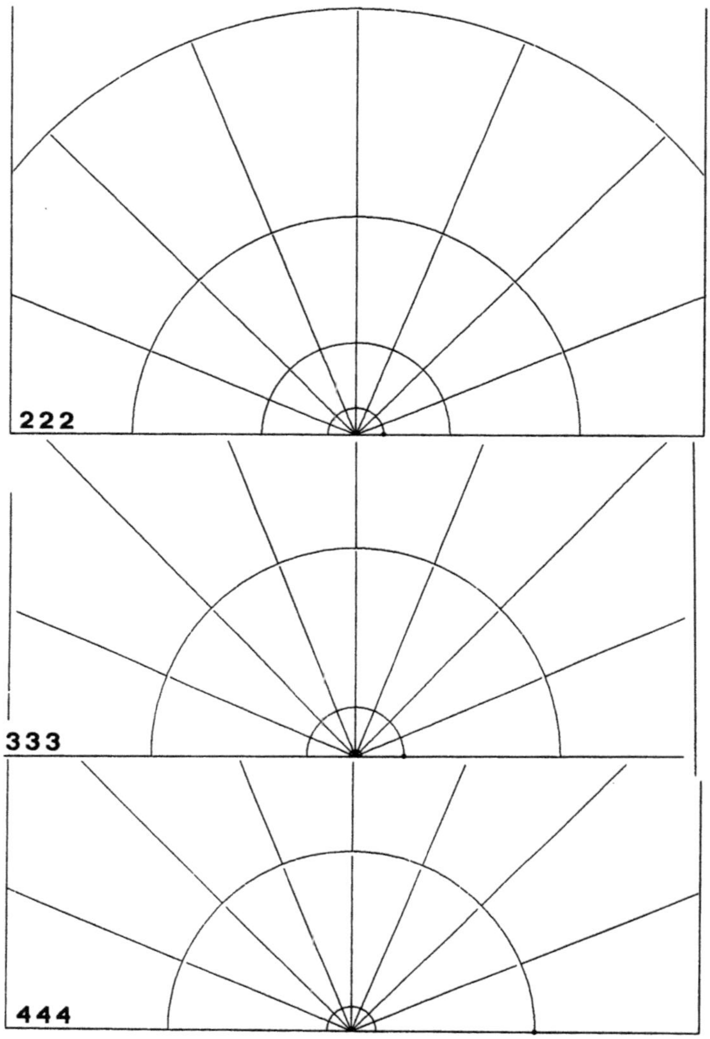

Figures 6.1 Spherical aberrations $\odot_k$ generated by ${}^k\chi_k^k$.

Circular coma: $\circ\!\!\triangleright_k$ Coma is a most fascinating aberration that is quite pernicious to image quality, since it is asymmetric, and its asymmetry visibly radiates and increases from the optical center. There is a single $\circ\!\!\triangleright_2$, but for higher orders, all aberrations with $j-m$ odd are comet-shaped, albeit they are "elliptical" and denoted $\ominus\!\!\triangleright$ and $\triangleright$. Only ${}^k\chi^k_{k-1}$ generates truly *circular* comas $\circ\!\!\triangleright_k$, and the size of these spots is linearly proportional to q, the distance from the optical center. (We repeat the remark that first-order $\circ\!\!\triangleright_1$ is pure magnification, so the relativistic coma is indeed pure, global coma.) In Figures 6.2 we show the circular comas up to seventh order.

We can give the exact result for the third-order $\circ\!\!\triangleright_2$ exponential series generated by ${}^2\chi^2_1 = p^2\mathbf{p}\cdot\mathbf{q}$, and also the seventh-degree approximation we used for the spots in the figures. They are:

$$\begin{aligned}
\mathbf{q}' &= \exp(\alpha\,{}^2\hat{\chi}^2_1)\,\mathbf{q}\\
&= \sqrt{1-2\alpha p^2}\,\mathbf{q} - 2\alpha\sqrt{1-2\alpha p^2}\,\mathbf{p}\cdot\mathbf{q}\,\mathbf{p} && (6.1a)\\
&\approx \mathbf{q} - \alpha(p^2\mathbf{q} - 2\mathbf{p}\cdot\mathbf{q}\,\mathbf{p})\\
&\qquad - \alpha^2p^2(\tfrac{1}{2}p^2\mathbf{q} - 2\mathbf{p}\cdot\mathbf{q}\,\mathbf{p}) - \alpha^3(p^2)^2(\tfrac{1}{2}p^2\mathbf{q} - \mathbf{p}\cdot\mathbf{q}\,\mathbf{p}), && (6.1b)
\end{aligned}$$

$$\begin{aligned}
\mathbf{p}' &= \exp(\alpha\,{}^2\hat{\chi}^2_1)\,\mathbf{p}\\
&= \mathbf{p}/\sqrt{1-2\alpha p^2} && (6.2a)\\
&\approx \mathbf{p} + \alpha p^2\mathbf{p} + \tfrac{3}{2}\alpha^2(p^2)^2\mathbf{p} + \tfrac{5}{2}\alpha^3(p^2)^3\mathbf{p}. && (6.2b)
\end{aligned}$$

We note that the particular combination $Ap^2\mathbf{q}+B\mathbf{p}\cdot\mathbf{q}\,\mathbf{p}$ translates in the figures as a *circle* of radius $\frac{1}{2}Bp^2q$, $q=|\mathbf{q}|$, and center at $(A+\frac{1}{2}B)p^2q$. These circles, as can be seen in the figure for (2 2 1), are tangent to a sector with apex at $\mathbf{q}$ and half-angle of 30° at the apex. Careful examination of the figure will reveal that the image lines of the radii are not quite straight: they are slightly bent inwards for larger p due to the remaining terms of the series in (6.1*b*) that sum circles on their own accord. The closed expression in (6.1*a*), read as $A(p^2)p^2\mathbf{p} + B(p^2)\mathbf{p}\cdot\mathbf{q}\,\mathbf{p}$, shows that as p increases from 0 to $1/2\alpha$, the radii grow from zero to a maximum and then decrease back to zero and become imaginary (certainly beyond $p=n$ the Taylor series expansion makes no visible optical sense); the segment of circle *centers* starts at $\mathbf{q}$ and ends then at the optical center $\mathbf{q}=0$.

To find the caustics, we investigate the differentials of the exact $\circ\!\!\triangleright_2$ in (6.1*a*). We find that the optical momentum plane has two bent half-lines joining at the optical axis ($\mathbf{p}=0$), and making there angles of $\kappa_p^{(2)} = \pm\arctan\sqrt{3} = 60°$ with the direction to the optical center. There, differential movement $d\mathbf{p}$ across that line will result in $d\mathbf{q}'=0$, *i.e.*, these half-lines map onto the *caustics* in the image $\mathbf{q}$-plane. Increasing p, the bent lines are given by $p_y^2 = 3p_x(1-2\alpha p^2)$, with angles $\kappa_p^{(2)} = \pm\arctan\sqrt{3(1-2\alpha p^2)}$ to the axis. For these 'caustic' lines $\mathbf{p}_c$, $\mathbf{q}'(\mathbf{p}_c,\mathbf{q})$ will be the visible image of the full caustic line on the screen. The angle between $\mathbf{q}-\mathbf{q}'(\mathbf{p}_c,\mathbf{q})$ and the direction to the optical center is

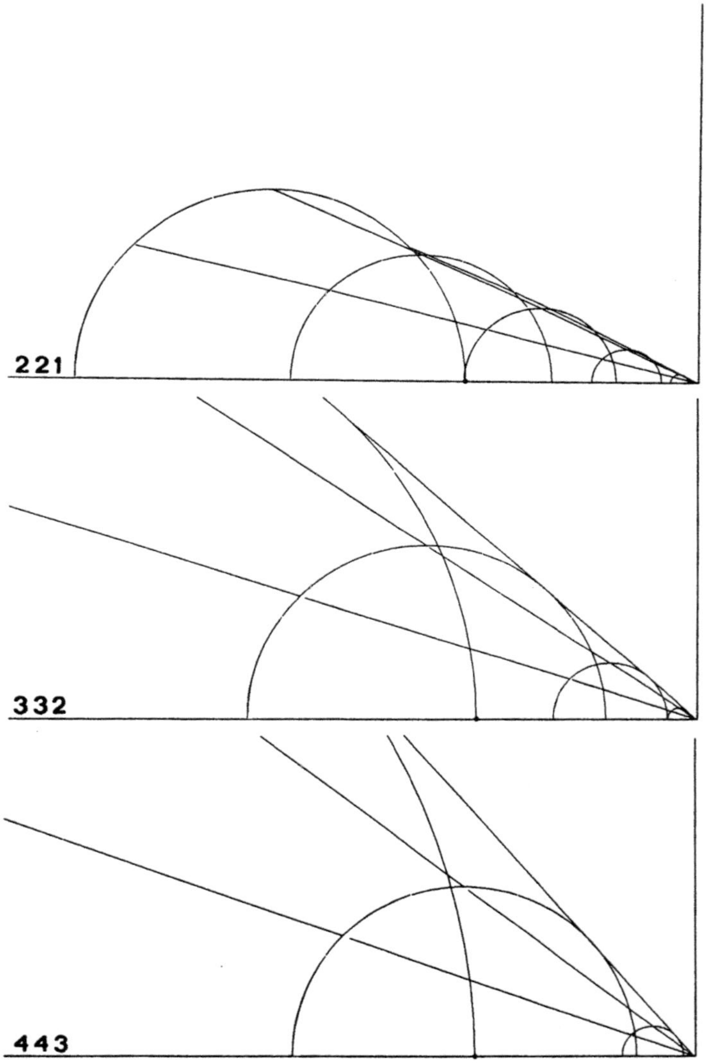

Figures 6.2 Circular comas $\mathcal{O}_k$ generated by ${}^k\chi^k_{k-1}$.

then $\pm\tan\kappa_q^{(2)} = \sqrt{3}R^2/[4R-3+2(1-R)/\alpha p^2]$, with $R^2 = 1-2\alpha p^2$. The half-angle of the $\mathbf{q}$-comet near the appex $[R \approx 1,\ (1-R)/\alpha p^2 \approx 1]$ is $\kappa_q^{(2)} = \pm\arctan 1/\sqrt{3} = \pm 30°$ as can be measured in the figure. This is a classical result [22].

The circular comas of rank k, $\mathrm{O\!\!>}_k$, generated by ${}^k\chi^k_{k-1} = (p^2)^{k-1}\mathbf{p}\cdot\mathbf{q}$ exhibit lines in momentum space of half-angle $\kappa_p^{(k)} = \pm\arctan\sqrt{2k-1}$ at the apex. This is 65.9052° for $\mathrm{O\!\!>}_3$ and 69.2952° for $\mathrm{O\!\!>}_4$. In position space —on the screen— the apex half-angles are $\kappa_q^{(k)} = \pm\arctan[(k-1)/\sqrt{2k-1}]$, namely 41.8104° for $\mathrm{O\!\!>}_3$ and 48.5904° for $\mathrm{O\!\!>}_4$. These results are also known [22] and can be measured in the figures.

Distortion: $\rightharpoonup_k$ This class of aberrations is the Fourier conjugate of circular coma, and has been commented upon above. It keeps images sharp, but distorts the picture. The mapping of phase space of $\rightharpoonup_2$ may be obtained from that of coma, Eqs. (6.1)–(6.2). Thus, under $\mathbf{q} \mapsto \mathbf{p}$, $\mathbf{p} \mapsto -\mathbf{q}$, we find

$$\mathbf{q}' = \exp(\alpha\, {}^2\hat{\chi}^2_{-1})\,\mathbf{q} = \mathbf{q}/\sqrt{1+2\alpha q^2}, \tag{6.3}$$

and a corresponding transformation for $\mathbf{p}'$, that undergoes circular comatic aberration. This duality was seen in relativistic coma. For general $\rightharpoonup_k$, generated by ${}^k\chi^k_{1-k} = \mathbf{p}\cdot\mathbf{q}(q^2)^{k-1}$ we give the result the result

$$\mathbf{q}' = \exp(\alpha\, {}^k\hat{\chi}^k_{1-k})\,\mathbf{q} = [1+2\alpha(k-1)(q^2)^{k-1}]^{-(k-1)/2}\mathbf{q}. \tag{6.4}$$

This effects an Escher-like map of the $\mathbf{q}$-plane into the circle $q^2 \le 1/2\alpha(k-1)$. The above four aberrations are the only nondegenerate ones in the aberration multiplets.

6.4. Degenerate families

Curvatisms/Astigmatures: $\bigcirc_k$—$|_k$ and $\bowtie_k$—8_k, constitute the next doubly-degenerate lines of aberrations inside the coma-distortion lines of the table, coalescing at the apex $\bigcirc_2$—$|_2$ that is well known in terms of third-order Seidel curvature of field and astigmatism. They are placed at the lines $m = k-2$ (generally called oblique spherical aberration in the classical literature) and $2-k$ (classically denoted with the blanket names of curvature of field and astigmatism) and, owing to their general peanut-shaped appearance, we may extend the class for the single $m = k-4 = 4-k$ case $\bowtie_4^{(m=0)}$—$8'_4$—$|_4$, for seventh aberration order. They are twofold or more *degenerate*. In Figures 6.3 we show the spot diagrams for the degenerate $\bigcirc_k$—$|_k$ pairs.

For third aberration order, $k=2$, $\bigcirc$ and $\bowtie$ coalesce into $\bigcirc_2$ generated by ${}^2\chi^2_0 = \frac{1}{3}[p^2q^2 + 2(\mathbf{p}\cdot\mathbf{q})^2]$; this we have called *curvatism*, a linear combination of the commonly known curvature of field (generated by the monomial p^2q^2) and astigmatism (generated by $(\mathbf{p}\cdot\mathbf{q})^2$). The aberration $j=0$ singlet $|_2$, degenerate with $\bigcirc_2$ is generated by ${}^2\chi^0_0 = (\mathbf{p}\times\mathbf{q})^2 = p^2q^2 - (\mathbf{p}\cdot\mathbf{q})^2$. Indeed, the general (A,B) linear combination of the two effects, to third aberration order, is

$$\exp(\alpha[Ap^2q^2 + B(\mathbf{p}\cdot\mathbf{q})^2]\hat{\ })\,\mathbf{q} = \mathbf{q} - 2Aq^2\mathbf{p} - 2B\mathbf{p}\cdot\mathbf{q}\,\mathbf{q}, \tag{6.5}$$

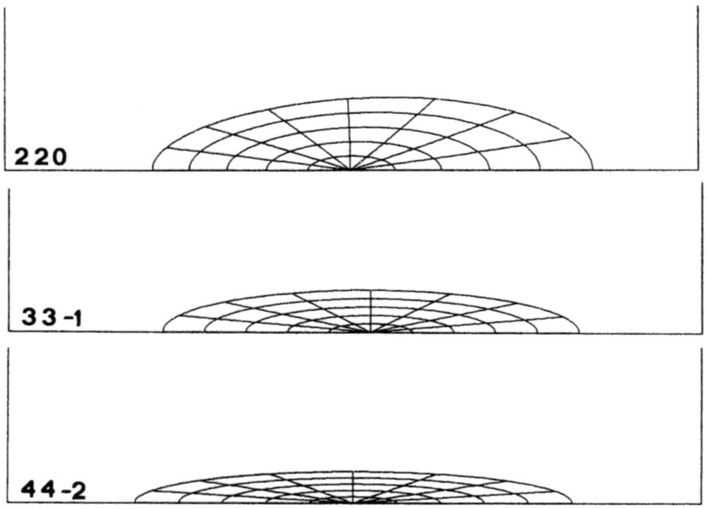

Figures 6.3 The curvatism-astigmature pairs $\bigcirc_k$—$|_k$ generated by ${}^k\Upsilon^k_{k-2}$ and ${}^k\Upsilon^{k-2}_{k-2}$.

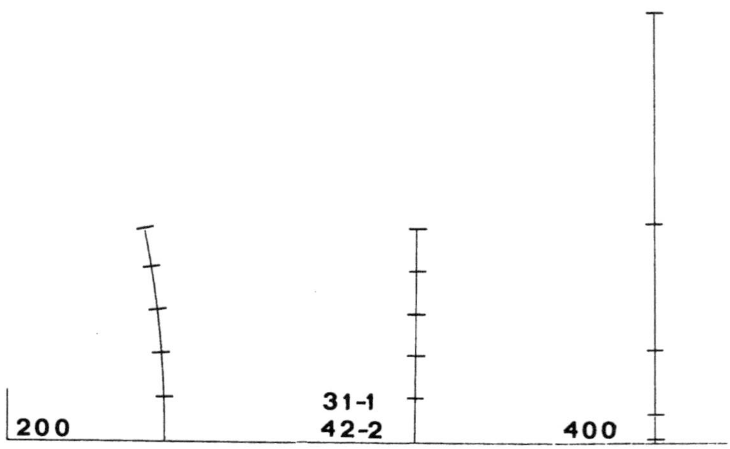

gives rise to spot diagrams that are ellipses centered on the object point $\mathbf{q}$ with horizontal half-axis $(A+B)q^2p$ and vertical axis Aq^2p. The size of the spots will thus characteristically increase as q^2, the square of the distance to the optical center, and linearly with ray angle p.

Traditional curvature of field gives rise to circular spots (○), and astigmatism to segments (degenerate ellipses) directed towards the optical center (—). *Curvatism* (⬭) yields elliptical spots with a 3:2 ratio of horizontal to vertical axes, and *astigmature* (|) provides vertical degenerate segments at right angles to the direction of the optical center. When we include higher terms in the exponential series, ellipses (and circles) become distorted, as is apparent in Figure 6.3—(2 2 0). Astigmatism may be integrated in its Lie series and seen to remain a segment $\mathbf{q}\exp(-2\alpha\mathbf{p}\cdot\mathbf{q})$, although no longer symmetric about the point.

The Lie series of astigmature $|_2$ may be integrated. Using an obvious matrix notation, and calling $\mathcal{X} = {}^2\mathcal{X}^0_0 \geq 0$, we may write

$$\{\mathcal{X}, \begin{pmatrix}\mathbf{p}\\ \mathbf{q}\end{pmatrix}\} = \mathbf{X}\begin{pmatrix}\mathbf{p}\\ \mathbf{q}\end{pmatrix}, \qquad \mathbf{X} = \begin{pmatrix} -\mathbf{p}\cdot\mathbf{q} & p^2 \\ -q^2 & \mathbf{p}\cdot\mathbf{q}\end{pmatrix}, \tag{6.6}$$

where the 2×2 matrix $\mathbf{X}$ is such that $\mathbf{X}^2 = -\mathcal{X}\mathbf{1}$ and $\det\mathbf{X} = \mathcal{X}$. From here,

$$\exp\alpha\hat{\mathcal{X}}\begin{pmatrix}\mathbf{p}\\ \mathbf{q}\end{pmatrix} = [\cos(2\alpha\sqrt{\mathcal{X}}\mathbf{1}) + 2\alpha\,\mathrm{sinc}(2\alpha\sqrt{\mathcal{X}}\mathbf{X})]\begin{pmatrix}\mathbf{p}\\ \mathbf{q}\end{pmatrix}. \tag{6.7}$$

For the conditions of the figure, $\mathbf{p}\cdot\mathbf{q} = p_xq$, $\mathbf{p}\times\mathbf{q} = \sqrt{\mathcal{X}} = -p_yq$, and $\phi = 2\alpha p_yq$,

$$\begin{pmatrix}p'_x\\ p'_y\end{pmatrix} = \begin{pmatrix}\cos\phi & -\sin\phi\\ \sin\phi & \cos\phi\end{pmatrix}\begin{pmatrix}p_x\\ p_y\end{pmatrix}, \qquad \begin{pmatrix}q'_x\\ q'_y\end{pmatrix} = q\begin{pmatrix}\cos\phi\\ \sin\phi\end{pmatrix}. \tag{6.7}$$

We see thus that the vector $\mathbf{p}'$ is rotated on a circle by an angle $\phi = 2\alpha p_yq$ and that the aberrated image spot $\mathbf{q}'$ forms into an arc of a circle, of which the third-term approximation is given in Figure 6.3—(2 0 0). Similar considerations apply for rank k to $|_k$, since ${}^k\mathcal{X}^0_0 = \mathcal{X}^{k/2}$.

In Ref. [10], while studying third-order aberrations in fibers, we noted that through the choice of profile parameters and multiples of a chosen fiber length, we could make all quintuplet aberrations zero; the aberration singlet, astigmature, would not vanish but steadily increase, as the skew rays spiral the fiber. Rays issuing in all directions from $\mathbf{q}$ thus distribute, for those lengths, along a growing arc of a circle centered on the optical axis. At those points along the fiber, we have an optical system exhibiting pure astigmature aberration.

The three ⋈ peanut-shaped aberrations are degenerate with the 8_k shapes. We show the corresponding spots in Figures 6.4. Ahead, we suggest they are "repeaters" of ⊙.

'Elliptical' comas: ⬭▷$_3$, ⬭▷$_4^{(m=1)}$, and ⬭▷$_4^{(m=-1)}$, are two-fold degenerate with ▷$_3$, ▷$_4^{(m=1)}$, and ▷$_4^{(m=-1)}$, respectively. Figures 6.5 show these degenerate pairs. They

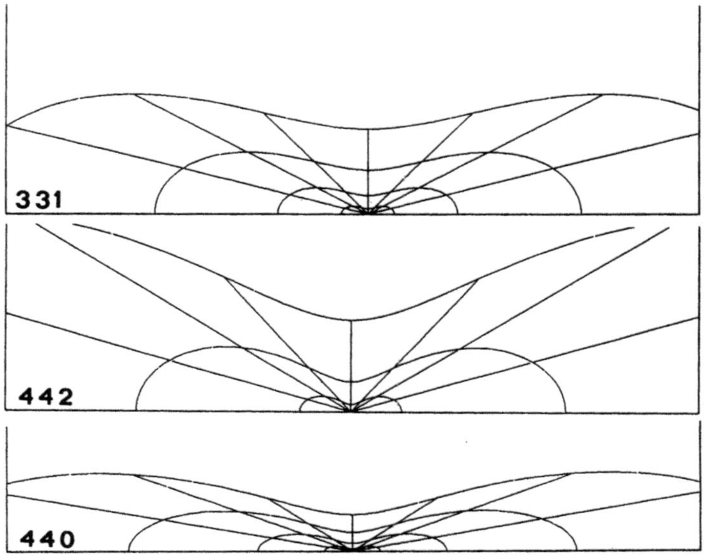

Figures 6.4 The degenerate pairs $\bowtie_k$—8_k, to seventh order.

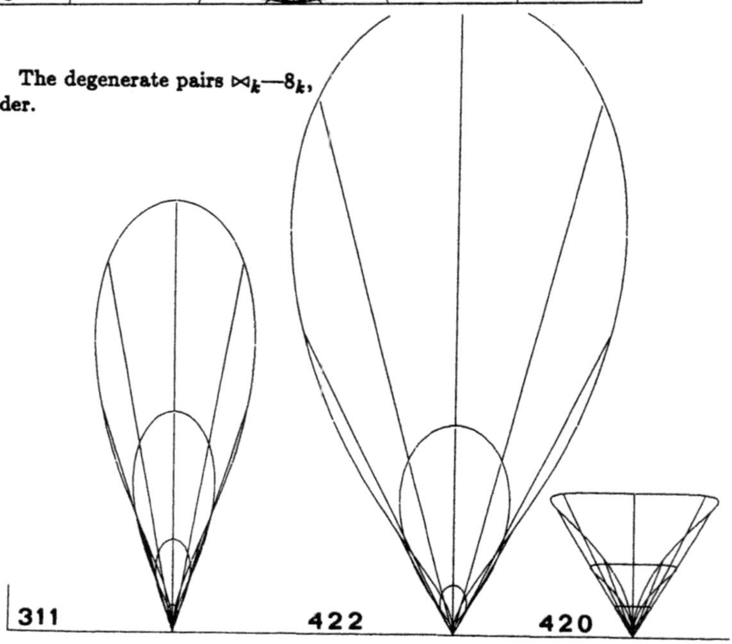

lie inside the previous aberration type. We may find analytically, as above, the caustic angles at the apex. At fifth aberration order, ${}^{3}\mathcal{X}_0^3$ generates ⬭$\triangleright_3$ with a pair of half-lines in momentum space, with half-angles $\kappa_p^{(3)} = \pm\arctan\sqrt{5} \approx \pm 65.9052°$ (the same as ⊃$\triangleright_3$ for *p*-space), mapping on the screen image caustics with $\kappa_q^{(3)} = \pm\arctan 1/\sqrt{5} \approx \pm 24.0983°$ at the apex (*different* from ⊃$\triangleright_3$ *q*-space). Similarly, ${}^{4}\mathcal{X}_1^4$ generates ⬭$\triangleright_4^{(m=1)}$ and maps $\kappa_p^{(4)} \approx \pm 68.4020°$ rays onto the caustic at $\kappa_q^{(4)} \approx \pm 39.3483°$ (no exact expression), and ${}^{4}\mathcal{X}_{-1}^4$ generates ⬭$\triangleright_4^{(m=-1)}$ mapping $\kappa_p^{(4)} = \pm\arctan\sqrt{7} \approx \pm 69.2952°$ rays to the caustic with half-angle $\kappa_q^{(4)} = \pm\arctan 1/\sqrt{7} \approx \pm 20.7048°$. The closed curves in the diagrams are *not* ellipses, in fact, so we placed the quotes around the name, to be less precise.

6.5. Repeater patterns

The seventh-order aberration quintuplet is a *repeater* of the third-order one, in the sense that they are generated by ${}^{4}\mathcal{X}_m^2 = \mathcal{X}\ {}^{2}\mathcal{X}_m^2$ out of the quintuplet in multiplication with $\mathcal{X} = {}^{2}\mathcal{X}_0^0$, the generator of astigmature. This *repetition* "8-shapes" the quintuplet (*viz.*, ⊙ → 8, ⊃$\triangleright$ → $\triangleright$, ⬭ → 8′, ↔ → $\triangleright$, • → |). Similarly, the fifth-order aberration triplet repeats the three Gaussian generators of free propagation (linearly growing circles), magnification (first-order coma and distortion), and | itself is the repeater of the unit map of q-space. We note that $(\mathbf{q}_a)_| = \{\mathcal{X}, \{a, \mathbf{q}\}\} = \{a, \{\mathcal{X}, \mathbf{q}\}\} = (\mathbf{q}_|)_a$ for any Gaussian or aberration generator function a, and astigmature | generated by $\mathcal{X}$ producing, from (6.5), $\mathbf{q}_| = \mathbf{q} - 2q^2\mathbf{p} + 2\mathbf{p}\cdot\mathbf{q}\,\mathbf{q}$, a vertical line.

A *repeater* aberration $\mathcal{X}a$ of a thus acts as

$$\begin{aligned} \mathbf{q}' &= \exp(\alpha\mathcal{X}a)\hat{}\,\mathbf{q} \\ &= \mathbf{q} + \alpha(\mathcal{X}\mathbf{q}_a + a\mathbf{q}_|) \\ &\quad + \tfrac{1}{2!}\alpha^2(\mathcal{X}^2\mathbf{q}_{aa} + 2\mathcal{X}a\mathbf{q}_{a|} + a^2\mathbf{q}_{||}) \\ &\quad + \tfrac{1}{3!}\alpha^3(\mathcal{X}^3\mathbf{q}_{aaa} + 3\mathcal{X}^2a\mathbf{q}_{aa|} + 3\mathcal{X}a^2\mathbf{q}_{a||} + a^3\mathbf{q}_{|||}) + \cdots, \end{aligned} \tag{6.8}$$

where $\mathbf{q}_a$ is the a-acted coordinate and $\mathbf{q}_|$ the astigmature spot. To first degree in α, compare the figures of the third- and seventh-order aberration quintuplets. In the formula above, we have $\mathcal{X}\mathbf{q}_a + a\mathbf{q}_|$; both $\mathcal{X}$ and $\mathbf{q}_|$ are zero for meridional rays (*i.e.*, in a plane with the optical axis) that come from the horizontal axis of the figures. Hence, when a is a single-valued (*i.e.*, non-comatic) aberration, $\mathcal{X}a$ will be zero on the waist, like 8 and 8′. When a is comatic, it sends rays both meridional and saggital (*i.e.*, across the plane with the optical axis) towards the optical center. The former are multiplied by the zero of $\mathcal{X}$ and regressed to $\mathbf{q}$, albeit retaining their vertical departure from meridionality (this explains the accumulating tangency of curves at the spot origin). Saggital rays remain in their general position, but the two-valued comatic 'cone' is now seen from a generator line, showing us the inside of the aberration.

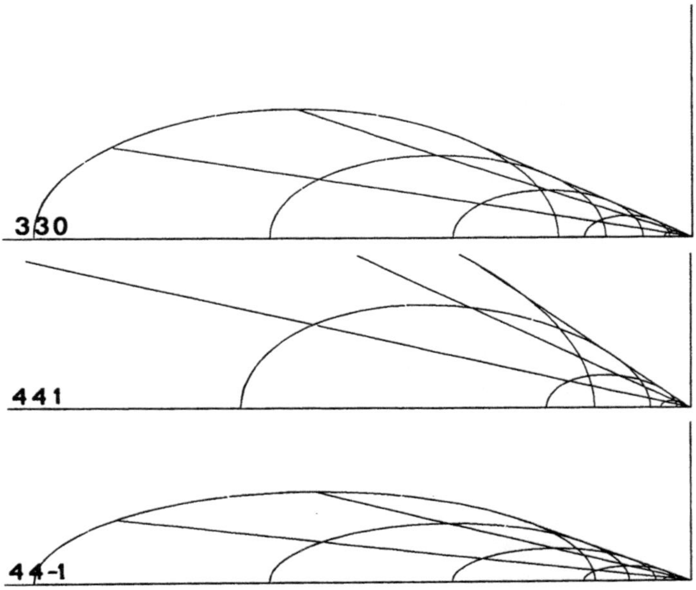

Figures 6.5 Elliptical comas and their degenerate partners $⊂\!⊃_k$—$\triangleright_k$, minus one *repeater*, to seventh aberration order.

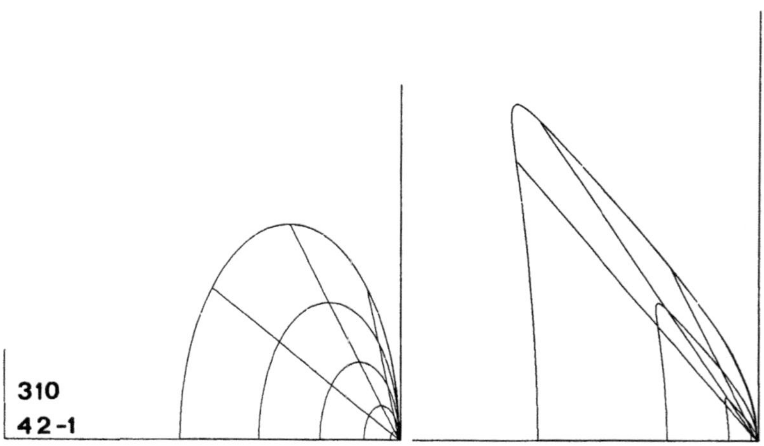

Another kind of repetition may detected between sequences of aberrations whose spot diagram has the *same shape*; in the figures we have only shown them once, therefore. They are $(3\ 1\ -1) = (4\ 2\ -2)$ in Fig. 6.3 and $(3\ 1\ 0) = (4\ 2\ -1)$ in Fig. 6.5. We shall now detail this *second* repetition mechanism that takes place for the two lowermost diagonals of each 'sector' of aberrations, and repeaters. Note that for ${}^{j+n}\chi^{j}_{-j} = (\mathbf{p} \times \mathbf{q})^{2n}(q^2)^j$, $\{{}^{j+n}\chi^{j}_{-j}, (q^2)^{n'} n\mathbf{q}\} = (q^2)^j \{{}^{n}\chi^{0}_{0}, (q^2)^{n'}\mathbf{q}\} = (q^2)^{j+n'}\{{}^{n}\chi^{0}_{0}, \mathbf{q}\}$ extracts the space-dependent factor to all orders, and that tells us how the spot increases with distance to the optical center, while the spot *shape* $\{{}^{n}\chi^{0}_{0}, \mathbf{q}\}$ is • for $n = 0$, | for $n = 1$, the cubic-growth | starting at $k = 4$, etc.

All | spot diagrams, therefore, including that of astigmature ${}^{2}\chi^{0}_{0}$, are straight lines *to first exponential term*; thereafter they become *circle segments*, as we saw above. When the growth function $\phi(q)$ of an astigmature segment with distance q from the optical center is $\phi(q) = \phi_2 q^2 + \phi_3 (q^2)^2 + \cdots + \phi_k (q^2)^{k-1} + \cdots$, then the relative contributions of kth rank astigmature may be determined from the coefficients ϕ_k, up to some aberration order. The same remarks become trivial for the shape of poci •, and apply for higher-order |'s as well in the general case $(j + n\ j\ -j)$, the lower fringe of the aberration sectors.

A similar argument holds for Lie transformations generated by ${}^{j+n}\chi^{j}_{1-j} = (\mathbf{p} \times \mathbf{q})^{2n}\mathbf{p}\cdot\mathbf{q}(q^2)^{j-1}$ and the series cut to the linear term $\{{}^{j+n}\chi^{j}_{1-j}, \mathbf{q}\} = (q^2)^j\{{}^{n+1}\chi^{1}_{0}, \mathbf{q}\}$ is independent of j. The lowest case $n = 0$ corresponds to distortion, ↼, where $\{{}^{1}\chi^{1}_{0}, \mathbf{q}\} = -\mathbf{q}$ (pure magnification) allows for the integration of the full series to (6.4). In repeaters with powers of $(\mathbf{p} \times \mathbf{q})^{2n}$ this accounts for the equality (to seventh order) of the $\triangleright_3$ and $\triangleright_4^{(m=-1)}$ degenerate with elliptical comas ⊂⊃. The general case involves $(j + n\ j\ 1 - j)$, the lower fringe of the aberration. For this reason $(3\ 1\ 0) = (4\ 2\ -2)$ in Figure 6.5.

6.6. Concluding remarks

Notwithstanding the explicit results of this section, it should be clear that our purpose is to treat group-theoretically the aberration phenomenon itself, because it is nonlinear. The optics of lenses and mirrors is only a particularly transparent application of the theory. In this context, we should admit to a grievous omission in these notes: the treatment of optical *fibers*, *i.e.*, systems whose evolution is governed by the Lie transformation $\exp -z\hat{H}(\mathbf{p}, \mathbf{q})$, where H is z-independent. The abstract problem indeed models classical mechanics as well, when $H = \frac{1}{2}p^2 + V(\mathbf{q})$. To third aberration order, the particular fiber problem (with nonflat interfaces) was treated in [10], and its general solution is provided (in principle) in Ref. [8], §3.5.3. We decided to exclude it from presentation for reasons of length, and for the purpose of exploring further the possibility of integrating efficiently in inhomogeneous media through integration of their corresponding group representatives in the 31-parameter space of seventh-order aberration theory.

As the presentation of tables and figures here suggests, the developments we pursue are geared to their use with personal computers, without the pressure of specific applications that have motivated much of the important work of our colleagues. We have tried to make a virtue out of these weaknesses by keeping some generality of results and computability by modest means. Alexander von Humboldt's dictum quoted at the beginning of this section is a reason to feel encouraged. The set of muSIMP functions to handle aberrating systems is being prepared and will be open. Perhaps it should be called MEXLIE.

Acknowledgement. I would like to thank Mr. Guillermo Correa for his collaboration in the effort to produce *visible* results —the figures. His SPOT_D system will read muSIMP output files and return spot diagrams that can appear on the monitor screen and on the IIMAS/DF laser printer. It works through a very handsome PASCAL-based interactive system. This too, will be open and reported elsewhere.

References

I must apologize for drawing too many references from my own work. A fuller account is in preparation, and may be found in the bibliography collected in *Lie methods in optics*, Lecture Notes in Physics Vol. 250 (Springer Verlag, Heidelberg, 1986), ed. by J. Sánchez Mondragón and the author.

[1] O. Stavroudis, *The Optics of Rays, Wavefronts, and Caustics*, (Academic Press, New York, 1978).

[2] A.J. Dragt, Lie-algebraic theory of geometrical optics and optical aberrations. *J. Opt. Soc. Am.* **72**, 372–379 (1982).

[3] K.B. Wolf, A group-theoretical model for Gaussian optics and third order aberrations. In *Proceedings of the XII International Colloquium on Group-theoretical Methods in Physics*, Trieste, 1983, Lecture Notes in Physics Vol. 201 (Springer Verlag, Heidelberg, 1984), pp. 133-136.

[4] M. Navarro-Saad and K.B. Wolf, Factorization of the phase-transformation produced by an arbitrary refracting surface , *J. Opt. Soc. Am. A* **3**, 340–346 (1986).

[5] M. Navarro-Saad and K.B. Wolf, The group-theoretic treatment of aberrating systems. I. Aligned lens optics in third aberration order. *J. Math. Phys.* **27**, 1449–1457 (1986).

[6] M. Navarro-Saad, *Cálculo de aberraciones en sistemas ópticos con teoría de grupos*. Thesis, Facultad de Ciencias, Universidad Nacional Autónoma de México, 1985.

[7] M. Navarro-Saad and K.B. Wolf, Applications of a factorization theorem for ninth-order aberration optics. *J. Symbolic Computation* **1**, 235–239 (1985).

[8] S. Steinberg, Lie series, Lie transformations, and their applications. In *Lie Methods in Optics*, Lecture Notes in Physics, Vol. 250 (Springer Verlag, Heidelberg, 1986), pp. 45–103.

[9] A.J. Dragt, E. Forest, and K.B. Wolf, Foundations of a Lie algebraic theory of geometrical optics. In *Lie Methods in Optics*, *op. cit.*, pp. 105–157.

[10] K.B. Wolf, The group-theoretic treatment of aberrating systems. II. Axis-symmetric inhomogeneous systems and fiber optics in third aberration optics. *J. Math. Phys.* **27**, 1458–1465 (1986).

[11] K.B. Wolf, Symmetry in Lie optics. *Ann. Phys.* **172**, 1–25 (1986).

[12] K.B. Wolf, The group-theoretic treatment of aberrating systems. III. The classification of asymmetric aberrations. *J. Math. Phys.* **28**, 2498–2507 (1987).

[13] H. Goldstein, *Classical Mechanics*, 2nd edition (Addison Wesley, Reading, Mass., 1980).

[14] A. Katz, *Classical Mechanics, Quantum Mechanics, Field Theory*. (Academic Press, New York, 1965).

[15] V. Bargmann, Irreducible unitary representations of the Lorentz group. *Ann. Math.* **48**, 568–640 (1947).

[16] C.P. Boyer and K.B. Wolf, Deformations of inhomogeneous classical Lie algebras to the algebras of the linear groups. *J. Math. Phys* **14**, 1853–1859 (1973); K.B. Wolf and C.P. Boyer, The algebra and group deformations $I^m[SO(n) \otimes SO(m) \Rightarrow SO(n,m)$, $I^m[U(n) \otimes U(m) \Rightarrow U(n,m)$, and $I^m[Sp(n) \otimes Sp(m) \Rightarrow Sp(n,m)$, for $1 \le m \le n$. *J. Math. Phys* **15**, 2096–2101 (1974).

[17] A. Atakishiyev, W. Lassner, and K.B. Wolf, The relativistic coma aberration. I. Geometrical optics. Comunicaciones Técnicas IIMAS, preprint No. 509, 1988 (submitted).

[17] A. Atakishiyev, W. Lassner, and K.B. Wolf, The relativistic coma aberration. II. Helmholtz wave optics. Comunicaciones Técnicas IIMAS, preprint No. 512, 1988 (submitted).

[18] H.M. Nussenzveig, Complex angular momentum theory of the rainbow and the glory. *J. Opt. Soc. Am.* **69**, 1068–1098 (1979).

[19] O. Castaños, E. López Moreno, and K.B. Wolf, Canonical transforms for paraxial wave optics. In *Lie Methods in Optics*, *op. cit.*, pp. 159–182.

[20] L. Seidel, Zur Dioptrik. *Astronomische Nachrichten* N° **871**,105–120 (1853).

[22] H. Buchdahl *An Introduction to Hamiltonian Optics.* (Cambridge University Press, 1970).

[23] M. Moshinsky, Canonical transformations in quantum mechanics. *SIAM J. Appl. Math.* **25**, 193–212 (1973).

[24] A.J. Dragt and J. Finn, Lie series and invariant functions for analytic symplectic maps. *J. Math. Phys.* **17**, 2215–2227 (1976); S. Steinberg, Factored product expansions of nonlinear differential equations. *SIAM J. Math. Anal.* **15**, 108–115 (1984).

[25] K.B. Wolf, Symmetry adapted classification of aberrations. Comunicaciones Técnicas IIMAS, preprint N°. 440, 1986. (*J. Opt. Soc. Am. A*, to appear.)

[26] V. Bargmann, Group representations in Hilbert spaces of analytic functions. In: *Analytical Methods in Mathematical Physics.* (Gordon and Breach, New York, 1970), pp. 27–63.

[27] K.B. Wolf, *Integral Transforms in Science and Engineering* (Plenum Publ. Corp., New York, 1979).

[28] K.B. Wolf, The group-theoretic treatment of aberrating systems. IV. The seventh-order aberration group and optical elements. Comunicaciones Técnicas IIMAS, preprint N° 490, 1987. (unpublished).

[29a] A.J. Dragt, *Lectures on Nonlinear Orbit Dynamics*, AIP Conference Proceedings N° 87 (American Institute of Physics, New York, 1982); *ibid.*, Elementary and advanced Lie algebraic methods with applications to accelerator design, electron microscopes, and light optics. *Nucl. Instr. Meth. Phys. Res.* **A 258**, 339–354 (1987).

[29b] A.J. Dragt, F. Neri, G. Rangarajan, D.F. Douglas, L.M. Healy, and R.D. Ryne, Lie algebraic treatment of linear and nonlinear beam dynamics (CTP-UM preprint, March 1988, *Annual Rev. Nucl. Part. Science*, to appear); E. Forest, M. Berz, and J. Irwin, Normal form methods for complicated periodic systems: A complete solution using differential algebra and Lie operators (LBL, Berkeley preprint, March 1988).

THE GENERALIZED HAMMERSTEIN INTEGRAL OPERATORS ON HILBERT SPACES

Mario Zuluaga

Departamento de Matemáticas, Universidad Nacional de Colombia, Bogotà, D.E. Colombia

Abstract

We study the equation $u \in sB\,\partial f(u)$, where $s \in R$, B is a linear selfadjoint and completely continuous operator defined on a real Hilbert space H and $\partial f(u)$ is the generalized gradient of a function $f : H \rightarrow R$ at u. Conditions are given for the existence of multiple solutions. The main analytical tools are the Liusternik-Schnirelmann theory and a reduction approach due to Lazer, Landesman and Meyers.

1. - Introduction -

A common type of Hammerstein's integral equation is

$$u = BF(u) \tag{1.1}$$

where $B:H\rightarrow H$ is a linear operator, H being a real Hilbert space and $F:H\rightarrow H$ is a potential operator, i.e., $F = \nabla f$, where f is a real valued function on H. The purpose of this work is to study the related equation where $F = \partial f(u)$ is the generalized gradient of a function $f:H\rightarrow R$ as defined in ref. (4). In this case $F : H\rightarrow 2^H$, and thus

$$u \in BF(u) \tag{1.2}$$

is meaningful. If, for example, $f \in C^1$, it follows from Proposition 4 in ref. (4) that (1.2) reduces to (1.1). The novelty of our approach in studying (1.2) lies in the use of variational methods. This approach unifies the well-know smooth cases. The work has been motivated by results in ref. (2) and ref. (8).

2. - Preliminaries -

Let H be a Banach space (in our case, H will be a real Hilbert space), and let f be a real valued, locally Lipschitz function (loc-lip, for short). The last requirement means that any point $u \in H$ has a neighborhood U such that, for some constant k and all y and z in U

$$|f(y) - f(z)| \leq k \ \|y - z\|$$

Definition 2.1 (F.H. Clarke, (4)). Let f be a loc-lip function. For each v in H the generalized directional derivative $f^{\circ}(u, v)$ in the direction v is

$$f^{\circ}(u, v) = \overline{\lim_{\substack{h\to 0\\ \lambda\to 0^+}}} \frac{f(u+h+\lambda v) - f(u+h)}{\lambda},$$

$h \in H$ and $\lambda \in (0,\infty)$.

Definition 2.2 (F.H. Clarke (4)). If $f{:}H\to R$ is loc-lip, the generalized gradient of f at u, denoted with $\partial f(u)$, is

$$\partial f(u) = \{z \in H \mid \langle z,v\rangle \leq f^{\circ}(u,v), \ \ \forall v \in H\}.$$

Definition 2.3 (K.C.Chang (2)). A loc-lip function f satisfies the Palais-Smale (P.S.) condition, if any sequence $\{u_n\}$ along which $|f(u_n)|$ is bounded and $\lambda(u_n) = \min_{w \in \partial f(u_n)} \|w\| \to 0$, has a convergent subsequence.

Let $f{:}H\to R$ be an even loc-lip function. Let

$$i_1(f) = \lim_{a\to 0^-} \Gamma(f_a)$$

and

$$i_2(f) = \lim_{a\to -\infty} \Gamma(f_a)$$

where $f_a = \{u \in H;\ f(u) \leq a\}$ and $\Gamma(f_a)$ is the genus of f_a. The following theorem is an extension to loc-lip functions of Theorem 9 of ref. (3).

Theorem 2.1. Let $f{:}H\to R$ be an even loc-lip function which satisfies the (P.S.) condition and $f(0) = 0$. Then, for each $m \in N$ such that $i_2(f) < m \leq i_1(f)$, there is at least a couple $\{u_m, -u_m\}$ of antipodal critical points of f such that

$$f(u_m) = \operatorname*{Inf}_{\Gamma(A)\geq m} \ \sup_{u \in A} f(u).$$

Here we say that $u \in H$ is a critical point of f if $0 \in \partial f(u)$.

Let $B: H \to R$ be a linear, continuous and selfadjoint operator. Let $\sqrt{B^2}$ be the positive square root of B^2. We define $B^+, B^-: H \to H$ as

$$B^+ = \frac{\sqrt{B^2} + B}{2} \tag{2.1}$$

and

$$B^- = \frac{\sqrt{B^2} - B}{2} \tag{2.2}$$

It is easily shown that B^+ and B^- satisfy

$$B^+ \cdot B^- = B^- \cdot B^+ = 0 \tag{2.3}$$

and

$$B = B^+ - B^- \tag{2.4}$$

We will write $A^+ = \sqrt{B^+}$ and $A^- = \sqrt{B^-}$. From (2.3) we have

$$A^+ \cdot A^- = A^- \cdot A^+ = 0 \tag{2.5}$$

and the operator $A = A^+ - A^-$ will then be called the principal square root of B. In all what follows, the operator B will be continuous and selfadjoint.

Lemma 2.1. Let $P_2: H \to H$ be orthogonal projection on $\mathrm{Ker}(B^+)$, and let $P_1 = 1 - P_2$. Then

$$A^- \cdot P_1 = P_1 \cdot A^- = 0 \tag{2.6}$$

$$A^+ \cdot P_2 = P_2 \cdot A^+ = 0 \tag{2.7}$$

$$A^+ \cdot P_1 = P_1 \cdot A^+ = A^+ \tag{2.8}$$

$$A^- \cdot P_2 = P_2 \cdot A^- = A^- \tag{2.9}$$

Proof: It is shown in Spectral theory (see ref. (1) chapter 27) that A^+, A^-, P_1, P_2 commute. From (2.3) we get that $B^-(u) \in \mathrm{Ker}\,(B^+)$ for all $u \in H$, so that

$$P_2 \cdot B^- = B^-, \tag{2.10}$$

and from (2.10), that

$$P_1 \cdot B^- = 0 \tag{2.11}$$

By taking square roots of eqs. (2.10) and (2.11), eqs. (2.9) and (2.6) are readily obtained.

Also

$$B^+ \cdot P_2 = 0 \tag{2.12}$$

and

$$B^+ \cdot P_1 = B^+, \tag{2.13}$$

and by taking square roots of eqs. (2.12) and (2.13) we obtain eqs. (2.7) and (2.8).

3. - Basic results -

Lemma 3.1. Let $F{:}H \to 2^H$. Then eq. (1.2) has a solution if and only if

$$(P_1 - P_2)(u) \in AFA(u) \tag{3.1}$$

also does.

Proof: Let $u \in H$ be a solution of (3.1), so that

$$(A^+ + A^-)(P_1 - P_2)(u) \in (A^+ + A^-)\,AFA(u) \tag{3.2}$$

From eq. (2.4) and eqs. (2.6) - (2.9) we get that $A(u) \in BF(A(u))$, so that $A(u)$ is a solution of eq. (1.2).

Now, let $u \in H$ be a solution of eq. (1.2). Then

$$u = B(v) \tag{3.3}$$

for any $v \in F(u)$, and from eqs. (2.4) - (2.6) we obtain that

$$AF(u) = AF(B(v)) = AFA\,(A^{+}(v) + A^{-}(v)). \tag{3.4}$$

Since $v \in F(u)$, it follows from eq. (3.4) that

$$A(v) \in AFA(A^{+}(v) + A^{-}(v)), \tag{3.5}$$

and from eqs. (2.6) - (2.9), that $A(v) = (P_1 - P_2)\,(A^{+}(v) + A^{-}(v))$. From eq. (3.5) we obtain that $A^{+}(v) + A^{-}(v)$ is a solution of eq. (3.1).

Lemma 3.2. Let $f\colon H \rightarrow R$ be a loc-lip convex function. Then

$$\partial(f \circ A)\,(u) = A\partial f(A(u)). \tag{3.6}$$

Proof: Clearly $f \circ A$ is loc-lip and convex. Let $w \in \partial(f \circ A)(u)$. From definition (2.1),

$$(f \circ A)^{o}\,(u,v) \leq f^{o}(A(u), A(v)), \tag{3.7}$$

for all $v \in H$, and from eq. (3.7) and definition (2.2)

$$\langle w,v \rangle \leq f^{o}\,(A(u), A(v)), \tag{3.8}$$

for all $v \in H$.

We claim that $w = A(z)$ for any $z \in \partial f(A(u))$. Lets proof it ab absurdo. By proposition (1) in ref. (4), $\partial f(A(u))$ is a convex, and weakly compact subset of H. Thus, so is $A\partial f(A(u))$. By the separation theorem, there is a $z_o \in H$ such that

$$\langle w,z_o \rangle > \langle A(z), z_o \rangle = \langle z, A(z) \rangle \tag{3.9}$$

for all $z \in \partial f(A(u))$. Thus

$$\langle w,z_o \rangle > \max_{z \in \partial f(a(u))} \langle z, A(z_o) \rangle. \tag{3.10}$$

Proposition (1) in ref. (4) states that

$$f^{o}(A(u), A(z_o)) = \max_{z \in \partial f(A(u))} \langle z, A(z_o) \rangle, \tag{3.11}$$

and eqs. (3.10) and (3.11) ensure that

$$\langle w, z_o \rangle > f^{o}(A(u), A(z_o)).$$

This inequality contradicts eq. (3.8). Then $w \in A\partial f(A(u))$, i.e. $\partial(f\circ A)(u) \subseteq A\partial f(A(u))$. To prove the other inclusion, let $z \in A\partial f(A(u))$. Then $z = A(w)$ for any $w \in \partial f(A(u))$. By proposition (3) in ref. (4), $\partial f(A(u))$ coincides with the subdifferential of f in the sense of convex analysis, so that

$$f(v) - f(A(u)) \geq \langle w, v-A(u) \rangle \tag{3.12}$$

for all $v \in H$. Also

$$f(A(v)) - f(A(u)) \geq \langle w, A(v) - A(u) \rangle = \langle A(w), v-u \rangle. \tag{3.13}$$

Taking into account that $f\circ A$ is convex, it then follows from eq. (3.13) that $A(w) = z \in \partial(f\circ A)(u)$.

Lemma 3.3. Let $f:H\rightarrow R$ be a loc-lip convex function. Let $J: H\rightarrow R$ be defined by

$$J(u) = 2f(A(u)) - \|P_1(u)\|^2 + \|P_2(u)\|^2 \tag{3.14}$$

If $0 \in \partial J(u)$ then eq. (3.1) has a solution, and so does eq.(1.2).

Proof: Clearly $J:H\rightarrow R$ is a loc-lip function and by proposition (4) and (8) of ref. (4) we have

$$\partial J(u) \subset 2\partial(f\circ A)(u) - 2\cdot(P_1 - P_2)(u). \tag{3.15}$$

If $0 \in \partial J(u)$, eq. (3.15) and Lemma (3.2) guaranty that

$(P_1 - P_2)(u) \in A\partial f(A(u)).$

Let's remark that it is natural to define $u \in H$ as a critical point of J if $0 \in \partial J(u)$. In the following let's define

$$X = P_1 H$$

$$Y = P_2 H.$$

The sets X and Y are closed subspaces of H, and $H = X \oplus Y$.

Lemma 3.4. Let $f{:}H \to R$ be such that
a) f is convex
b) for all $(\alpha, c, d) \in R^3$, with $\alpha > 0$, there is an $M > 0$ such that if $\|u\| \geq M$ then

$$|f(u)| \leq \frac{\alpha}{2} \|u\|^2 + c\|u\| + d$$

then, for all $x \in X$, the map $J_x{:}Y \to R$ given by $J_x(y) = J(x+y) = 2\, fA(x+y) - \|x\|^2 + \|y\|^2$, satisfies

1) $J_x(y) \to +\infty \qquad$ if $\|y\| \to \infty$

2) J_x is strictly convex.

Proof. Let $x \in X$ be fixed. From condition b) it follows that for $\alpha > 0$ such that $\alpha \|A\| < 1$ and for any $c, d \in R$ there exists $M > 0$ such that if $\|x+y\| \geq M$, then

$$J_x(y) \geq (1 - \alpha \|A\|^2)\|y\|^2 - 2c\|A\|\,\|x+y\| - (1 + \alpha\|A\|^2)\|x\|^2 - 2d. \tag{3.16}$$

Since $1 - \alpha\|A\|^2 > 0$, from eq. (3.16) it follows that $J_x(y) \to +\infty$ if $\|y\| \to \infty$. Finally, since $f \circ A$ is convex and $\|y\|^2$ is strictly convex, then J_x is strictly convex.

As it is well-known (see for example ref. (7)), consequences 1), 2) of Lemma (3.4) imply for each $x \in X$ the existence of an $\hat{y} \in Y$ such that $J_x(\hat{y}) = \min_{y \in Y} J_x(y)$, and such y is unique. Letting $y = T(x)$, this defines a map $T{:}X \to Y$. Given $x \in X$, $T(x)$ is the unique member of Y such that

$$J_x(T(x)) = \min_{y \in Y} J_x(y)$$

Lemma 3.5. Let f:H→R be a loc-lip function satisfying conditions a), b) of Lemma (3.4). Then if $x \in X$ and $y_1, y_2 \in Y$,

$$<\partial J_x(y_1) - \partial J_x(y_2), y_1 - y_2> \geq 2 \|y_1 - y_2\|^2, \tag{3.17}$$

where the left handside of eq. (3.17) is to be understood in the sense that the inequality holds for all $z \in \partial J_x(y_1)$ and $\theta \in \partial J_x(y_2)$.

Proof. Since J is a loc-lip function, so is J_x. From Proposition (4) and(8) in ref. (4) we obtain

$$\partial J_x(y_i) \subset 2\partial(f \circ A(x+.))(y_i) + 2y_i, i = 1, 2. \tag{3.18}$$

Let $z_i \in J_x(y_i)$, i = 1, 2. From eq. (3.18)

$$z_i = 2\theta_i + 2y_i, \tag{3.19}$$

where $\theta_i \in \partial(f \circ A(x + .))(y_i)$, i = 1, 2. Since f∘A is convex, it follows from Proposition (3) in ref. (4) that, for all $w \in Y$

$$f \circ A(x + w) - f \circ A(x + y_i) \geq <\theta_i, w - y_i>, i = 1, 2. \tag{3.20}$$

As a consequence of eq. (3.20) we have that

$$<\theta_1 - \theta_2, y_1 - y_2> \geq 0 \tag{3.21}$$

and eq. (3.17) now follows from eq. (3.21).

Lemma 3.6. Let f:H→R, and assume that f satisfies all conditions of Lemma (3.5). Let B:H→H be a linear selfadjoint and completely continuous operator. Then

a) The map $x \to \|T(x)\|^2$ is Weakly Upper Semi Continuous (W.U.S.C.), i.e. if $x_n \rightharpoonup x_o$ (weakly), then

$$\|T(x_o)\|^2 \geq \overline{\lim_{n\to\infty}} \|T(x_n)\|^2. \tag{3.22}$$

b) If $x_n \rightharpoonup x_o$ then, for any weakly convergent subsequence $\{T(x_{n_k})\}$ of $\{T(x_n)\}$, $T(x_{n_k}) \rightharpoonup T(x_o)$.

c) The map $T:X \to Y$ is continuous.

Proof. Take $y_1 = 0$ and $y_2 = T(x)$ in eq. (3.17). Since $T(x) \in Y$ is a minimum of J_x, Proposition (6) in ref. (4) ensures that $0 \in \partial J_x(T(x))$ and, from eq. (3.17)

$$\|\partial J_X(0)\| \geq 2\, \|T(x)\|. \tag{3.23}$$

Since J_X is convex, Proposition (3) in ref. (4) guarantees that

$$\partial J_x(0) = \left\{\theta \in Y;\, J_x(y) - J_x(0) \geq \langle\theta, y\rangle, \forall y \in Y\right\}.$$

Let $\theta \in \partial J_x(0)$. Then

$$\langle\theta, y\rangle \leq 2\,|f\circ A(x+y)| + 2\,|f\circ A(x)| + \|y\|^2, \tag{3.24}$$

and also

$$\|\theta\| = \sup_{\|y\|\leq 1} |\langle\theta,y\rangle| \leq \max\left\{2.\sup_{\|y\|\leq 1} |f\circ A(x+y)|,\ 2.\sup_{\|y\|\leq 1} |f\circ A(x-y)|\right\} + 2\,|f\circ A(x)| + 1, \tag{3.25}$$

as follows from eq. (3.24).

It is well-known (see for example ref. (6)) that if B is a completely continuous operator then so is $A = A^+ - A^-$, Let W be a bounded subset of X, then $\{f\circ A(x);\, x \in W\}$ and $\{f\circ A(x\pm y);\, x \in W, \|y\| \leq 1\}$ are also bounded, and from eq. (3.25) we also see that $\partial J_x(0)$ is bounded for $x \in W$. Hence inequality (3.23) implies that $T:X \to Y$ is bounded on bounded sets.

Let $\{x_n\} \subset X$, and assume that $x_n \rightharpoonup x_o$, $x_o \in X$, then $\{x_n\}$ is bounded. This implies that $\{T(x_n)\}$is bounded, and thus there exists $\{T(x_{n_k})\} \subset \{T(x_n)\}$ such that $T(x_{n_k}) \rightharpoonup y_o$, $y_o \in Y$. Since $0 \in \partial J_{x_{n_k}}(T(x_{n_k}))$,

$$J_{x_{n_k}}(y) - J_{x_{n_k}}(T(x_{n_k})) \geq 0, \text{ for all } y \in Y.$$

For $y = y_o$ we have

$$2\left\{ f\circ A(x_{n_k} + y_o) - f\circ A(x_{n_k} + T(x_{n_k}))\right\} + \|y_o\|^2 \geq \|T(x_{n_k})\|^2, \tag{3.26}$$

and since A is completely continuous, $A(x_{n_k} + T(x_{n_k})) \rightarrow A(x_o + y_o)$. Hence from eq. (3.26)

$$\|y_o\|^2 \geq \overline{\lim_{n_k \rightarrow \infty}} \|T(x_{n_k})\|^2. \tag{3.27}$$

We now show that $y_o = T(x_o)$. Clearly, it is sufficient to establish that

$$J_{x_o}(z) \geq J_{x_o}(y_o) \tag{3.28}$$

for all $z \in Y$. Now

$$J_{x_{n_k}}(z) \geq J_{x_{n_k}}(T(x_{n_k})), z \in Y, \tag{3.29}$$

and eq. (3.29) is equivalent to

$$2.f\circ A(x_{n_k} + z) + \|z\|^2 \geq 2.f\circ A(x_{n_k} + T(x_{n_k})) + \|T(x_{n_k})\|^2. \tag{3.30}$$

Since A is completely continuous, $f\circ A(x_{n_k} + z) \rightarrow f\,A(x_o + z)$ and $f\circ A(x_{n_k} + T(x_{n_k})) \rightarrow f\circ A(x_o + y_o)$. On the other hand the map $\| \;\|^2$ is W.L.S.C., i.e. if $w_n \rightharpoonup w_o$ then $\underline{\lim} \|w_n\|^2 \geq \|w_o\|^2$ (see, for example ref. (7)). Passing to the lower limit on both sides of eq. (3.30), we see that

$$2.f\circ A(x_o + z) - \|x_o\|^2 + \|z\|^2 \geq 2.f\circ A(x_o + y_o) - \|x_o\|^2 + \|y_o\|^2, \tag{3.31}$$

and eq. (3.31) is equivalent to eq. (3.28). This proves eq. (3.28) and, as a matter of fact, establishes b). Assertion a) now follows from b) and eq. (3.27). It remains to prove that T is continuous. Assume that this is not the case for $x \in X$. Then there exist $\varepsilon > 0$, a sequence $\{x_n\} \subset X$ such that $x_n \to x$, and a subsequence $\{T(x_{n_k})\} \subset \{T(x_n)\}$ such that

$$\|T(x_{n_k}) - T(x)\| > \varepsilon.$$

Since T is bounded on bounded sets, $\overline{\{T(x_{n_k})\}}$ is closed and bounded, hence weakly compact. On the other hand, J_x is convex, and thus W.L.S.C.. Since any W.L.S.C. map reaches its minimum on any weakly closed set, there exists $y_1 \in Y$ such that

$$\|y_1 - T(x)\| > \varepsilon \tag{3.32}$$

and for all n_k, $k = 1, 2, \ldots,$

$$J_x(T(x)) < J_x(y_1) \leq J_x(T(x_{n_k})). \tag{3.33}$$

On the other hand, since J is continuous, there is $N \in \mathbb{N}$ such that

$$J_{x_n}(T(x)) < J_x(y_1) \tag{3.34}$$

for all $n > N$. From eqs. (3.32) - (3.34) we see that for all n_k large enough and all $n > N$, $J_{x_n}(T(x)) < J_{x_{n_k}}(T(x_{n_k}))$. Hence for $n = n_k$, we have that $J_{x_{n_k}}(T(x)) < J_{x_{n_k}}(T(x_{n_k}))$. This is in contradiction with the definition of T.

Definition 3.1. Let $J:H \to R$. For all $x \in X$ and $y \in Y$ we define $J_x, J_y:H \to R$ by

$$\begin{aligned} J_x(h) &= J(x + P_2 h) \\ J_y(h) &= J(y + P_1 h) \end{aligned} \tag{3.35}$$

Then we have the following

Lemma 3.7. Let $J:H \to R$ be a loc-lip function and let J_x, J_y be defined by eq. (3.35). If $0 \in \partial J_x(y) \cap \partial J_y(x)$, then $0 \in \partial J(x+y)$.

Proof. It is clear that if J is a loc-lip function then so are J_x, J_y. Let $0 \in \partial J_x(y) \cap \partial J_y(x)$. Since $0 \in \partial J_x(y)$ then

$$J_x^o(y,v) = \overline{\lim_{\substack{h\to 0\\ \lambda\to 0^+}}} \left[\frac{J(x+y+P_2h+\lambda P_2v) - J(x+y+P_2h)}{\lambda}\right] \geq 0$$

for all v H, so that

$$\overline{\lim_{\substack{h\to 0\\ \lambda\to 0^+}}} \left[\frac{J(x+y+h+\lambda P_2v) - J(x+y+h)}{\lambda}\right] \geq 0 \tag{3.36}$$

In particular,

$$\overline{\lim_{\substack{h\to 0\\ \lambda\to 0^+}}} \left[\frac{J(x+y+h+\lambda P_1v+\lambda P_2v) - J(x+y+h+\lambda P_1v)}{\lambda}\right] \geq 0 \tag{3.37}$$

On the other hand, since $0 \in \partial J_y(x)$,

$$J_y^o(x,v) = \overline{\lim_{\substack{h\to 0\\ \lambda\to 0^+}}} \left[\frac{J(x+y+P_1h+\lambda P_1v) - J(x+y+P_1h)}{\lambda}\right] \geq 0. \tag{3.38}$$

Also

$$\overline{\lim_{\substack{h\to 0\\ \lambda\to 0^+}}} \left[\frac{J(x+y+h+\lambda P_1v) - J(x+y+h)}{\lambda}\right] \geq 0, \tag{3.39}$$

as follows from eq. (3.38). Now,

$$J^{o}(x+y,v) = \overline{\lim_{\substack{h\to 0\\ \lambda\to 0^+}}} \left\{ \left[\frac{J(x+y+h+\lambda P_1 v+\lambda P_2 v) - J(x+y+h+\lambda P_1 v)}{\lambda} \right] + \right.$$

$$\left. + \left[\frac{J(x+y+h+\lambda P_1 v) - J(x+y+h)}{\lambda} \right] \right\}.$$

From eqs.(3.37) and (3.39) it then follows that $J^{\circ}(x+y,v) \geq 0$ and thus $0 \in \partial J(x+y)$.

4. - The main theorem -

Theorem 4.1. Let $B:H\to H$ be a linear, selfadjount and completely continuous operator. Let $f: H\to R$ be such that

a) f is a loc-lip, convex function
b) For each $(\alpha,c,d) \in R^3$ $\alpha > 0$ there exists $M > 0$ such that if $\|u\| \geq M$ then

$$|f(u)| \leq \frac{\alpha}{2}\|u\|^2 + c\,\|u\| + d. \tag{4.1}$$

Then eq. (1.2) has a solution.

Proof. Let $J:H\to R$ be defined by eq. (3.14), and let J_x, J_y: $H\to R$ be defined by eq. (3.32). If f is a loc-lip function, so are J, J_x, J_y. Since T(x) is the minimum of J_x on Y, it is easy to see that $J^{\circ}_x(T(x), v) \geq 0$ for all $x \in X$ and $v \in H$. Also because of Definition (2.2) $0 \in \partial J_x(T(x))$. We will prove that there exists $x_o \in X$ such that $0 \in \partial J_{T(x_o)}(x_o)$. From this fact, the preceeding conclusion, and Lemma (3.7) we have that $0 \in \partial J(x_o+T(x_o))$. From Lemma (3.3) eq. (1.2) has a solution.

First we consider the map $G:X\to R$ defined by

$$G(x) = J_x(T(x)) = J(x+T(x)). \tag{4.2}$$

From the definition of T(x),

$$G(x) = \min_{y \in Y} J_x(y). \tag{4.3}$$

Now, from eqs. (4.2) and (4.3) we obtain that

$$G(x_1) - G(x_2) \le J(x_1 + T(x_2)) - J(x_2 + T(x_2)), \tag{4.4}$$

for all $x_1, x_2 \in X$. From eq. (4.4) and the fact that J is a loc-lip function we have that

$$|G(x_1) - G(x_2)| \le k \,||x_1 - x_2||, \tag{4.5}$$

for some $k > 0$ and x_1, x_2 in some neighbourhood U. Relation (4.5) guarantees that G is a loc-lip function. Now we claim that if $0 \in \partial G(x)$ then $0 \in \partial J_{T(x)}(x)$. In fact, from definition (2.3)

$$\overline{\lim_{\substack{h \to 0 \\ \lambda \to 0^+}}} \left\{ \frac{J(x+P_1(h+\lambda v) + T(x+P_1(h+\lambda v))) - J(x+P_1 h+T(x+P_1 h))}{\lambda} \right\} \ge 0. \tag{4.6}$$

For all $v \in H$, let

$$I(v) = \overline{\lim_{\substack{h \to 0 \\ \lambda \to 0^+}}} \left\{ \frac{J(x+T(x)+P_1(h+\lambda v)) - J(x+T(x)+P_1 h)}{\lambda} \right\}.$$

By the definition of T(x) we have,

$$I(v) \ge \overline{\lim_{\substack{h \to 0 \\ \lambda \to 0^+}}} \left\{ \frac{J(x+P_1(h+\lambda v) + T(x+P_1(h+\lambda v))) - J(x+T(x)+P_1 h)}{\lambda} \right\}.$$

In particular, if $h \in Y$ then $P_1 h = 0$, and we have

$$I(v) \geq \overline{\lim_{\lambda \to 0^+}} \left\{ \frac{J(x+\lambda P_1 v + T(x+\lambda P_1 v)) - J(x+T(x))}{\lambda} \right\}. \tag{4.7}$$

We claim that the right hand side of eq. (4.7) is greather than zero. Otherwise, there would be an $\alpha < 0$ and $\{\lambda_n\} \subset R^+$ such that $\lambda_n \to 0$ and that

$$\left\{ \frac{J(x+\lambda_n P_1 v + T(x+\lambda_n P_1 v)) - J(x+T(x))}{\lambda_n} \right\} \leq \alpha < 0,$$

for all $n \in \mathbb{N}$.Since J and T are continuous, then for all $h \in H$ with $\|h\|$ sufficiently small,

$$\left\{ \frac{J(x+P_1(h+\lambda_n v)+T(x+P_1(h+\lambda_n v))) - J(x+P_1 h+T(x+P_1 h))}{\lambda_n} \right\} < 0. \tag{4.8}$$

But eq. (4.8) contradicts eq. (4.6). Hence $I(v) \geq 0$. This is equivalent to $0 \in \partial J_{T(x)}(x)$.

Our next step is to show that there exists an $x_o \in X$ such that $0 \in \partial G(x_o)$. In fact, by the definition of T(x), it is clear that $G(x) \leq J(x) = 2f{\circ}A(x+T(x)) - \|x\|^2$. From eq. (4.1) with $\alpha > 0$ and $\alpha \|A\| < 1$ we get that $J(x) \to -\infty$ if $\|x\| \to \infty$. Then $G(x) \to -\infty$ if $\|x\| \to \infty$. Now,

$$G(x) = 2f{\circ}A(X+T(x)) - \|x\|^2 + \|T(x)\|^2.$$

Since A is completely continuous, it follows from b) of Lemma (3.6) that $2f{\circ}A(x+T(x))$, for $x \in X$ is W.U.S.C.. Since it is well-known that $-\|x\|^2$ is W.U.S.C. and also, by a) of Lemma (3.6), $\|T(x)\|^2$ is W.U.S.C., then G(x) is W.U.S.C.. Hence there exists an $x_o \in X$ such that $G(x_o) \geq G(x)$ for all $x \in X$. From Proposition (6) in ref. (4) we, now, obtain that $0 \in \partial(-G)(x_o)$, and from Proposition (4) in ref. (2), that $0 \in \partial G(x_o)$.

5. - Multiple solutions -

In this section we consider the problem

$$u \in s\, B\, \partial f(u), \qquad s \in R. \tag{5.1}$$

We have the following:

Theorem 5.1. Assume all conditions of Theorem (4.1) hold. Also assume that
a) f is strictly convex
b) f is even
Then, for all $n \leq \dim X$, there is an $s(n) \geq 0$ such that if $s > s(n)$ then eq. (5.1) has, at least, $2n + 1$ solutions.

Proof. The proof is based on Theorems (2.1) and (4.1). As pointed out in Theorem (4.1), if $x_0 \in X$ is a critical point of $G:X \to R$ then $x_0+T(x_0)$ is a critical point of the functional $J:H \to R$, and $A(x_0+ T(x_0))$ is a solution of problem (1.2).

Without loss of generality we may assume that $f(0) = 0$. Since f is even, and convex as well, it is easily shown that $0 \in \partial f(0)$. Thus, for all $s \in R$, 0 is solution of problem (5.1). Because of Proposition (4) in ref. (2), problem (5.1) becames

$$u \in B\,\partial\,(sf)\,(u), \qquad s \in R. \tag{5.2}$$

Now let's observe that, for all $s > 0$, the map $sf:H \to R$ satisfies all conditions of Theorem (4.1) and let's consider the functional $G_s:X \to R$ given by

$$G_s(x) = 2(sf) \circ A(x+T_s(x)) - \|x\|^2 + \|T_s(x)\|^2, \tag{5.3}$$

where $T_s(x)$ is the minimum of the functional

$$J_{s,x}(y) = 2(sf) \circ A(x+y) - \|x\|^2 + \|y\|^2.$$

From eq. (3.23) we get that

$$\|\partial J_{s,0}(0)\| \geq 2\,\|T_s(0)\|. \tag{5.4}$$

Now, from $J_{s,0}(y) = 2(sf) \circ A(y) + \|y\|^2$ being strictly convex and even it easily follows that $0 \in \partial J_{s,0}(0)$. From eq. (5.4) we see that $T(0) = 0$ and from eq. (5.3), that $G_s(0) = 0$. Now, if $G:X \to R$ is a loc-lip function then, so is G_s. Now, we prove that G_s is even. In fact, since f is even, so is $J_s(u) = 2(sf) \circ A(u) - \|P_1u\|^2 + \|P_2u\|^2$. On the other hand,

$$G_s(x) = J_s(x+T_s(x)) = \operatorname*{Inf}_{y\in Y} \; J_s(x+y)$$
$$= \operatorname*{Inf}_{y\in Y} \; J_s(-x-y)$$
$$= \operatorname*{Inf}_{y\in Y} \; J_s(-x+y)$$
$$= G_s(-x).$$

The next step is to prove that G_S satisfies the P.S. condition of definition (2.3). Let $\{x_n\}\subset X$ be such that $G_s(x_n)$ is bounded and $\min_{w\in\partial G_s(x_n)} \|w\| \to 0$. Since $G_s(x_n)\to -\infty$ if $\|x_n\|\to\infty$, then $\{x_n\}$ is bounded.

Now we are going to prove that

$$\partial G_s(x) \subset 2\left\{P_1 A\partial(sf)\, A(x+T_s(x)) - x\right\}, \tag{5.5}$$

for all $x \in X$. In fact, since T_s is continuous

$$G_s^o(x,v) = \overline{\lim_{\substack{h\to 0\\ \lambda\to 0^+}}} \left\{\frac{J_s(x+h+\lambda v+T_s(x+h+\lambda v)) - J_s(x+h+T_s(x+h))}{\lambda}\right\}$$

$$\leq \overline{\lim_{\substack{h\to 0\\ \lambda\to 0^+}}} \left\{\frac{J_s(x+h+\lambda v+T_s(x+h)) - J_s(x+h+T_s(x+h))}{\lambda}\right\}$$

$$= \overline{\lim_{\substack{h\to 0\\ \lambda\to 0^+}}} \Bigg[\left[\frac{2(sf)\circ A(x+T_s(x) + h+T_s(x+h) - T_s(x) +\lambda v)}{\lambda}\right] -$$

$$-\left[\frac{2(sf)\circ A(x+T_s(x) +h+T_s(x+h) - T_s(x))}{\lambda}\right] - \frac{(\|x+h+\lambda v\|^2 - \|x+h\|^2)}{\lambda} \Bigg].$$

$$\leq 2s(f\circ A)^o(x+T_s(x) - 2 \langle x,v\rangle \tag{5.6}$$

for all v, x $\in$ X.

If $z \in \partial G_s(x)$, it follows from eq. (5.6) that

$$<z,v> \leq 2s(f\circ A)^{o}\,(x+T_s(x),v) - 2<x,v>, \tag{5.7}$$

for all $v \in X$, and if $z \in 2s\partial\,(f\circ A)(x)$ then

$$<P_1z,w> \;=\; <z,P_1w> \;=\; <P_1z,P_1w> \leq 2s(f\circ A)^{o}(x,P_1w), \tag{5.8}$$

for all $w \in H$. Thus, if $z \in \partial\, G_s(x)$, then, from eqs. (5.7) and (5.8) we obtain that $z - 2x \in 2sP_1\partial(f\circ A)(x)$. Assertion (5.5) now follows from Lemma (3.2) and Proposition (4) in ref. (2).

Now, from Proposition (1) in ref. (4), $\partial G_s(x_n)$ is weakly compact, and since $\|\;\|$ is W.L.S.C. then, $\|\;\|$, reaches a minimum on $\partial\, G_s(x_n)$. Let $z_n\in \partial G_s(x_n)$ be such a minimum. By assumption, $z_n \to 0$. Also, from eq. (5.5)

$$z_n = 2(P_1\theta_n - x_n) \tag{5.9}$$

where $\theta_n\in A\partial(sf)\circ A(x_n+T_s(x_n))$. Since $\{x_n\}$ is bounded then $\{T_s(x_n)\}$ is bounded and since A is completely continuous, there exists a convergent subsequence of $A(x_n+T_s(x_n))$, which we label in the same form. This implies that $\bigcup_{x_n} \partial(sf)\,(\,A(x_n+T_s(x_n))$ is bounded. Hence, from eq. (5.9), $\{\theta_n\}$ and $\{x_n\}$ have convergent subsequences. We have proved that G_s satisfies the P.S. condition.

We know that $-\,G_s(x)\to+\infty$ if $\|x\|\to\infty$. Then $i_2(-G_s) = \infty$. Now let $X_n\subset X$ be a subspace such that $\dim(X_n) = n$. Write $S_{r,n} = S_r \cap X_n$, where $S_r = \{u \in H;\ \|u\| = r\}$, and let $M\subset S_{r,n}\times\mathbb{R}^+$ be such that $\|T_s(x)\|^2 > r^2$, for all $(x,s) \in M$. Let

$$d = d(r,n) = \inf_{(x,s)\in M}\ f\circ A(x+T_s(x)). \tag{5.10}$$

Since f is strictly convex, even, and $f(0) = 0$, then $f(u) > 0$ if $u \neq 0$.

This implies that $d \geq 0$. Let us see that $d > 0$. Otherwise, there would be a sequence $\{x_n+T_{s_n}(x_n)\}$ such that $f\circ A(\,x_n+T_{s_n}(x_n))\to 0$ and $x_n+T_{s_n}(x_n) \rightharpoonup x_o+y_o$ for some $x_o\in X$, $y_o\in Y$. Since in X_n weak convergence and strong convergence coincide, then $\|x_o\| = r$. Since A is completely continuous, we have that $0 = f\circ A(x_o+y_o) > 0$. This contradiction proves that $d > 0$.

Now, if $s > r^2/d$ then $G_s(x) > 0$ for $x \in X_n$, $\|x\| = r$. In fact, if $(x,s) \in M$ then $\|T_s(x)\|^2 > r^2$, and so $G_s(x) > 0$. If $(x,s) \in M$ then $2sf \circ A(x+T_s(x)) > r^2$, and also $G_s(x) > 0$. It is well known that the genus of $S_{r,n}$ is n. Thus, for $s > r^2/d = s(r,n)$, $i_1(-G_s) \geq n$. Hence, if $s(n) = \operatorname{Inf}_{r>0} s(r,n)$, we have that $i_1(-G_s) \geq n$ for $s > s(n)$. Thus we have seen that G_s satisfies all conditions of Theorem (2.1), and thus, G_s possesses 2n critical points $\{x_k, -x_k\}$, $k=1,2,\dots n$, for $s > s(n)$. As pointed out in Theorem (4.1) $\{\pm x_k\, T_s(\pm x_k)\}$ $k = 1, 2, \dots n$, are critical points of J_s, and $A^+(\pm x_k) - A^-(T_s(\pm x_k))$ are solutions of eq. (5.2). Since A^+ is injective on X then problem (5.2) has 2n solutions for $s > s(n)$, and since 0 is also solution, the theorem has been proved.

Remark

Since $-\partial f(u) = \partial f(-u)$, problem (5.1) can be stated in the form

$$-u \in (-s)\,(-B)\,\partial f(-u), \tag{5.11}$$

$$-B = B^- - B^+.$$

If E is the orthogonal projection over Kernel (B^-), essentially the same argument as in the proof of Theorem (5.1) shows that for all $n \leq \dim (I - E)(H)$ there exists $s(n) \leq 0$ such that for $s < s(n)$ problem (5.11) has, at least, 2n+1 solutions.

References

(1) G.Bachman and L. Narici: *"Functional Analysis"*, Academic Press, New York, 1966.

(2) K.C.Chang: "Variational methods for nondifferentiable functionals and their applications to partial differential equations", J. of Math. Anal. and Appl., 80 (1981) 102-129.

(3) D.Clark: "A variant of the Liusternik-Schnirelman theory", Ind. Univ. Math. J. 22 (1972) 65-74.

(4) F.H.Clarke: "A New approach to Lagrange multipliers", Math. Operation Research, 1 (1976) 165-174.

(5) A.Lazer, E.Landesman and D.Meyers: "On saddle point problems in the calculus of variations. The Ritz algorithm and the monotone convergence", J.Math. Anal. and Appl. 53 (1975) 549-614.

(6) A.Krasnoselskii: *"Topological Methods in the theory of nonlinear integral equations"*, Pergamon Press, New York, 1964.

(7) M.Vainberg: *"Variational method for the study of nonlinear operators"* Holden Day, San Francisco, 1964.

(8) M.Zuluaga: "Operadores integrales de Hammerstein, su espectro y aplicaciones" Rev. Colombiana de Matemáticas, 17 (1983) 73-98.

Acknowledgements

I acknowledge with thanks professor Jairo Charris for his help in reading and correcting the language of early drafts of this manuscript.

INDEX

A

B

C

D

E

F

G

H

T

V

W

X

Y

Z

www.ingramcontent.com/pod-product-compliance
Ingram Content Group UK Ltd.
Pitfield, Milton Keynes, MK11 3LW, UK
UKHW020403150926
8624IPUK00004B/327

9 789971 506636